WESTON D. GARDNER, M.D.

Professor of Anatomy, Medical College of Wisconsin;
Medical Director, Curative Workshop of Milwaukee;
Formerly, Director of Medical Education, Deaconess Hospital
Milwaukee, Wisconsin

WILLIAM A. OSBURN, M.M.A.

Associate Professor and Chairman,
Department of Medical Art and Visual Education,
The University of Texas Southwestern Medical School,
Dallas, Texas

P9-CQV-995

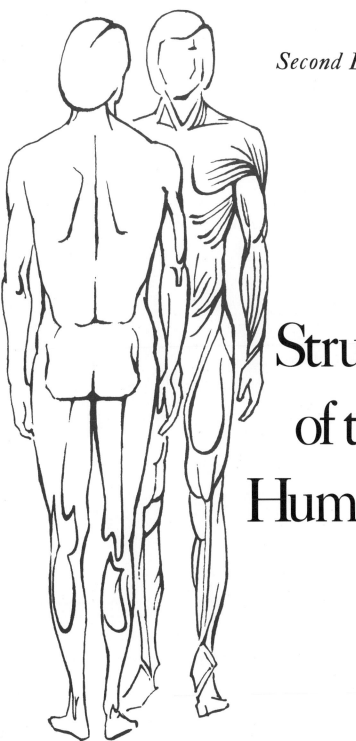

Second Edition

Structure
of the
Human Body

W. B. Saunders Company   *Philadelphia/London/Toronto*

ART DIRECTION BY WILLIAM A. OSBURN
Contributing Artists
    Alice Aubuchon
    James Bonner
    Weston Gardner
    Patricia Kenny
    William Osburn
    William Poinsett
    Kay Reinhardt
    Ruth Coleman Waterlin
    Don Johnson

Illustrations marked "from Healey" are used by permission of John E. Healey, M.D., from his book, *Synopsis of Clinical Anatomy*, Philadelphia, W. B. Saunders Co., 1969.

Illustrations marked "from Bloom and Fawcett" are from: Bloom, W., and Fawcett, D.: *A Textbook of Histology*. 9th ed. Philadelphia, W. B. Saunders Co., 1968.

Illustrations marked "from Leeson and Leeson" are from: Leeson, T. S., and Leeson, C. R.: *Histology*. 2nd ed. Philadelphia, W. B. Saunders Co., 1970.

Illustrations marked "from Huffman" are from: Huffman, J.: *Gynecology and Obstetrics*. Philadelphia, W. B. Saunders Co., 1962.

Listed here is the latest translated edition of this book together with the language of the translation and the publisher.

Spanish (2nd Edition) — NEISA, Mexico 4 D.F., Mexico

W. B. Saunders Company: West Washington Square
    Philadelphia, PA 19105

    1 St. Anne's Road
    Eastbourne, East Sussex BN21 3UN, England

    1 Goldthorne Avenue
    Toronto, Ontario M8Z 5T9, Canada

Structure of the Human Body   ISBN 0-7216-4021-4

© 1973 by W. B. Saunders Company. Copyright 1967 by W. B. Saunders Company. Copyright under the International Copyright Union. All rights reserved. This book is protected by copyright. No part of it may be reproduced, stored in a retrieval system, or transmitted in any form or by any means, electronic, mechanical, photocopying, recording, or otherwise, without written permission from the publisher. Made in the United States of America. Press of W. B. Saunders Company. Library of Congress catalog card number 72-82804.

Print No: 9   8   7   6

Preface to the
Second Edition

The authors are pleased that the first edition, including its Asian offprint and its Spanish and Portuguese translations, has been useful to faculty and student consumers to the point that a second edition has been indicated. Our objectives in this edition are the same as in the first, although we recognize the extension of usefulness to a larger reader group. The preface to the first edition, on the pages following, states that the reader objective of this book is the college and university student who is studying gross human anatomy for the first, and perhaps only, time. The fact that some medical and dental professors have found this text useful, as the time allotted to anatomy in those professional curricula has been reduced, is gratifying to both the authors and the publisher.

The second edition has been expanded to present more information which the authors feel will be useful to both teachers and students.

A section on the basic steps in human development has been added to Chapter 1 to set the scene for the many embryological correlations which occur throughout the text. The careful scrutiny of manuscript by, and the suggestions of, Dr. Stanley Kaplan of The Medical College of Wisconsin are acknowledged, with special appreciation for his contributions as an embryologist and teratologist.

Although this text is devoted to systematic gross anatomy, the first edition presented microscopic correlations where they were believed to be helpful. These have been extended in this edition to broaden the coverage of microscopic anatomy in the respiratory, alimentary, genitourinary, endocrine, and integumentary systems. The critical reading of the manuscript by, and suggestions of, Dr. Lowell A. Sether of The Medical College of Wisconsin are gladly acknowledged in this regard.

Especially new are the ultrastructural correlations which are placed in special indented paragraphs of smaller type so that they may be used by the teacher in accord with his course objectives. The ultrastructural passages have been reviewed by, and the electron photomicrographs of Chapter 1 have been provided by, Dr. Kenneth A. Siegesmund and Patricia J. Delaney of The Medical College of Wisconsin. Ultrastructural illustrations in later chapters have been borrowed with permission from the publications of Dr. Don W. Fawcett of Harvard Medical School, Dr. C. R. Leeson of the University of Missouri, Columbia School of Medicine, and Dr. T. S. Leeson of the University of Alberta Faculty of Medicine.

The chapter on the nervous system has been improved by bringing the section on functional nerve components into exact coincidence with present teaching in neuroanatomy. In addition, the viewpoints on visceral sensory components, the autonomic nervous system, and the rami communicantes have been tuned to contemporary teaching with the assistance of Drs. Robin L. Curtis and F. David Anderson of The Medical College of Wisconsin.

Passages on the gamma efferent system, muscle spindles, and related proprioceptive pathways have been added in response to the use of these concepts by physical and occupational therapists.

Many new drawings have been added to the list of illustrations to accompany new portions of the text. Many illustrations have been enlarged to full column size in response to suggestions from teachers and students. A few of the illustrations from the previous edition have been modified to reflect current information and terminology. The color atlas plates have been divided for greater readability and increased in number in this edition.

The contributions of certain faculty members who have used the first edition are especially acknowledged. Dr. Frances H. Higginbotham of West Virginia University School of Medicine marked author and typographical errors, page by page, in the first printing. She has also developed a student workbook designed to parallel this text for use in elementary anatomy courses. Dr. Richard W. Coleman of Upper Iowa College, Fayette, Iowa, has shown by course outlines, study sheets, and examinations how a college course sequence can be designed around this text. Both scientist-teachers, we are sure, would welcome correspondence about their innovations in the use of this book. Dr. W. Clyde Skillen of West Chester State College, West Chester, Pennsylvania, has presented viewpoints about the functional attachments of the abdominal muscles, which are incorporated as a footnote in the text. Finally, Dr. Ronan O'Rahilly has convinced the text author that the "upper extremity is the superior pole of the skull, and the lower extremity of the body is the sole of the foot." We have changed every "extremity" reference to "limb," in deference to his anatomic purism, but seriously doubt if physical and occupational therapists will pay much attention, as they use the term "extremities" daily in their work.

The continued rhetorical assistance of Barbara P. Gardner, R.N., is once again warmly acknowledged.

The authors have had three successive W. B. Saunders Company editors whose help has been invaluable. Tyler Buchenau retired with the first edition. Carl May ensured the continuance of this text and helped set the tone for changes in the second edition. Richard Lampert has seen this edition through to completion. Lorraine Battista has again created the typographic design which, in itself, ensures the excellence of appearance and readability of a text.

Students at the Medical College of Wisconsin in using this book have also brought shortcomings to the attention of the author of the text. In the future, as in the past, the authors will welcome teacher and student suggestions arising from experience with this edition.

WESTON D. GARDNER

WILLIAM A. OSBURN

Preface to the First Edition

Anatomy has different meanings to persons who prepare themselves to work in each of the many facets of the health sciences. Many advances made since the Second World War have created the need for a vast corps of teachers, specialists, and technicians to assist the physician and nurse in providing total health and therapeutic care. Social changes, spurred by recent legislative trends, have accentuated this need and have produced even greater demands for trained personnel in the health professions.

The skills required in the rapidly advancing health fields must be built upon a more extensive foundation of knowledge of human structure and function than that provided by introductory courses of the past. Occupational standards have caused curricular requirements to become more stringent. Well trained and research minded faculty members now direct introductory courses of anatomy in the colleges and universities. Since students come into these collegiate courses well prepared by the science courses of modern high schools, it is obvious that students and their teachers require a great deal from their texts.

This book has been written to meet the needs of the student who, in preparing for one of the modern health fields, is making what may be his only exploration into the study of anatomy. Its objective is to provide a verbal and visual description of the structure of the human body in which only the facts of human morphology which can be related to the living body are included. These facts are centered about the structure of cells, tissues, and organs, and their interaction with each other as the person lives, works, and plays. Although the emphasis is upon fibrous, skeletal, muscular, and nervous tissues, all systems of the human body are described in their roles in making up the human body.

Two persons have collaborated to create this book: a physician anatomist who is actively engaged in teaching and illustration, and a medical illustrator whose academic preparation included biology, anatomy, and other medical sciences. As persons involved in the educational preparation of students who are entering the health professions, the authors realize that introductory courses in human anatomy vary in content and in length. It is recognized that teachers will present material selectively and assign text readings in accordance with the objectives of their own courses and the needs of their own academic environment. This can be done only if the text is adequate and if the student realizes that his instructor preplans the assignments in accord with the particular objectives of the course.

With these factors in mind a number of unique presentations have been designed to facilitate the learning required to achieve a broad understanding of human anatomy. A unique verbal-visual correlation is basic to this book and results in illustrations which are part of its plan for learning. Throughout the book the text and illustrations are inseparable. The illustrations were designed to depict single concepts or a few pertinent facts in the simplest possible manner. The book has been designed so that these illustrations are placed exactly where they may be read along with the text not only to offer visual reinforcement but also to supply information necessary to the understanding of the text. Especially unique are the tabular presentations of muscles, nerves, and blood vessels, in which illustrations are incorporated with the tabular text so that each structure is described visually as well as verbally. This style of presentation should provide a useful guide for the student of this text and even for those pursuing more complex professional studies.

Since the learning of structure of complicated adult organs is facilitated by knowing how they developed from much simpler forms, developmental anatomy has been included wherever the facts of embryology will contribute to interest and understanding.

Many of the readers of this book will go on to study the effect of injuries, disease, and aging upon the human body. Mention is made, therefore, at appropriate points of the application of basic human morphology to function or to disturbances of human structure. Paramount at the outset, however, is the need to master the elements of human structure as a sound foundation for future studies and work. This need has been the criterion for the inclusion and arrangement of subject matter.

The latest revision of official anatomic terminology, *Nomina Anatomica* anglicized, as recommended to the Sixth International Congress of Anatomists, Paris, 1955, and adopted by the Seventh International Congress of Anatomy, New York, 1960, has been followed in the text, but older, more familiar terms and even words in lay usage are included where they are useful in description.

It is impossible today for anyone to write or to illustrate human anatomy without being influenced by and indebted to the great teachers of anatomy and the masters of illustration of the past and present. Both fields continue to progress with new developments in knowledge and techniques. The authors are mindful of their indebtedness to both their predecessors and their peers.

The contributions of Dr. Lowell Sether, Dr. Kenneth Siegesmund, and Mr. Alan Becker of the Department of Anatomy of Marquette University Schools of Medicine and Dentistry are acknowledged. As experienced medical-dental-paramedical teachers they have served as consultants and as readers of the text and illustrations.

Few books today carry a dedication. Barbara Peever Gardner, R.N., has had a previous book in anatomy dedicated to her. That she has served once again as critical reader, rhetorical consultant, typist, and source of inspiration must be recorded here.

The special and unusual correlation of text and illustration required over 625 new, specifically designed drawings, obviously a job of such magnitude that it demanded the cooperative efforts of a sizeable group of artists working under unified direction. The authors wish to gratefully acknowledge the services of these talented medical illustrators whose names are listed on the title page, and those of Miss Mary Ellen Durning, who contributed intelligently and artistically to the page design, incorporating text and illustration.

The index for this book was made thoughtfully by the person who actually wrote the words of the text. As far as possible the index follows the anticipated thoughts and needs of the readers. It is a part of the learning plan of the book, intended to be used frequently to expedite the finding of the facts contained.

It is the sincere hope of the authors that this book will serve the purposes of the student with a strong background in biology who needs to master the elements of human structure as a sound foundation for further studies and future work in biological and health sciences. Suggestions for improvement or additions will be welcome always from both student readers and faculty members.

Milwaukee, Wisconsin   WESTON D. GARDNER

Dallas, Texas   WILLIAM A. OSBURN

Contents

Chapter 3

THE SKELETAL SYSTEM AND ITS JOINTS.............................. 77

Chapter 4

Chapter 7

THE RESPIRATORY SYSTEM

Chapter 8

THE DIGESTIVE SYSTEM

Chapter 9

THE GENITOURINARY SYSTEM

Chapter 10

THE ENDOCRINE SYSTEM

Chapter 11
THE INTEGUMENTARY SYSTEM

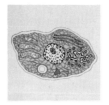

1

Introduction

Anatomy is the study of the structure of living things. This definition includes the structure of the myriads of viruses and microorganisms, the simplest and most complex plants, and all the phylogenetic ranks of animal life which precede man in the science of biology. Human anatomy, with which this book is concerned, is the study of the structure of the living human body. Since structure is the framework that makes function possible, knowledge of human anatomy provides the basis upon which an understanding of function depends.

Approaches to Anatomy

Because human anatomy has been one of the oldest objects of man's curiosity it has developed into one of the most extensive bodies of knowledge. There are several ways to approach the facts of human structure which have been accumulated for centuries and have been refined or added to by modern developments in tools and techniques.

One approach to human anatomy is to realize that the human body is composed of cells, which combine to form tissues, which in turn combine in a variety of ways to form specific organs. Similar organs or even different ones, which unite either in location or in function to perform specific roles for the body as a whole, form organ systems. The human body may be studied by learning about one system at a time in its entirety, no matter into what portions of the body it may extend. Although parts of a particular system may come into relation with another, the major focus is upon the system being studied. This approach to the human body is known as *system-*

atic anatomy. This book is based largely upon such a systematic approach.

Another approach to human anatomy is found when the human body is looked upon externally. Systems are not seen, but the body appears to be made up of characteristic parts, such as the head, neck, trunk, and limb. One of these regions may be studied at a time in its entirety, taking notice of the appearance of structures of any systems which may be present. Attention is paid to the relationship of one component to another and its depth or extent. This approach is termed *regional anatomy*. From time to time in this book regional information will be presented.

Special Fields of Anatomical Study

It has become customary to divide the extensive subject matter of anatomy into special categories according to the concepts contained or the methods of study.

Gross anatomy is concerned with the structure of the human body which may be observed by the unaided eye. It is historically much the oldest field of anatomy, for its findings are largely derived from external observation or from dissection which did not have to await the development of special tools or techniques. This book is concerned mainly with gross anatomy.

Microscopic anatomy is devoted to the finer details of structure which can be recognized with the microscope. This body of knowledge is sometimes called *histology* or the study of tissues. The field has progressed with refinements in microscopes and the techniques of preparing tissues for study. Although the light microscope is the classic tool of microscopic anatomy, new knowledge is constantly being acquired or old facts revamped by phase microscopes, ultraviolet and x-ray diffraction techniques, and the electron microscope. Investigation of the structure of tissues by a combination of microscopic and biochemical techniques has brought about the development of the new field of *histochemistry*. When the findings of microscopic anatomy are pertinent, they are mentioned in this book.

Developmental anatomy is the study of the human organism from its conception until its mature adult form is attained. One phase of the study of the development of the new individual up to the time of birth is a separate science, *embryology*. Developmental anatomy is concerned also with the growth of human structure throughout childhood and adolescence.

Topographic anatomy is limited to the conformation of the surface of the human body and includes the changes in form created by the action of muscles and changes in posture. Since these changes require life and motion, this study is often called *living anatomy*.

Neuroanatomy is a special study of the nervous system and sense organs of the body. It combines knowledge gained by gross and microscopic means. Today experimental approaches based upon electronics, physiology, and surgery are important in this field.

Radiological anatomy contributes knowledge which may be disclosed by x-ray methods. It reveals the pattern of structure in tissues or organs by means of shadows cast on a photographic film when x-rays are directed through the body.

THE ORGANIZATION OF LIVING MATTER

LIFE AND LIVING MATTER

One who begins the study of anatomy has already gained insight into the origins and characteristics of life during an earlier study of biology. The quality of life, which escapes easy definition, has been the climactic result of a number of processes in the physical world which began with the proper combination of elements under suitable environmental conditions in past eons of time. Compounds resulting from the combination of essential elements were forged into ever more complex molecules under the influence of catalysts and enzymes. The process continued with the formation of macromolecular complexes. Many similar complexes may be synthesized in modern biochemical and biophysical laboratories, but the products do not have the qualities associated with living matter or protoplasm. Protoplasm is differentiated from nonliving combinations of similar chemical complexes by its possession of characteristics considered to be the *attributes of life*. The attributes of living matter are:

1. Organization into units of specific size and shape.

2. Ability to enter into chemical activities of a constantly changing nature, resulting in the building or maintenance of protoplasm and the transformation of energy.

3. Movement.

4. Irritability, or the response to stimuli from the external environment.

5. Growth through an increase in the size and number of its structural units.

6. Ability to reproduce.

7. Adaptation to changes within the external environment.

CELLS

Protoplasm has been defined as the living substance of which an organism is composed. At low levels of magnification protoplasm appears to be structureless, i.e., an amorphous, viscous mass. This mass, however, is organized into minute bits of the living substance called *cells*. Cells are the units of structure of the living body. The pioneer of microscopy, Van Leeuwenhoek, saw cellular forms with his crude lenses. Hooke described them in the plant structure of cork. Schleiden and Schwann enunciated the cell theory which stated that cellular units were independent organisms which gathered together in complex ways to form plants and animals. Later came the dictum that all cells originate from pre-existing cells by division of their central nuclei. This law, *omnia cellula e cellula*, put to rest previous theories of the origin of life such as spontaneous generation. The later work of Virchow, which elaborated the theory that disease occurs by alterations in normal cells, established the basis for medicine as we know it today.

There are millions of cellular units in the human body. Each cell possesses the attributes of life and manifests the form and structure which is characteristic of the function it performs in the body.

Cell Structure in General

With the development of staining methods to reveal cell structure, light microscopy showed that a cell is surrounded by an envelope or membrane.

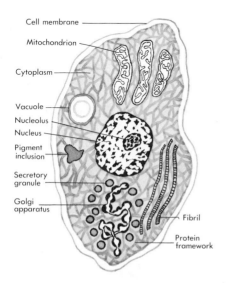

Cell membrane
Mitochondrion
Cytoplasm
Vacuole
Nucleolus
Nucleus
Pigment inclusion
Secretory granule
Golgi apparatus
Fibril
Protein framework

A central structure, the *nucleus,* is imbedded in a viscous mass of protoplasm which contains other structures, living and nonliving. Such a view of the cell was held until the 1950's. We know that strides in acquiring knowledge are greater following the development of new tools for research. Among several new techniques for cell observation was the electron microscope. This instrument has taken human vision beyond the limit obtained with the light microscope. For instance, light microscopy can disclose structures measured in *microns* (thousandths of a millimeter), but the electron microscope reveals forms measured in *angstroms* (ten-thousandths of a micron). Therefore, the knowledge gained by the electron microscope is not a separate body of facts but an extension of the previous knowledge of the cell into a new frontier in which protoplasm is composed of molecules and macromolecular complexes, which in turn are recognized as the structures found by light microscopy.

In the brief account of cell structure which follows, those factors which are common to both light and electron microscopy are mentioned first. Accessory facts, related to newer knowledge of "ultrastructure" gained by biochemical or electron microscopic study, are presented in smaller type for those readers who desire this further information.

Overview of the Cell

The cell, without the aid of any staining, looks like a transparent, slightly refractile, jellylike sphere. When treated with chemicals in preparation for staining, the cell protoplasm undergoes precipitation of its electrolytes and coagulation of its proteins. After being stained with dyes, the cell appears to have a dense external membrane, an amorphous or delicately granular protoplasm, and a very dense, central body, the nucleus. The appearance of the protoplasm is deceptive. The granularity hides a host of submicroscopic structures and a lacy protoplasmic maze upon which, and within which, a number of complex proteins, carbohydrates, and fats intermix and interact. The myriad physical interfaces made possible by the lacy protoplasmic framework allow many metabolic reactions to proceed simultaneously, speeded by the catalytic force of enzymes. Such reactions are related to the nourishment of the cell, its energy reactions, or its special products.

The Chemical Basis of Protoplasm

The complex chemical structure of the cell may be reduced to very simple substances of molecular nature. These combine to form larger aggregates or *macromolecules.* There are three basic kinds of macromolecules in the cell: proteins, nucleic acids, and polysaccharides. It is important to understand these before going on to further study of the cell.

Proteins are made available to the cells by the digestion and absorption of foods and their transport into the tissues of the body. Proteins must be reduced to their simplest form, molecules of *amino acids,* before they can be used by the cell. These simple molecules are building blocks, or *monomers,* for cell activity. Cells use energy to chemically combine amino acids into macromolecular chains called *peptides.* These chains lengthen to form *polypeptides* by the process of *polymerization.* Many living structures of cells and tissues are formed by these macromolecules or *polymers*.

Nucleic acids are more complex macromolecules in which amino acids, sugars, and phosphates are combined. Nucleic acids have been shown to be of great importance in the synthesis of proteins which form many parts of cells and tissues. Nucleic acids are also responsible for the storage and transfer of genetic information about an individual and his cells during reproduction and development. There are two types of nucleic acids. For simplicity, they are referred to as DNA (deoxyribonucleic acid) and RNA (ribonucleic acid).

The *polysaccharides* are macromolecules in which many simple sugars are polymerized to produce more complex substances. These substances will be mentioned as they participate in the structure of cells and tissues.

The cell is made up principally of water. In watery solution are many electrolyte ions, which are of great import to the life and integrity of the cell. Sodium, potassium, magnesium, calcium, and iron may be singled out, as well as the chloride, phosphate, and carbonate ions.

Cell Membrane

At the periphery of the cell, the protoplasm may be thickened to form a limiting membrane.

This may be punctured by microscopic manipulation to collapse the cell and to permit less viscous portions of the protoplasm to escape. In other cells the surface protoplasm may act as a membrane, rather than forming a definite boundary. In either case the cell is given form, its contents limited, and the diffusion of substances into and out of the cell permitted.

The cell membrane, also called the *plasmalemma,* usually consists of three layers forming a trilaminar boundary membrane at the surface of the cell. This *unit membrane* is the standard

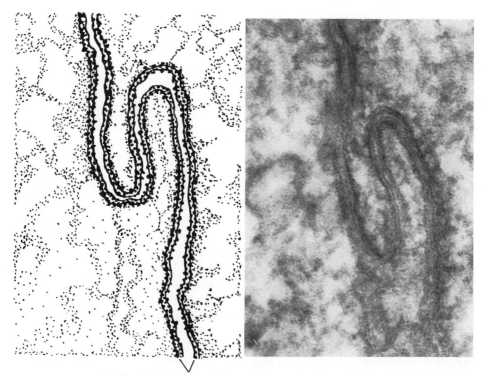

Trilaminar unit membrane
X80,000

construction form at the surface of all cells and often makes up the surface of other structures which require a limiting membrane. The macromolecules comprising the unit membrane are made up of a combination of proteins, lipids (fats), and phosphate ions. The macromolecules aggregate in such a way that openings occur between them. The *pores* thus formed permit selective passage of elements into and out of the cell by diffusion. Sometimes polysaccharides are found coating the unit membrane. These are a defense against bacteria or toxic substances, for the coating must be dissolved before entry can be gained to the cell.

Nucleus

The structural and functional center of activity in a cell is to be found in a dense viscous body composed of complex proteins: it is termed the nucleus. The nucleus contains the chromosomal material responsible for the preservation and reproduction of genetic traits (genetic coding) and is believed to regulate the metabolism and function of the cell. The nucleus is surrounded by its own nuclear membrane.

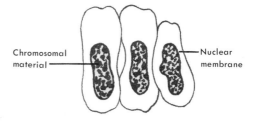

The nucleus is usually located in that part of the cell where metabolic activity is greatest. In tall gland cells, for instance, the nucleus is found near the base where secretory activity is high.

The control of the cell's activity in metabolism and in the synthesis of macromolecules is believed to be the function of the nucleus. The regulation of genetic processes also is believed to be at the direction of the nucleus, because of the *chromosomes* which have been found to be tightly coiled helices of the nucleoprotein DNA. DNA in a dispersed form is the *chromatin material* seen within the nucleus of a nondividing cell.

The *nuclear membrane* is composed of an inner and outer layer with a clear space between.

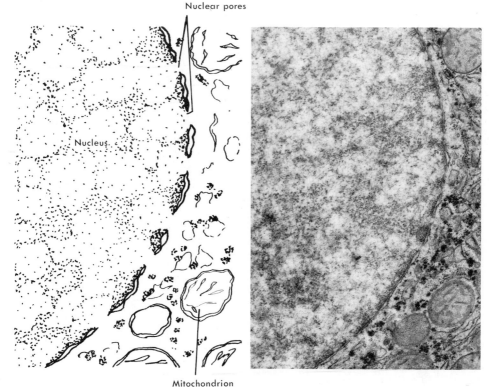

Nuclear pores

Nucleus

Mitochondrion

(X28,000)

There are gaps, *nuclear pores,* in the nuclear membrane, which are 1000 angstroms across. This makes continuity possible between the interior of the nucleus and the protoplasm of the cell surrounding it.

Within the nucleus is a very dense spherical body, the *nucleolus,* which is believed to be made up of the other nucleoprotein, RNA. The nu-

cleolus is prominent in cells which are synthesizing protein. RNA has been observed entering the nucleolus and then departing from the nucleus via nuclear pores into the protoplasm of the cell as if it were a messenger. It is probable that the nucleolus has DNA nucleoprotein-bearing genetic information, which is picked up by such messenger RNA for use outside the

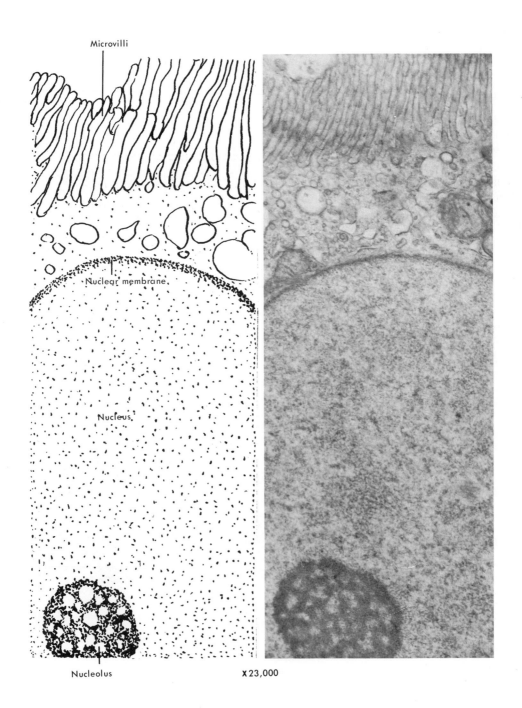

Microvilli

Nuclear membrane

Nucleus

Nucleolus   X 23,000

nucleus. The nucleolus is less dominant in inactive cells and actually shrinks in cells which are dying.

Cytoplasm

The remainder of the protoplasm of the cell is called the *cytoplasm*. It has the characteristics of protoplasm described previously, is limited by the cell membrane, and may be thought of as containing the nucleus.

Contents of Cytoplasm. Formed structures are contained within the cytoplasm, in addition to the submicroscopic protein framework. The cytoplasmic contents take many forms. Some are living portions of the cytoplasm (*organelles*), whereas others are products of cell activity or substances which have passed through the cell membranes (*inclusions*). Mitochondria and the Golgi apparatus are types of organelles.

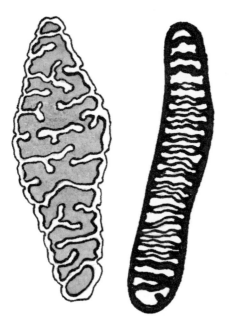

The *mitochondria* are complexly folded rods or cylinders, which are composed of fats or of combinations of fats and proteins. These are the site of enzyme activities which catalyze chemical reactions in the cytoplasm. The *Golgi apparatus* is a spirally arranged, tubular structure located near the nucleus; it is concerned with the formation of secretions by the cell.

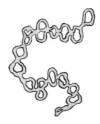

Inclusions take the form of *vacuoles, granules, pigments,* and particulate matter. Nutritive substances recently absorbed by the cell are frequently of a density different from that of the cytoplasm. These substances collect as fluid droplets called vacuoles. Secretions manufactured by the cell appear first as minute droplets, which become larger as secretory activity increases. Although liquid in nature, they can be precipitated by using the staining techniques of light microscopy; thus they are commonly referred to as granules. The activity of certain cells is associated with the presence or the formation of pigmented materials which give their natural color to particles within the cytoplasm. Other particulate matter within some cells may result from the scavenging of debris and bacteria or from deterioration of the cell.

Electron microscopy has brought much new information about the rather granular, amorphous cytoplasm seen by light microscopy. It is now known that the lacy submicroscopic protein framework of the cytoplasm is actually a reticulated network of tubules or flattened sacs called the *endoplasmic reticulum*. These channels are continuous with the nuclear membrane at the nuclear pores. It is believed that such a tubular network has several roles which go far to explain some previous mysteries of cell function.

1. A tremendous surface area is provided by a myriad of anastomosing interfaces on which chemical reactions can occur simultaneously.

2. Separate locations are possible at which different enzyme reactions can occur.

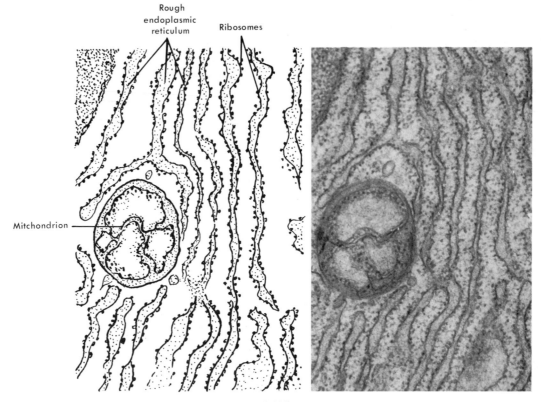

Rough
endoplasmic
reticulum

Ribosomes

Mitchondrion

(X40,000)

3. A channel through this tubular network of the cytoplasm makes it possible for large molecules to pass through the cytoplasm from the nucleus to the tissue spaces outside of the cell without contacting other reacting substances during passage.

Two kinds of endoplasmic reticulum are observed. The *granular* or rough form has small spherical bodies attached to the external surface of the channel (See illustration above.) These bodies are *ribosomes* which, in their attached forms, are manufacturing proteins. It is believed that *messenger RNA* carries genetic code information from the nucleus into the cytoplasm. Another RNA form, *transfer RNA,* picks up a single amino acid from the cytoplasm and takes it to the surface of a ribosome, where other amino acids are polymerized with it to form a peptide chain and subsequently a protein. (See illustration at top of the following page.) The *smooth* form of endoplasmic reticulum lacks ribosomes. Fat metabolism and the storage of carbohydrate as glycogen are ascribed to this form.

In some cells disconnected channels make up a system of *microtubules,* i.e., projecting *cilia* and internal *filaments.* (See illustration at bottom of page 10.)

A type of cell organelle which is very active in secretory activities is the *Golgi complex* or *Golgi apparatus.* (See illustration at bottom of following page.) With electron microscopy the copy is seen to be more like a system of flat sinuses, which are stacked like saucers near the nucleus. *Vacuoles* are frequently found in association with this complex when a cell is actively secreting. Vacuoles appear to bud from the end of the Golgi sacs like small bubbles. This apparatus is continuous with both forms of endoplasmic reticulum. It is believed that raw materials travel through the endoplasmic channels to the Golgi apparatus where enzymes or secretions are synthesized and then released in packaged form as vacuoles or secretory granules. Most of the syntheses of carbohydrates and fats occur in this manner. The packaged enzymes for fat formation appear in a dense vacuole form called *liposomes.*

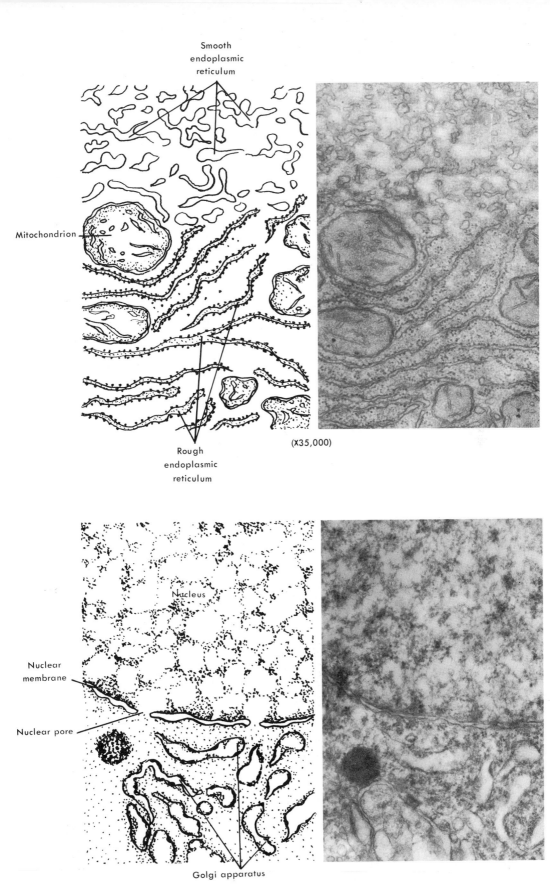

Smooth
endoplasmic
reticulum

Mitochondrion

Rough
endoplasmic
reticulum

(X35,000)

Nucleus

Nuclear
membrane

Nuclear pore

Golgi apparatus

(X35,000)

9

Mitochrondria, depicted previously, appear in the cytoplasm as complexly folded rods which have a typical trilaminar unit membrane. The

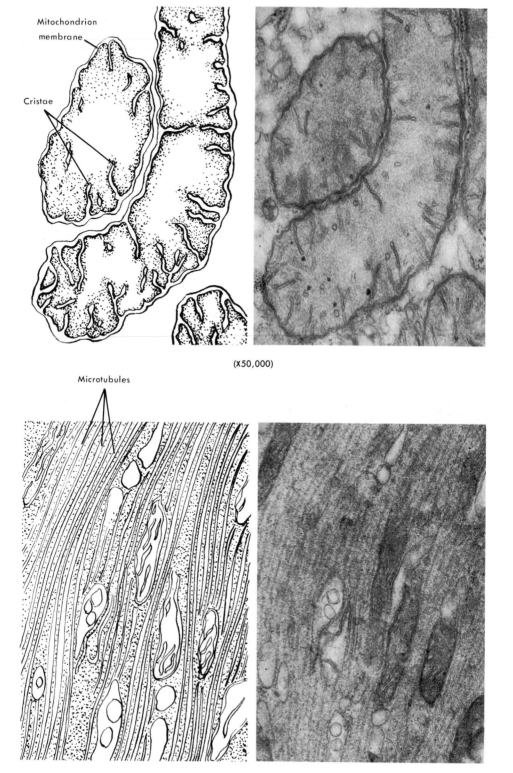

Mitochondrion membrane

Cristae

(X50,000)

Microtubules

(X35,000)

rods of mitochondria are hollow, and the membrane projects inward as *cristae*. Mitochondria, as living organelles, appear to be the cell's machinery for producing energy from food elements such as adenosine triphosphate (ATP). They do this through enzyme reactions which use oxygen, which is passed through the membrane of the mitochondrion. Mitochondria are also active in producing enzymes related to fat metabolism. As living organelles within the cell, mitochondria have their own DNA and reproduce by budding. They are very sensitive to any process which affects a cell adversely, and they react quickly by degenerative changes in themselves.

TISSUES

In living organisms composed of one or a few cells, the cells are remarkably alike and carry on activities which may be repeated from cell to cell. A corollary of the increasing complexity of living organisms is that cells lose this self-sufficiency. A cell becomes specialized to perform a specific role. All cells of a particular type acquire structural characteristics which are adapted for their function. As a consequence of their specialization these cells lose their ability to perform other functions which are now done best by other cells which have specialized differently. The result is that the human body develops as a community of mutually interdependent cells which, in a state of health, work in their own ways for the benefit of the organism as a whole.

A *tissue* is a collection of similarly specialized cells and their products which are united in the performance of a specific function. A tissue has three components: the cells which are characteristic of that tissue, an intercellular medium or tissue fluid, and intercellular products of cellular activity. These components are seldom seen in an exact balance. One is frequently dominant at the expense of the others. In the lining of membranes or organs the cells predominate. In tendons the intercellular product (fibers) is the major component. In the fluid tissues (blood and lymph) the intercellular medium, or plasma, assumes the leading role.

Intercellular Medium

All cells exist in an intercellular medium which may vary in consistency from a watery hydrosol to a viscid hydrogel. Even if the cells are closely packed so that they touch each other, or are compressed by the bulk of their intercellular products, there is always at least a film of intercellular medium about the cells. This tissue fluid bathes the cell and permits the diffusion of substances to and from the cells and the capillaries of the circulatory system. This medium has been called the interior environment of the body. There is a balance between the chemical composition of the intercellular medium and that of the blood. The intercellular medium is the depot, constantly replenished by the diffusion of substances from the capillaries, from which the cells draw the materials necessary for life, nourishment, and energy.

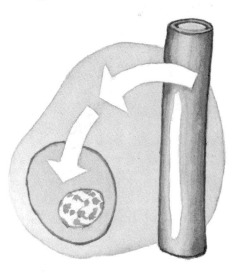

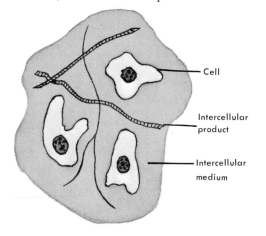

Cell

Intercellular product

Intercellular medium

Intercellular products are formed within this medium or are added to it by the function of the cells. It may be extensively modified by the cells in certain tissues so that resiliency or rigidity becomes its characteristic (as in cartilage and bone).

Intercellular Products

In some tissues the cells and the intercellular medium may be sufficient for function. This arrangement is found, for example, where the cells are aligned to form membranes, ducts, or glands.

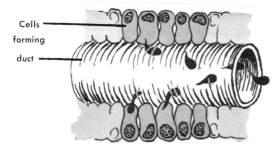

Cells forming duct

If products of secretion are formed which are conveyed to other areas of the body by the blood, these substances will be present only transiently in the intercellular medium as they diffuse toward the capillaries.

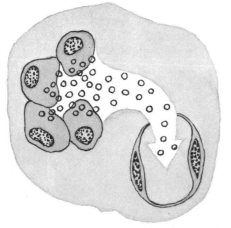

In contrast to these the term intercellular products is applied to substances or structures, formed in the intercellular medium as the result of cell activity, which are usually fixed in their position and are essential to the function

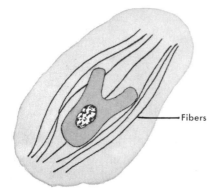

Fibers

of the tissue. The fibers of a tendon are an example of a true intercellular product.

THE FUNDAMENTAL TISSUES

At first thought it might seem that in an organism as complicated as the human body there would be a relatively large number of tissues. Analysis of the construction of the body, however, reveals that, although many tissues are widespread and occur in many different combinations, there are but four basic or *fundamental tissues.* Despite structural or functional modifications, each tissue of the body can be assigned to one of the fundamental tissues, and every organ can be analyzed into a characteristic combination of several or all of them. The fundamental tissues are epithelium, connective tissue, muscle tissue, and nerve tissue.

EPITHELIUM

Epithelial tissue is a membrane which covers the surface of the body and lines the cavities or tubes within the body. Epithelium is primarily a cellular tissue with few intercellular products and a minimal amount of intercellular medium. Fundamentally this tissue is a series of cells which abut upon each other or interlock to form a sheet.

A film of intercellular fluid between the cells provides a medium for the exchange of nutritive and waste substances by diffusion to and from the capillaries.

The capillaries occupy an adjacent tissue and are not components of an epithelium. It was believed formerly that in some epithelia the cells are united by an intercellular cement substance which they elaborate.

The junctions between epithelial cells must be secure if the cells are to maintain their alignment as a sheet covering a surface. The epithelial covering of the body (skin) or epithelial sheets lining passageways within the body must be flexible enough to provide for stretching, contraction, or movements of the organs of which they are a part. These factors are reason enough for the reduction in intercellular material or fluid, since their presence would reduce the adhesion of cells.

Between some epithelial cells an adhesive may be formed of polysaccharide macromolecules reminiscent of intercellular cement. Electron microscopic study reveals, however, that the dense lines and bars seen between some adjacent face, the cells may part slightly, with a widening of the intercellular space and a small increase in intercellular material (the former intercellular cement and terminal bars between cells seen by light microscopy).

3. *Desmosomes* occur where the surface cytoplasm becomes very tense at a point of considerable widening of the intercellular space. Very dense dots appear at the cell surface (desmosomes), and these are associated with cytoplasmic *filaments*. The latter are aligned perpendicular to the cell surface, as if to strengthen the cell surface or to provide for surface tension of the cytoplasm.

4. In places where cells are subject to a shearing force which might pull them apart, the

Intercellular bridges

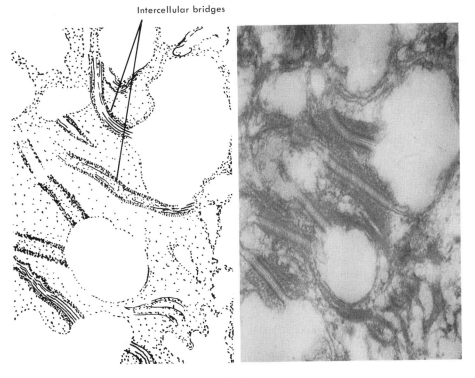

(X60,000)

epithelial cells are more often formed by arrangements of the unit membrane and the surface cytoplasm than by true adhesive material.

1. At the surface of adjacent cells, the inner lamina of the membranes may fuse to form a *tight junction,* so that surface materials cannot penetrate between the cells.

2. Below the tight junction at the cell sur-surfaces are formed into minute interlocking projections.

Arrangements of Epithelial Cells

Epithelial cells are arranged according to their function. Those which line an internal cavity where protection is not important form a sheet

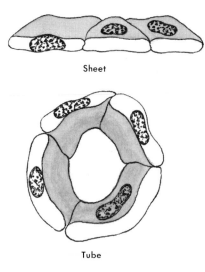

Sheet

Tube

of very flat cells, one layer in thickness, called a *simple squamous epithelium*. The sheet of cells may extend across a surface or may be rolled to form the inner lining of a tube. Some epithelial cells may have more than a simple lining or covering function. If the membrane requires more substance, or if either secretion of cell products or absorption is its role, the single layer of cells thickens. In this way *simple cuboidal* or *simple columnar epithelia* are formed.

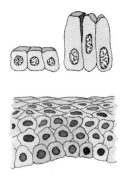

If an epithelium is more than one cell in thickness it is called a *stratified epithelium*. Such a membrane is better suited for protection and is found covering the external surface of the body as part of the skin. The lining membranes of organs which may be subjected to friction (esophagus) or to damage by harsh agents in the environment (respiratory system) become stratified. A simple epithelium cannot stretch or retract well enough to accommodate changes in the surface area of organs that periodically distend.

In such an organ, as for instance the urinary bladder, a special form of stratified epithelium is found which permits cells to slip past each other as changes occur in the surface area. In stratified epithelia, new cells are continually formed by the division of cells in the deepest layers to replace aging or injured cells at the surface.

Special Epithelia

Certain groups of simple epithelia are noted for their performance of special functions.

Endothelium. Endothelium is a simple squamous type of epithelium in which the membrane consists of a single layer of flattened cells rolled into a tube. Endothelium forms the inner lining of the vessels of the circulatory and lymphatic systems and the internal lining of the heart. The thinness of the epithelium permits diffusion of substances through capillary walls. Blood cells can migrate through the endothelial walls by insinuating themselves between the cells. The smoothness of endothelium enhances the propulsion of blood through the vessels.

Mesothelium. Mesothelium is a type of simple squamous epithelium which forms the smooth and delicate internal lining of the closed body cavities in which the heart, the lungs, and the organs of the abdomen are located.

Mesenchymal Epithelium. Mesenchymal epithelium is also a simple squamous epithelium. This type lines the internal surface of small sacs and cavities of the body. Mesenchymal epithelium is found forming lubricating sheaths about tendons. It makes up friction- and pressure-relieving sacs (bursae) at points where tendons or muscles cross over bony prominences. The membranes which enclose joints are lined by this form of epithelium.

Membranes

A membrane is a sheet of tissue which covers a surface, lines a cavity or passageway, or subdivides a portion of the body. Some membranes which require strength to protect or to enclose an organ must be of a fibrous nature, but many other membranes of the body are primarily epithelial in nature.

Mucous Membranes. Early anatomists found many membranes which were covered by a film of mucus and named them *mucous membranes*. A more accurate description is that mucous membranes are those which line the internal surfaces of organs or cavities which communicate with the outside of the body. When reinforced by a layer of connective tissue, the epithelial mucous membrane is termed a *mucosa.* Mucous membranes are found in the alimentary, respiratory, urinary, and reproductive systems.

Serous Membranes. The older anatomists also noted that other membranes were associated with a watery surface film and called them *serous membranes.* Today this term is applied to epithelial membranes lining the internal surfaces of the large and small closed body cavities which do not communicate with the external environment. The combination of the epithelium with a strengthening layer of fibrous tissue is a *serosa.* The special epithelia of mesothelium and mesenchymal epithelium are found in serous membranes.

Glands

Some epithelial cells have a more specialized function than that of lining a surface. Although these cells may be situated in a lining membrane, they form substances within their cytoplasm which pass through the cell membrane to be used on the surface. These substances are secretory products or *secretions.* The process of their formation and release by the cell is called *secretion* and the cells are known as *secretory* or *glandular cells.*

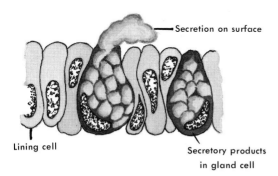

Lining cell

Secretion on surface

Secretory products in gland cell

Examples of secretions are the mucus of the nasal passages, saliva, and the digestive enzymes.

In some epithelia the secretory cells may become so numerous that they form most of the membrane and combine both lining and secretory functions. If a greater mass of secretory cells is needed, they clump together in separate units. Space requirements may force these cells to move

away from the surface of the membrane into other tissues of an organ. The clump of secretory cells is then called a *gland.*

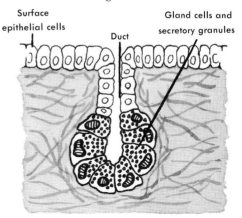

Surface epithelial cells

Duct

Gland cells and secretory granules

Glands remain connected to the surface by a cord of cells which hollows out to form a tube, the *duct* of the gland, through which the secretions are conveyed to the surface. Large glands, which require more space than the original organ can provide, move beyond its borders to form separate organs. Only the duct connecting this gland to its parent organ remains to indicate the point from which the gland developed. The liver is an example of such a large gland.

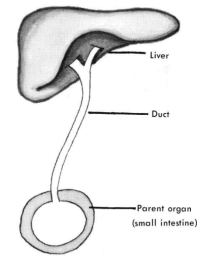

Liver

Duct

Parent organ (small intestine)

The internal structure of glands varies from simple tubes or sacs to complex branching ducts and units of secreting cells. Glands with ducts to convey secretions to a specific location are called *exocrine* glands. Some glands have no ducts; their products are needed by many or all cells of the body and require a greater, or even a universal, distribution. These glands are permeated by capillaries of the circulatory system which provide for the distribution of the secretory products; such glands are called *endocrine* glands. The salivary glands are examples of exocrine glands, whereas the thyroid gland is an endocrine gland. The pancreas is an example of a gland which combines both endocrine and exocrine functions. Its product, insulin, is endocrine in nature, but the elaboration of digestive enzymes for use in the alimentary system is its exocrine function.

CONNECTIVE TISSUE

Connective tissue is a fundamental tissue which is widespread throughout the body. It serves many functions which may be classified very generally as packing and filling, uniting dissimilar structures, and resisting physical forces. These many functions result in tissues of varied appearance. There are basic characteristics, however, which are shared by all connective tissues: (1) The intercellular products are dominant in the tissue. (2) The products take the form of fibers. (3) The cells regulate the activity of the tissue and form the fibers in the intercellular medium.

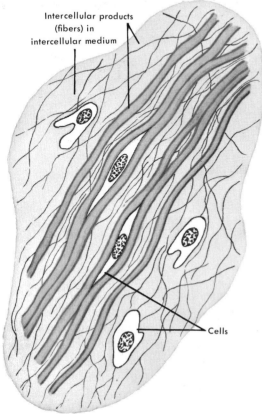

Intercellular products (fibers) in intercellular medium

Cells

Variation in the appearance and function of the various types of connective tissue are due to differences in the form and arrangement of the fibers, differences in cell activity, or modifications of the intercellular medium. These variations are described in Chapter 2, which is devoted to the connective tissues.

FLUID CONNECTIVE TISSUE

In this form of connective tissue the cells are suspended in a large amount of intercellular medium in which they are free to move. The intercellular medium is a watery liquid in which many substances may be dissolved or carried in a colloidal suspension.

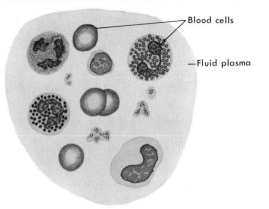

Blood cells

Fluid plasma

The fluidity of this tissue requires vessels composed of other tissues to convey it. Common fluid tissues are blood, lymph, and cerebrospinal fluid.

MUSCLE TISSUE

Muscle is primarily a cellular tissue in which the cytoplasm of the cells is composed largely of minute fibrils. The fibrils possess the quality of contractility. There are several types of muscle tissue in the human body. One type, smooth muscle, enables hollow organs to change their size and volume or to move the substances they contain.

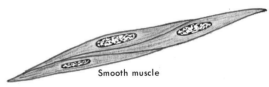

Smooth muscle

Another type, voluntary muscle, makes up the muscular system that produces movement of the body or its parts and maintains the position of the body in space.

Voluntary muscle

A third type of muscle tissue, cardiac muscle, makes possible contraction of the heart and propulsion of blood through the circulatory system.

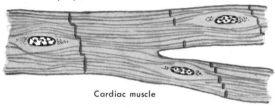

Cardiac muscle

In muscular tissues the intercellular medium is reduced in amount. There are no true intercellular products, although portions of other tissues and systems, such as nerves or capillaries, may be found between the cells.

NERVE TISSUE

Nerve tissue is another primarily cellular tissue in which the cells possess the properties of irritability and conductivity.

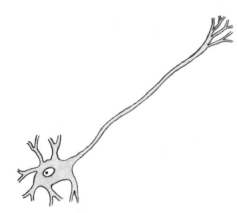

These qualities endow it with the ability to initiate and transmit the electrical impulses which regulate the activities of the body, provide it with awareness of the environment, and permit adaptation to environmental changes. See color plate 12 for a detailed illustration.

ORGANS AND SYSTEMS OF THE BODY

The human body is an extremely complicated arrangement of cells that are assembled into fundamental tissues. Like any complex body of knowledge, however, the study of the human body is facilitated when we realize that it is composed of basic units that may exist in a variety of combinations. Once the elementary units are recognized, they become the basis for understanding the whole. When the study of the complex human body is undertaken on the basis of either systematic or regional gross anatomy, it is found that the fundamental tissues described are the elementary units.

TISSUE COMBINATIONS

Although a tissue consists of a group of similarly specialized cells united in the performance of a specific function, these cells are often combined with other tissues. The combination of tissues may be necessary to the function of the dominant cell type. For example, connective tissue is combined with an epithelium in order to strengthen the sheet of epithelial cells and to bring capillaries into proximity with these cells.

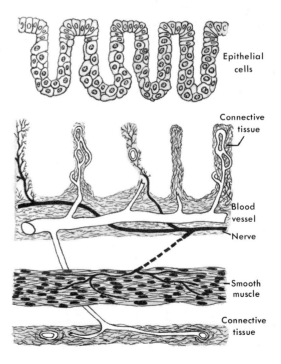

Epithelial cells

Connective tissue

Blood vessel

Nerve

Smooth muscle

Connective tissue

In another instance, the columnar cells of the epithelial lining of the stomach secrete mucus, hydrochloric acid, and digestive enzymes. The epithelial membrane forms a delicate sac which initiates some of the steps in the digestion of food. Connective tissue strengthens the membrane, provides space for its glands, and brings the blood supply needed by the epithelial cells. The food, however, must be churned in order to be mixed with the glandular secretions, and it must be moved along to other areas of the alimentary tract. These processes are facilitated by the addition of smooth muscle to the epithelial and fibrous sac. Still another fundamental tissue is added by the nerves which are necessary to regulate the secretion of the glands and stimulate the contraction of the muscle. A wrapping of more connective tissue completes the structure of one of the large functioning units of the organism—the stomach.

controlled. No single cell or tissue type can perform all these related functions but a group of related fundamental tissues can. Endothelium provides the smooth internal lining. An elastic form of connective tissue serves to resist the physical force and contribute to the wave motion. The size of the vessel is changed by the contraction and relaxation of smooth muscle.

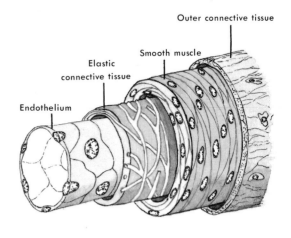

ORGANS

An organ may be defined simply as a unit or structure of the body which performs a specific function. In view of preceding descriptions, however, this definition may be given more substance. In the sense of the specialization of cells, an organ is a group of similar cells, or often several such groups, which have specialized to perform a specific function or related functions for the benefit of the organism. From the standpoint of the concept of fundamental tissues, it is appropriate to consider an organ as the combination of tissues into a unit for the performance of a specific function or a series of related functions. This concept has been illustrated in the description of the tissues forming the stomach.

The tubelike structure of an artery may be taken as a typical example of an organ. The specific function of an artery is to distribute blood to the tissues of the body. Certain requirements are imposed by this function. The organ must be a tube with an extremely smooth lining to reduce friction. It must be elastic to receive the jetlike spurt of blood and to pass it along in a pulsing, wavelike motion. The vessel must have a mechanism for changing its diameter so that the amount of blood flowing into part of the body can be adjusted and the pressure of its flow

The combination of these elements forms the organ, and their individual actions are coordinated to make its function possible.

ORGAN SYSTEMS

All the specific structures and organs of the body may be grouped according to similarities in structure or general function. Each group is an *organ system*. The grouping of organs into systems is in recognition of the fact that a number of organs may be involved in the performance of a general function. For instance, a number of muscles of widely different sizes, shapes, and locations possess the common property of contractility. They are united as an organ system by their function of moving parts of the body. On the other hand, a number of organs varying in size, structure, and location are grouped together as the alimentary system because of their role in related phases of the digestion of food.

The grouping of organs into systems also provides an approach for the study of the structure and function of the human body. This approach has already been described as systematic anatomy.

The systems of the body are as follows:

Skeletal system: the arrangement of bones and cartilage for rigidity and protection of the body.

Articular system: the means by which bones are joined together or move upon one another.

Muscular system: the grouping of muscles to move parts of the body and to preserve or to change the body's position in space.

Nervous system: the varieties of nerve tissue that make possible awareness of changes in the external environment, control of body activities, and adaptation.

Circulatory system: the blood and the organs which pump and distribute it to the tissues of the body.

Integumentary system: the skin and its modified structures which cover the body and perform special functions in secretion, excretion, and reception of stimuli from the external environment.

Alimentary system: the organs and passages which receive, digest, and absorb food. Some of the organs form digestive enzymes. Other organs excrete the unusable parts of food and other body waste products.

Respiratory system: the passages and organs which conduct environmental air into a close relation with the circulatory system for the exchange of oxygen and carbon dioxide.

Urinary system: a system of organs which remove waste products from the blood and eliminate them from the body.

Reproductive system: organs which form the sex cells in preparation for fertilization of the ovum. Organs of this system in the female nurture and protect the embryo from its conception until the time of birth.

Lymphatic system: a system of organs combining a tubular network for the return of some of the tissue fluid to the venous system with organs which filter the fluid and form certain blood cells.

Endocrine system: a group of widely scattered glands which distribute their products, hormones, by means of the circulatory system. Hormones may be needed for the metabolism of all body tissues (thyroid) or for the regulation of the development or activity of certain organs (pituitary).

THE LANGUAGE OF ANATOMY

The living human body is active and dynamic. Even in sleep it is rarely quiescent. Its position is subject to change in accordance with the will of the individual. In addition the body undergoes a continual series of muscular readjustments, many of which are reflex in nature, in order to maintain balance and posture. If the body as a whole is continually changing its position, so do the structures within it, for they are either producing movement or affected by it. Some organs possess mobility or are subject to changes in size or position as they function. Other organs may be loosely fixed within the body and may shift their position in response to movement of the body or changes in its center of gravity. Still other organs may remain in a relatively constant position but will displace other organs as they increase or decrease in size. Therefore, it is important to realize that the relationship of body components to each other may frequently change.

In order to study the organization of the human body it is necessary to arrest it in some arbitrary position. Then the structures of the body can be described in relation to that position. When information has been gained by reference to a standard static position, the study of functional changes can be undertaken. This standard *anatomical position* is the basis for all anatomic description. It is necessary also to employ accurate and precise terms in anatomic description so that what one states about a body part conveys the same meaning to everyone. These terms, which are used also in connection with the anatomical position, are grouped together as *anatomical terminology.*

THE ANATOMICAL POSITION

The human body is described and studied as if it were standing completely still in an erect or upright position. The head is positioned squarely upon the neck, untilted; the eyes stare directly forward; the chin is neither elevated nor depressed; and the nose is in the center line of the body. The weight of the body is transmitted downward along an imaginary line which is perpendicular to the floor. This line extends from just behind the ear, through the hip joint, grazing just in front of the knee joint, to a point on the foot in front of the ankle joint. The feet are directed forward. The arms are placed so that the elbow joints are fully straightened and the hands are rotated so that the palms face forward. This position of the forearm and hands is an admittedly unnatural one in life but makes accurate description possible.

PLANES OF THE BODY

The human body is a functioning unit, but many descriptions are made in terms of imaginary planes passing through it. A plane passing downward through the body which divides it into right and left portions is a *sagittal plane*. If such a plane passes directly along the exact midline of the body, perfectly symmetrical right and left halves of the body are created; this is the *median sagittal plane*. Any other sagittal planes are parallel to this and are termed *lateral sagittal planes*. Lateral sagittal planes are usually identified by a point of reference, such as "through the midpoint of the right inguinal ligament" or "through the left shoulder joint."

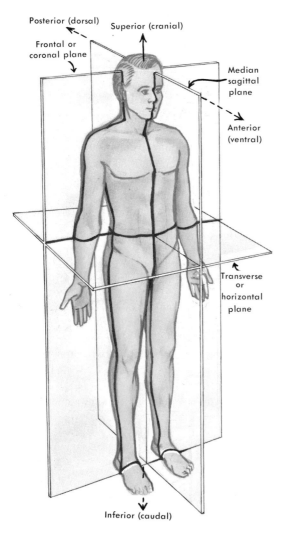

Posterior (dorsal)
Superior (cranial)
Frontal or coronal plane
Median sagittal plane
Anterior (ventral)
Transverse or horizontal plane
Inferior (caudal)

If a plane is passed horizontally through the body (in the anatomical position) so that it is divided into upper and lower portions, reference is made to the *transverse* or *horizontal plane*. A transverse plane is identified specifically by a reference point, such as "through the umbilicus" or "at the level of the iliac crests."

A plane dividing the body into front and back portions is referred to as a *frontal* or *coronal plane*.

TERMS OF POSITION

In anatomic description, specific terms are used to refer to the general location of a structure or to its position relative to other structures. Use of these terms avoids ambiguities which might arise if conventional lay terms were employed. *Medial* indicates that the structure at point lies nearer the median plane than another, whereas *lateral* designates a position farther from the median plane. Examples: "the medial end of the clavicle" or "the acromion process of the scapula lies lateral to the coracoid process." Occasionally some other central point of reference is used instead of the median plane. When this is done the reference point is always indicated in the description, as in the following: "At this point the ulnar nerve is medial to the brachial artery."

The terms "front" and "back" are usually taken by the layman to mean the corresponding surfaces of the body. Other terms are more acceptable in anatomy. *Anterior* may be used to refer to the front surface of the body, but it has a broader meaning in the sense of a structure's being *nearer* to the front of the body. Examples: "the anterior abdominal wall" or "the subclavian vein lies anterior to the subclavian artery." In the same way, *posterior* may refer to the back surface of the body or it may mean that a structure is nearer to the back of the body. Examples: "the posterior surface of the arm" or "the esophagus is posterior to the heart." The word *ventral* is used interchangeably with "anterior" and has the same meanings. *Dorsal* is linked similarly with "posterior."

In making the transition from the study of biology, one should avoid the use of "anterior" when referring to the head end of the body. Although this might be appropriate for the study of a four-footed animal, it can be seen from the previous description that such a use will introduce confusion when descriptions are based upon the anatomical position. *Superior* is used to indicate the head end of the human body or a position of relative nearness to it. Examples: "the superior limb" or "the apex of the lung is superior to the first rib." *Cranial* is also used in this sense, particularly in connection with the nervous system. *Inferior* is used to indicate the tail or lower end of the body and the relative nearness of a structure to it. "Posterior" is not used for this meaning because of the restricted meaning already described. *Caudal* as the opposite of "cranial" is in common use. "Above" and "below" are to be avoided whenever possible because the true meaning will be obscure if the position of the body changes.

Many structures must be described in relation to the depth of their location in the body. When a structure being described is located nearer the center of an organ, cavity, or part of the body, *internal* is the word of choice. If it is farther from the center, *external* is used. When reference is made to the surface of the body, the structure is said to be *superficial* or *deep,* depending upon the relative depth involved. Examples: "the endocardium is the internal lining of the heart" or "the periosteum is the external covering of a bone." In terms of depth, "the median cubital vein is superficial to the bicipital tendon" or "the peritoneum lies deep to the muscles of the abdominal wall."

Special terms are used in describing the limbs or structures related to their long axis. Only two of these are general enough to mention now. *Proximal* suggests that the item of concern is *nearer* to the attached end of the limb or thus the trunk of the body. A *distal* structure is *farther* from the attached end of the limb, and therefore it is toward the free end. Examples: "the artery enters the proximal end of the bone" or "the distal attachment of the muscle is to the tibial tuberosity."

USE OF TERMS

The study of anatomy is rich in the number of new terms which will be encountered. Ones which are related particularly to certain systems or organs will be explained in those sections of the book. New terms should be pondered until their meaning is clear. A medical dictionary is an invaluable aid to this end. It is important to remember that anatomic terminology is a language in itself. Certainly the subject must be studied, but to do so, its language must also be mastered. It is not sufficient to read this language and to understand. It must be written and spelled correctly. It must be spoken and pronounced correctly. This language is to be a tool of future studies and of later professional work. Therefore, it is suggested that when the principles of a unit of study have been grasped, an attempt should be made to write short descriptions or outlines in which an active use of the language of anatomy is made. Then speak it in recitations or in conversations with teachers and colleagues. Knowledge is rarely complete until it can be communicated in a professional manner.

THE NECESSITY FOR CLASSIFICATION

It has been said that nature is not arbitrary and does not classify the structures of which it is composed. If a portion of an organ is examined under a microscope, it will be seen that different cells exist in characteristic arrangements which blend or merge smoothly with each other without sharp demarcations. When a portion of the body is dissected, a variety of structures are revealed in their natural association. Muscles of varying shapes and sizes stretch between their attachments, enclosed or separated by sheets of connective tissue, with arteries and veins coursing between them. All are bound smoothly into a functioning complex by more connective tissue so that the individual components cannot be revealed until the uniting tissue is removed. Nothing is named, tabulated, or sorted out until man desires to learn about it. It is the nature of man and of his search for knowledge that the unknown must be identified, organized, and classified. Truly, classification is a device man uses to speed his learning. Things must be taken out of their natural setting to be examined singly. They must be named and their identifying qualities described. Order and understanding are brought out of ignorance. The danger exists that the knowledge which ensues may be but an extensive listing of descriptions of isolated entities. To counter this, one must learn to correlate new information with that which has been already learned. It is necessary to restore classified knowledge to nature's state of the functioning whole if learning is to be complete.

DEVELOPMENT OF THE HUMAN BODY

ORIGINS OF THE HUMAN BODY

"In the beginning" is just as important to the understanding of the human body as it is to the history of man or of the physical world. Indeed beginnings are highly important to the student of anatomy, for the complexities of human structure are more easily understood if one knows just how these complexities developed. This section is concerned with the very simple but very significant origins of human development. It is not concerned with the study of developmental processes for their own sake, nor does it attempt to cover all of the period prior to birth. Rather this section seeks to provide an understanding of certain early stages of human development in order to assist the learning of human body structure. It will be helpful to return to this section from time to time as new organs or systems are studied.

FERTILIZATION AND CLEAVAGE

Fertilization

The sex cells or *gametes* are the spermatozoa in the male and the ovum in the female. A number of anatomic mechanisms are necessary in both parents in order that a new individual may be conceived and begin its development toward the dramatic point where a newborn infant enters the world. The parental mechanisms are merely summarized here; they are described more fully in Chapter 9 in the section on adult reproductive organs.

1. Ovulation is the release of an ovum from the female ovary. This ovum passes into the uterine tube (oviduct) of the female and progresses toward the uterus, the organ in which the new individual will develop. This process takes from 4 to 7 days, but the ovum retains its physiological capacity to be fertilized for only about 24 hours. If conception is to occur, the ovum and the male sex cells, the spermatozoa, must meet

GENITOURINARY SYSTEM

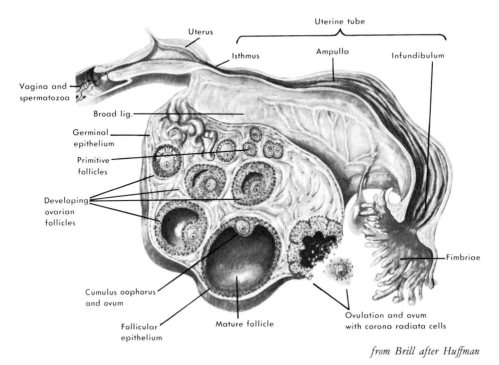

from Brill after Huffman

in the upper portion of the uterine tube within 24 hours after ovulation.

2. Mature spermatozoa are extremely delicate, free-swimming forms which lose their ability to fertilize the ovum after one or two days. Furthermore, the acidity of the female genital tract presents them with an adverse environment. The spermatozoa are suspended in the seminal fluid, which is produced by a number of glands associated with the male reproductive and urinary passageways. The seminal fluid adds volume to the spermatozoa and assists in countering the adverse vaginal conditions. The addition of spermatozoa to the seminal fluid at the time of ejaculation makes up the male copulatory fluid, the *semen*.

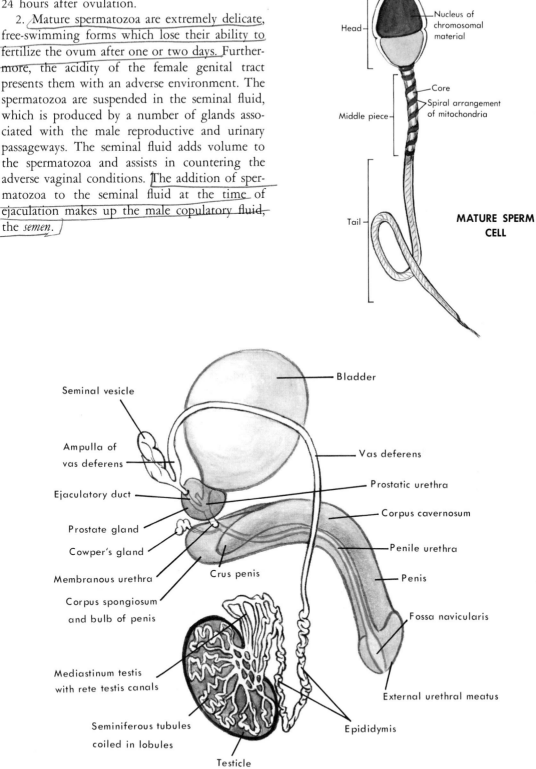

MATURE SPERM CELL

MALE REPRODUCTIVE ORGANS

3. A few hundred thousand of the millions of spermatozoa in ejaculated semen survive to reach the upper portion of the uterine tube. Although only one is necessary for fertilization, all are thought to be required to assist that one in liquefying a cementlike substance between the cells of a protective layer, the *corona radiata,* which enshrouds the ovum.

Fertilization of the ovum brings about the union of the chromosomes contained within each spermatozoon and ovum. The chromosomes of the new individual represent both parental lines, so that ancestral traits influence the form and personality of each of us. The future person, at this moment, is but a single cell, the *zygote.* At a diameter of 150 microns, the zygote is larger than any potential embryonic or adult cell. The zygote soon begins to divide ceaselessly (cleavage), progressing toward the point of human development when a fully developed, independent human organism is formed.

First Divisions of the Zygote

Cleavage begins at about 36 hours of post-fertilization age, during which the zygote is still in the upper third of the uterine tube. In cleavage the single cell first divides to form two *daughter cells.*

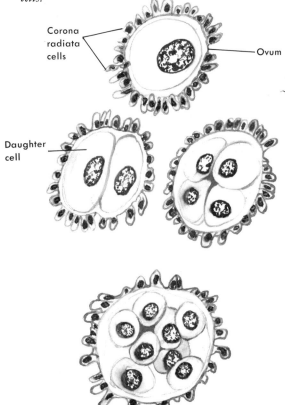

Corona radiata cells

Ovum

Daughter cell

Time lapse motion pictures, taken through a microscope, reveal that the zygote scarcely pauses before further division of the two daughter cells occurs. The cells resulting from cleavage are termed *blastomeres.* The blastomeres are smaller than the ovum and, with succeeding cell divisions, become still smaller until the normal cell size for the human is reached (7 micra). In the early stages of cleavage, this partitioning process results in a mass of cells which approximate the fertilized ovum in size.

At one time it was believed that one of the two original daughter cells would be set aside at the moment of the first cleavage of the zygote. This *gonadal cell* would pursue an independent course of cell divisions in anticipation of the formation of the reproductive cells of the new individual in the future. The other daughter cell, the *somatic cell,* was believed to form all of the other tissues and organs of the body.

Recent information suggests, however, that the DNA nucleoprotein in the nucleus of *all* cells, including both daughter cells, bears the genetic characteristics of the individual. These characteristics are transmitted during the reproductive process. During the formation of the spermatozoa and ova, the number of chromosomes in each cell is reduced from 46 to the haploid number, 23. This form of division, in which the number of chromosomes in each cell is reduced from 46 to 23, is known as meiosis. When the chromatin material of the zygote is assembled after fertilization, it is found that 46 chromosomes, 23 from the spermatozoon and 23 from the ovum, are present.

All cells resulting from the first division of the zygote bear the identification of the new individual's sex in special chromosomal material of the nucleus. It is possible to determine the true sex of any person by analyzing the nuclear chromatin in a very simple microscopic examination of many types of body cells. According to the newer belief, it is only in the second month of development that certain cells of the early embryo become set aside to form the basic *primordial germ cells* of the reproductive system. All other cells continue to develop as somatic cells.

Formation of the Morula. During the first moments of the 4 to 5 day journey of the zygote to its destination in the uterus, the original daughter cells divide a number of times. The two-cell stage gives way to four cells, then sometimes by an unequal rate of daughter cell division, to six cells, but with an eight-cell stage coming almost immediately. The cells at this period of their development are the *blastomeres.* The blastomeres divide repeatedly to form a spherical mass of cells, which seems to grow before one's eyes in a popcornlike eruption.

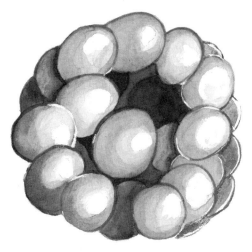

At about the end of the fourth postfertilization day, the solid mass of cells, now called the *morula,* is still draped by the remaining cells of the corona radiata. A delicate, nonliving membrane, the *zona pellucida,* covers the morula thinly. The morula has reached the point at which persistence in a mulberry or raspberrylike form is incompatible with the life of the cells lying deepest within the ball. In this form these cells could not nourish themselves, and they would be endangered by the pressure of adjacent or overlying cells. The solid sphere

of cells commences a hollowing-out process, in which the pressure of fluid accumulating between the cells causes the peripheral cells to spread out over a larger surface area, followed by their deeper counterparts. This process forms what is to become a central cavity. The entire hollow mass is now called the *blastula.*

Development of the Blastula.   The hollowed ball of cells is now nearing the uterine end of the uterine tube. The hollow center, the *blastocele,* becomes filled with a greater amount of secretion from the blastomeric cells.

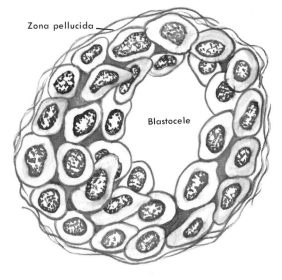

Zona pellucida

Blastocele

The blastomeres continue to divide and to spread out over an everenlarging surface area underneath the zona pellucida. The saclike structure is now termed the *blastocyst* because of its form. It is nourished at

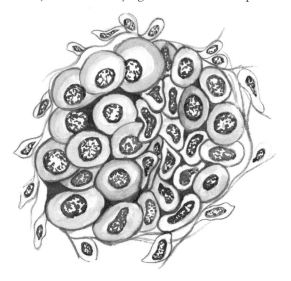

this stage by secretions of the epithelial cells of the uterine tube; this nutriment is called *embryotroph.*

Implantation of the Blastocyst.   Between the fifth to seventh postfertilization days the blastocyst, always growing by cell division, is propelled into the uterus. The inner lining of this organ has been prepared to lodge the blastula by hormonal influences which originated when the ovum was ready to be released from the ovary. The remainder of embryonic development takes place in the uterus. Only the innermost lining of the uterus, the *endometrium,* is of concern during this time. It is epithelial in nature and contains many glands in its supporting connective tissue. Hormonal action has thickened this layer and developed its tubular glands to a succulent, secretory stage.

The blastocyst rolls onto the surface of the endometrium as it emerges from the uterine end of the uterine tube. Secretions of uterine endometrial glands are added to the embryotroph to make "uterine milk" to nourish the blastocyst. As the blastula comes to rest, it sheds the remaining, deteriorated cells of the corona radiata; it now consists of a hollow mass of cells which has developed a dominant group. The dominant cells are eccentrically placed in relation to the rest of those making up the sphere. The dominant group

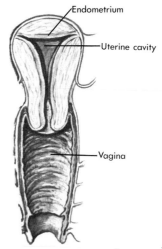

from Healey

is called the *inner cell mass,* and its cells are collectively called the embryoblast since the embryo will develop from them. The other remaining cells, called *trophoblast cells,* will form the membranes which will surround the embryo and become part of the *placenta,* an organ which establishes the connection between the embryo and the mother. The blastocyst, now composed of these two parts, sinks into the softened surface of the endometrium. The blastocyst could be

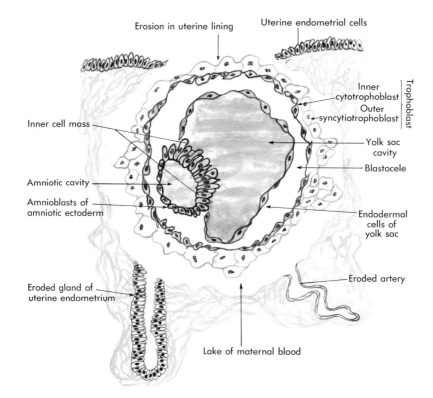

called a parasite upon the maternal host at this time, for it proceeds to invade the connective tissue and glands of the endometrium. The trophoblastic cells increase in number. The endometrial blood vessels surrounding the blastocyst are opened up to provide a lake of maternal blood, and the endometrial glands of the site of implantation are destroyed. The disruption of cells, glands, and blood vessels in the area provides a mixture of materials which now nourishes the blastocyst until the trophoblastic cells establish permanent relations with the maternal (uterine) circulation. The damaged endometrial surface is closed temporarily by a type of clotting and, later, by the growth of epithelial cells over the site of implantation.

These cells make up the *ectoderm,* the germ layer from which the skin, the nervous system, and many of the sense organs of the body will develop. Other flattened cells grow out from the margins of the ectoderm toward the internal aspect of the cytotrophoblast. These cells, the *amnioblasts,* further obliterate the blastocele as they form a sac of their own, the *amnion,* within which is the *amniotic cavity.* The amniotic membrane will come to envelop the entire fetus. It will retain a fluid, the *amniotic fluid,* which bathes and cushions the fetus until the time of birth. The layman calls it the "bag of waters." Its rupture and consequent gush of fluid through the birth canal is one announcement of the impending birth of the child.

Conversion of the Blastocyst into the Embryo

The simple blastocyst passes through a number of steps in order to form the fundamental tissues.

Splitting of the Trophoblast. Two layers of cells develop from the trophoblast. The inner layer continues to function as the envelope of the blastocyst. It is now called the *cytotrophoblast.* The outer group of cells forms a loose arrangement of cells around the blastocyst. These invasive cells are instrumental in destroying the maternal endometrial tissue, glands, and arteries to form the maternal lake of blood. They also excite a reaction in maternal cells surrounding the blastocyst, which stimulates the development of maternal tissue contributions to the placenta and to the membranes which will surround the growing embryo.

Formation of the Germ Layers. The inner cell mass has consisted of only a single layer of cells of similar, columnar appearance. Now these cells divide and specialize. Three distinct layers of cells develop from the embryoblast portion of the blastocyst. These layers are called the *primary germ layers,* because all of the varying types of cells of the adult body can be traced to their origins from them. The stages of development of these layers overlap but occur in the following sequence.

ECTODERM. The embryoblastic cells form a plate of columnar cells on the side of the embryonic disc which is toward the amniotic cavity.

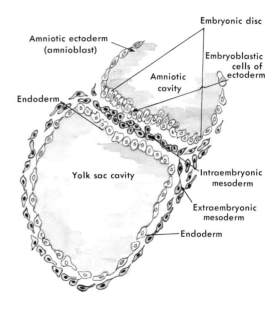

ENDODERM. On the side of the inner cell mass which is toward the amniotic cavity, the columnar embryoblastic cells again divide to develop a new form of cuboidal cells which give the embryo a two-layered appearance. This new layer is the *endoderm.* The cells of the endoderm are the forerunners of the alimentary and respiratory systems. Further cell division in the endoderm also forms a delicate membrane, the *yolk sac,* which has only a transient existence in the human. The embryoblastic cells of the inner cell mass have formed now a two-layered plate, the *embryonic disc.*

MESODERM.   The embryo rapidly proceeds to convert itself from a bilaminar disk to a triple-layered form. The third layer, the *mesoderm,* forms by cell division of ectodermal cells, in which stellate cells may split away (delaminate) from the ectoderm to intrude between the endoderm and the ectoderm. Mesodermal cells have the potentiality to form many types of cells and tissues—the precursors of certain blood cells, phagocytic (scavenging) cells, the basic cell types of connective tissues, as well as cells of fat, cartilage, bone, and even of epithelial membranes. Muscle cells also come from the mesoderm, but in their present state the proliferating mesodermal cells form only a flat sheet of cells between the other two layers of the embryonic disc.

TOTIPOTENTIALITY.   The embryo is in its third week of postfertilization development when the three-layered disc becomes evident. The cells of all three layers are highly totipotential at this stage. This means that the cells are not only capable of forming their own derivatives by specialization but also of developing cells characteristic of a different environment to which they might be moved. In such transplantations, the timing of embryonic development determines whether cells of a transplanted germ layer will develop their own type of tissue or will be induced to form a tissue which is characteristic of the area of the embryo into which they have been moved.

HENSEN'S NODE AND THE PRIMITIVE STREAK.   When mesoderm appears between the other two cell layers, an early indication of organization can be noted in the embryonic disc. If one were able to look "down" on the disc from the amniotic cavity, a small circular thickening with a pit at its center could be seen. This struc-

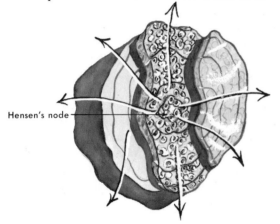

Hensen's node

ture, *Hensen's node,* is a small knot of ectoderm, mesoderm, and endoderm. Trailing away from one point of Hensen's node is a narrow band, darker and denser than the rest of the disc, which, as the *primitive streak,* first establishes a longitudinal axis in the embryonic disc.

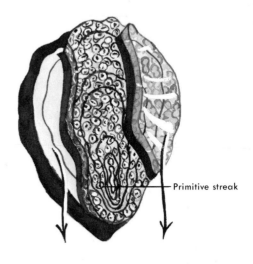

Primitive streak

The primitive streak is also composed of highly potential cells of all three germ layers. In the third week of development it is not possible to distinguish in terms of future importance between the endoderm, the ectoderm, and the mesoderm. The derivatives of the first two are primarily epithelial in nature. The mesoderm itself gives rise to a vast group of derivatives which become universally distributed throughout the body. However, a great proportion of the structure and function of tissues and organs derived from ectoderm and endoderm is contributed by the mesoderm and its own specialized products, which become structurally associated with the other two germ layers.

EARLY DEVELOPMENT OF THE EMBRYO

A three-layered embryo has been created by the development of the primary germ layers from the inner cell mass. The embryo has settled into the uterine endometrium surrounded by its own protective membranes. A temporary arrangement for nourishment has been established through the activities of the trophoblast and its derivatives. The trilaminar embryonic disc is enlarging as the result of continuing, rapid cell divisions in all germ layers. It is beginning to elongate under the influence of the primitive streak in a caudal (tail-ward) direction. The next step is to understand the main stages of development in certain structures which are essential to the embryo. It may be helpful first, however, to stop and examine the following diagrammatic table, which is used by Kaplan* to explain the timing of activities in terms of the perspective of the entire prenatal (prior to birth) period.

The entire developmental time period discussed in the text and illustrations up to this point is represented by the pre-embryonic period on the table. In the next five weeks of the embryonic period which follow, the foundations for the development of organs are laid down. Afterwards, all is growth.

Immediate Essentials
for Embryonic Growth

The embryo has some immediate needs. If they are not met, it will develop into a shapeless,

---

*Kaplan, Stanley: Unpublished lectures in human development and teratology.

disorganized mass of cells, or it will die from the lack of nourishment and oxygen required for the metabolism of the growing and dividing cells. Similarly, the waste products of cellular activity must be removed. The needs of the embryo are:

1. Establishment of a longitudinal axis for the future body, and the determination of dominance in its growth pattern.

2. Development of a system of support for the expanding mass of derivatives from the germ layers.

3. Provision of nutrition and of a means of elimination by the formation of a network of vessels which will establish connections with the maternal circulation of the uterus. These vessels must be formed before the early cellular absorption of nutriments from the maternal blood lake becomes inadequate to support life.

4. Coordination of the activities of a complex organism by the early formation of a central nervous system. The embryo does not require much in the way of awareness of its external environment. This becomes an immediate need, however, at birth when the newborn infant is separated from the maternal environment. Therefore, sense organs of many types must be formed

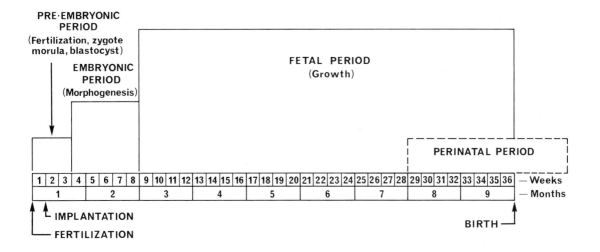

early, so that they may be developed to a stage of function by the time they are needed.

Axis and Dominance. Cell growth and division causes the embryonic disc to spread peripherally in all directions. If unchecked by some force for organization, the disc would come to resemble nothing more than a three-layered flat plate. From this point on, however, the main direction of growth is toward the caudal end, under the influence of Hensen's node and the primitive streak.

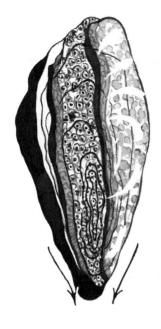

The forces set in motion by this direction of growth reshape the germ layers into an oval, and then along a distinctly longitudinal axis. The growth center is always made up of young, unspecialized cells. Therefore, the germ layer components which lie anteriorly or to the head end of the disc will always be older and, hence, more specialized. The head or cranial end of the embryo assumes dominance over the rest of the body, and structures at this end will always be in a more advanced state of development and specialization. Even at birth the baby's head, neck, and upper trunk are larger than his abdomen, hips, and legs.

Support of the Body. The embryo is soft and predominantly cellular. The establishment of a longitudinal axis sets off even more growth. Cells spread apart and can no longer support each other by mutual pressure. Physical forces induce responses by the mesoderm. Some of the meso-

dermal cells begin to specialize to support the increasing bulk of cellular masses. They spread apart, but since they are stellate in form (star-shaped with several cell processes or arms on each cell), they can maintain contact with their neighbors.

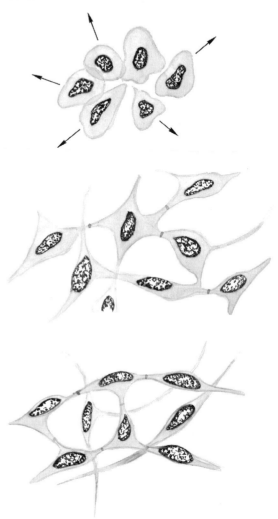

A meshwork or network of mesodermal cells fills in under the ectoderm and endoderm. Now the mesodermal cells function as the major supporting tissue of the embryo and are given a new name, the *mesenchyme.* The mesenchyme will go through progressive specializations as the organism needs more support or greater resistance to physical forces, which only more specialized structures can provide.

Without some means of stiffening along the longitudinal axis, the elongating body would not be able to maintain its form. The embryo makes use of a structure which is seen in chordate animals, the *notochord,* to provide this type of support, although it is soon replaced. Cells of Hensen's node form a plate of densely clustered cells in the wake of the caudally moving primitive streak.

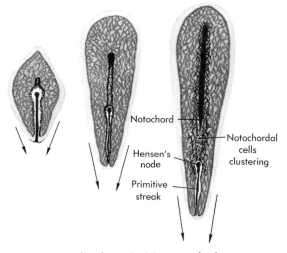

As the primitive streak draws away caudally these cells follow, and more are formed in the space previously occupied by the primitive streak. The *notochordal plate* does not spread out into the adjacent mesodermal cells but becomes condensed into a tight cylinder of cells which form a stiff rod. The surrounding mesoderm will form its successor later. The cells of the notochord undergo an internal change and form a solid, intercellular material which increases the strength of the rod. The role of the notochord is limited, however, because it cannot support a ceaselessly growing embryo beyond the limit of its own strength. The mesoderm about it forms more specialized tissues—cartilage first (which also proves inadequate), and then the bony vertebrae. This much attention is given to a transient structure to emphasize the fact that adult structure is attained only by the modification or replacement of simple ones which were perfectly adequate for the purposes of the embryo at the time that they were developed. The creative force which organizes and guides the stages of development of the human body is economical. It sees that what is needed develops, but with an eye to future needs. It ensures that these needs will be fulfilled at the proper time by modification, specialization, or by the timely associated growth of a replacement.

Provisions for Nutrition, Elimination, and Fluid Transport. The early embryo cannot nourish itself, but it does expend energy in growth and forms waste products. These wastes increase in amount as cell division and specialization proceed. The human embryo has no vast amount of food materials stored in a yolk. The formation of layers of cells and thicker membranes makes the simple diffusion of nutrients and waste products between the uterine lake and the embryo difficult. Organization of a better means of absorption and elimination, as well as of a vascular system, is essential to the continued life of the embryo. Simple systems to satisfy these needs develop sooner than almost any other organ systems of the body.

FORMATION OF ENDOTHELIAL TUBES. Some mesodermal cells in the embryo and in its covering membranes cluster together in tight little knots or islands. These are *angioblastic cell clusters* or *blood islands.* They form the simple *endothelial tubes* which will transport blood.

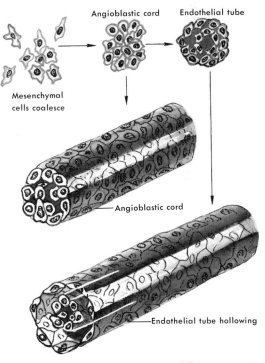

This process is an early indication that the cardiovascular system derives from the mesoderm. The blood islands coalesce into solid cords of cells, which frequently branch or anastomose with each other. The cords hollow out as the outer cells flatten down to form a continuous tube. The more centrally located cells form clumps within the lumen of the tube. They become converted into the earliest blood cells and are believed to secrete a fluid which is the first blood plasma. Therefore, the first blood cells and blood plasma are also products of the mesoderm. The development of the blood vessels and their contents goes on within the mesoderm throughout the embryo.

CONNECTION WITH THE MATERNAL CIRCULATION. The *chorion,* the grown-up inner layer of the trophoblast, has developed from an original sac of flattened and cuboidal cells. It is now a complexly folded membrane which has formed many fringelike processes. These are the *chorionic villi,* the covering cells of which are supported by mesenchyme in which the endothelial vascular tubes are prominent. The villi project into uterine blood lakes.

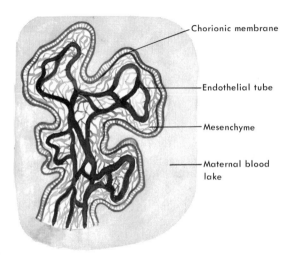

Substances can diffuse in both directions. When circulation is established in the embryo, substances passing from the maternal blood through the surface cells of the chorionic villi may diffuse into the endothelial vascular tubes, after seeping through the meshwork of supporting mesenchyme. Waste substances from the embryo are removed from its primitive vascular system by diffusion in the opposite direction.

DEVELOPMENT OF EMBRYONIC VASCULAR CHANNELS. Absorption and excretion over a broad surface are all that are required to establish a proper exchange with the maternal circulation. The same process must occur in the tissues of the embryo, but the blood must be moved through the distributing and collecting channels

of the embryo itself, as well as to and from the chorionic villi. The mother's heart moves the maternal blood under pressure to the uterine blood lakes.

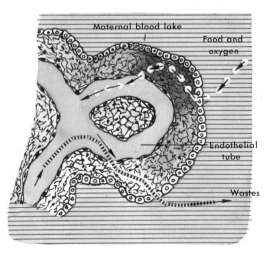

Osmotic phenomena and active transport induce the exchange of substances across the chorionic villi and into the endothelial tubes of the membranes of the embryo. It is up to the embryo to move its own blood through its own developing vascular system.

The method by which this is accomplished involves first the simple anastomosing network of endothelial tubes, which has already been described. Initially no part of the network is dominant.

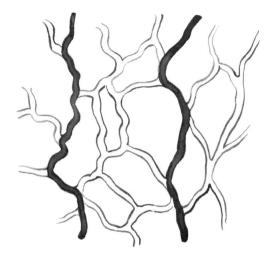

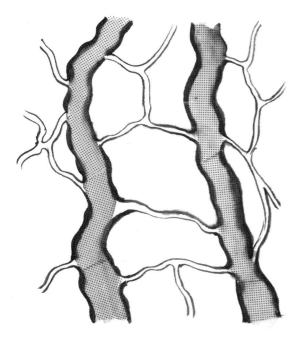

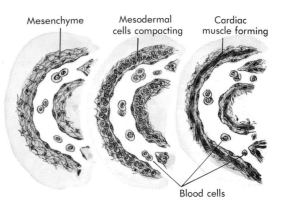

Mesenchyme — Mesodermal cells compacting — Cardiac muscle forming

Blood cells

Gradually, however, one main pathway develops on each side of the notochord, as parts of the earlier networks contribute to the two main channels by increasing their diameter and by straightening to align themselves with the longitudinal axis of the embryo. The two endothelial tubes join at the head end of the embryo.

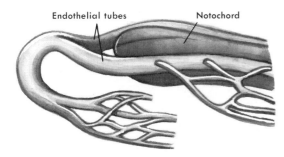

Endothelial tubes — Notochord

The junction, ahead of the notochord, forms a single tube which connects the dominant parts of the network just described to other portions as they develop.

APPEARANCE OF THE CARDIAC TUBE. Soon there is a compacting of mesodermal cells about the now-single *cardiac tube*. This mantle of mesoderm demonstrates the totipotentiality concept as it develops into an entirely new tissue, primitive cardiac muscle, which combines with the endothelial tube to become the forerunner of the heart.

The first stirrings of contractility can be observed in the cardiac tube as the primitive muscle becomes established. At first mere random twitchings occur. Gradually rhythmicity is achieved. Contractility and rhythmicity are inherent properties of cardiac muscle which occur long before any nerves reach the developing heart. With a simple pump operating, the primitive blood plasma containing the rudimentary blood cells begins to move in the endothelial channels. At first the movement is only a to-and-fro series of waves without much progress in direction.

ESTABLISHMENT OF CIRCULATION IN THE EMBRYO. As the rhythm of the cardiac tube's contraction becomes more regular, the blood is propelled from the dominant channels through the cardiac tube and out into other parts of the vascular network. The inflow channel will develop into a system of veins, whereas the outflow channels will become the future aorta and large arteries.

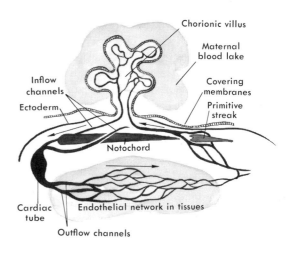

Inflow channels — Ectoderm — Chorionic villus — Maternal blood lake — Covering membranes — Primitive streak — Notochord — Cardiac tube — Endothelial network in tissues — Outflow channels

Distribution of the blood is now possible. Vascular channels have formed in the body stalk of the embryo which attaches it, via the covering membranes, to the wall of the uterus. These channels now connect the circulatory system of the embryo to the endothelial tubes of the chorion. Oxygen and nutrient substances entering the chorionic villi from the mother's circulation can be transported to all parts of the rapidly growing body. Waste products of metabolism can be picked up for transport to the chorion where they are eliminated, across the placenta, into the maternal circulation. The genesis of the cardiovascular system has been emphasized, for it is important to realize that this system does much for the early embryo that other systems will do later in the adult body. It moves blood. This will always be its role; but right now it also forms the blood. It absorbs food in a utilizable form. It breathes for the embryo when the lungs have not formed and could not operate in the fluid environment of the amniotic cavity if they had formed. In addition it excretes waste substances. This very simple system is far from perfect. Like so many other tissues and organs of the embryo, it is adequate for the purposes of the embryo at an early stage, but it will become insufficient to provide for the continued growth and increasing complexity of a larger embryo. This system, however, is not replaced. It must continue to function to maintain life. Modification occurs as the system continues to function, until the complicated heart of the adult has been developed.

Nervous System.   The first indication of the formation of cells which will establish a means of control over the activities of the systems of the body is found when the embryo is about three weeks old. The beginnings of the nervous system, therefore, coincide with the appearance of the first endothelial tubes. The nervous system and the cardiovascular system develop concurrently, although the latter becomes functional first.

Following the establishment of a longitudinal axis, a plate of ectodermal cells thickens along the upper surface of the embryonic disc. The *neural plate* lies above the earliest notochordal cells. The two structures develop simultaneously. Cells along the edges of the neural plate multiply faster than those in the midline, with the result that the more lateral cells pile up into *neural folds* with a *neural groove* in between.

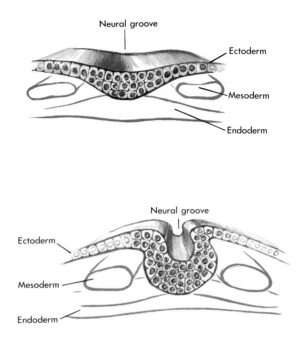

of this special ectoderm parallels that of the notochord as it grows caudally, following the primitive streak. Therefore, while the plate is becoming a groove at the head end of the embryo, the more caudal part of the ectoderm is just forming into a neural plate in the lower and younger regions.

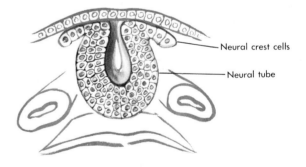

The special *neurectoderm* continues to grow in length and in thickness of the neural folds. The laterally placed neural folds finally topple toward each other in a rolling manner much like the breaking of a long wave.

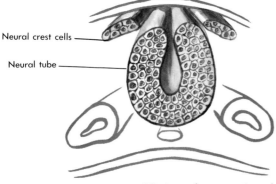

Their surfaces touch and fuse above the neural groove and convert it into the *neural tube,* whose lateral walls will always be thicker than its roof or floor.

During the thickening of the neurectoderm, some of the cells are left behind in long strips and cords. These are *neural crest* cells, which do not accompany the neural folds as they form the neural tube. The neural tube and the neural crests sink deeper into the mesoderm of the embryonic disc. The ectoderm grows over the neural tube and the neural crest to cover the gap left in the midline as these become internal organs. The neural tube will undergo many specializations to become the brain and spinal cord. The neural crest will develop into collections of nerve cells located outside of the central nervous system (*ganglia*) and will participate in the formation of more peripheral parts of the nervous system and pigment cells of the skin. The processes of cells in the wall of the neural tube become nerve fibers of the brain and spinal cord. Some of the processes, especially those from the ganglia, will grow out toward the structures of the body which need a nerve supply as development proceeds. In this way the growing brain and spinal cord will assume control of body activities, initiate and coordinate responses, and provide the body with an awareness of changes or dangers in the external environment.

Segmentation of the Embryo

A distinguishing characteristic of all vertebrates is segmentation or metamerism of the body. In the lower vertebrates the body is built up of units which are similar in structure, and each unit, or metamere, is provided with simple organs and musculature which make it relatively self-sufficient. As the vertebrate becomes more complex, organ systems develop which cross the metameric lines to serve all of the units of the organism. Primitive nerve networks, vascular and digestive channels, and simple systems of excretion and reproduction are examples of such progress. The human embryo passes through similar stages, but the steps are not as sharply delineated. Some steps may be quite transient and, therefore, slurred or obscured by more obvious events occurring at the same time.

Changes in the Mesoderm. While the primitive streak retreats steadily in a caudal direction to stimulate the organization of lower areas, mesodermal cells spread out laterally to fill in all areas of the embryonic disc between the ectoderm and the endoderm.

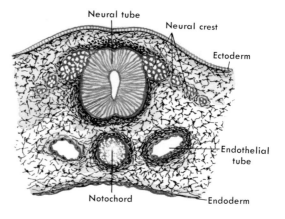

Distribution of Unspecialized Mesoderm

Unspecialized mesoderm is everywhere—about the neural plate and neural tube, around the first vascular channels and cardiac tube, and abutting the ectodermal and endodermal surface. Many mesodermal cells act as a general supporting and filling network, but other groups of cells begin to cluster together in certain areas to perform specialized roles.

LATERAL PLATE MESODERM. This term is applied to the mesodermal cells which have spread out under the endoderm and ectoderm. The lateral plate mesoderm will develop into the connective sheets under the skin (*fascia*) and between major structures. It will ensheath blood vessels and form their outer tunics, surround the growing nerves, and develop the outer layers of the alimentary and respiratory organs which arise from the endoderm.

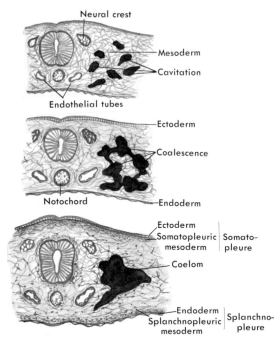

Splitting of Lateral Plate Mesoderm

Cavitation within the lateral plate mesoderm begins as a series of bubblelike spaces amidst the cells, which become larger and coalesce. Soon an irregular cavity is found on each side of the longitudinally oriented notochord and neural tube. The *intraembryonic coelom* is formed by these two cavities, which are connected at the cephalic end of the embryo ahead of the blind end of the neural tube. The coelom is the first indication of a body cavity. In its early stages, it separates the lateral plate mesoderm into two main sheetlike masses. The portion which lies between the coelom and the ectoderm is now renamed the *somatopleuric mesoderm.* It will remain in association with the ectoderm. These tissues comprise the *somatopleure* which forms the body wall of the embryo and from which will develop the variety of structures of the body wall of the adult. The portion of the mesoderm between the coelom and the endoderm is termed the *splanchnopleuric mesoderm.* When the endoderm begins to form the alimentary tract, this part of the mesoderm forms a mantle around the early digestive tube and any outpouchings from it. The mesodermal part of the *splanchnopleure* resulting from this relationship will form the muscle, connective tissue, and outer covering (serosa) of the alimentary organs. The endodermal part will form the lining epithelium (mucosa) and glands.

CARDIOGENIC MESODERM.   The mesoderm which forms the early cardiac tube is classified as cardiogenic mesoderm because it performs this single function. The cephalic connection of the two limbs of the coelom comes to surround the cardiac tube.

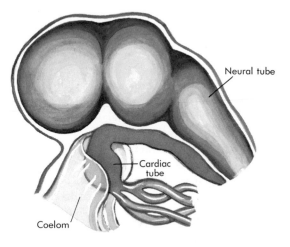

When the coelom divides later to form separate cavities within the abdomen and the thorax, this portion will become the pericardial cavity surrounding the heart.

INTERMEDIATE MESODERM.   The splanchnopleuric and somatopleuric mesoderm are continuous along the lateral border of the longitudinal structures where the coelom does not intrude. Some of the mesodermal cells here will specialize to form the reproductive organs, the kidneys, and (as believed by some embryologists) part of the adrenal glands.

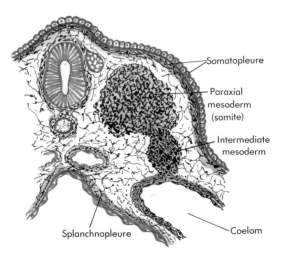

PARAXIAL MESODERM. Dense columns of mesodermal cells become evident beside the notochord and neural tube. These columns are very significant because of the variety of structures they will produce. Segmentation is most marked here. The bulky columns divide into a series of blocks of mesodermal cells whose regular arrangement along the longitudinal axis is reminiscent of metamerism. Each block is a *somite*. The formation of somites is progressive and follows the gradient of development from the cephalic to the caudal end of the embryo until about 44 pairs are formed.

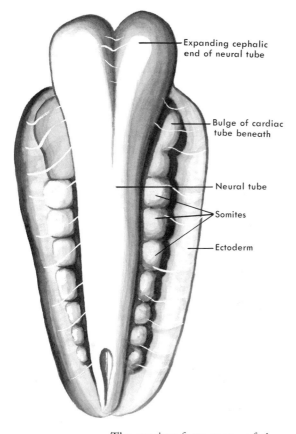

The somites form many of the elements of the skeleton, musculature, and skin of the segmental areas under which they lie.

Developments Concurrent with Somite Formation. It cannot be said that the early somite embryo has acquired organs or organ systems. One may recognize, however, many structures which have been observed already, as well as note those appearing for the first time.

CIRCULATION. The cardiac tube and the network of embryonic vascular channels move and distribute blood. The pericardial cavity is forming about the primitive heart as an early subdivision of the coelom.

NERVOUS SYSTEM AND SENSE ORGANS. The formation of neural and cardiac tubes, the primitive coelom, and the somites occurs simultaneously. The neural tube at this point shows a spurt of development which is characteristic of structures located cephalically in the embryo. The entire head end is expanding, and with it there is expansion of the neural tube. Its lateral walls thicken and bulge, as the neurectodermal cells grow and divide rapidly. The internal cavity of the neural tube dilates in a similar fashion. A form of segmentation develops, in which a series of constrictions appears in the neural tube at locations where growth does not occur. The constrictions mark off the division of the central nervous system into portions which will persist as major divisions of the adult nervous system. The bulging parts with dilated neural cavities are

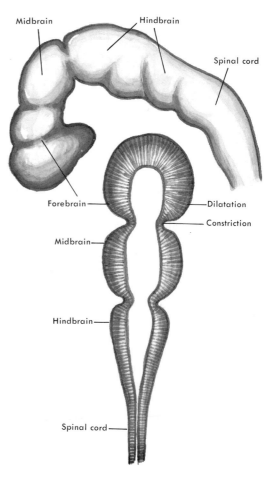

maternal relationship through the placenta. Reproduction is far in the future. It is not surprising that the appearance of genitourinary organs has been delayed. But the embryo does develop three sets of urinary organs. The first, the *pronephros,* is quite primitive, never functions, and soon vanishes.

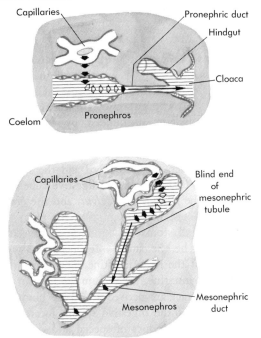

The second kidney, the *mesonephros,* is more advanced in structure and functions for a time before being replaced by the definitive kidney, the *metanephros,* which becomes part of the urinary system of the adult. The structure of these kidneys, as well as that of the adult organ, is described in Chapter 9.

What of the beginnings of the reproductive system? In the second month of embryonic development, certain cells come to the fore as primordial cells of the genital system. These have been scattered in a quiescent state under and amidst the endodermal cells of the yolk sac. They migrate to a position under the mesodermal lining of the primitive coelom near the nephrogenic cord. When finally clumped together, they gather both in the mesenchymal connecting cells of the area and in the mesodermal coelomic lining overlying them to form a long *genital ridge.* The mesonephric ducts of the second kidney, no longer needed for excretion, are incorporated into a swelling, elongated body which is termed the *gonad.* There is no indication at this time of the actual sex of the future person. The gonad and the embryo are said to be sexually undifferentiated. Many further changes must go on in the gonad, genital ducts, and the genitourinary outlet from the body, the *cloaca,* before the internal and external sex organs will be recognized as male or female. The genetic sex has been determined since fertilization, however, although it is not discernible now on inspection of the embryo.

LATER CHANGES IN THE EMBRYO

A more extensive excursion into even the basic elements of the further formation of organs and organ systems would not be profitable now. The stage has been set, but the next steps will be more meaningful if they are considered in company with the adult form of the systems. Therefore, a section devoted to developmental considerations will be found as part of the introduction to each system or as part of the description of individual organs, in order to aid in the understanding of how such complex structures developed.

The table on the next page is a partial review of the principles which have been presented in this section.

ORGAN OR STRUCTURE	EARLIEST INDICATION	PRODUCT OF FURTHER DEVELOPMENT
Entire body	Primitive streak	Organs and organ systems
Notochord	Mesoderm following primitive streak	Temporary axial support; part of intervertebral disks
Blood Vessels	Angioblastic mesoderm	Endothelial tubes; arterial and venous channels
Heart	Endothelial tubes; cardiogenic mesoderm	Cardiac tube
Placenta	Chorionic ectoderm	Chorionic villi
Central nervous system	Ectodermal neural plate	Neural tube; primitive vesicles of brain
Ganglia	Neural crests	Root and autonomic ganglia
Nerves	Neural tube; neural crest	Sensory and motor nerves
Alimentary system	Endoderm; splanchnopleuric mesoderm	Lining, muscle, glands of foregut, midgut, and hindgut derivatives
Voluntary muscles	Paraxial mesoderm	Somites; myotomic hypomere and epimere
Bone and cartilage	Paraxial mesoderm	Somites; sclerotome; somatopleure; unspecialized limb bud mesoderm
Skin	Ectoderm; somatopleuric mesoderm	Somatopleure
Urinary system	Intermediate mesoderm	Pronephros; mesonephros; metanephros
Genital system	Unspecialized cells in yolk sac endoderm; mesoderm of coelomic lining; mesonephric ducts	Gonad and genital ducts
Body cavities	Lateral plate and cardiogenic mesoderm	Coelom; pericardial cavity

2

The

Connective

Tissues

CONNECTIVE TISSUE IN GENERAL

Connective tissue alone does not form an organ or an organ system, although it is the paramount tissue of the skeleton and of the articulations of the human body. Its importance lies in the fact that it is a widely distributed fundamental tissue which, in its connecting or uniting function, is absolutely essential to the structure and function of all other tissues and organs. Without connective tissue, organs would be collapsed and shapeless, lacking interior frameworks, external wrappings, or protection. Strength, where needed, would be lacking and the body as a whole would be devoid of an axis or interior support. Vital organs could not be protected. The familiar features of a person's face or body form would not exist. It would be impossible for a person to sit, to stand, or to alter his position. Walking and physical activity would be beyond one's ability. The human body would be a mass of quivering protoplasm. Therefore a consideration of the many forms of connective tissue is important as a prelude to the study of the systems of the human body.

FUNCTIONS OF CONNECTIVE TISSUE

Before encountering the bewildering variety of the connective tissues it will help to reflect upon some of the principles which underlie the grouping together of such widely distributed and greatly varying tissues. If the fundamental premise that these tissues connect and unite is kept in mind, the many modifications can be understood and more specific functions can be determined. Connective tissue:

1. Binds together structures which need to be held in an intimate working association or which ensure the integrity of an organ.

2. Supports structures of the body and provides rigidity where needed.

3. Protects organs with coverings, sheaths, and capsules or by surrounding vital areas with bony or cartilaginous portions of the skeletal system.

4. Subdivides organs or regions of the body by sheets, partitions, or membranes of varying strengths and thicknesses.

5. Separates groups of structures which have a common function from others which have a different function.

6. Unites dissimilar tissues, such as muscle and bone, which by their nature or the vigor of their action cannot join successfully themselves.

7. Forms the material which packs and fills the many odd crevices and crannies between and around the structures of the body.

8. Provides the unit of tissue construction wherever tensile strength is required.

9. Provides the framework through which vessels and nerves may proceed to their destinations and may be distributed to the organs they supply.

10. Modifies its cells in development to form specialized tissues, such as fat, cartilage, and bone.

Origin of Connective Tissues

The human embryo grows as a rapidly burgeoning mass of cells which originates from the fertilized ovum. Very early in development three layers of cells (the germ layers) can be identified as the first evidence of specialization. The outer group of cells forms the *ectoderm,* which specializes to form the epithelium of the outer covering of the body and to develop the central nervous system and parts of the sense organs. The inner group of cells forms the *endoderm* from which will come the epithelial and glandular portions of the alimentary and respiratory systems and parts of the urinary system. A group of cells appears between these two layers of cells; this third group, the *mesoderm,* is the forerunner of all of the connective tissues in the body.

two of the requirements of a connective tissue in that it possesses cells and an intercellular medium. This arrangement soon becomes inadequate as the embryo continues to grow and physical forces become evident. An early change is that the physical contact of the mesenchymal cells is broken as the mesenchymal cells spread apart to keep pace with the growing embryo. Support is provided by the appearance of delicate fibers amidst the cells and by the formation of a mucoprotein material in the intercellular fluid which gives it a viscous nature. This more advanced tissue is called *embryonal* or *mucous connective tissue.*

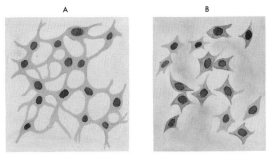

Mesenchyme   Breaking of cell contact

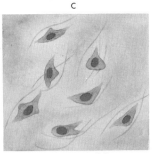

First appearance of fibers
around fibroblasts

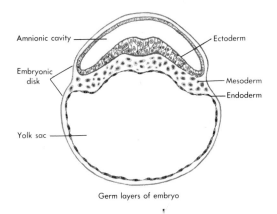

Germ layers of embryo

Mesoderm.   Mesoderm possesses no particular strength in itself, for it is as soft and cellular as the other embryonic germ layers. It gives no indication of being able to perform all the functions ascribed to connective tissues. Its importance lies in the ability of its immediate offspring, the *mesenchymal cell,* to specialize into many different types of cells which can perform the roles of connective tissue.

Mesenchyme.   Mesodermal cells undergo their first specialization to produce stellate cells with branchlike processes which establish physical contact with other similar cells. The loose meshwork which is formed provides a rudimentary form of support and union in the growing embryo which is called *mesenchyme.* An intercellular fluid forms within the meshwork so that mesenchyme fulfills

It provides most of the support of the body of the embryo until more adult tissues appear, and it persists at birth as *Wharton's jelly* in the umbilical cord. The jelly-like vitreous humor of the posterior chamber of the eye is akin to mucous connective tissue.

Totipotentiality of Mesenchyme.   Of greater importance than the provision of temporary support in the embryo is the role of mesenchyme in forming other tissues by its specialization. So many adult tissues can be traced to origins from mesenchyme that this embryonic tissue may be termed "totipotential" in its ability to specialize into a variety of forms. The following chart indicates how many and how significant these specialized forms of tissue are.

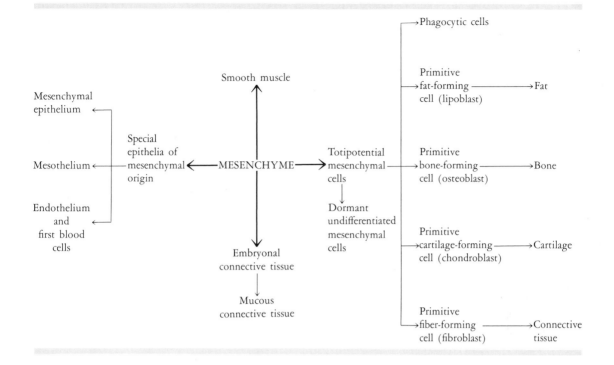

BASIC FORM OF CONNECTIVE TISSUES

Now it can be seen why the mesenchymal cell is described as totipotential. The many routes open to it in specialization result in a variety of adult tissues, all of which are grouped together as connective tissues, which are of great significance to the body. It is important to understand, however, that once a mesenchymal cell has embarked upon a particular specialization, such as forming cartilage or fat, it loses its power of totipotentiality. It cannot change its characteristics to form some other type of connective tissue. Associated with this fact is the lessened ability of the cells to reproduce as they become older and more specialized.

Dormant Mesenchymal Cells. Some of the mesenchymal cells in the embryo do not specialize in accord with their totipotentiality. They remain in an undifferentiated dormant state as the body grows and develops into its adult form. These dormant cells are located wherever the adult forms of connective tissue or its modifications are found. These are the cells which, unspecialized and possessing full totipotentiality, can develop into any of the connective tissues to meet the needs of growth, maintenance, or repair.

All connective tissues have the same basic pattern of structure, no matter how varied their appearances may be. The differences between them can be attributed to changes or modifications in one or more of the components in the basic pattern. In turn, the structural modifications permit the gradations and varieties of function in this group of tissues. Three structural components are characteristic of the basic pattern of all connective tissues. These are the connective tissue cells, fibers, and intercellular medium. All may be modified. In some instances one may be so dominant as to obscure the others. But all are present in every type of connective tissue.

Connective Tissue Cells

In even the most highly modified connective tissues, the cells were first fiber-forming cells, and the modified tissue is first organized as some form of fibrous connective tissue before the modifications occur. Therefore, the main type of cell in connective tissues is the fiber forming cell or *fibroblast.* This cell, in development, results from the breaking of physical contact by mesenchymal cells in which the primitive cells separate from each other and their branching processes retract to form tapering pseudopodic extensions of the cytoplasm.

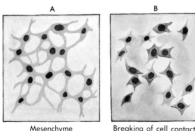

Mesenchyme   Breaking of cell contact

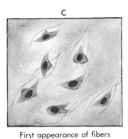

First appearance of fibers   Fibrocytes and fibers
around fibroblasts

The fibroblast is responsible for the formation of connective tissue fibers, for regulating the metabolism of the tissue, and for modifications in the intercellular medium. When the fibroblast has produced a characteristic type and number of fibers, or has acted to modify its medium, it becomes fixed in position. The definitive, adult connective tissue cell is sometimes termed a *fibrocyte,* but most histologists retain the term *fibroblast* for an adult, fiber-forming cell.

Other types of cells may be found in varying numbers in a connective tissue. Some mesenchymal cells specialize into cells that are not unlike fibroblasts but that do not form fibers. Although in a state of health, these cells attach themselves passively to the surface of fibers; if the tissue becomes inflamed, amoeboid motion is a prominent characteristic. These restless cells, termed connective tissue *macrophages* or *histiocytes,* move amidst the other components of the tissue. They are noted for their ability to ingest foreign material, bacteria, and cellular debris. This phagocytic function, coupled with the universal distribution of the macrophages through the widespread reaches of connective tissue in the body, makes them important in the front line defenses of the body against infection and inflammation. Other cells found in connective tissues are *plasma cells,*

which are believed to form antibodies, *mast cells,* which are associated with anticoagulant substances, and varying numbers of white blood cells. The last migrate into all connective tissues from the blood vessels and represent another segment of the defense forces of the body. Some are always on patrol in connective tissues. If infection strikes, large numbers of white blood cells stream into the affected area and are temporarily the dominant cell type. Later they are succeeded by large numbers of macrophages, which move in, or develop from dormant mesenchymal cells, to clean up the debris of the infection.

Connective Tissue Fibers

Fibers are the outstanding characteristic of the connective tissues and in most, but not all, are most directly concerned with function of the tissue.

Types of Fibers. Fibers of some type are always present even if they are not dominant or are obscured by a modification of the intercellular medium. Frequently in connective tissues the type of modification or the function depends directly upon the kind of fiber which is present. All fibers are protein in nature. Three types of fibers are found.

Collagenous fibers are formed from complex polypeptide molecules which are linked together to form minute fibrillae. The fibrils run a slightly wavy course and are grouped together into branching bundles by a cement substance. The protein collagen of which they are formed is extremely strong and is well suited to resist physical forces.

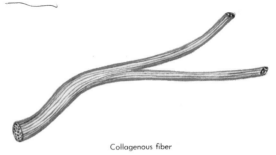

Collagenous fiber

Collagenous fibers, therefore, may be expected where strength, support, or firm union is required.

Elastic fibers, formed from the protein elastin, are more delicate fibers which tend to course separately, to branch frequently, and to form networks.

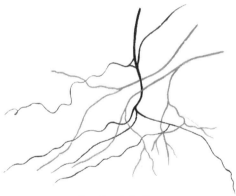

Branching elastic fibers

Elasticity, rather than unyielding strength to oppose forces, is their functional characteristic. Where this property is needed, elastic fibers predominate in the connective tissue. Some degree of elasticity is desirable in situations in which connective tissues wrap or separate structures which move or change their size. In these situations elastic fibers are frequently intermixed with collagenic fibers.

Reticular fibers are the most delicate of connective tissue fibers. The protein *reticulin* of which the reticular fibrillae are composed is similar to collagen but with fewer peptides making finer fibers. They branch frequently and join with others to form fine networks. This quality suits them to unite or fasten coarser connective tissue fiber arrangements to cellular membranes, to group individual muscle fibers into functioning arrangements, and to hold or group masses of cells in cellular organs or glands.

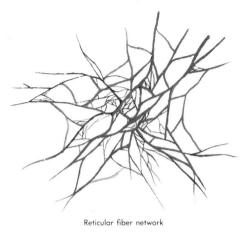

Reticular fiber network

Formation of Fibers. There are no fibers in mesenchyme, but soon after mesenchymal cells separate to specialize into fibroblasts, delicate fibrils appear in the intercellular medium adjacent to the cells. Electron microscopic studies and the observation of cultures of embryonic connective tissue have made it evident that the fibers form from the amalgamation of protein molecules in the intercellular medium and that the fibroblasts regulate the process.

Amino acid molecules are absorbed by the fibroblast through its unit membrane from the surrounding intercellular medium. The amino acids are assembled into polypeptides on the ribosomes of the fibroblast endoplasmic reticulum. The polypeptides traverse the channels of the endoplasmic reticulum to reach the Golgi complex. Here three polypeptide chains become coiled about each other to form a helix or spiral called *tropocollagen.* This is secreted by the Golgi complex to form vacuoles, which move to the surface of the fibroblast. The tropocollagen passes through the unit membrane into the intercellular medium, where the macromolecular helices become polymerized into delicate fibrils, which become oriented according to physical forces in the forming tissue.

Intercellular Medium

The intercellular medium of the connective tissues varies in its components and in its complexity according to the degree to which the basic form of connective tissue is modified. In the simplest and most elemental form, the intercellular medium consists of a watery *tissue fluid,* which resembles blood plasma in its composition. This form is found in mesenchyme where the cells are bathed in a liquid medium containing proteins, electrolytes, and nutritive substances through which oxygen and carbon dioxide diffuse. As modifications and specializations occur, other substances appear as parts of the intercellular medium. It has been seen that a mucoprotein material is added in embryonal or mucous connective tissue. At the time that fibers appear in a connective tissue, a *ground substance* is added.

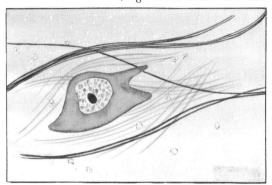

Adult connective tissue with ground substance and fibers in intercellular medium

The ground substance converts the almost aqueous tissue fluid into a colloidal mixture which has more "body" to it although the degree of viscosity varies. The ground substance is a mixture in which complex molecules, formed by the combination of carbohydrates and proteins, are the predominant components. These have been identified as various forms of protein polysaccharides (formerly called mucopolysaccharides). Differences in their chemical form result in morphological characteristics which are frequently keys to the more advanced modifications.

The most common protein polysaccharide ground substance is hyaluronic acid combined with proteins to form *hyaluronate*. It may be the only ground substance intermixed with tissue fluid in the more elemental form of loose fibrous connective tissue which is universally distributed throughout the body as a wrapping and packing material. This useful substance is able to bind water into its colloidal framework. In the process the viscosity of the intercellular medium is increased so that the connective tissue acquires the greater density and firmness known as tissue turgor. An important benefit of the presence of hyaluronic acid is that the intercellular medium becomes more resistant to the passage of fluid and fluid-borne objects through it. This state of altered density enables the intercellular medium of connective tissues to resist the spread of invading bacteria or noxious substances. They must first secrete an enzyme, *hyaluronidase*, to liquefy the ground substance before they can spread through a tissue.

THE FIBROUS CONNECTIVE TISSUES

Most of the functions of connective tissue which were listed are performed by tissues that are predominantly fibrous. The arrangement of the fibers produces characteristic gross anatomical units. These fibrous structures are classified according to their gross appearance, their fiber arrangement, and their function. They are listed in order of progressing strength or ability to resist physical forces.

LOOSE, IRREGULARLY ARRANGED CONNECTIVE TISSUE

This arrangement of connective tissue is the type universally encountered in the body and fulfills the previous descriptions of the basic form of connective tissue. It is a loose association of fibrocytes, fibers, and intercellular medium in almost equal proportions. The fibers are rather widely separated into an irregular interlacing network. Both collagenous and elastic fibers are present. The gross appearance

of this type of connective tissue is that of delicate interlacing strands resembling a dense cobweb. Tissue fluid and a minimal amount of ground substance fill the interstices of the network of fine fibers.

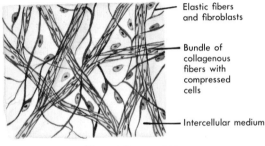

Elastic fibers and fibroblasts

Bundle of collagenous fibers with compressed cells

Intercellular medium

Microscopic appearance

The early anatomists, dissecting unembalmed bodies, called this tissue arrangement *areolar tissue* (foamy or gassy tissue) because gaseous products of decomposition bubbled amidst the fiber network. Microscopists today refer to such a tissue pattern as *areolar connective tissue*. It is, however, described more accurately as a loose irregular arrangement in which both collagenous and elastic fibers are present. This type of connective tissue is adapted for packing and filling between structures and for wrapping or loosely binding them together. Loose wrapping rather than strength is its major characteristic, but areolar connective tissue does provide for general regional support or for broad subdividing planes in the body within the limits of a loose meshwork.

Fascia

A sheet or broad band of loose fibrous tissue which covers the body underneath the skin or invests the muscles and organs is called fascia. It is an areolar mixture of interlacing collagenous and elastic fibers. When such a mixture is encountered in dissection or in surgery it is called a fascia. This is a very general term and, as will be seen, several types can be described.

Tela Subcutanea

The fascia under the skin is modified in development by the presence of many mesenchymal cells which specialize as fat-forming cells. The result is a loose connective tissue with a high fat content which lies as a continuous sheet (tela) underneath the skin (subcutanea) in all regions of the body.

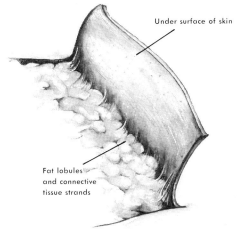

Under surface of skin

Fat lobules
and connective
tissue strands

Tela Subcutanea gross appearance

It was formerly called the *superficial fascia* because of its location and to distinguish it from deeper-lying sheets of connective tissue.

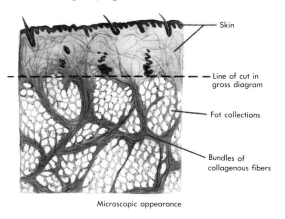

Skin

Line of cut in
gross diagram

Fat collections

Bundles of
collagenous fibers

Microscopic appearance

The tela subcutanea possesses insulating and temperature-regulating qualities owing to its fat content. Its looseness permits a flexible union of the skin to deeper structures and provides for the passage of nerves and blood vessels to the skin.

DENSE, IRREGULARLY ARRANGED CONNECTIVE TISSUE

There are many places in the body where a loose meshwork of connective tissue fibers cannot provide a wrapping or a sheet that is dense enough to enclose, restrain, or separate functioning structures. A somewhat stronger type develops in which there are more fibers which are grouped together into larger bundles. These appear to the eye as large coarse strands which are gathered into a tight network with smaller interstices than are found in a loose connective tissue. The proportion of elastic fibers diminishes, and the areolar mixture becomes predominantly collagenous to provide for a measure of strength.

Deep Fascia

This type of tissue arrangement is adapted for enclosing individual muscles in a dense, flexible meshwork which wraps their contractile elements into a functioning unit. In contrast to the superficial fascia or tela subcutanea, these connective tissue envelopes are devoid of fat and lie deeper in the body, thereby receiving the name *deep fascia.* Other uses of deep fascia are to assemble muscles of similar function into related groups or to make thin flexible sheets which separate several such functioning groups from each other. In this way a part of the body, such as the forearm, can be subdivided without sacrificing movement or flexibility of the contained structures.

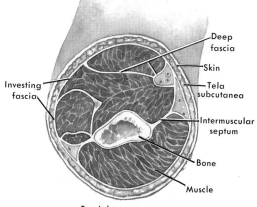

Deep
fascia

Skin

Tela
subcutanea

Investing
fascia

Intermuscular
septum

Bone

Muscle

Fascial arrangements

Therefore, great strength is not necessary, and during surgery or in dissection deep fascia may be lifted, picked away, or separated from the structures it encloses.

Investing Fascia

In many regions of the body whole groups of different structures are disposed along a single axis. The linear arrangement may include groups of muscles, a bone or bones, and many vessels and nerves. The form of the particular body part where this arrangement occurs, such as the arm, thigh, or neck, tends to be cylindrical. It is desirable that the functioning components be restrained within the columnar limits of body form. The skin and tela subcutanea are too soft and flexible a covering to perform such a role. A dense interlacing arrangement of coarser collagenous fiber bundles develops into a sleevelike arrangement between the more delicate deep fascia and the fatty tela subcutanea; this is an *investing fascia.*

Intermuscular Septa

Sometimes muscles or muscle groups of the limbs are so numerous and so powerful that the deep fascia between them becomes developed into strong, heavy fibrous partitions. These *intermuscular septa* extend from the bony framework of the limb outward to blend with the sleeve of investing fascia. The many septa consist of even denser and coarser collagenous fiber bundles in which elastic fibers become quite sparse. Strength is a necessary quality, because in some cases there are more muscle fibers than there is bony surface upon which they can secure attachment. The heavy intermuscular septa, which are themselves attached to these bones, provide surfaces near the bones that are strong enough to extend the area for muscular attachment.

Fibrous Organ Capsules

All organs are wrapped with a layer of connective tissue which protects the soft parts within, contributes to the form of the organ, and separates it from adjacent structures. These roles may be performed adequately by the looser areolar type of connective tissue. Some organs, such as the liver, spleen, and brain, are extremely soft and highly cellular.

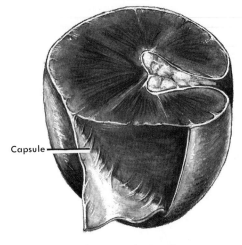

Capsule

They need the firmer case provided by a thick envelope of closely interlacing collagenous fibers similar to investing fasciae and intermuscular septa.

REGULARLY ARRANGED CONNECTIVE TISSUES

The irregular matted meshwork of connective tissue is not strong enough for all situations in the living body. The contraction of muscles exerts a tremendous pull upon the bones to which they attach. The combined pull of a number of muscles produces movements of the joints of the body, but at the same time applies forces of a degree such that disruption of the bony and cartilaginous elements of the joints may occur unless they are bound together by tissues which can resist these forces. An irregular, matted fiber network might shear, tear, or separate from the structures it attempted to unite. In its place, as in tendons, is found a very dense, completely collagenous connective tissue in which the fiber arrangement is distinctly different. The tissue is so dense that the fibroblasts are compressed between the massive fiber bundles. The intercellular tissue fluid is reduced to a thin film over the surface of the flattened cells and between the fibers. The fiber bundles no longer form an interlacing network but course in a close, compact parallel arrangement. The fibers are arranged in line with the direction of forces operating in the tissue. This type of connective tissue is eminently suited for situations in the body in which great tensile strength is necessary.

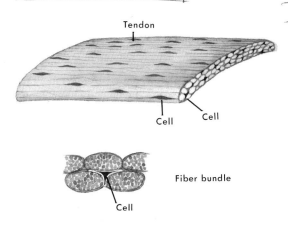

Tendon

Cell   Cell

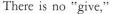

Fiber bundle

Cell

There is no "give," no elasticity to the gross structures which are formed by such connective tissue. Its design is for the firmest possible union and for the resistance of forces that might tend to disrupt the structures which are joined.

Tendons

Muscle fibers are capable of great contractile force but yet are soft and cannot themselves adhere to the hard bones upon which they attach. *Tendons* provide an intermediate connection by which the muscle fibers may be securely attached to a bone but yet through which the pull of the muscle may be transmitted undiminished. Tendons appear as either cylindrical cords or flattened bands extending directly from their muscles to the bones which are to be moved. The heavy collagenous fibers are bundled together and wrapped by a network of areolar connective tissue.

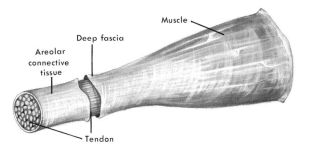

Muscle

Deep fascia

Areolar connective tissue

Tendon

In certain locations free movement of the tendon in response to muscle contraction is facilitated by *tendon sheaths*. These are fibrous sleeves,

lined by mesenchymal epithelium, which surround the tendon when it is subjected to pressure or friction, such as in its passage between heavy muscles, between muscles and bones, or through a tunnel between bones. A film of fluid is formed by the epithelial lining of the tendon sheath to act as a lubricant.

Biceps tendonitis

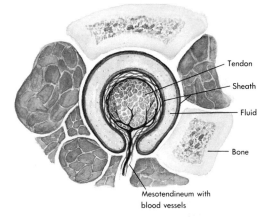

Tendon

Sheath

Fluid

Bone

Mesotendineum with blood vessels

Tendons in their course may have to cross the surface or margin of a bone. The rubbing of the collagenous cable upon the bone would impede the movement of the tendon. Ultimately injury to the tendon would occur. At such points the cells of mesenchyme develop a delicate sac or *bursa* of fibrous tissue between the two structures.

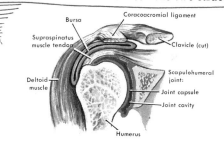

Bursa   Coracoacromial ligament

Supraspinatus muscle tendon   Clavicle (cut)

Deltoid muscle   Scapulohumeral joint:

Joint capsule

Joint cavity

Humerus

A lining of mesenchymal epithelium produces a lubricating fluid. In this case, movement of the tendon produces movement of the walls of the bursa, which slide smoothly upon each other to facilitate passage of the tendon and to protect it from the bone.

The nature of the connection between a muscle and its tendon and of the attachment of a tendon to bone will be discussed with the muscular system and with the consideration of bone.

Aponeuroses

An *aponeurosis* is a broad flat tendinous sheet. The dense collagenous fibers are arranged in a parallel fashion but in the form of thin, extremely strong ribbons. These may be separated, at intervals, by a delicate areolar network which conveys fine blood vessels. Aponeuroses are found in several situations serving to attach muscles to bones. In one case a large powerful muscle may arise from an extensive bony surface or from a series of bones. The attachment is made by a great number of thin, short collagenous ribbons. Since the pull exerted by the many contractile fibers of a large broad muscle is tremendous, only the parallel collagenous arrangement can provide the tensile strength required. In another case, aponeuroses serve to attach large muscles which migrate or tend to grow away from their bones as they develop. Here broad, flat tendinous sheets form to link the muscle to its origin.

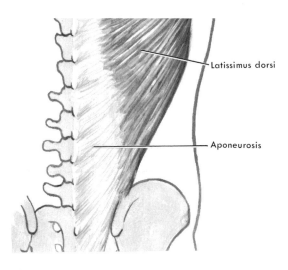

Latissimus dorsi

Aponeurosis

In other cases a muscle or group of muscles may spread out beyond the limits of bone available for attachment. A fanshaped, aponeurotic sheet or a collagenous shelf adjacent to the bone serves to extend or broaden the area for muscle attachment.

Aponeuroses also serve to provide support for organs of the body or to form a base for muscular attachment where there is no portion of the skeletal system but yet great strength is needed. An example is found in the anterior wall of the abdomen. Broad aponeurotic sheets permit the abdominal muscles, which approach from both sides, to interlace and find attachment. At the same time both the muscles and the aponeuroses restrain the abdominal organs against the forces of gravity in the erect position.

Ligaments

Bones which fit together to form joints must be free to move. Yet they must be held together against forces which would tend to disrupt their working association. In most joints, motion must be free in certain directions but restricted in others to avoid instability or uncoordinated action. The form of the articulating surfaces, the pull of muscles, the pressure of tendons crossing over the joint, and the sleeve-like capsule which surrounds the joint contribute to the stability of the joint. A great deal of the support of any joint is provided by *ligaments,* which are short, flat, tough bands of fibrous tissue which connect or restrain bones that must remain together for function. As in tendons and aponeuroses, closely packed collagenous fibers are arranged in a parallel manner to give the tensile strength needed.

Types of Ligaments. The most common ligaments are separate fibrous bands which appear on the external surfaces of joints. They are firmly united to the fibrous covering of the bones or send anchoring fibers into the substance of the bone itself.

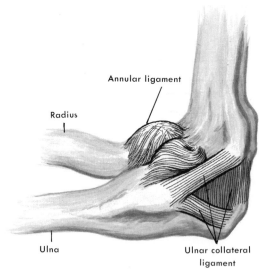

Annular ligament

Radius

Ulna

Ulnar collateral ligament

Externally located ligaments may reinforce one surface of a joint or may act as a fibrous wrapping which holds a group of bones together. Other ligaments may be located within the cavity of a joint where they run a very short course in acting to hold the articulating bones together as they move upon each other. Collagenous ligaments are unyielding and inelastic. But they may be disposed in such a manner that they may be loose while the joint moves in one direction but become taut if it attempts to move in another direction. In such cases the bands function as *check ligaments* in that they permit a desired movement but prevent those which should not occur.

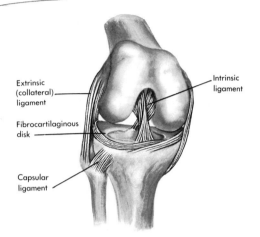

Extrinsic (collateral) ligament

Intrinsic ligament

Fibrocartilaginous disk

Capsular ligament

A *joint capsule* is a thin fibrous envelope which encloses the articulating ends of the bones that form a joint and the cavity between the bones.

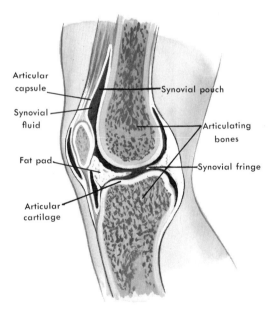

Articular capsule

Synovial fluid

Fat pad

Articular cartilage

Synovial pouch

Articulating bones

Synovial fringe

The capsule is lined by a mesenchymal epithelium, the *synovial membrane,* which forms a lubricating fluid within the joint. The capsule, therefore, is an enclosing and lubricating membrane which is not strong. At points where the capsule might tear if undue force were applied to the joint, reinforcing bands develop as thickenings of the fibrous envelope. These *capsular ligaments* strengthen the capsule but alone would be insufficient to provide much stability to a joint.

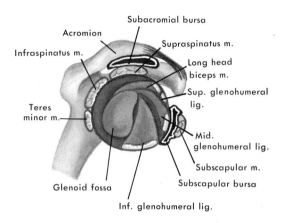

Acromion

Subacromial bursa

Infraspinatus m.

Supraspinatus m.

Long head biceps m.

Sup. glenohumeral lig.

Teres minor m.

Mid. glenohumeral lig.

Subscapular m.

Glenoid fossa

Subscapular bursa

Inf. glenohumeral lig.

Elastic Ligaments. In all the situations described the function of the ligament requires tensile strength and inelasticity. The many joints between the bones of the spinal column, however, require a firm form of union which must be somewhat resilient. Flexibility of intricate trunk movements must be preserved at the same time that sufficient rigidity of the column of vertebrae is ensured. An unusual form of connective tissue develops in which large elastic fibers predominate. These are bound together by a network of fine collagenous fibers to form *elastic ligaments.* The

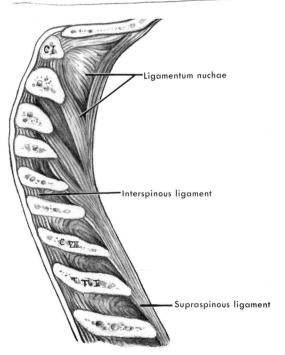

CI

Ligamentum nuchae

Interspinous ligament

C VII

T I

Supraspinous ligament

high elastic content of the tissue gives these ligaments a characteristic yellow color which contrasts with the glistening white aponeuroses and fibrous ligaments. Elastic ligaments are found in the neck region (ligamentum nuchae) and in connecting portions of adjacent vertebrae (ligamenta flava).

Other Forms of Ligaments. Other tissues that may be called ligaments do not fulfill the descriptions which have been given for true ligaments. Capsular ligaments fall partly in this category. Notches in the margins of bones are sometimes roofed over by connective tissue. These strands convert the notches to shallow tunnels which hold blood vessels and nerves in place as they pass over the margin of the bone; they are often referred to as ligaments. A still looser usage of the term ligament is found in reference to folds of the lining of the abdominal and pleural cavities. These folds represent the draping of the lining from one organ to another or provide a loose attachment of one organ to another or to the body wall. Organs are only relatively restrained by these folds and freedom of movement is characteristic. The use of the term ligament, though common with surgeons, is not in keeping with its meaning.

MODIFIED CONNECTIVE TISSUES

The changes in the fibrous connective tissues which produce many different structures are mainly changes in degree which principally affect the type and arrangement of the fibers. The modified connective tissues are highly specialized tissues in which the cells, the intercellular products, and the intercellular medium are extensively altered. The basic form of a connective tissue may no longer be apparent, or it may be so masked that an entirely different tissue seems to exist. Analysis of the composition of such a tissue and the manner of its development reveals that the modified forms are basically connective tissues in which the new forms are associated with special functions. In each type it will be recognized that none of the more simple arrangements of connective tissue studied so far could provide a structural basis for the more highly developed func-

tion. The modified connective tissues are fat, cartilage, and bone.

FAT

Fat is a modified connective tissue in which the cells become dominant as the result of chemical changes in the cytoplasm of mesenchymal cells. A specialized tissue forms which is recognized microscopically as *adipose connective tissue* but which appears grossly to the eye in the familiar form known as *fat*.

Origin of Fat

There are many totipotential mesenchymal cells present in areas where fat will be formed. The cells enlarge as minute droplets of fatty substances appear within the cytoplasm. When the cells are established in their specialization as fat-forming cells, they are known as *lipoblasts* or *steatoblasts*. The fat droplets continue to form and quickly coalesce.

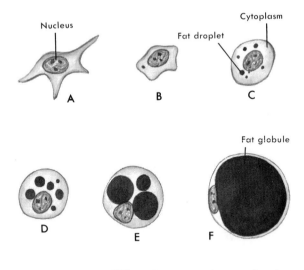

When the process is completed, one large fat droplet occupies most of the space within the cell membrane and displaces the remaining rim of cytoplasm, compressing the nucleus to one side of the cell. Its characteristic appearance suggests the name, *signet ring cell*.

Adult Gross Form of Fat

The clusters of swollen fat cells have developed within the fibers of an areolar connective tissue which are now adapted to form fat. The fatty substances within the cells are semiliquid at body temperature, and the cells are closely packed in a slippery mass which would be displaced easily if not restrained. Reticular connective tissue fibers become developed into transparent gossamer membranes which surround groups of fat cells and convey capillaries into the scanty intercellular medium. Each group of fat cells is a *fat lobule,* which may be distinguished by the eye as a glistening white or yellow sphere or polyhedron.

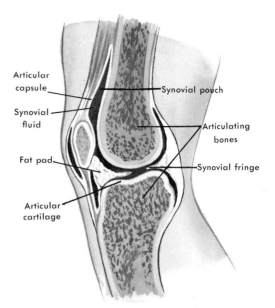

Tela Subcutanea gross appearance

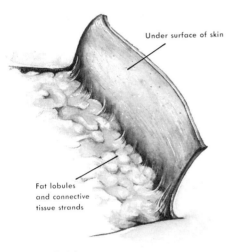

Pressure applied to fat lobules ruptures the delicate connective tissue wrapping and a number of the cell membranes. The escape of the fatty materials gives a typical greasy or fatty "feel."

When fat appears in an adult areolar connective tissue, the cell groups are smaller and are interspersed among the arrangements of elastic and collagenous fibers, blood vessels, and nerves as yellow patches. The amount of fat varies in location in the body, but it is generally distributed in the tela subcutanea, around internal organs, and in the folds of the abdominal lining membranes (mesenteries) which attach abdominal organs to the posterior body wall. Fat smooths the contours of the body, rounding bony or muscular surfaces and filling pockets or spaces between other structures as in the arm pit.

Despite its cellularity, fat possesses considerable resiliency owing to the combined effect of the pressure of lipoid materials against the membranes of a large number of cells. *Fat pads* are

masses of adipose cells which are sheathed by strong *envelopes* of connective tissue fibers. They appear in joints where they serve to take up the shock of weight-bearing or to ease the impact of one bone upon another. A packing of fat tissue between the bony walls of the orbit and the eyeball facilitates eye movement and cushions this vital organ. Fat pads appear in areas such as the sole of the foot which are subject to shock and to friction.

CARTILAGE

Cartilage is a modified connective tissue in which the intercellular medium is altered extensively by the addition of substances which change its consistency to provide resiliency, greater strength, and some rigidity.

Formation of Cartilage

In areas where cartilage will develop, mesenchymal cells are already present. These totipotential cells divide rapidly and cluster compactly together in a cellular mass which approximates the shape of the structure to be formed. This may be a rod, a ring, a crescent, or an irregular plate or block. The mesenchymal cells then begin to separate and function briefly as fibroblasts. A mixture of fine collagenous and elastic fibers is formed in the intercellular medium, although in some of the areas where particular types of cartilage are developed, one fiber type may be formed at the exclusion of the other. Cartilage, therefore, is organized initially as a connective tissue.

The totipotentiality of the mesenchymal cells is seen as they cease the formation of fibers and proceed to modify their intercellular medium. The cells either add complex polysaccharide substances (chondroitin sulfate and chondromucoid) to the intercellular medium, or they induce their formation from simpler materials which have diffused into the medium from the capillaries. At this stage the intercellular medium begins to become firm and the cells are renamed *chondroblasts* because of their role in forming cartilage.

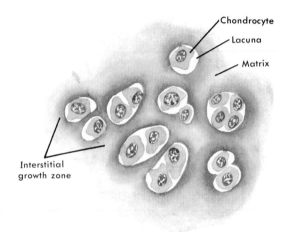

Chondrocyte
Lacuna
Matrix
Interstitial growth zone

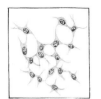

Mesenchymal cells   Separation

Fibroblasts   Collagenous and elastic fiber   Chondroblast and hardening
                        formation   intercellular substances

The substances added to the intercellular medium engulf the connective tissue fibers and incorporate them into a stiffening cartilaginous *matrix*. As the chondroblasts spread apart the matrix forms around and between them. The cells become imprisoned within minute pockets of the hardening matrix which are known as *lacunae*. The quality of the matrix is more than gelatinous but less than glasslike in nature; it is best described as being rubbery and resilient. When the cartilaginous matrix has been formed, the cells within the lacunae are termed *chondrocytes*.

Growth and Metabolism of Cartilage

Since cartilage forms first within the embryo, many of the structures of cartilaginous nature increase in size with growth of the body. In young cartilage the chondrocytes divide to form new cells within the lacunae which become en-

larged to accommodate several inhabitants. These cells form more matrix between them and develop their own lacunae. In this way groups of cells develop which are separated from other groups by an increasing amount of matrix. This *interstitial growth* of cartilage ceases as the matrix becomes more rigid. Later growth of cartilage occurs at the borders of a cartilaginous mass where there are mesenchymal cells in the adjacent connective tissue. These cells produce *appositional growth* at the periphery by becoming chondroblasts which develop new cartilage.

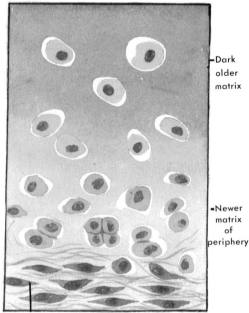

Dark older matrix

Newer matrix of periphery

Chondroblasts

The stiff consistency of the cartilaginous matrix makes difficult the growth of blood vessels.

Cartilage, therefore, is devoid of blood vessels. Nutriments seep through the matrix slowly from blood vessels in adjacent tissues. Consequently the metabolism of cartilage is low and, if the tissue is injured, its repair is slow.

Adult Forms of Cartilage

Cartilage is classified according to the kind of connective tissue fibers present in the matrix and the proportion of the fibers to the matrix. *Hyaline, elastic,* and *fibrous* types of cartilage are recognized.

Hyaline Cartilage. This is the most common cartilage in the body and is a modified areolar connective tissue. Both collagenous and elastic fibers may be present as a fine feltwork which is masked by the cartilaginous matrix. Hyaline cartilage appears to the eye to be a firm smooth substance which is bluish white and translucent.

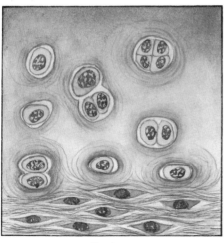

Hyaline

A familiar form is the gristle of meat. The skeleton of the fetus is composed of this type of cartilage. It is replaced in the formation of the long bones of the body. Hyaline cartilage forms the smooth articular surfaces found where bones move upon each other in the joints. It forms the cartilages of the respiratory system and the costal cartilages that form the flexible union of the ribs to the breast bone.

Elastic Cartilage. Elastic fibers are dominant in this form of cartilage, which resembles a modified elastic connective tissue. The elastic fibers form a network in the matrix that masks them.

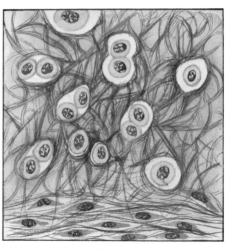

Elastic

The fibers are more numerous and coarse than in hyaline cartilage. They impart their yellow color to the matrix and make it more fibrous and opaque in appearance. Elastic cartilage is found where movement of the cartilaginous structure occurs, as in the epiglottis, which guards the entrance to the larynx during swallowing, and in the cartilages of the larynx, which are moved to produce tension upon the vocal cords. The framework of the external ear is of elastic cartilage, as is that of the auditory (eustachian) tube, which connects the middle ear with the nasal passages.

Fibrous Cartilage. A distinctly tendinous character is found in fibrous cartilage which may be considered to be a modified dense collagenous connective tissue. Heavy bundles of collagenous fibers are imbedded in a cartilaginous matrix.

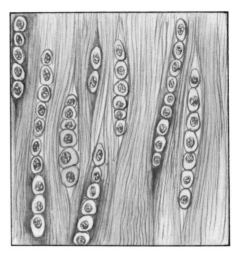

Fibrous

Cartilage lacunae containing chondrocytes are arranged in the matrix between the fiber bundles. The fiber content is quite high and frequently exceeds the amount of matrix. Fibrous cartilage, therefore, is an exceedingly tough material. It is useful in uniting bones at joints where movement is limited, as between the bones of the skull. Fibrous cartilage withstands compression and thus may be found making up the cushioning pads between the vertebrae (intervertebral disks) and reinforcing the hyaline articular cartilages at the knee and hip. If a tendon or ligament must attach to hyaline cartilage, the union is often made through the medium of fibrous cartilage, which, in composition, is a blend of the two structures.

Perichondrium

A structure formed of cartilage is enclosed within an envelope of dense fibrous connective tissue, the *perichondrium*. This envelope protects the cartilage and contains blood vessels for its nutrition. An important feature of the perichondrium is the presence of mesenchymal cells which develop into chondroblasts when needed for appositional growth of the cartilage or for its repair. The fibers of the membrane blend into the connective tissue adjacent to the cartilage. Only the opposing surfaces of articular cartilages which move upon each other are devoid of the covering of a perichondrium.

BONE

The greatest modification of connective tissue occurs in bone or osseous tissue. The structural alteration forms discrete units called bones, which form most of the skeletal system of the adult body. Bone is characterized by strength and rigidity, but the structures of which it is composed are also considerably elastic. The addition of calcium salts to the intercellular medium is responsible for the quality of rigidity. These minerals are deposited in characteristic patterns conforming to the direction of physical forces in the tissue. The structural pattern is responsible for the great strength of bone. The underlying fibrous framework of connective tissue contributes elasticity to bone which would otherwise be quite brittle.

Origin of Bone

Prior to the formation of bone the physical support and protection of any part of the body are furnished by mesenchymal cells or by hyaline cartilage. In the first case bone may be formed directly, whereas in the second case the cartilage must be removed as bone is formed to take its place. In either case the general steps in the formation of bone are the same.

Connective Tissue Framework. In an area where bone will be formed, mesenchymal cells develop into fibroblasts. By the secretion of tropocollagen, these form a delicate embryonal connective tissue in which fine collagenous fibers predominate as the intercellular product. Many capillaries grow into the tissue to provide it with a rich vascular supply.

Formation of Osteoid. Some of the fibroblasts enlarge and line up in parallel rows amidst the fibers and are joined by large numbers of new cells which originate from other mesenchymal cells. The cells in these ranks first modify the connective tissue by altering the intercellular medium. The cells, now known as *osteoblasts,* cause the addition of viscous protein polysaccharides to the intercellular medium of the connective tissue. The connective tissue fibers are incorporated into the viscous intercellular medium,

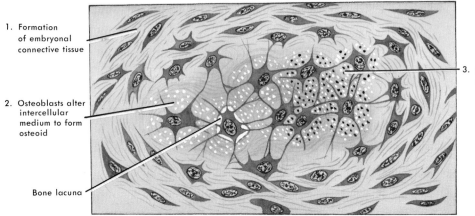

1. Formation of embryonal connective tissue

2. Osteoblasts alter intercellular medium to form osteoid

3. Calcification of osteoid to form bone

Bone lacuna

which acquires a rubbery elastic consistency. In the process of developing this new substance, which is known as preosseous tissue or *osteoid,* the cytoplasm of each cell spreads out into many highly branched processes. The processes of an osteoblast stretch out to maintain continuity with those of other cells. The osteoid matrix forms around the osteoblast and its processes. The central part of the cell becomes enclosed within a lacuna. This cellular pocket in the matrix differs from the lacuna of cartilage because many minute channels are formed about the osteoblastic processes which radiate from the central part of each cell. Each lacuna in the osteoid is, therefore, connected by *canaliculi* with a number of other lacunae. In this manner the osteoblasts maintain contact with each other and the diffusion of nutriments is facilitated.

Just prior to being imprisoned in the matrix they have formed, osteoblasts may divide to form new cells which arrange themselves outside the new osteoid. They are joined by new osteoblasts which have differentiated from nearby mesenchymal cells. Growth and extension of the osteoid occur as the new cells modify their own area in the stages which have been described.

Osteoid is not bone but rather an intermediate stage of bone formation. It is the living organic part of bone which endows this tissue with its toughness and elasticity. About one third of the total weight of a bone is represented by its osteoid content; the remainder is due to the mineral content which is yet to be added. A bone may be reduced to its osteoid form by decalcification with acids. When this is done, the shape and structural pattern remain. The difference is that the rigidity is gone, and the demineralized osteoid can be bent, twisted, or slightly stretched.

Calcification of Osteoid. The final stage in the formation of bone occurs as the osteoblasts draw upon mineral ions which diffuse into the intercellular medium of osteoid from the blood vessels. The mineral salts—calcium carbonate and calcium phosphate—are deposited within the intercellular medium and upon the surface of the collagenous fibers as very finely dispersed crystals. The matrix acquires rigidity and osteoid becomes bone. The osteoblasts now are adult bone cells, the *osteocytes,* which having formed bone now regulate its metabolism.

Bones as Living Tissue. It would be erroneous, in studying a skeleton or individual bones, to consider these dried, rocky structures as typical of living bone. They are merely mineral shells within which are dead osteocytes, shriveled blood vessels devoid of blood, and shrunken osteoidal components. Living bone courses with circulating blood whose mineral ion content is at equilibrium with that of the bone matrix. The mineral content of bone tends to be constant in the healthy person, but there is a continual adjustment in keeping with the needs of other tissues for these ions. If disease, vitamin deficiency, or faulty intake or utilization of calcium occurs, excessive amounts of the minerals may be withdrawn from bone. Softening of the bones or imperfect bone formation may result. After birth, much of the formation of blood cells occurs in the interior marrow of some bones. Bones, therefore, are living and are important for many reasons other than their rigidity.

Types of Bones

Bone tissue, like cartilage, is organized into structural units each of which is called a bone. Wherever bone tissue is found, its underlying composition is the same but there are differences in form, structural pattern, and means of development or growth which have led to various classifications.

Bone Shape. The bones of the body may be grouped as long, short, irregular, or flat. These terms will be applied in the descriptions of the units of the skeletal system.

Long bones are the largest bones in the body and include the prominent bones of the limbs.

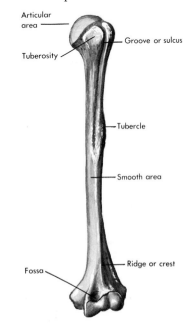

Articular area

Groove or sulcus

Tuberosity

Tubercle

Smooth area

Ridge or crest

Fossa

A long bone has a longitudinal axis of considerable length. At each end there is an expanded portion, known as the *epiphysis,* which usually articulates with a similar part of another bone at a joint. The shaft of the bone between the two epiphyses is called the *diaphysis.*

Short bones differ from long bones only in being smaller and having less prominent ends.

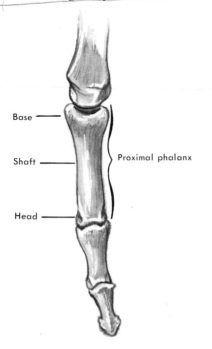

Base —

Shaft —

Head —

— Proximal phalanx

Irregular bones are so named because of their blocklike, irregular shape. Several surfaces of an

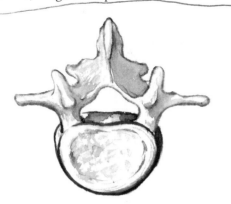

irregular bone may articulate with several other bones, as in the wrist bones. Other irregular bones may have processes of varying sizes or shapes for muscle attachments or for articulation with other bones, as in the bones of the spinal column.

Flat bones are platelike bones that usually have

two broad surfaces (*tables*) with narrow edges by which one flat bone articulates with another. The bones of the vault of the skull are examples of this type.

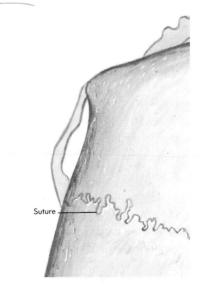

Suture —

Structural Pattern.   Bones, or portions of a bone, are also classified according to the density of their structural pattern into *compact* and *spongy bone.*

Compact bone makes up the tubular shaft of a long bone and is continued onto the epiphysis as its extremely thin outer shell. The inner and outer tables of a flat bone are also of this type.

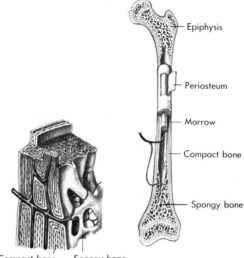

— Epiphysis

— Periosteum

— Marrow

— Compact bone

— Spongy bone

Compact bone   Spongy bone

Compact bone is a dense homogeneous arrangement of osseous tissue. It presents a solid, ivory-like appearance with no interruptions in its substance. The physical principles underlying the structure of any rigid tube apply to the shaft of a long bone composed of compact bone. It is as strong as a solid cylinder but is lighter and withstands bending, compression, and weight-bearing equally well.

Spongy bone has the appearance of a latticework. It is constructed of many small, splinterlike, irregular rods (*trabeculae*), which are formed into a network with spaces between. The trabeculae are reworked during the growth of spongy bone to align themselves with the lines of force that are imposed upon the whole bone. The appearance, therefore, is not unlike the arrangement of beams and girders in the steelwork of a complicated bridge or heavy machine. Spongy bone is light but strong. It is found in the epiphyses of long and short bones and adjacent to the marrow cavities of their shafts. The center of irregular bones may be of this type, and the bony spicules between the tables of flat bones (*diploë*) resemble spongy bone in construction.

Method of Formation. Bones may be further classified according to the manner by which their bony tissue is formed.

Membrane bones are the flat bones. These bones develop directly in the mesenchyme without the presence of a preceding skeletal member formed of hyaline cartilage. The manner in which they form is termed *intramembranous bone formation,* and it is identical to the processes described already for the origin of bone. The formation of membrane bone is continuous, but flat bones must change their size and shape constantly in order to keep pace with the steady growth of the brain which they surround and protect. One surface of the bone is enlarged while bone is subtracted from the other surface. A special type of cell, the *osteoclast,* appears wherever bone must be removed. These multinucleated cells, which may be but a different form of the original osteoblast, are found arranged along the surface of a bone which is undergoing resorption or along trabecula within it. It is common, during growth of the skull, for osteoclasts to remove bone on the concave inner surface of a flat bone while osteoblasts are laying down new bone on the convex outer surface.

Cartilage bones or *replacement bones* are the long, short, and irregular bones of which most of the skeleton is composed. These bones are preceded

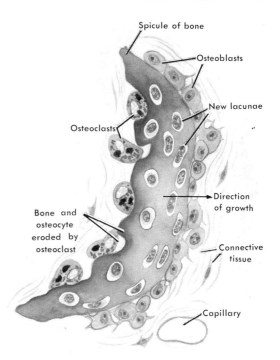

in the body of the embryo and the child by miniature skeletal structures which have a similar form but which are composed of hyaline cartilage. The cartilage models protect the growing body and give it rigidity but must be replaced by osseous tissue which can withstand better the forces of active adult life. *Endochondral* or *intracartilaginous bone formation,* therefore, involves the simultaneous removal of hyaline cartilage and the formation of bone while all the functions and growth of the skeletal system are being carried on. The actual process of bone formation is not unlike that described previously, but it presents a more complicated picture because of the presence and removal of cartilage. (See color plate 2.)

At the beginning of endochondral bone formation the cartilage in a central location softens and degenerates. Under the perichondrium, dormant mesodermal cells form new mesenchyme, which invades the degenerating cartilage as the *periosteal bud.* Simultaneously other mesodermal cells become transformed into osteoblasts, which lay down a thin shell of bone under the periosteum at the surface to strengthen the area of degeneration and invasion. Other mesenchymal cells, swept in with the periosteal bud, develop into phagocytic cells, which commence the removal of the dying cartilage. Other osteoblasts develop from the mesenchymal cells of the enlarging periosteal bud and start to form trabeculae of bone amidst the young connective tissue and degenerating cartilage. Such an area of bone formation is called an *ossification center.* The first such center to appear is in the diaphysis of what will become a long bone or at the center of an irregular bone; it therefore is the *primary ossification center* of that bone.

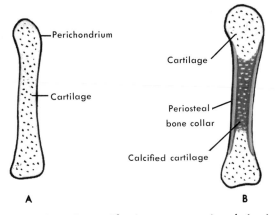

During the ossification process the skeletal member continues to enlarge and much of the growth is cartilaginous. New *secondary ossification centers* appear in the epiphyseal ends of the bones and in particularly large projections where several muscles attach. The process of bone formation gradually catches up until eventually the location of new cartilage formation is restricted to narrow disks between the diaphysis and the epiphysis. Growth of the bone during childhood occurs at these *epiphyseal plates* whose location, when bone growth ceases in early adult life, is marked by *epiphyseal lines* on the external surface of the bones. (See color plate 2.)

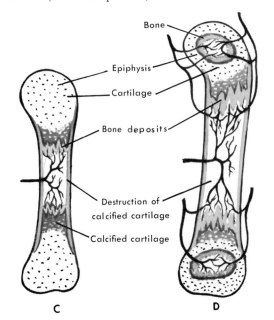

The bone laid down by the ossification centers is not permanent. It is spongy bone and in places must be reworked to align the trabeculae with new physical forces imposed by muscular activity and weight-bearing. In some locations the original bone must be converted into compact bone. A marrow cavity replaces the early bone at the center of the shaft. Therefore, osteoclasts are constantly at work to remove or to remodel much of the bone. In their wake, osteoblasts lay down more permanent bone.

In adult bone, particularly the compact type, the proportion of osseous tissue is increased markedly over the proportion of connective tissue and vascular elements. An adequate blood supply is necessary, but it will be recalled that the osteocytes are imprisoned in their lacunae with only delicate processes maintaining contact through the canaliculi of the calcified osteoid. Therefore, in the definitive reworking of bone, the osseous tissue is organized into a series of units, the *haversian systems* or *osteons,* which are oriented to provide strength and also an adequate blood supply to the osteocytes.

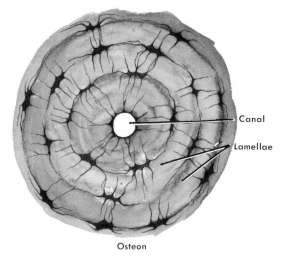

Osteon

When an area of original bone is remodeled, a long tunnel is created by the absorption of existing bone. Connective tissue containing sprouting capillaries and mesenchymal cells migrates into the excavated bone. Osteoblasts develop and move to the periphery of the tunnel. Concentric lamellae of bone are laid down, each succeeding lamella having a smaller circumference than its predecessor. Finally a narrow cavity is left at the center. This *haversian canal* forms the center of an haversian system.

Branches of blood vessels, nerves, and lymphatics are contained within its connective tissue. Nutriments diffuse from the vessels into the canaliculi of the lamellae to maintain the metabolism of the osteocytes.

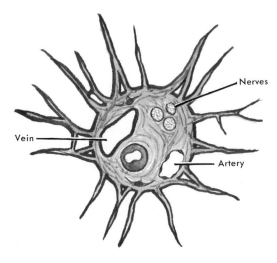

Once an haversian system has been formed, it may be involved in a subsequent revision of the bone structure. It is not strange, therefore, to find irregular fragments of old haversian systems filling the spaces between the concentric patterns of newer ones.

Bone is reworked continually until ossification extends into the epiphyseal plates and bone growth ceases. Even in an adult bone, however, the coordinated activities of osteoclasts and osteoblasts may be resumed to produce minor reorganizations of bone structure in response to new physical forces or to carry out the repair of fractures. Dormant undifferentiated mesenchymal cells stand ready under the periosteum and in the connective tissue of the marrow to respond to these needs.

THE SKELETAL SYSTEM IN GENERAL

All the bones in the human body, when assembled in their relationships to each other, comprise the skeletal system. To view a mounted, articulated skeleton in the classroom is only a beginning of the study of the living person. The study of the skeletal system is a valid beginning, however, for it is the framework of the body and of each of its parts. Muscles and fascia clothe the skeletal system and are related vitally to its functions. The joints join together portions of the skeletal system and permit them to move upon each other. The skeletal system provides rigidity, support, protection, muscle attachment, and leverage. Bones also contribute to the formation of blood cells and storage of mineral salts.

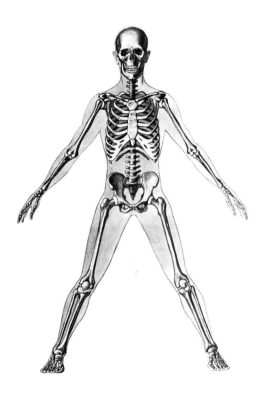

FUNCTIONS OF THE SKELETAL SYSTEM

Rigidity. There must be a means of providing the body as a whole, and many of its component parts, with a form of stiffening. Most of the structures of the body are soft, pliable, and compressible. Muscles are firm only when contracting, and the various forms of connective tissue (other than cartilage and bone) are stiff only when resisting physical forces. The other elements of the body would clump together in a shapeless mass if a rigid framework were not present. The skeletal system provides an internal mechanism of rigidity upon which other systems can act or be aligned.

Support.   In the same way, the skeletal system provides a framework from which elements of other systems of the body may be suspended, fixed, or derive support.

Protection.   Vital organs of the body are encased by the skeletal system, either completely or in association with elements of connective tissue and the muscular system. The brain is completely surrounded by the bones of the skull. The spinal cord is almost completely protected by bony parts of the vertebrae, although some apertures are filled in by strong connective membranes or ligaments. The thoracic organs are partly enclosed by the vertebrae, ribs, and sternum, but much of their protection is derived from muscles and fasciae which complete gaps in the thoracic cage.

Muscle Attachment.   The contraction of a muscle would be ineffectual if one end were not firmly attached to a fixed portion of the skeletal system and the other end equally firmly attached to the skeletal member of the body part which is to be moved.

Leverage.   The arrangement of bones or projections from the mass of a particular bone provide the leverage required for many muscles to act. Bones act as levers in typical physical systems whereby the power of muscular contraction is transformed into motion and work through the function of the joints.

Blood Formation.   In late fetal life and in childhood the richly vascular marrow tissue within the marrow cavities of most bones takes over the production of blood cells from the liver. During late childhood and early adult life the red marrow of the long bones of the limbs is gradually replaced by the fatty yellow marrow with consequent cessation of blood formation. Red marrow persists, in the adult, in portions of the hip bone, vertebrae, sternum, and ribs, which remain active in the manufacture of blood cells.

Mineral Salt Storage.   The osseous tissue of all parts of the skeletal system provides the body with a vast depot of mineral salts as described previously.

NUMBER OF BONES

While it may seem an easy matter to count the number of bones in a skeleton, anatomists do not like to be explicit about the total, usually considered to be about 206. Additional bones are

found, the number varying among individuals. Some of these are small *sesamoid bones,* which develop within tendons either to reduce friction as the tendons cross over bony prominences or to provide the additional leverage which may be gained by a slight change in the course of the tendon. *Accessory bones* (supernumerary bones) occur where there are several ossification centers in a developing bone. The osseous tissue formed by one center may fail to fuse with that of the remainder of the bone. A separate, additional bone results. Ordinarily the flat bones of the skull meet at irregular *sutures* which are bridged by fibrous tissue which ossifies late in life. It is not uncommon for ossification of a flat bone to fall short of the suture point. In such cases, a wide fibrous membrane (and a potential weak spot in the skull) is avoided by the development of a small, irregular, accessory bone as a bony island in the suture line. These *sutural* (wormian) *bones* further compound the difficulty of listing an exact number of bones which may comprise the skeletal system of any one person.

SKELETAL SYSTEM CLASSIFICATION

The skeleton, while a functional entity, is divided for purposes of morphological description into the axial skeleton and the appendicular skeleton.

Axial Skeleton

The axial skeleton is made up of the bones which establish the longitudinal axis of the body, protect its vital organs, and offer attachment or support to structures of the head, neck, and trunk. The *vertebral column,* composed of the individual vertebrae and the intervertebral discs between them, is the central pillar of the axial skeleton. The *skull,* balanced upon the vertebral column, encloses and protects the brain. Most of its bones are closely fitted together to form the cranial vault and the face. Separate bones of the skull are the *mandible,* which is the skeleton of the lower jaw, the *hyoid bone,* and the *ear ossicles.* In the thoracic region the *ribs* articulate with the vertebral column posteriorly and the *sternum* or breast bone anteriorly to form the thoracic cage.

Pectoral Girdle. The erect position of man has freed the upper limb from weight-bearing. The upper limb has become highly movable and dexterous. Its freedom of movement would be restricted by too close an apposition to the thoracic wall or too firm an articulation. The main connection to the axial skeleton is via a mobile ball-and-socket joint with the shoulder bone or *scapula.*

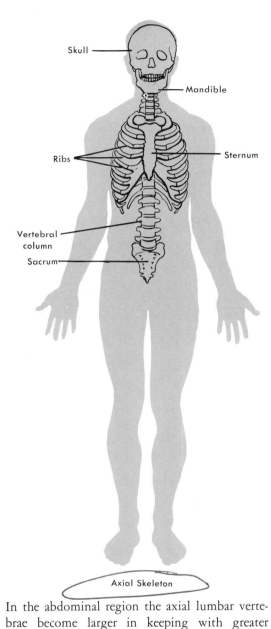

Axial Skeleton

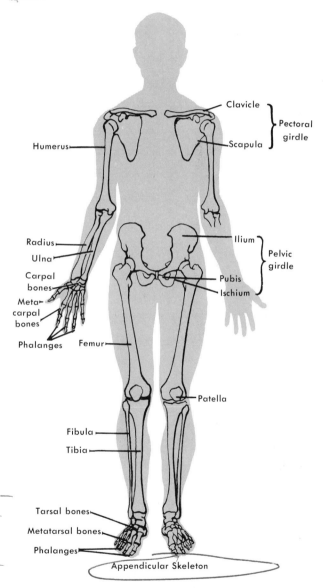

Appendicular Skeleton

In the abdominal region the axial lumbar vertebrae become larger in keeping with greater stresses of support and weight-bearing, whereas in the pelvic region the *sacrum* represents the fusion of the axial sacral vertebrae into one bone which forms the keystone of the bony pelvis. The small and rudimentary coccygeal vertebrae which comprise the *coccyx* complete the axial skeleton inferiorly.

Appendicular Skeleton

The appendicular skeleton consists of the bones of the limbs plus the intermediate girdle bones through which the limbs are connected to the axial skeleton.

The scapula is not directly joined to the thoracic cage upon which it lies but is attached to it and to the vertebral column by muscles. The pectoral girdle is completed anteriorly by the *clavicle,* which is a strut interposed between the sternum and the scapula.

Upper Limb.   The upper limb articulates with the pectoral girdle at the shoulder joint. Here the *humerus,* the bone of the upper arm, moves upon the scapula. The forearm has two bones, the *radius* and the *ulna,* which meet the humerus in the elbow joint. The lower ends of these bones form the wrist joint with the small *carpal bones.* The *metacarpal bones* form the skeleton of the hand from which the *phalanges* extend distally to form the framework of each finger and the thumb.

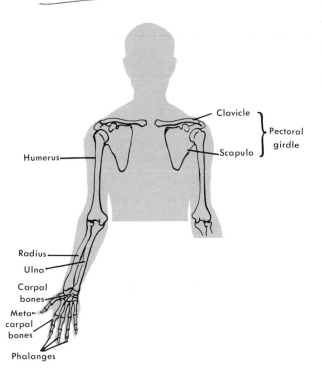

Pelvic Girdle.   The range of motion of the lower limb is less extensive, and its articulation

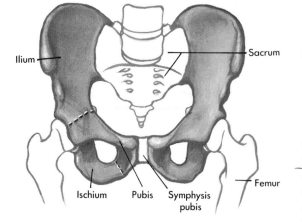

with the axial skeleton is directly concerned with weight-bearing, posture, locomotion, and the forces of gravity. The pelvic girdle, therefore, is more substantial and is both directly and firmly attached to the axial skeleton. Three component bones, the *pubis, ischium,* and *ilium,* are fused in the adult to form the *hip bone.* The components of the two hip bones form the pelvic girdle, which is united to the sacrum of the axial vertebral column at the sacroiliac joints. The bony *pelvis* is formed by the hip bones, sacrum, and coccyx. It supports and protects the pelvic organs.

Lower Limb.   The bone of the thigh, the *femur,* articulates with the pelvic girdle at the hip joint. The two bones of the leg, the *tibia* and *fibula,* meet the femur at the knee joint. The lower ends of these two bones articulate with the more superior of the bones of the foot at the ankle joint.

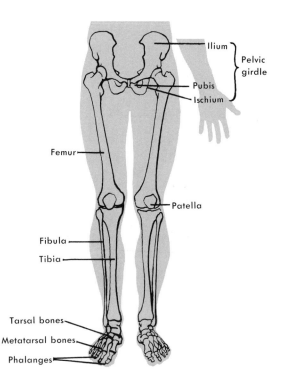

The *tarsal bones* are only somewhat analogous to the wrist bones because of modification for weight-bearing and walking. The *metatarsal bones* resemble the skeleton of the hand in making up the arches of the foot. Distal to these are the *phalanges* of the toes.

THE STUDY OF BONES

The text descriptions and illustrations of bones should be supplemented, if possible, by the study of the skeleton or of individual bones in the classroom. Reference to a skeleton is particularly helpful because the bones are mounted and oriented in their usual position. They may be moved in a manner which approximates their movements in the living body.

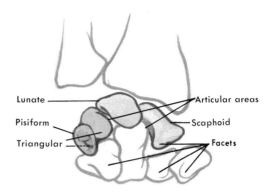

General Suggestions

Examine a bone for its general features. Is it part of the axial or of the appendicular skeleton? Is it a long, short, irregular, or flat bone? What is its size? Does it have a characteristic shape which will assist in its description or identification? Does it articulate with other bones? Which surfaces are concerned with articulations? How is the bone oriented in the anatomic position? With the bone properly oriented, methodically study its various surfaces and borders, observing the specific features which are encountered.

Bone Landmarks

Every bone presents characteristic features or markings in addition to the larger structural aspects which give it a recognizable shape.

Smooth Areas. An area of bone which is perfectly smooth frequently indicates that it was surrounded only by its periosteum and that no fibrous or muscular structure was attached to it. Sometimes "fleshy" muscle fibers are attached to such an area but without coarse tendinous or ligamentous connections. Smooth areas may have muscles adjacent to them which are running a parallel course. At other times smooth portions of bones lie near the surface of the body with only the tela subcutanea intervening between the bone and the skin. Such a part of the bone is described as the *subcutaneous portion.* A *facet* is a small smooth area which occurs between rough areas or projections. There may be nothing attached to a facet, or some adjacent structures coursing obliquely or at right angles to it may move across the surface. A facet may be capped with cartilage at the point where the bone articulates with another.

Elevations. The linear attachment of a sheet of fibrous tissue, such as an intermuscular septum or aponeurosis, may produce a raised *line* on a bone. A tougher fascia, the linear attachment of a muscle, results in a higher, wider, or rougher *crest.* More localized tendinous attachments produce blunt *processes* or drawn-out, sharper *spines.* The attachment of several tendons which compete for the same location on the bone forms a more prominent elevation. If small, the elevation is termed a *tubercle;* if larger, it is called a *tuberosity.*

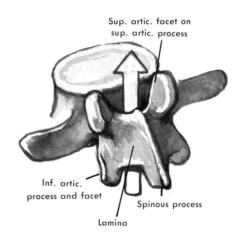

Depressions. Depressions on bones may be relative, because of the presence of adjacent elevations, or may be actual excavations of the bone. They usually indicate that some structure, or several structures, were lodged at the point or permanently fitted against the bone. A circumscribed depression of any size is termed *a fossa.* A linear groove along a bone usually indicates the course of a closely associated tendon which, in development produced this groove or *sulcus.*

Openings. A *foramen* is a hole made in a bone

by a structure, such as a vessel or nerve, which enters to supply it. If the bone is not too thick, the term foramen is also applied to a hole in a bone that is traversed by a structure entering or leaving the part of the body protected by the bone. A *canal* is a bony passageway of some length made by structures which course a considerable distance through a bone.

JOINTS IN GENERAL

The structural arrangements of tissues by which bones are joined together are called joints. The articular system is comprised of these widespread independent units and is a functional extension of the skeletal system.

Without some kind of stable union of the many separate bones the skeletal system would lose much of its effectiveness in providing rigidity and support for the body. However, an equally important function of the articular system is to make movement possible at many points of skeletal union. A few joints provide fixed union, but most joints are characterized by flexible union with varying degrees of movement allowed, depending on the nature of the joint surfaces and the arrangement of reinforcing structures.

CLASSIFICATION OF JOINTS

The classification of joints is burdened by the carryover of terms, largely based upon Greek descriptions, from older days of anatomy. Although newer, anglicized terms will be used in classification and descriptions in this text, some of the older terms will also be mentioned. The terminology to be used is easier to grasp if it is preceded by an explanation of the developmental stages through which the various types of bony union pass.

Development of Joints

The degree of specialization in the mesenchyme which loosely connects the ends of the developing bones determines the nature of the joint between them. If the mesenchyme develops into a fibrous connective tissue, the ends of the bones will be united by fibrous strands to form a *fibrous joint*. In other situations mesenchyme forms cartilage as the connection between the bones. Such joints are *cartilaginous joints*. The older term "synarthrosis" was applied to both of these types to indicate the absence of a cavity

within the joint and, therefore, either the absence or limitation of movement. If most of the mesenchyme between the ends of the bones disappears in development, a cavity appears between the bones. Mesenchyme remains only at the edges of the bones to develop into a fibrous capsule around the joint. The inner lining of this capsule elaborates a gelatinous lubricating fluid, the synovial fluid, which is the actual substance found between the bone ends. Such joints, permitting movement, make up the majority of those in the body and are called *synovial joints*. From their possession of a joint cavity came the older term of "diarthrosis."

Joint Classification and Types

Each of the three types of joints can be further classified according to modifications and variations.

Fibrous Joints. In fibrous joints the bones are united by fibrous strands of varying length. In these joints union is paramount and motion is either prohibited or severely limited.

SUTURES OR SUTURAL JOINTS. The bones of the skull are closely applied to each other along wavy or intricately fashioned borders. The connective tissue between them is reduced to the dense short strands of the *sutural ligament,* which, in the adult, forms the *sutural joints.*

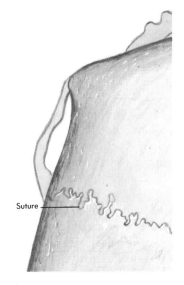

Suture

The ligament and the intricate, dove-tailed borders of the bones effectively prohibit movement. In adult life, dormant mesenchymal cells in the sutural connective tissue specialize into osteoblasts. The suture gradually becomes reduced in width or disappears as the sutural ligament ossifies. The ossified fibrous joint is termed a *synostosis*.

SYNDESMOSES.   A syndesmosis is a fibrous joint in which the fibrous strands are long enough to permit some movement while holding adjacent bones in firm approximation. The two bones are usually in a side-to-side relationship rather than end to end.

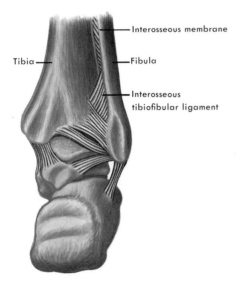

Tibia — Interosseous membrane
Fibula
Interosseous tibiofibular ligament

The binding of the lower end of the slender fibular bone of the leg to a notch in the side of the tibia by a short interosseous ligament is an example.

GOMPHOSES.   An unusual fibrous joint, the gomphosis, is found in the connective tissue intervening between a tooth and its socket in the jaw.

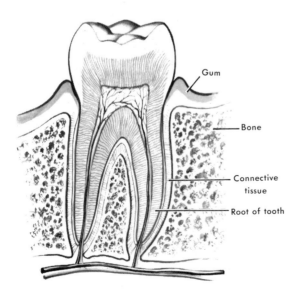

Gum
Bone
Connective tissue
Root of tooth

The fibrous tissue conforms to the irregularities between tooth and socket and holds the tooth firmly but permits almost imperceptible movements with the shocks of grinding and biting.

Cartilaginous Joints.   In cartilaginous joints the resilient nature of the cartilage matrix is used to advantage in providing a flexible connection between bones.

SYMPHYSES.   A symphysis is found where two bones, or a number of bones in a series must be united but with the possibility of slight flexible movement that results more from the pliability of the intervening matrix than from movement of the bones. A prominent example is the intervertebral disk whose fibrocartilaginous nature holds adjacent vertebrae firmly together yet permits a slight rocking.

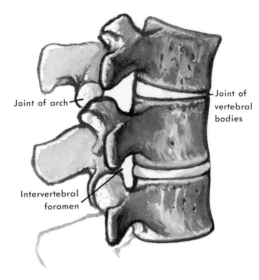

Joint of arch
Joint of vertebral bodies
Intervertebral foramen

A disk of fibrocartilage similarly unites the hip bones anteriorly in the pelvic girdle.

SYNCHONDROSES.   A close relationship between bone and cartilage is seen in the synchondrosis. One type is found in connection with the costal cartilage which intervenes between the bony part of a rib and its point of connection with the sternum. Whereas the joint of the cartilage with the sternum is movable, the connection between cartilage and bony rib is tight but the flexibility of the hyaline cartilage accommodates to movements of the rib and sternum with respiration. Similar hyaline joints occur between parts of the sternum. Both these types of synchondroses tend to ossify in early adult life with resultant reduction of flexibility. A temporary form of synchondrosis is represented by the epiphyseal plate of cartilage between the epiphysis and diaphysis of a developing long bone. This is technically a joint because it joins two parts of bone. Ordinarily no motion is possible, although epiphyseal separation is possible if undue force pulls upon the head of a long bone in a child. The temporary joint disappears when the epiphyseal plate ossifies at the end of bone growth.

Synovial Joints. Most of the joints of the body are of the synovial or diarthrodial type. Characterized by a joint cavity, an enclosing membrane, an internal lubricating fluid, and bone ends covered by articular cartilage, the synovial joint affords skeletal union while offering mobility. The division of these joints into subtypes depends upon the nature of the articulating surfaces and the kind of motion which can result.

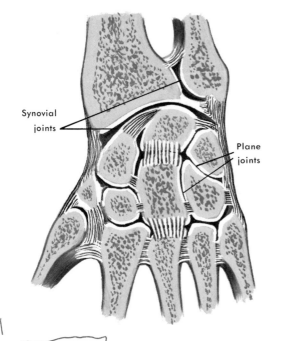

PLANE JOINTS. Many small joints of the body are located between small bones or between small processes of larger bones. The point of movement is usually marked by a flattened articular facet which is capped in life by a plaque of hyaline articular cartilage. Only simple gliding or short, slipping movements are possible. Examples are the joints between the bones of the wrist and between the articular processes of adjacent vertebrae.

PIVOT JOINTS.   A pivot joint exists where the only movement possible is one of rotation about a long axis. The head of the radius rotates upon

a pivot formed by part of the lower end of the humerus as the hand is turned from a palm up to a palm down position.

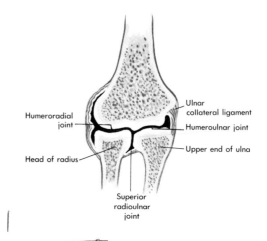

HINGE JOINTS. Movement in only one axis is possible also in hinge joints, but here the movement is at a right angle to the long axis of the participating bones, as in bending of the elbow or knee.

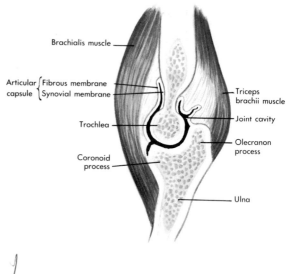

CONDYLOID JOINTS. Condyloid joints allow for movements along two axes but against a single elliptical surface of one bone. There may be a long range of motion along one axis and a shorter range along the other. A good example is found in the movements of the index finger. Bending can occur but sidewise movements are also possible.

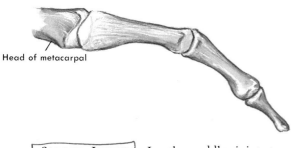

Head of metacarpal

STRUCTURE OF A GENERALIZED JOINT

Certain principles of structure are fundamental to all joints. These principles are exemplified by the characteristics of a generalized synovial joint.

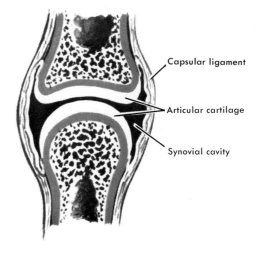

Capsular ligament

Articular cartilage

Synovial cavity

SADDLE JOINTS. In the saddle joint two different surfaces on one bone, whose end looks like a saddle, oppose another bone. In different movements one or another side of the saddle slides against the trough-shaped end of the other bone.

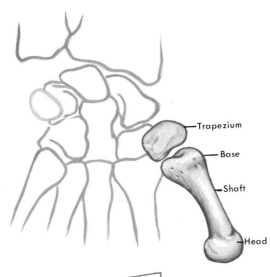

Trapezium

Base

Shaft

Head

The factors contributing to the construction of such a joint are:

1. Form of the articulating bones
2. Disposition of the articular cartilage
3. Form and disposition of the articular capsule and synovial membrane
4. Nature of the synovial cavity and synovial fluid
5. Kinds of reinforcing and stabilizing structures

BALL-AND-SOCKET JOINTS. Movement along many axes is possible in a joint where the end of one bone is shaped like a sphere to fit into a cuplike depression or socket of another. Maximum freedom of motion may be expected in such a joint of which the shoulder and hip joints are typical examples.

Articulating Bones

Some part of each bone must come into apposition with the other when they meet to form a joint. This part, called the *articular surface*, is shaped into a smooth facet. The point of articulation will be at the extremity of a long bone but it need not be on the actual end surface. The facet may be located on a side (fibular notch of the tibia) or extend onto the anterior or posterior surface of the end (femoral condyles). In the case of small irregular bones, such as the carpals, the articular facets may be located on any surface of the bones. Joints in the human body cannot be compared in all ways to bearings in a machine. They are alike in that they bring opposing sur-

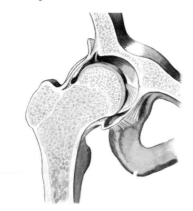

faces together, transfer loads, facilitate movement, and require lubrication. The articulating surfaces of bones in human joints differ from machine bearings in that not all portions of the articular surfaces are in contact at a given time. Even a ball and socket do not always make an exact fit with each other in human joints. Incongruities in the opposing surfaces make joint movement a succession of rolls and glides in which the actual point of contact at any moment is small and shifts almost immediately to another over a wide area. A more appropriate analogy would be to the shifting points of contact in the teeth of a gear.

Articular Cartilage

The articular facets on the opposing bones are capped by hyaline cartilage. Articular cartilage makes an absolutely smooth, bluish white, opalescent surface. It is upon these surfaces, lubricated by the synovial fluid, that bones glide, roll, and turn upon each other. The resiliency of the cartilage cushions joint movement and the slippery, lubricated surface minimizes friction. The articular cartilage follows the incongruities of the bone exactly so that the load is spread over many points of contact and little pools of synovial fluid can collect to aid in lubrication. Articular cartilage has neither a blood nor a nerve supply. Nourishment is by way of slow diffusion from the synovial fluid and from capillaries of the underlying bone. If the cartilage were supplied with nerve fibers, the pressure of weight-bearing and the motion of the surfaces would make the movement of joints a continually painful experience. The rich nerve supply to the joint capsule provides the central nervous system with adequate information about the status of the joint.

Articular Capsule

When the synovial cavity forms between the ends of the bones in a developing joint, only a collar of mesenchyme remains along their sides. This develops into a sleeve, the *articular capsule,* which encloses the ends of the participating bones. The articular capsule consists of two portions, an outer *fibrous membrane* and an inner *synovial membrane.*

Fibrous Membrane.   The fibrous membrane extends along all surfaces of the articulating bones to finally blend with the periosteum. In a synovial joint the collagenous fibers of the fibrous membrane are the primary mechanism in uniting the bones of the joint. In some small joints the fibrous capsule may be the sole structure responsible for the integrity of the joint. In other joints, localized collagenous thickenings appear in the fibrous capsule along lines of force to reinforce it against disruption or to strengthen it to check joint movement in a certain direction. These localized thickenings are called *capsular ligaments,* although their function is more to strengthen the capsule than, according to the definition of ligaments, to bind the bones together.

Synovial Membrane.   The synovial membrane underlies the fibrous capsule in bordering the joint cavity and extends with the capsule past the articular cartilage. At the line where the fibrous membrane blends with the periosteum, the synovial membrane turns back upon itself to closely invest the periosteum of the bone surface as far as the margin of the articular cartilage. The lining of the joint cavity between the load-bearing surfaces is formed only by the lubricated articular cartilage, for the delicate synovial membrane could never withstand the pressures involved.

Synovial Cavity

The presence of the ends of the bones covered with articular cartilage within the articular capsule reduces the actual joint or synovial cavity to a potential slit. There is room only for a thin film of synovial fluid between the cartilages at the actual point of contact. The small incongruities permit transitory pools of fluid to develop but space is at a premium. In the larger joints there are larger, semipermanent points of incongruity where, for instance, a rounded knob rolls upon a flatter plateau. The potential space that develops in these locations is usually filled by fringelike overgrowths of the synovial membrane, the *synovial fringes.* These extra tags of synovial membrane may contain fat. In this way larger *synovial* (haversian) *fat pads* are created within the synovial cavity. The fat pads act as lubricating cushions which can slide into a transient space one moment and equally easily slide out of the way as the bearing surface changes. There is more space for fluid beyond the actual line of a joint crevice where the synovial membrane follows the fibrous membrane along the edges of the bones before folding back upon itself. Such extensions of the synovial membrane form *synovial pouches* or *synovial recesses,* which frequently serve as bursae for tendons crossing the joint.

The *synovial fluid* or *synovia* is actually a gelatinous, viscous substance, the main ingredient of which is the protein polysaccharide hyaluronate, which is like the white of an egg. Such a viscid material is better suited than a watery fluid would be to penetrate between the bearing surfaces and to withstand their pressures. Lubrication of the articular surfaces, to reduce friction and enhance smooth movement, and nourishment of the articular cartilage are the functions of the synovia.

Reinforcing and Stabilizing Structures

The development of the cavity within a synovial joint enhances its mobility but reduces the stability that is contributed by intervening materials in the fibrous and cartilaginous joints. In the absence of accessory structures, the stability of a synovial joint will depend upon the shape of the articulating surfaces and the strength of the articular capsule. Frequently the *shape of one or both articulating surfaces* provides stability in some, if not all, of the directions the joint can move. For example, the wrenchlike form of the upper end of the ulna, conforming to the scroll-like lower end of the humerus, is quite efficient in limiting the movement of the hinge-type elbow joint to the single plane of bending and

straightening. In addition, as the forearm straightens, the posterior part of the ulna slides into a fossa on the back of the humerus to impinge finally upon a fat pad. The elbow joint is arrested in its movement at full straightening. Without this stablizing factor, the straightened elbow would be weak and in danger of bending backward unpredictably. The *strength of the articular capsule* is a stabilizing feature in some of the smaller joints. In addition, the length and arrangement of the collagenous fibers in the articular capsule are utilized to lend stability in one direction of movement but to enhance mobility in the opposite direction. Still using the elbow joint as an example, it will be found that the capsular fiber arrangement is such that the anterior part of the sleeve becomes taut as the joint straightens. The same part of the capsule readily relaxes as the joint bends.

In most joints, however, other devices are necessary to ensure stability or to reinforce the foregoing factors.

Capsular Ligaments. As previously mentioned, these localized collagenous thickenings of the capsule reinforce the sleeve along principal lines of force. They may become important in creating tautness of the capsule in particular directions of action.

Accessory Ligaments. Ligaments extending from one of the bones participating in the joint to the other act to bind them together into a more stable joint or to protect the joint against forces which might disrupt it. When the ligaments are external to the capsule of the joint, they are termed *extrinsic ligaments.* A ligament which crosses the joint cavity within the capsule is called an *intrinsic ligament.* These short bands are great contributors to stability in the direction of their alignment. Many of the extrinsic ligaments are also called *collateral ligaments,* because, located as they are at the side of a joint, they do not interfere with desirable movements but firmly resist movement at an angle to them. Other ligaments are aligned in the direction of a desirable movement, and do not resist it until the joint reaches either a point of maximum function or one beyond which the joint would become unstable. The ligament becomes taut at such a point and it prevents further movement. These ligaments may be either extrinsic or intrinsic in location and are appropriately called *check ligaments.*

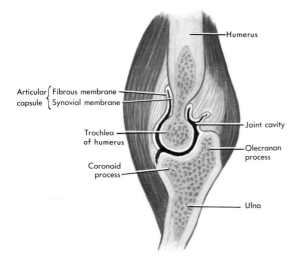

Humerus

Articular {Fibrous membrane / Synovial membrane} capsule

Trochlea of humerus

Coronoid process

Joint cavity

Olecranon process

Ulna

Environmental Forces. Weight-bearing joints are stabilized by the *force of gravity,* which aids in forcing the articulating bones together. This is particularly effective where the axis of the joint, as in the hip, knee, and ankle, corresponds to the line of the erect position. Gravity can work to the disadvantage of a joint if the body falls while one of the bones is caught in an abnormal movement or in a fixed position. Collateral ligaments may be overstretched or torn to result in a "sprain" or the bone may be forced out of normal apposition in a "dislocation." Less tangible as a stabilizing factor is the *atmospheric pressure,* which, always pressing upon the human body, tends to keep bones together.

Action of Muscles and Tendons. A most important stabilizing factor is found in the fact that a muscle arising from one bone, in passing to insert upon another bone, crosses the joint between them. The pull of the muscle is against not only the bone of insertion but upon its own bone of origin and, inevitably, the bones are drawn together at the joint. A group of muscle tendons, arranged about all surfaces of a joint, strengthens the capsule and aids the ligaments in furthering joint stability. It will be learned later, in the consideration of muscle action, that muscle groups opposite to those which produce a given action are in a continual state of tension which, by resisting the action, smooths the movement at the joint. This antagonistic action of opposite muscles gives them a check ligament function in that they assist in stopping the primary action at the point of maximum stability.

Fibrocartilaginous Structures. Plates or disks of fibrocartilage are occasionally found in joints to:

1. Build up the margins of a flat articular surface to receive the curved end of another bone (knee joint).

2. Deepen a shallow socket of one bone to receive the ball-like end of another (shoulder joint).

3. Fill the gap between two bones where one is too short to reach the other (ulna at wrist joint).

4. Divide a small joint into two cavities so that two sets of movements can be carried on (clavicle and sternum; skull and mandible).

Soft Tissues. A minor role in stabilizing joints is played by the soft tissues surrounding them. Although the skin is mobile and distensi-ble, it does mold the tela subcutanea about the joint. The deeper investing fasciae serve to restrain the tendons about a joint as well as the joint capsule.

JOINT MOVEMENTS

Bones, and thereby joints, move in many directions, along different axes, and through different planes of the body. These factors make it difficult to define standard terms of joint movement in a manner that will be applicable to each joint. The most used terms can be explained according to their broad, general meanings. When special situations require, a further explanation will be given with the joints concerned.

Flexion is the bending of one part of the body toward another, usually toward the ventral surface, so that the angle of the joint between the parts is decreased. Examples are bending the elbow, raising the thigh upon the abdomen, and moving the lower limb forward in walking.

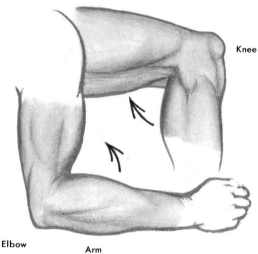

Knee

Elbow   Arm

Extension is the movement of one part of the body away from another, usually toward the dorsal surface, so that the angle of the joint between the parts is increased. Examples are straightening the fingers, bending the neck backward, and swinging the leg forward at the knee.

Abduction is the drawing of a limb away from the midline of the body (the leg at the hip joint as illustrated) or, in the case of the fingers, from the midline of the hand through the middle (third) finger.

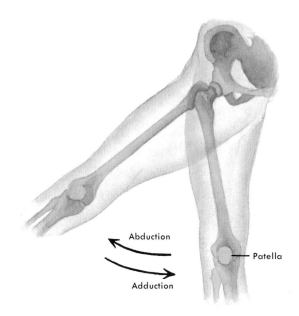

Abduction

Patella

Adduction

Adduction is the drawing of a limb toward the midline of the body or, in the case of the fingers, toward the midline of the hand through the middle (third) finger.

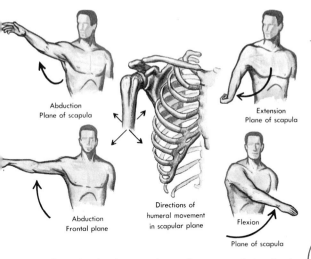

Abduction
Plane of scapula

Extension
Plane of scapula

Abduction
Frontal plane

Directions of
humeral movement
in scapular plane

Flexion
Plane of scapula

Rotation is the turning of a part of the body about its long axis. In the case of a limb the movement occurs at the proximal joint (shoulder or hip), but this motion is structurally impossible at the more distal joints and the limb rotates as a unit. In *medial rotation* the ventral surface of the limb turns toward the midline of the body. In *lateral rotation* the ventral surface turns outward, away from the midline. *Circumduction* is the combination successively of flexion, abduction, extension, and adduction in a limb. The distal portion of the limb describes a circular path while the head of the bone, at the proximal joint, serves as the pivot for the movement. Example: Raise the arm from the side of the body and cause the outstretched fingers to describe a circle. The head of the humerus is the apex of a cone and the pivot for the movement.

Combined Movements. Although it is possible to perform a single movement at but a single joint at one time, it should be readily apparent that such a pure movement rarely occurs in life, would be awkward, and would frequently serve no useful purpose. Most of the movements of the body are the combination of several directions of movement at one joint in company with the simultaneous or successive movements of other joints. The grace and ease of body movements depend on such combinations.

THE STUDY OF JOINTS

The study of joints necessarily is also one of the skeletal system so that the description of joints will be found following or in association with the description of the bones concerned. The orderly study of any joint demands the acquisition of the following information:

1. Location and type of joint.
2. Form and relations of the bony parts.
3. Structure of the joint.
4. Arrangement of the capsule.
5. Reinforcing and stabilizing mechanisms.
6. Structures near or crossing the joint.
7. Movements of the joint.

Know these factors for all Joints!

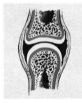

3

The Skeletal System And Its Joints

The joints are functional extensions of the morphological arrangements by which bones meet and move upon each other.

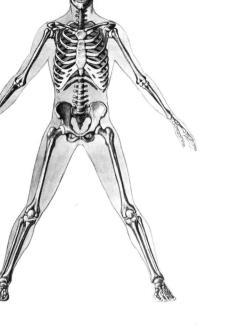

Therefore, the discussions of the skeletal system and the articular system are combined in this chapter. As the bones of the body are described, accounts of the joints between bones will be found in their logical anatomic locations.

THE AXIAL SKELETON AND ITS JOINTS

The axial skeleton consists of the vertebral column, the skull, and the thoracic skeleton.

THE VERTEBRAL COLUMN

The vertebral column establishes and maintains the longitudinal axis of the body, provides a pivot for the support and movement of the head, and gives the limbs a structural base for articulation and action through the pectoral and pelvic girdles.

GENERAL FEATURES

The vertebral column combines the strength of a rigid pillar with the flexibility of a multi-

. Axial support of the body in the
. .ition would be impossible without the
. .gth of the vertebral column. Without its
. xible joints the back would be poker stiff with
associated loss of agility.

The Vertebrae

The vertebral column consists of 33 individual
bones, the *vertebrae,* which are arranged serially,
one atop another, through the posterior part of
the neck, thorax, abdomen, and pelvis. The vertebrae are divided as follows: seven cervical, twelve
thoracic, five lumbar, five sacral, and four coccygeal.

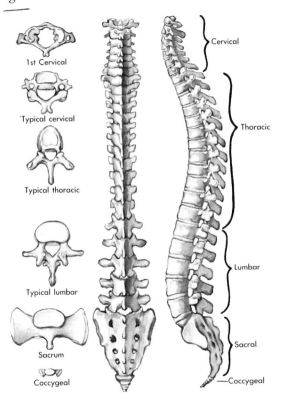

1st Cervical

Typical cervical

Typical thoracic

Typical lumbar

Sacrum

Coccygeal

Cervical

Thoracic

Lumbar

Sacral

Coccygeal

The sacral vertebrae are fused to make one
solid bone, the *sacrum,* which is the keystone of
the pelvis and which articulates with the hip bone
of the pelvic girdle. The coccygeal vertebrae are
irregular and rudimentary in the human in contrast to their importance as skeletal members in
species with a tail.

Vertebral Column as a Unit

The importance of the vertebral column to the
body lies not so much in the individual bones

as in their articulation to create the *spine* or back-
bone. The vertebrae are held together by strong
ligaments which unite adjacent vertebrae or may
extend for a considerable distance along the
length of the column. Similarly, short muscles
exert their pull between adjacent vertebrae or
across the span of several vertebrae. Although the
movement of one vertebra upon another may be
slight, the sum of many small movements along
the entire spine results in considerable movement
and flexibility. Larger muscles may produce the
movement of the vertebrae of one body region
upon those of another region. The combination
of vertebral, muscular, and ligamentous action
maintains the alignment of the vertebrae in the
spinal column against the tremendous forces of
gravity, weight-bearing, and the pull of the muscles of the trunk and limbs.

It would seem possible that the various forces
acting upon the vertebrae would exert such a
severe compression upon them that there might
be the danger of crushing together the superior
and inferior surfaces of the block-like bones, or
that adjacent vertebrae could grind upon each
other. The danger of crushing is averted by a
pattern of strong trabecular structure in the
spongy bone of the interior of the vertebrae. The
grinding of adjacent vertebrae is prevented by
strong, resilient *intervertebral disks* which are interposed between them.

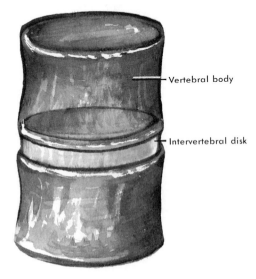

Vertebral body

Intervertebral disk

Curvatures of the Vertebral Column

The vertebral column occupies the midline of
the body but has a straight axis only when viewed

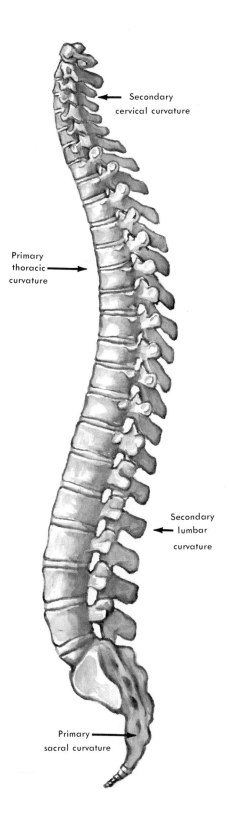

Secondary
cervical curvature

Primary
thoracic
curvature

Secondary
lumbar
curvature

Primary
sacral curvature

directly from in front or behind. Its axis in profile presents a series of gradual curves. Before birth the fetus is curled in the uterus with its head bowed upon the chest and its thighs folded upon the abdomen. The fetal vertebral column, therefore, is bent into a continuous gentle curve with its concavity facing forward. This is the C-shaped *primary fetal curvature* of the vertebral column. The superior end of the fetal vertebral curvature is straightened when the infant is able to raise its head. The continual holding of the head erect without support and its movement by the neck muscles reverse the primary curvature in the cervical portion of the vertebral column so that the concavity faces posteriorly. A similar reversal of the primary curvature occurs in the lumbar region of the spinal column to compensate for the forces of weight-bearing and locomotion when the young child struggles to stand and learns to walk. The lumbar portion of the adult vertebral column, therefore, also presents a curvature with a posterior concavity. The cervical and lumbar curvatures are termed the *secondary* or *compensatory curvatures* of the spine. The persistence of the primary curvatures of the thoracic and sacrococcygeal portions of the vertebral column in adult life maintains balance in the axial skeleton and also accommodates the organs of the thorax and pelvis.

The primary and secondary curvatures are normal curvatures in the anteroposterior plane of the body. An abnormal or accentuated thoracic curvature, resulting from occupation or disease, is known as *kyphosis* (hunchback). An accentuated lumbar curvature is the condition known as *lordosis*. In health there should be no lateral curvatures of the vertebral column and each vertebra should be in a midline position. Diseases of bone or of muscle may pull some vertebrae out of vertical alignment to create abnormal lateral curvatures or *scoliosis*. Such curvatures result in postural deformity and locomotor imbalance.

THE VERTEBRAE

The cervical, thoracic, and lumbar vertebrae form a series of vertically aligned, independent bones which change very gradually from the smaller, delicate upper cervical vertebrae to the large, coarse, heavy vertebrae of the lumbar region. Adjacent vertebrae in any one region may present only barely distinguishable differences whereas, taken as a group, the vertebrae of each region are characteristic. The sacrum, through the fusion of its component bones, does not resemble the other vertebrae.

General Vertebral Structure

A vertebra consists of an anterior block of bone, the *body,* and a group of bony bars, plates, and processes which jut posteriorly and then close together to form an irregular ring, the *vertebral arch.* The posterior margin of the body combines with the internal surfaces of the elements of the vertebral arch to bound a central aperture, the *vertebral foramen.*

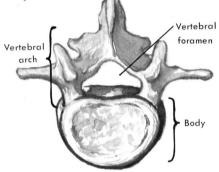

When all of the vertebrae are aligned and openings between adjacent vertebrae are closed by ligaments and periosteal membranes, the successive vertebral foramina combine to form the *vertebral canal,* in which the spinal cord, its nerves, and spinal blood vessels are located.

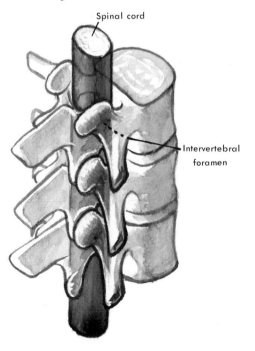

Vertebral Body. The vertebral body varies in shape from region to region. Its upper and lower surfaces are flattened. Lips around the margins enclose a broad central area which is roughened by the attachments of the intervertebral disks. The sides of the body present a pinched-in, concave appearance and are pitted by many minute foramina through which nutrient blood vessels enter the bone.

Vertebral Arch. Two short bars of bone, the *pedicles,* jut backward from each side of the vertebral body to form the beginning of the vertebral arch. The upper surface of the pedicle is shallowly indented and the lower surface is deeply indented to form the *superior* and *inferior vertebral notches.*

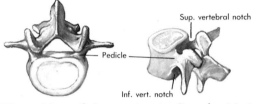

When either of the notches is aligned with its neighbor of the vertebra above or below, they form together an *intervertebral foramen* for the transmission of spinal nerves from the vertebral canal.

At the posterior end of each pedicle the vertebral arch flares out into broad plates of bone, the *laminae,* which complete the closure of the vertebral canal by fusing with each other at the midline.

A smooth transition from the pedicle into the lamina is interrupted by the formation of several *processes* at their point of junction. Rising upward on each side from the union of the pedicle with the lamina is a sharp oval plate of bone, the *superior articular process.* Jutting sharply downward from the same junction on each side is another thin oval plate of bone, the *inferior articular process.*

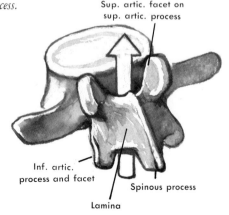

Each of these platelike processes bears a smooth *articular facet.* The superior articular facet faces backward; the inferior articular facet faces forward. When adjacent vertebrae are aligned, the facets, capped with a smooth articular cartilage, interlock.

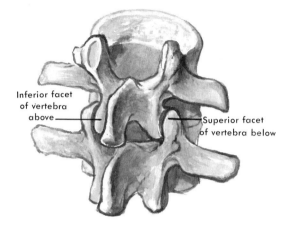

Inferior facet of vertebra above — Superior facet of vertebra below

The four articular facets of each vertebra plus the intervertebral disk comprise the points of articulation between adjacent vertebrae.

Transverse processes are short bars of bone which thrust out laterally on each side from the junction of the pedicle with the lamina. They are levers for the attachment of muscles to the vertebrae.

Spinous processes, one for each vertebra, represent the median continuation of the vertebral arch posteriorly after fusion of the laminae. The spine is directed downward as well as backward in the midline to end in a blunt expanded tip. This bony projection, which also serves for muscle and ligamentous attachment, can be felt and seen in the midline furrow between the back muscles except in the upper cervical region.

Joints of the Vertebral Column

These joints are concerned with the rigidity of the vertebral column, its transmission of weight, and its flexibility in providing for mobility of the trunk and for adjustments of position to maintain balance and posture. Joints of the vertebral column occur between the bodies of the vertebrae and between the articular processes of

the vertebral arches. Ligaments assist in maintaining alignment of the vertebrae.

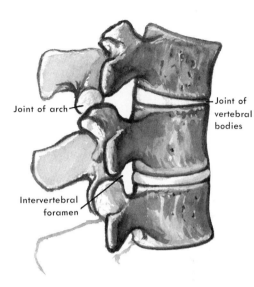

Joint of arch — Joint of vertebral bodies

Intervertebral foramen

Joints of the Vertebral Bodies. Union, alignment, and limited motion of adjacent vertebrae are accomplished by the *intervertebral disks.* The disks are resilient compressible cushions of fibrocartilage which are interposed between the bodies of adjacent vertebrae to unite them while allowing a slight rocking motion. They absorb the forces of muscle pull, gravity, and weight-bearing which would otherwise tend to grind the vertebrae together. Each disk is a rounded plate, approximately one quarter the height of the vertebrae between which it is placed, and fits into the flattened impressions on the opposing surfaces of the vertebrae. The disk is made up of two parts. The peripheral part or *annulus fibrosus* is composed of concentric rings which are more fibrous than cartilaginous. The central part of the disk or *nucleus pulposus* is eccentrically located toward the posterior margin. The substance of the nucleus pulposus is more cartilaginous than fibrous. It is a pulpy, highly elastic tissue which is normally highly compressed and, in tending to escape compression, acts as a shock absorber. The nucleus pulposus is held in place by the annulus fibrosus and by pressure of the vertebrae, but, with injury to the fibrocartilaginous ring around it, it may expand posteriorly because of its eccentric position to press upon the spinal cord and the nerve roots within the vertebral canal.

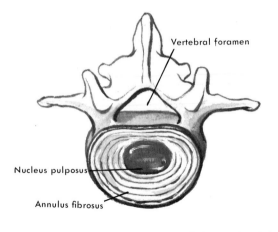

The rocking motion of adjacent vertebrae upon each other causes the portion of the annulus fibrosus and the edge of the nucleus pulposus toward the direction of movement to be compressed while the opposite side expands. Although the actual amount of movement between adjacent vertebrae is slight, the summation of movement of a series of vertebrae results in appreciable movement of a vertebral region or of the vertebral column as a whole. Thicker disks in the cervical and lumbar regions allow a greater compression-expansion, and thus greater intervertebral motion, than in the thoracic region where the disks are thinner.

Joints of the Vertebral Arches. The greatest movement between adjacent vertebrae occurs at the joints of the vertebral arch. The points of articulation are between the inferior articular processes of the vertebra above and the superior articular process of the vertebra below. These are synovial joints of the plane or gliding type. Small, loose articular capsules enclose the joints but do not limit their motion. The direction of motion is determined by the manner in which the articular facets of the processes, capped by hyaline articular plaques, face each other. The flat surfaces of the *cervical* articular processes are angled in their interlocking and permit relatively free motion in flexion, extension, and lateral bending. The forward and backward interlocking of the *thoracic* articular processes reduces these same motions to short shifting movements. The thinner thoracic intervertebral disks and the presence of the ribs contribute to the lesser mobility of the thoracic vertebrae.

The superior and inferior articular facets of the *lumbar* vertebrae meet in an almost anteroposterior plane. The smooth articular surfaces permit adjacent lumbar vertebrae to glide easily upon each other in flexion and extension which, abetted by thick intervertebral disks, are quite free movements in the lumbar region. The facets are also tilted just enough to allow limited lateral bending, but they interlock to prevent rotation. The joints between the vertebral arch of the fifth lumbar vertebra and the sacrum are an exception. The inferior articular facets of the fifth lumbar vertebra, instead of facing laterally, are turned forward to engage the posteriorly facing articular facets of the sacrum. The vertebral column is, therefore, hooked behind the facets of the sacrum and prevented from being displaced downward on the sacrum. Lateral rotation of the fifth lumbar processes upon the sacrum is free and increases the degree of lateral bending at these lowest synovial joints of the vertebral column.

Ligaments Affecting Vertebral Joints

Ligaments restraining or aligning the vertebrae are grouped into those related to the vertebral bodies and those of the vertebral arches.

Ligaments of the Vertebral Bodies. Groups of strong collagenous fibers are applied to the anterior surface of the vertebral column. Some of the fibers extend from one vertebral body across the vertebral disk, to which it is firmly attached, to the adjacent vertebral body. Other fibers extend across several vertebrae. All of the fibers are tightly interlaced into a broad ligamentous band, the *anterior longitudinal ligament,* which extends from the atlas to the pelvic surface of the sacrum. This strong ligament is prominent in maintaining the alignment of the vertebrae and in limiting extension of the vertebral column. On the opposite side of the vertebral bodies, facing the vertebral canal, is the *posterior longitudinal ligament.* This is not nearly so strong, for it is more a thickening of the connective tissue lining of the vertebral canal as it passes over the posterior aspect of the vertebral bodies and their disks. It assists in alignment and aids the ligaments of the vertebral arches in checking flexion of the vertebral column.

Ligaments of the Vertebral Arches. Short elastic ligaments, the *ligamenta flava,* extend from the lamina of one vertebral arch to the one above it. These ligaments unite the bony laminae to enclose the vertebral canal posteriorly. The ligamenta flava stretch with flexion of the vertebral column, but their elasticity checks this motion and brings the vertebrae back from a flexed position.

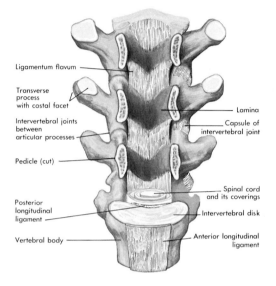

Thin collagenous *interspinous ligaments* run between the spinous processes to limit flexion.

The *supraspinous ligament* is a thin ribbon formed by a continuous series of collagenous fibers running along the tips of the spinous processes and blending laterally with the individual interspinous ligaments. In the cervical region the supraspinous ligament acquires an appreciable elastic tissue content and becomes heavier, to be called the *ligamentum nuchae.* In this region the ligament not only slows flexion but acts as do the ligamenta flava in bringing the neck back to a neutral position. Below the cervical region the supraspinous ligament, with the interspinous ligaments, resists flexion.

Regional Vertebral Characteristics

The cervical, thoracic, and lumbar vertebrae have the basic structural features just described but exhibit characteristic regional differences.

Thoracic Vertebrae. The thoracic vertebrae are intermediate in size between the smaller, lighter cervical vertebrae and the more massive lumbar vertebrae. The general description for vertebrae which has just been given applies specifically to the thoracic vertebrae. The only additions to this plan are occasioned by the articulation of a pair of ribs with each thoracic vertebra. The rib moves upon the vertebra at two points, the superior costal facet and the transverse costal facet.

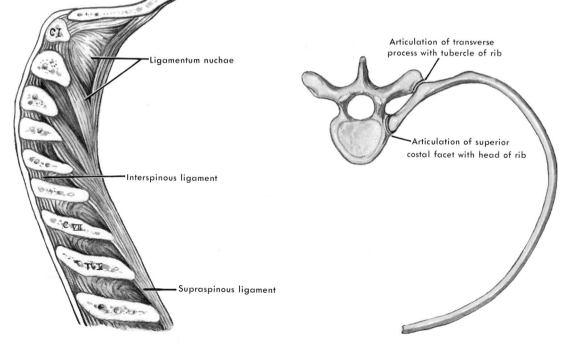

The *superior costal facet* is a slightly raised, smooth, oval area located on the body of the thoracic vertebra at its junction with the pedicle. The head of the rib overrides the border of the vertebral body so that a small inferior costal facet may be found on the lower border of the body of the vertebra above. The oval *transverse costal facet* on the anterior surface of the transverse process articulates with a tubercle on the same rib.

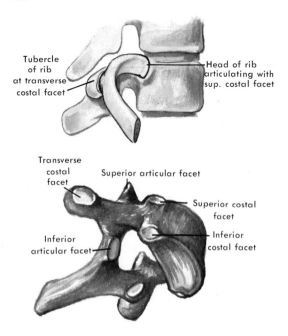

Cervical Vertebrae. The cervical vertebrae increase in size serially as they approach the thoracic region but are typically smaller than the other vertebrae. The vertebral bodies are quite small and are compressed anteroposteriorly. The cervical vertebral foramen is large in relation to the bone as a whole because the cervical portion of the spinal cord is enlarged. The pedicles of the vertebral arch are reduced almost to stublike insignificance by modifications of the transverse and articular processes. The laminae are long and narrow to encompass the larger vertebral foramen and they end in a short, divided or *bifid spine.* The short transverse processes are drawn out into shallow troughs which have sharp, thin forward edges, or *anterior roots,* and raised posterior walls, or *posterior roots.* Adjacent to the pedicle, the floor of the transverse process is pierced by the foramen transversarium or *transverse foramen* in the upper six cervical vertebrae. This foramen transmits the vertebral artery in its upward course to the brain.

The cervical spinal nerve emerging from the intervertebral foramen hugs the posterior root and passes behind the vertebral artery which is coursing vertically through the transverse foramen, and then passes laterally along the trough formed by the two roots. The articular processes form an irregular flattened column at the junction of the pedicle and the lamina just posterior to the transverse process. The upper part of this column is the *superior articular process* bearing the *superior articular facet.*

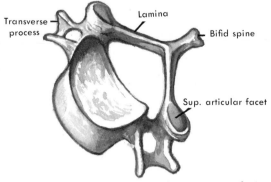

The lower part of the column is formed by the *inferior articular process* with its *inferior articular facet.*

The first and second cervical vertebrae present modifications in structure in connection with their function in the support and movement of the head. They are described with the atypical vertebrae. The seventh cervical vertebra is larger than the others and possesses a longer prominent spinous process which can be seen and felt at the base of the neck posteriorly. For this reason it is frequently referred to as the *vertebra prominens.*

Lumbar Vertebrae. The lumbar vertebrae are the largest of the individual vertebrae, each structural feature being stronger and more massive than its counterpart in other regions. The large, thick vertebral body is compressed anteroposteriorly, and its oval contour is indented into a concavity which faces posteriorly on the vertebral foramen so that the body is bean-shaped in appearance.

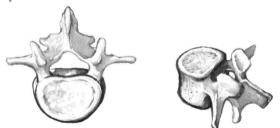

The triangular vertebral foramen is not as large as that of a cervical vertebra, although it is larger than the round foramen of the thoracic region. Short rugged pedicles thrust out posteriorly from the body. A sharply indented superior vertebral notch depresses the superior border of the pedicle below the level of the thick lip on the upper surface of the body. The inferior vertebral notch is a long, vertically slanting groove on the medial surface of the thick pedicle so that the inferior surface of the pedicle is almost straight. The junction between the pedicle and the lamina is interrupted by the origins of the superior articular and transverse processes. The coarse laminae fuse to form a short oblong spinous process which projects horizontally. The superior articular processes rise abruptly from the junctions of pedicle and lamina and are twisted posterolaterally so that the inferior articular facets, which face anteriorly and laterally, are received into their embrace. The transverse process curves out posterolaterally on each side at the same level as the pedicle.

Atypical Vertebrae

Certain vertebrae do not conform to the general or regional types because they are specially modified (cervical vertebrae 1 and 2), are fused into a single large bone (sacral vertebrae), or are rudimentary (coccygeal vertebrae).

First Cervical Vertebra. This bone is appropriately named the *atlas* because it supports the head, just as the mythological figure held up the earthly sphere. In development, the body of the atlas is detached and fuses to the superior surface of the second vertebra.

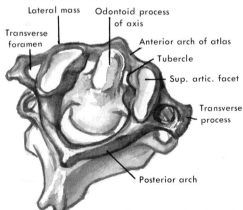
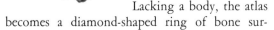

Lacking a body, the atlas becomes a diamond-shaped ring of bone sur-

rounding a large vertebral foramen. At the lateral extremities of the diamond the bone is thickened into *lateral masses* which are connected by anterior and posterior bars of bone, the *anterior* and *posterior arches*. At a medial location on each lateral mass is the troughlike *superior articular facet* which receives the occipital condyle of the skull somewhat like a fist that fits into a cup. The two articular facets are tilted medially toward each other. This disposition indicates that the skull can rock in a nodding motion anteriorly and posteriorly or from side to side, but the tilting of the articular facets prohibits rotation of the skull upon the atlas. The *transverse process* extends laterally from the superior facet but without the trough of other cervical vertebrae. It is pierced by the transverse foramen through which the vertebral artery ascends to turn medially to groove the posterior arch before turning upward again to enter the skull. Medial to the superior facet the lateral mass bears a prominent tubercle to which is attached the important *transverse ligament* of the atlas which is described with the second cervical vertebra. The anterior arch closes the ring anteriorly and bears a striplike facet on its posterior surface for articulation with the odontoid process of the second cervical vertebra. The inferior surface of each lateral mass presents a flat *inferior articular facet* which meets the superior facet of the second cervical vertebra.

Second Cervical Vertebra. This bone is called the *axis* because it forms a pivot around which the atlas bearing the skull may rotate. The axis is much more like a cervical vertebra in appearance although its superior surface is modified. The separated body of the atlas forms the *odontoid process* of the axis which projects upward into the anterior portion of the vertebral foramen of the atlas. The *superior articular facets* are found on the superior surface of the body on each side of the odontoid process. They are large and rounded with a slightly domed surface upon which the inferior facets of the atlas may turn.

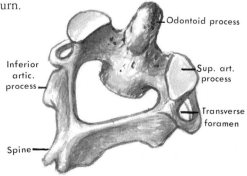

The odontoid process carries an oval facet on its anterior surface where it articulates with the facet on the posterior surface of the anterior arch of the atlas. Alignment of the anterior arch with the odontoid process is maintained by the strong transverse ligament of the atlas. This is strung from the tubercles of the lateral masses to hug the posterior surface of the odontoid process.

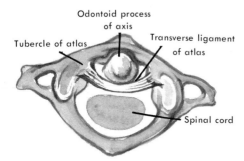

The transverse ligament of the atlas divides the vertebral foramen of the atlas into a smaller anterior compartment and a larger posterior one. The odontoid process of the axis projects superiorly into the anterior compartment between the anterior arch and the transverse ligament. The spinal cord traverses the posterior compartment. The rotation which is not permitted between the skull and the atlas is facilitated by the pivot arrangement by which the atlas may turn around the odontoid process upon the broad superior facets of the axis.

Special Vertebral Joints

The unusual structures of the atlas and the axis are associated with special joints for the support and movement of the skull.

Atlanto-occipital Joints. These synovial joints are located on each side of the superior aspect of the vertebral column. Here the occipital condyles of the base of the skull articulate with the superior articular facets of the atlas. The meeting of the rounded condyles with the concave articular facets forms a condyloid type of joint in which motions can be performed in two planes. The anteroposterior motions of flexion and extension occur easily as in nodding of the head. Lateral bending is also free. Rotation of the skull upon the atlas is impossible, because the oblique placement of both the condyles of the skull in the medially tilted facets of the atlas prevents turning in either lateral direction. The

gap between the anterior arch of the atlas and the skull is closed by the *anterior atlanto-occipital membrane* which is a curtain-like upward continuation of the anterior longitudinal ligament to the anterior margin of the foramen magnum. The posterior longitudinal ligament cannot similarly continue upward to the base of the skull because the atlas has no vertebral body. Instead ligamenta flava from each side of the posterior arch of the atlas fuse and extend to the posterior margin of the foramen magnum as the *posterior atlanto-occipital membrane.*

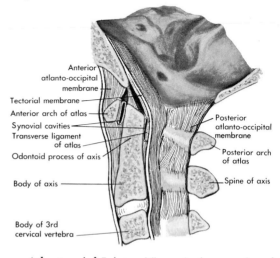

Atlantoaxial Joints. The articular capsules of three synovial joints shroud the median articulation of the odontoid process of the axis with the anterior arch of the atlas and the laterally placed plane joints between the articular facets of the two bones. The odontoid process is held in contact and alignment with the anterior arch of the atlas by the *transverse ligament of the atlas.* As the atlas, bearing the skull, moves upon the broad surfaces of the superior articular facets of the axis, its anterior arch rotates about the pivot made by the odontoid process. The transverse ligament of the atlas, draped around the posterior surface of the odontoid process, completes the ring about the pivot. This ligament must be strong, for if it should rupture, the odontoid process could shift posteriorly to squeeze the soft spinal cord immediately behind against the posterior arch of the atlas. Vertical extensions of the transverse ligament suggest the synonym *cruciate ligament.* Other than this ligament, the spinal cord has only the slight protection of the *tectorial membrane,* a weak upward continuation of the posterior longitudinal ligament from the body of the axis

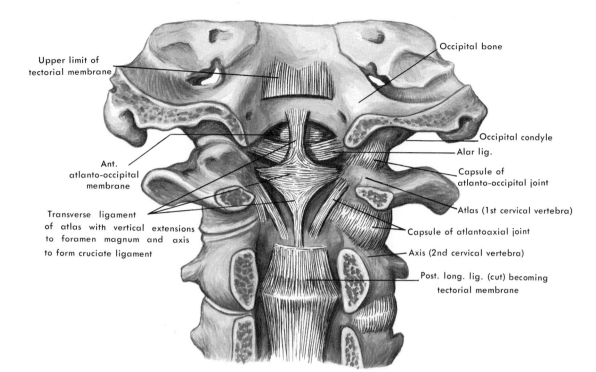

Upper limit of tectorial membrane

Occipital bone

Occipital condyle

Alar lig.

Capsule of atlanto-occipital joint

Ant. atlanto-occipital membrane

Atlas (1st cervical vertebra)

Transverse ligament of atlas with vertical extensions to foramen magnum and axis to form cruciate ligament

Capsule of atlantoaxial joint

Axis (2nd cervical vertebra)

Post. long. lig. (cut) becoming tectorial membrane

through the foramen magnum to the floor of the skull. This membrane is a sling to support the spinal cord, not a barrier to ward off a rampaging odontoid process. *Alar ligaments,* extending between the odontoid process and the occipital condyles, check rotation movements.

Sacrum.   The sacrum, formed by the five sacral vertebrae fused together by the four ossified intervertebral disks between them, is an irregular, spade-shaped bone which is curved with a concavity facing anteriorly; it is wedged between the two hip bones to complete the pelvic girdle posteriorly. The sacrum presents a smooth anterior or pelvic surface which retains the sacral curve, a persistence of the primary fetal curvature of the spinal column. The convex posterior surface, roughened by many muscle attachments, can be felt below the "small of the back." The bone is widest above where its central portion articulates with the fifth lumbar vertebra and tapers inferiorly to a narrower apex which articulates with the coccyx. (See opposite illustration.)

The anterior or *pelvic surface* may be divided into three zones: a central bar of bone, a series of foramina, and lateral masses of bone. The central bar of bone is composed of the five fused vertebral bodies. (See opposite illustration.) The superior end of the bar, representing the first sacral body, appears much like a lower lumbar

vertebral body and articulates with the fifth lumbar vertebra via a prominent lumbosacral intervertebral disk. The anterior lip of this surface is greatly thickened and with the prominent disk makes up a landmark of the pelvis known as the *sacral promontory.* Below this point the anterior surface follows the concave sacral curvature.

Four *transverse ridges* cross the surface to mark the location of the ossified sacral intervertebral

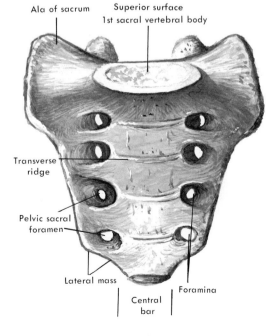

Ala of sacrum

Superior surface 1st sacral vertebral body

Transverse ridge

Pelvic sacral foramen

Lateral mass

Foramina

Central bar

disks. Four pairs of *pelvic sacral foramina* are found, one at each end of the transverse ridges. These foramina represent only the anterior end of a sacral intervertebral tunnel into which the sacral nerves course from the vertebral canal. Division of the spinal nerve into anterior and posterior rami, which occurs just lateral to other vertebrae, occurs inside this tunnel. Only the ventral divisions of the first four sacral nerves are transmitted through the pelvic sacral foramina. Lateral to the central bar of bone on each side, five heavy bony processes surround the foramina and then fuse with each other to form the *lateral masses* which make up the lateral parts of the sacrum. The portions of the lateral masses which make up the pelvic surface would be ribs in the thoracic region. The superior margin of the lateral mass adjacent to the body of the first sacral vertebra is broad and smooth. It is termed the *ala* or wing of the sacrum and forms, with the sacral promontory, part of the superior border of the pelvis. The side of the ala and of the lateral mass corresponding to the level of the first and second sacral vertebrae is flat and roughened into the *auricular surface* where the sacrum on each side fits against the hip bone at the sacroiliac joint.

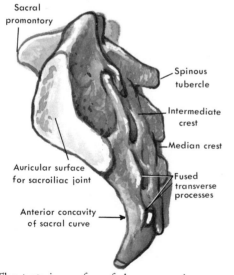

Sacral promontory

Spinous tubercle

Intermediate crest

Median crest

Auricular surface for sacroiliac joint

Fused transverse processes

Anterior concavity of sacral curve

The *posterior surface* of the sacrum is as rough and convex as the anterior surface is smooth and concave. Fusion of sacral vertebral elements hides the *sacral* (vertebral) *canal*, which can be seen clearly only at the inferior tip where failure of the fourth and fifth laminae to fuse leaves an open triangular interval, the *sacral hiatus*. The small triangular sacral canal contains only the lowest spinal nerves, spinal membranes, and fat.

It ends at the sacral hiatus which in life is covered by a fibrous membrane. The true intervertebral foramina open out laterally to provide egress of the sacral nerves into the tunnel previously described. The central part of the posterior surface is rough and nodular. It is made up largely of the fused vertebral laminae.

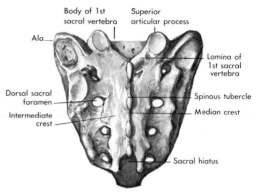

Body of 1st sacral vertebra

Superior articular process

Ala

Lamina of 1st sacral vertebra

Dorsal sacral foramen

Spinous tubercle

Intermediate crest

Median crest

Sacral hiatus

A *median crest* is formed by the fused spinous processes and bears four *spinous tubercles* which represent the tips of the upper four spinous processes, the fifth being lacking in the formation of the sacral hiatus. On each side of the median crest there is a lower *intermediate crest,* which represents the fused articular processes and bears small *articular tubercles*. Just lateral to the articular tubercles are the openings of the *dorsal* or posterior *sacral foramina* through which pass the dorsal or posterior rami of the first four sacral nerves. The fifth sacral dorsal ramus passes out through the sacral hiatus onto the posterior surface of the tip of the sacrum. Lateral to the foramina the posterior surface of the lateral masses is formed by the fusion of the transverse processes of the sacral vertebrae. The superior portion of the posterior surface resembles a vertebra more than do other portions of the bone. It is made up of elements of the first sacral vertebra. Its laminae are more distinct. Prominent *superior articular processes* thrust up superiorly on each side above the first posterior sacral foramina. These processes face posteriorly to interlock firmly with the inferior articular processes of the fifth lumbar vertebra above.

Coccyx. The four coccygeal vertebrae are fused in the human to form an irregular, slender, and tapering bone, the *coccyx*. This inferior extremity of the vertebral column is the vestige of a tail. The fused bones are vertebrae only in a very rudimentary sense and only the first coccygeal bone has structural features to be noted. This part of the coccyx articulates with the sacrum via an intervertebral disk. The pelvic surface is smooth, continues the sacral concavity, and is covered by pelvic ligaments which attach to it.

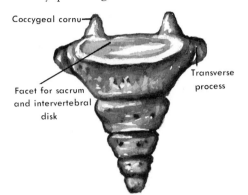

Stubs of transverse processes project laterally, whereas the only vestiges of a vertebral arch are found on the posterior surface in a pair of *coccygeal cornua* which point upward toward the lowest of the sacral articular tubercles. The lower three coccygeal bones, each smaller than the one above, form the tip of the coccyx which can be felt in the gluteal fold between the buttocks just above the anus.

Summary of General Movements of the Vertebral Column

The summation of small movements between adjacent vertebrae results in a considerable range of motion of the vertebral column as a whole. In addition, movement is particularly free where the vertebrae of one region meet those of another, as at the lumbosacral junction and the junction of the twelfth thoracic with the first lumbar vertebra. The movements of the vertebral column are flexion, extension, lateral flexion, and rotation.

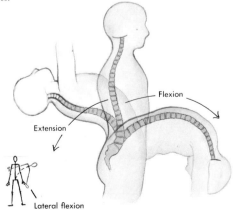

Flexion is forward bending of the spinal column. This movement is quite free in the cervical region. Flexion is reduced in the thoracic region by the presence of the ribs, sternum, and clavicles. The restriction of flexion here protects the thoracic organs from compression. Flexion is free in the lumbar region and particularly at the lumbosacral junction. Much of the forward bending is performed at the hip joints rather than in the vertebral column.

Extension is backward bending of the vertebral column as in arching of the back. This movement is free in the cervical region, restricted in the thoracic region, and is greatest in the lumbar region and at the junction of the last lumbar vertebra with the sacrum.

Lateral flexion is lateral bending of the vertebral column. It is the least free of the motions of the cervical vertebrae but is facilitated if combined with rotation of the head. Lateral flexion is greatest in the lumbar region and, though possible, in the thoracic vertebrae is impeded by the ribs.

Rotation is a twisting or turning motion of the vertebral column which is a summation of the amount of turning of adjacent vertebrae permitted by their intervertebral disks and the nature of their articulations with each other. Articulations which will facilitate the other movements of the vertebral column do not allow significant rotation, so that the amount of rotation of the vertebral column is less than might be expected when one considers the freedom of the hands-on-hips turning of the body in physical education exercises. Rotation combined with lateral flexion is freest in the cervical region and present in the thoracic vertebrae, but is limited in the lumbar region by the forces of weight-bearing. Much of the freedom of the trunk to turn is due to movements of the pelvis and the hips rather than of the vertebral column.

THE SKULL

The skull is composed of 22 separate bones united by saw-toothed fibrous joints which ossify during adult life. The primary function of the skull is to protect the brain and its associated sense organs for vision, hearing, taste, and smell. Apertures of the skull admit light rays to the eyes, sound waves to the hearing mechanism, air to the respiratory system, and food to the digestive system. A separate bone, the mandible, forms the framework of the lower jaw. Fourteen irregular bones of the skull, including the mandible, form the *facial skeleton* to which attach muscles which guard the apertures of the head and provide for facial expression. Eight flat bones of the *cranial skeleton* form the *cranium,* which encases the brain.

THE CRANIAL VAULT

The flat bones of the cranium are heavier posteriorly and toward the brows but thinner laterally where heavy muscles provide protection. The cranium is spherical, a shape which serves well in protecting the brain, for blows to the head tend to glance off a smooth spherical surface. The portions of the skull which sweep upward from the sides and over the superior surface of the skull form the *cranial vault.* Four bones enter into the formation of the cranial vault. One *frontal bone* presents its curved convex aspect from the level of the eyebrows upward as the anterior part of the cranium. The frontal bone is separated from the two *parietal bones* by the *coronal suture* which runs transversely up and over the vault of the cranium just in front of the midpoint or vertex of the skull. The midline *sagittal suture* runs anteroposteriorly between the parietal bones. Posteriorly, two *lambdoidal sutures* slant upward and anteriorly to meet the end of the sagittal suture. The lambdoidal sutures separate the parietal bones from the *occipital bone* which forms the posterior aspect of the cranial vault.

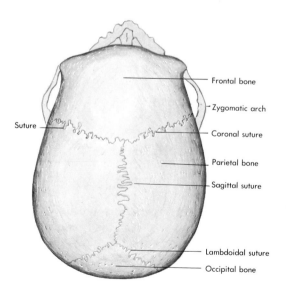

	Frontal bone
	Zygomatic arch
Suture	Coronal suture
	Parietal bone
	Sagittal suture
	Lambdoidal suture
	Occipital bone

Fontanelles

At birth the cranial bones do not meet and their fibrous joint membranes have not ossified to form the sutures. In addition, right and left frontal bones have not yet fused into one at the midline. A diamond-shaped aperture covered by a fibrous membrane lies at the intersection of the frontal and parietal bones. This aperture, the *anterior fontanelle,* is used as a landmark of the fetal skull in obstetrical diagnosis. A triangular aperture, the *posterior fontanelle,* is located at the intersection of the lambdoidal and sagittal sutures.

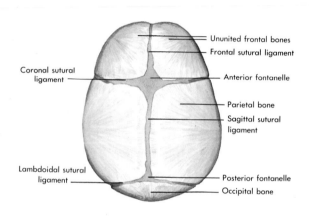

	Ununited frontal bones
	Frontal sutural ligament
Coronal sutural ligament	Anterior fontanelle
	Parietal bone
	Sagittal sutural ligament
Lambdoidal sutural ligament	Posterior fontanelle
	Occipital bone

Both fontanelles close as the bones grow together and the sutural ligaments ossify during early childhood.

Tables of the Skull

The flat bones of the cranium consist of two plates or *tables* of compact bone which are separated by spongy bone and marrow spaces called the *diploë.* The external (outer) table is smoothly convex and forms the surface of the cranium. The internal (inner) table is the irregularly concave receptacle for the brain. Severe direct force to the

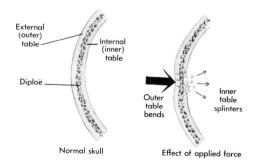

External (outer) table	Internal (inner) table
Diploë	
	Outer table bends
	Inner table splinters
Normal skull	Effect of applied force

skull may cause the spherical outer table to bend and absorb much of the force, but the concave inner table may break or splinter to injure the brain, blood vessels, or nerves of the skull.

ANTERIOR ASPECT OF THE SKULL

When viewed from the anterior aspect the skull presents the osteological features which underlie the forehead and the face. The anterior aspect may be divided into frontal, maxillary, nasal, orbital, and mandibular areas.

Frontal Area

The frontal bone of the cranial vault underlies the frontal or forehead region. As the frontal bone slopes downward it is bulged slightly on each side above the orbit by the *frontal eminence* below which a flattened ridge, the *superciliary arch,* curves transversely across the bone at the level of the eyebrows. A depressed flat area between the frontal eminences, the *glabella,* may be traced inferiorly to the root of the nose where the frontal bone meets the paired nasal bones. Below the superciliary arches the frontal bone thickens into the *supraorbital margin* as it turns inward to form the roof of the orbit. The *supraorbital foramen* pierces the junction of the medial and middle thirds of the supraorbital margin. Nerves and blood vessels to the forehead pass through the foramen.

Maxillary Area

The upper jaw is formed by the two *maxillary bones* (maxillae) which meet at the midline beneath the nasal opening. The center or *body* of each maxilla is a concavely depressed area under each orbit. The body is hollowed by the *maxillary sinus,* one of a series of outpouchings from the nasal cavity. The *frontal process* of the maxilla projects upward from the body of the bone as a buttress which supports the nasal bone of its side. In so doing the frontal process forms most of the medial boundary of the orbit and the upper lateral boundary of the nasal opening. The *alveolar process* of the maxilla projects inferiorly from the body of the bone and medially to meet its fellow from the opposite side. The two alveolar processes form the inferior and lower lateral

boundaries of the nasal opening. The upper teeth are borne in sockets which are deeply indented and form a scalloped margin on the inferior border of the alveolar process. The *zygomatic process* juts out laterally from the body of the maxilla to meet the zygomatic bone which with a portion of the body forms the *infraorbital margin.* The *infraorbital foramen* perforates the body of the maxilla below the margin for the transmission of nerves and blood vessels to the face below the orbit.

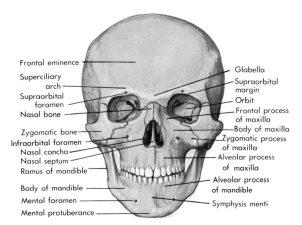

Nasal Area

Most of the external nose of the living person has a cartilaginous framework, missing in the dried skull, which is affixed to the nasal bones superiorly and to the bony margins of the nasal opening laterally and inferiorly. The two nasal bones, united at the midline, support the upper part of the bridge of the nose. Within the triangular nasal opening the *bony nasal septum* hangs downward in the midline. This partition, frequently angulated or deviated to one side, is defective anteriorly near the floor of the nasal cavity. The defect is filled and a partition for the cartilaginous framework of the external nose is made by a cartilaginous plate, the *septal cartilage.* The nasal cavity is thus divided into two sides from the lateral margins of which project irregular curved shelves of bone, the *nasal conchae.*

Orbital Area

The *orbits* are pyramidal recesses in the skull which contain the eyeballs and the muscles, vessels, and nerves associated with them, as well as nerves and blood vessels in transit to the face.

The *base* of the orbital pyramid is the open orbital aperture whose supraorbital and infraorbital margins have been described. The *roof* of the orbit is made by the *orbital plate* of the frontal bone which turns posteriorly at the supraorbital margin to separate the orbit from the cranial cavity above. The *medial wall* of the orbit is made up of several bones. Anteriorly is the frontal process of the maxilla which is succeeded posteriorly by the small *lacrimal bone.* The posterior part of the medial wall is formed by the orbital portion of the *ethmoid bone,* a bone of the floor of the skull which also makes up much of the roof and part of the lateral wall of the nasal cavity. The *lateral wall* of the orbit is made up anteriorly by the zygomatic (cheek) bone and posteriorly by a portion, the greater wing, of the sphenoid bone.

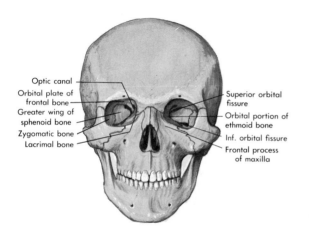

Optic canal
Orbital plate of frontal bone
Greater wing of sphenoid bone
Zygomatic bone
Lacrimal bone
Superior orbital fissure
Orbital portion of ethmoid bone
Inf. orbital fissure
Frontal process of maxilla

The *floor* of the orbit is triangular because the lateral wall runs anterolaterally from the apex while the medial wall parallels the midline. The maxilla forms the medial part of the floor and the zygomatic bone forms the lateral part. An irregular crevice, the *inferior orbital fissure,* runs along the floor, separating the bones of the floor and the lateral wall. Many nerves and blood vessels, not concerned with the orbit, traverse this fissure to reach a deep region of the face, the infratemporal fossa. The boundaries of the orbital pyramid meet posteriorly at the *apex* which presents two openings. Medially on the apex the *optic canal* transmits the optic nerve and ophthalmic artery from the cranial cavity through the lesser wing of the sphenoid bone into the orbit. This canal is separated by a thin bar of bone from the laterally placed *superior orbital fissure* which transmits other blood vessels and nerves into the orbit.

Mandibular Region

The separate *mandible* completes the facial skeleton, participates in the formation of the floor of the mouth and oral cavity, and bears the lower teeth. The bone consists of the *body,* which arches laterally and posteriorly under the maxillae, and the *rami,* which rise vertically on each side to articulate with the skull. Only the body is seen well anteriorly. This curved bar of bone develops in two halves which fuse at the midline in a thickened vertical ridge, the *symphysis menti.* The inferior border of the body is round and, centrally, forms the *mental protuberance* which is the framework of the chin. The superior border of the body is scalloped into the *alveolar process* which is indented by the sockets of the lower teeth.

LATERAL ASPECT OF THE SKULL

The spherical surface of the skull permits many of the osteological features of the anterior aspect of the face to be viewed from the side. Other areas are also found on the lateral aspect.

Zygomatic Area

The *malar* (cheek) *surface* of the central part of the *zygomatic bone* forms the prominence of the cheek. Medially the zygomatic bone meets the zygomatic process of the maxilla and slides over it to form much of the infraorbital margin. The *frontal process* rises to meet the frontal bone and forms the lateral orbital margin and part of the lateral orbital wall. Behind the frontal process is a deep groove beside the cranial vault behind the orbit. This is the *temporal fossa* which lodges the temporalis muscle. The posterior margin of the malar surface flares out into the *temporal process* of the zygomatic bone. This process passes posteriorly, bridges the temporal fossa, and meets a long narrow bar of bone, the *zygomatic process of the temporal bone,* which comes forward from the side of the skull to form the *zygomatic arch.*

Cranial Vault

The lateral aspect of the cranial vault is divided into *temporal* and *auditory-mastoid areas.*

Temporal Area. The *superior temporal line* arches across the side of the cranial vault from the zygomatic process of the frontal bone. The line curves posteriorly and then inferiorly before arching forward onto the zygomatic process of the temporal bone. The *temporal fossa* between these limits is flat superiorly but deepens into a trough under the zygomatic arch. The *squamosal suture* forms a roughly concentric semicircle within the temporal fossa parallel to the superior temporal line which outlines the squamous portion of the temporal bone. Above this suture the vault of the cranium is formed by the parietal bone.

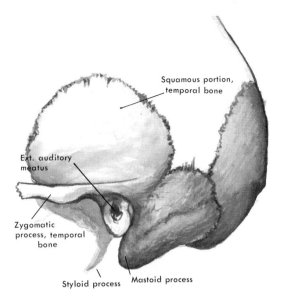

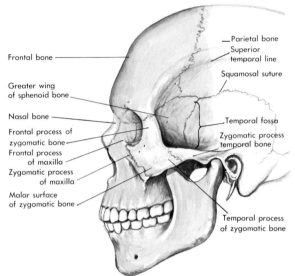

Auditory-Mastoid Area. An imaginary line projected posteriorly from the zygomatic arch traverses a broadened shelf of temporal bone over the articulation of the mandible with the skull. Continuing posteriorly the line coincides with the *supramastoid crest* on the bulkier portion of the temporal bone. The line finally intersects with the lambdoidal suture behind which is the occipital bone. The blunt triangular *mastoid process* projects downward from the main part of the temporal bone. The interior of the mastoid process is hollowed out by the mastoid *air cells* which connect with the cavity of the middle ear. Anterior to the mastoid process and inferior to the supra-meatal crest is the opening of the *external auditory meatus*, a bony canal which leads into the middle ear cavity of the temporal bone. The *styloid process* projects anteroinferiorly as a slender bar of bone from the floor of the meatus to provide attachment for certain muscles of the tongue, pharynx, and neck. The *mandibular fossa* is a hollow anterior to the meatus which is part of the articulation of the mandible with the skull.

Mandible. The *ramus* of the mandible rises as a quadrangular plate from the posterior end of the body where the inferior border of the two parts forms the squared *angle* of the mandible.

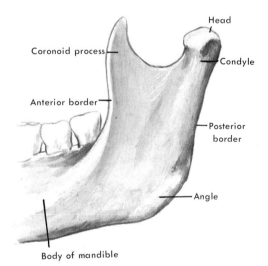

The anterior border of the ramus becomes a thin triangular plate, the *coronoid process,* which receives the attachment of the temporalis muscle behind the zygomatic arch. The posterior border of the ramus continues superiorly as the more tubular *condyle* to enter into the articulation with the base of the skull by means of a rounded *head.*

Temporomandibular Joint. This synovial joint is between the head of the mandible and an articular eminence on the mandibular fossa of the skull. The head of the mandible is oriented transversely across the mandibular fossa with an articular disk interposed between them. The articular eminence is convex but the rest of the fossa is concave. The superior surface of the articular disk is reciprocally shaped to fit them. The inferior surface of the disk is concave to receive the rounded head of the mandible. The synovial membrane forms separate synovial cavities on each side of the disk but a loose fibrous *common articular capsule* encloses the entire joint. The front and sides of the joint are strengthened by the *temporomandibular ligament.*

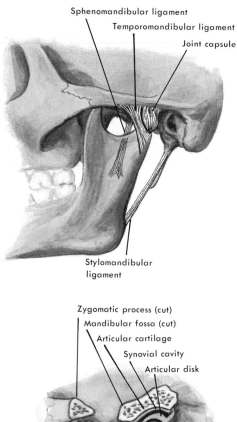

Sphenomandibular ligament

Temporomandibular ligament

Joint capsule

Stylomandibular ligament

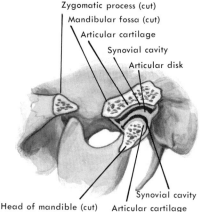

Zygomatic process (cut)

Mandibular fossa (cut)

Articular cartilage

Synovial cavity

Articular disk

Synovial cavity

Head of mandible (cut)   Articular cartilage

Opening and closing of the jaws are primarily hinge movements in which the head of the mandible moves on the inferior surface of the articular disk. In *protrusion,* the disk, carrying the head of the mandible with it, is pulled forward onto the articular eminence. In *retraction,* the disk

and head are moved backward together into the mandibular fossa. *Lateral movement* of the jaws occurs by the movement of the disk and head as a unit across the mandibular fossa. These movements are combined in grinding, biting, and chewing.

BASE OF THE SKULL

The rough base of the skull is formed into a series of projections, depressions, and foramina. These are concerned with the attachment of neck and tongue muscles, the suspension of the pharynx, the respiratory apertures, and the entrance or exit of blood vessels and nerves en route to or from the head.

Mandible from Below

Viewed from below the mandible has a horseshoe shape. Its inferior border is smooth and rounded, but the internal surface is marked by tubercles for the attachment of the muscles of the floor of the mouth and the tongue. A fossa just lateral to the symphysis lodges the sublingual salivary gland, and one on the internal surface of the ramus accommodates the submandibular gland. The *mandibular foramen* slants into the center of the ramus to carry the inferior dental nerve and vessels into the lower jaw.

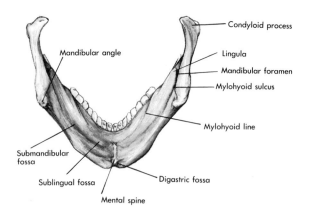

Mandibular angle

Condyloid process

Lingula

Mandibular foramen

Mylohyoid sulcus

Mylohyoid line

Submandibular fossa

Sublingual fossa

Digastric fossa

Mental spine

Base of the Skull

With the mandible removed, the alveolar processes of the maxillae bearing the upper teeth can be seen surrounding a depressed shelf of bone, the *hard palate.* The hard palate forms the roof of the mouth and the floor of the nasal cavity. It ends abruptly posteriorly at the margin of the two posterior openings of the nasal cavity, the *posterior choanae,* which are separated by the posterior end of the nasal septum. The base of the skull posterior to the nasal openings is made by the sphenoid bone from which project on each side the *lateral* and *medial pterygoid plates.* These limit the nasal openings laterally, offer points for suspension of the pharynx, and provide attachment for some of the masticatory muscles. The sphenoid bone is succeeded more posteriorly in the midline by the tonguelike basilar portion of the *occipital bone.* A large opening, the *foramen magnum,* appears in the occipital bone. This huge foramen provides passage for the brain stem from the cranial cavity into the vertebral canal where it becomes the spinal cord. Two large bony masses, the *occipital condyles,* located anterolaterally on the margins of the foramen, bear facets for the articulation of the skull with the superior articular facets of the atlas. The posterior aspect of the base of the skull is formed by the occipital bone, which meets the temporal bone laterally. A number of large openings of the base of the skull are traversed by vessels and nerves which enter or leave the cranial cavity. These are described with the floor of the cranial cavity.

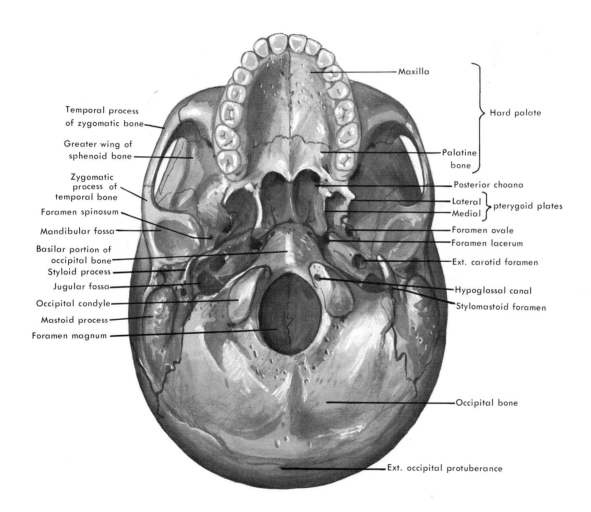

CRANIAL CAVITY

The internal surface of the cranial vault is concave in all directions and molded in relief to the surface of the brain. The *dura mater,* the outer covering of the brain, is affixed to bones of the vault and carries both arteries and venous passageways, called *venous sinuses,* which groove the bones.

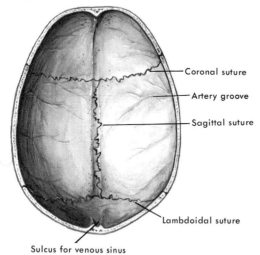

- Coronal suture
- Artery groove
- Sagittal suture
- Lambdoidal suture

Sulcus for venous sinus

One of these, the *superior sagittal sinus,* courses along the midline of the vault leaving a shallow trough to mark its passage. This is surrounded by shallow fossae holding outpouchings of the wall of the sinus. These are related to structures through which cerebrospinal fluid about the brain is drained into the venous system. Branches of the *middle meningeal artery,* which supply the coverings of the brain, form grooves and tunnels which radiate across the inner surface of the cranial vault from the squamous portion of the temporal bone.

Three large portions of the brain rest upon the *floor of the skull* which is shaped into three cuplike depressions or *cranial fossae* to hold them. The frontal lobe anteriorly, the temporal lobe laterally, and the cerebellum and brain stem posteriorly rest upon the anterior, middle, and posterior cranial fossae respectively.

The *anterior cranial fossa* is formed by the orbital plates of the frontal bone on each side between which are the thin *cribriform plates* of the ethmoid bone, perforated by the passage of filaments of the olfactory nerve from the nasal cavity beneath. A sharp projection, the *crista galli,* rises upward between the troughs of the cribriform plate to provide an attachment for the dural membrane of the brain. The posterior part of the fossa and its sharp posterior margin are formed by the butterfly shape of the body and lesser wings of the sphenoid bone. This margin is drawn out medially on each side to form the anterior clinoid process. This process overhangs the middle cranial fossa and is another point of dural attachment. The process is above the *optic canal.*

The body of the sphenoid bone rises up as a squared block in the midline of the *middle cranial fossa.* This block is also named the *sella turcica* from a fancied resemblance to a Turkish saddle. It presents a central depression which lodges the pituitary gland and is termed the *pituitary fossa.* The anterior margin of the fossa bulges upward into the *tuberculum sellae.* The posterior margin or *dorsum sellae* bears the *posterior clinoid processes.* The floor of the middle cranial fossa drops off rapidly on each side of the midline into deep cups which join the lateral wall of the cranial vault. The medial part of the floor is made by the

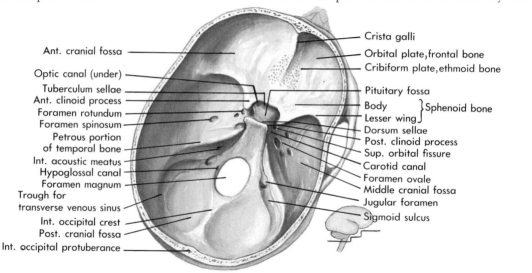

Ant. cranial fossa

Optic canal (under)
Tuberculum sellae
Ant. clinoid process
Foramen rotundum
Foramen spinosum
Petrous portion
of temporal bone
Int. acoustic meatus
Hypoglossal canal
Foramen magnum
Trough for
transverse venous sinus
Int. occipital crest
Post. cranial fossa
Int. occipital protuberance

Crista galli
Orbital plate, frontal bone
Cribiform plate, ethmoid bone

Pituitary fossa
Body   } Sphenoid bone
Lesser wing
Dorsum sellae
Post. clinoid process
Sup. orbital fissure
Carotid canal
Foramen ovale
Middle cranial fossa
Jugular foramen
Sigmoid sulcus

sphenoid bone and the lateral part is formed by the squamous part of the temporal bone. Major foramina are found in the floor and walls of the middle cranial fossa. The internal carotid artery rises from the neck through the *carotid canal* beside the posterior clinoid process. The *superior orbital fissure* opens into the orbit through the anterior wall under the anterior clinoid process. The *foramen rotundum,* just behind the fissure, transmits the maxillary division of the trigeminal nerve. Farther posterolaterally is the *foramen ovale* which is traversed by the mandibular division of the same nerve. Lateral to the foramen ovale is the *foramen spinosum* through which the middle meningeal artery enters the skull.

The *posterior cranial fossa,* the deepest and largest of the three fossae, is distinguished by the large *foramen magnum.* The occipital bone slopes upward from the foramen in the midline to meet the dorsum sellae. The anterior margin of the fossa on each side is formed by the petrous portion of the temporal bone which contains the internal ear mechanism. The *internal acoustic meatus* is a large foramen on this margin which carries the vestibulocochlear nerve into the internal ear. The facial nerve and internal auditory vessels also traverse this foramen. Shallow depressions beside and behind the foramen magnum hold the lobes of the cerebellum. The *internal occipital crest* rises in the midline between these depressions onto the posterior wall of the fossa to terminate in the *internal occipital protuberance.* A trough runs laterally across the occipital bone from this eminence conveying the *transverse venous sinus.* The same sinus turns inferiorly as the *sigmoid sinus* in a deeper trough, the *sigmoid sulcus.* These continue medially and forward under the overhanging petrous bone to the large jagged *jugular foramen* through which the sinus leaves the skull to become the internal jugular vein of the neck. The glossopharyngeal, vagus, and accessory nerves pass through this foramen to reach the neck. Another important foramen, the *hypoglossal canal,* opens on the wall of the foramen magnum above the occipital condyle. The hypoglossal nerve leaves the skull for the tongue through this aperture.

THORACIC SKELETON

The thorax is the region of the trunk between the root of the neck and the diaphragm. Its skeletal members are the sternum, ribs, and thoracic vertebrae. The limited motion in the thoracic section of the vertebral column and the thoracic primary vertebral curvature serve to protect the thoracic organs and to increase the capacity and dimensions of the thorax to accommodate them.

STERNUM

The vertebral column is a protective pillar for the great vessels, esophagus, and vital elements of the nervous system which lie just anterior to it, deep in the thorax. The heart, the great vessels springing from it, and the anterior surfaces of the lungs face toward the anterior surface of the thorax where they are exposed to possible trauma. They are protected by the *sternum* or breast bone. This long, narrow, shieldlike plate of bone, centered in the anterior midline of the thorax, provides a sturdy barrier against injury and affords a flexible anterior attachment for the ribs which also allows for change in the dimensions of the thorax with breathing.

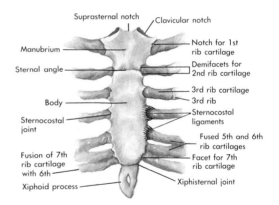

The sternum develops as three portions which fuse ultimately into the single bone. The three parts of the bone are the manubrium, body, and xiphoid process. The *manubrium,* about 2 inches long, is the superior part of the sternum. It makes up the anterior midline portion of the common boundary between the superior aperture of the thorax and the root of the neck. The manubrium meets the body of the bone at the prominent *sternal angle* (of Louis), a transverse ridge across the sternum which marks the junction of the cartilage of the second rib with the sternum and is a landmark for counting the ribs. The superior surface of the manubrium is indented by a median *suprasternal notch.* This is easily felt in the living person in the suprasternal fossa between the knoblike medial ends of the collar bones and the tendons of the sternocleidomastoid muscles of the neck. The superior surface of the manubrium is indented at each side by the *clavicular notches* which present articular facets to receive the medial ends of the clavicles. Immediately below these notches, the lateral borders of the manubrium are notched to lodge the cartilages of the first ribs. It should be noted that although all ribs articulate directly with their vertebrae posteriorly, none join the sternum directly. Their anterior articulation is via a costal cartilage which provides a flexible but strong union. The manubrium narrows below the notches for the first ribs to widen again at the sternal angle where it possesses a half notch and *demifacet* for the upper part of the second costal cartilage.

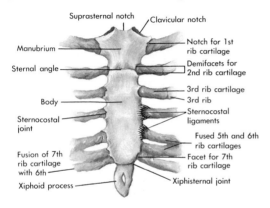

The *body* of the sternum is the longest portion of the bone. It varies in its width, owing to scalloping of the lateral borders to receive the costal cartilages, but gradually tapers inferiorly. The lateral border is immediately indented superiorly by a half notch and demifacet for the lower part of the articulation of the second costal cartilage. The lateral border below this point presents successively five more notches and facets for the cartilages of the third to seventh ribs. The eighth, ninth, and tenth ribs curve about the thorax at a level lower than the sternum and fall short of attaining the midline. Their cartilages fuse with each other and then with the cartilage of the seventh rib to acquire an indirect attachment to the sternum and to complete the *"thoracic cage."*

The *xiphoid process* is the inferior tip of the sternum. Like the coccyx, it is rudimentary and is frequently imperfect in development. It may be felt as a beady process in the soft tissues below the manubrium in the *infrasternal angle* between the upward converging margins of the lower ribs.

RIBS

The ribs are arclike ribbons of bone which generally follow a curving course from their junction with the vertebral column to the anterior aspect of the trunk wall. The first seven ribs are called the *true ribs* because they articulate with the sternum through their costal cartilages. The eighth, ninth, and tenth ribs are termed the *false ribs* because they join the sternum only indirectly through fusion of their cartilages with each other and then fusion of the joint cartilage with the seventh costal cartilage. The eleventh and twelfth ribs, the *floating ribs,* are so named because they are so short that they end amidst muscles of the lateral abdominal wall. The ribs vary in appearance according to their level. A typical rib will be described and then the differences displayed by the others will be explained.

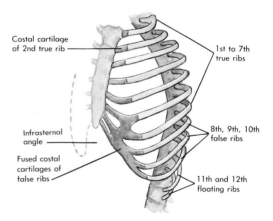

Typical Ribs

Although the first and second ribs are true ribs, they are so short that the third to seventh ribs are more typical of ribs in general. A typical rib possesses a slightly knobby *head* posteriorly by which it articulates with the vertebral column. The surface of the head is divided. A larger *inferior facet* is for articulation with the superior costal facet on the body of the particular vertebra to which the rib belongs. The smaller *superior facet* articulates with the inferior costal facet on the body of the vertebra above. The rib starts its curving course from the head as the narrow *neck* inclines posterolaterally toward the transverse process of its vertebra. The neck bears a *tubercle* for articulation with the transverse costal facet on the vertebral transverse process.

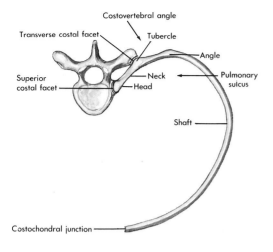

In front of the transverse process the neck of the rib curves laterally to become the smooth *shaft*. At the *angle* of the rib the bone abruptly changes its direction to descend while curving laterally and then anteriorly to follow the surface of the thoracic wall. Several factors influence the direction of each rib.

The Form of the Superior Aperture of the Thorax. The thorax is cagelike in its skeletal elements. Its narrow superior aperture, corresponding to the base of the neck, is made up of the body of the first thoracic vertebra, the first rib, and the superior border of the manubrium of the sternum. Arteries destined for the arms must first rise through the superior aperture of the thorax and turn laterally across the base of the neck to pass through the armpit.

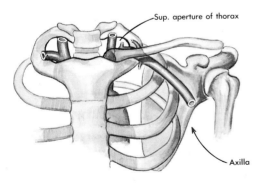

A sharp angulation of the arteries is avoided by the oblique plane of the superior thoracic aperture in which the superior margin of the sternum is set lower than the first thoracic vertebra and actually corresponds with the level of the body of the second thoracic vertebra. The first rib de-

scends throughout its short curved course to reach this level. The second and third ribs descend similarly.

The Length of the Sternum. The sternum not only is set lower but is shorter than the length of the 12 thoracic vertebrae. Whereas the upper three ribs descend to meet the sternum through their cartilages, the fourth and fifth ribs, descending laterally to follow the oblique course of the ribs above, must straighten anteriorly and approach the sternum almost horizontally. The lower true ribs and the false ribs descend far below the level of the sternum laterally and must ascend anteriorly to reach the sternum through their cartilages.

The Form of the Thorax. The thorax is conical. It tapers from a narrow superior aperture to a broad inferior aperture. The lower true ribs and the false ribs flare farther outward in their lateral course and descent than the ones above. Most of the surface of the shaft is smooth, although attachment is afforded on the lateral surface to the serratus anterior muscle coming from the scapula and to the external oblique muscle of the abdominal wall. The intercostal muscles attach to the superior and inferior borders. The *costal groove* on the inferior border gives protected passage to the intercostal vessels and nerve. The anterior end of a typical true rib is hollowed out to receive its costal cartilage in the *costochondral junction*. The flexible cartilage takes a direction appropriate for the level to join the rib to the sternum at the *sternochondral junction*.

The posterolateral angulation of the neck of the rib forms the *costovertebral angle* between the posterior surface of the rib and its vertebra. This angle is filled in by deep muscles of the back. The same angulation produces, on the thoracic surface, an increase in the anteroposterior axis of the thoracic cavity and an accompanying concavity of the posterior thoracic wall. The *pulmonary sulcus* thus formed on each side provides more space to accommodate the lungs.

Atypical Ribs

Certain ribs need special mention because of their form or unusual characteristics.

First Rib. The first is the shortest of the true ribs. It describes a tight arc in bounding the superior thoracic aperture. The bone is broader than most ribs and is flat. It is important because of the many structures which cross it or attach to it. The apex of the lung rises into the base of the neck medial to the first rib. The subclavian artery and vein groove the outer surface which is also crossed by nerves descending from the neck to the arm. The outer surface and internal border receive attachments of the scalene muscles of the neck, whereas the external border provides attachments for the scalene muscles of the neck, the intercostal muscles, and the serratus anterior muscle of the chest wall.

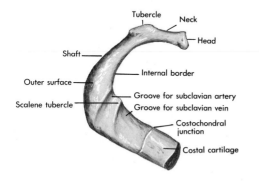

Eighth to Tenth Ribs.   The false ribs terminate almost as soon as they turn medially. Therefore, they are shorter than the typical ribs. Their upward and medial course is completed to the sternum by fused cartilages which with those of the opposite side form the *infrasternal angle.* The infrasternal angle outlines the anterior portion of the inferior thoracic aperture.

Eleventh and Twelfth Ribs.   The floating ribs are short imperfect bars which end amidst the muscles of the lateral abdominal wall in blunt cartilaginous tips.

JOINTS OF THE THORACIC SKELETON

Joints occur in the thoracic skeleton between the ribs and the vertebral column, between the costal cartilages and the sternum, and between the parts of the sternum.

Costovertebral Joints

The ribs articulate posteriorly with the vertebral column both at their heads and at their tubercles. Most of the ribs present a divided synovial joint between the head and the side of the body of their corresponding vertebra and of the one above. The synovial cavity is divided by an intra-articular ligament which attaches the head

to the intervertebral disk between the two vertebrae. The articular capsule is strengthened by the *radiate ligament* which spreads along the capsule from the head to the vertebrae and their disks.

The synovial joint between the tubercle of the rib and the transverse process is termed the *costotransverse joint* to distinguish it from the other costovertebral joint. The small filmy articular capsule is weak but it is shrouded and the joint is strengthened by three *costotransverse ligaments.*

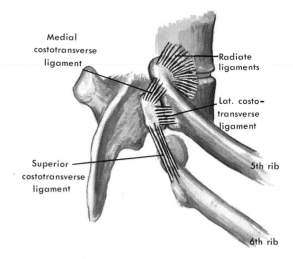

The lateral costotransverse ligament extends from the tip of the transverse process to the lateral part of the tubercle of the rib. Its medial counterpart sends collagenous fibers from the neck of the rib to the transverse process. The superior costotransverse ligament is a vertical sheet of collagenous fibers running from the neck to the transverse process of the vertebra above. These ligaments stabilize the posterior end of the ribs as they move.

Sternochondral Joints

The end of the bony part of the rib is hollowed out for a tight fit with its costal cartilage at the costochondral joint. The continuity of the fibrous periosteal-perichondrial covering strengthens this union. The costal cartilages articulate with the sternum in several ways. The first rib, which undergoes no movement as it borders the superior thoracic aperture, is joined firmly to the manubrium sterni by a true cartilaginous joint. The cartilage of the second rib, because it articulates with demifacets on both the manubrium and the body of the sternum at their junction, is a synovial joint with two synovial cavities divided by an intra-articular ligament. The cartilages of ribs three through seven have single small synovial joints with their corresponding facets on the sternum. *Sternocostal ligaments* radiate from each cartilage onto the anterior and posterior surfaces of the sternum, the collagenous fibers from one ligament blending with those of adjacent ligaments to form a continuous sternal membrane. Small *interchondral* (but synovial) *joints* occur where the eighth, ninth, and tenth cartilages fuse with each other and where the common cartilage of these ribs unites with the seventh costal cartilage.

Joints of the Sternum

The cartilaginous *manubriosternal joint* between the manubrium and body of the sternum has been cited as an example of a synchondrosis. The resiliency of the cartilaginous matrix makes this angulated union strong though flexible. This joint is subject to calcification later in life with lessened flexibility. The *xiphisternal joint* is also a cartilaginous joint, but in early adult life the ossification center of the xiphoid process extends into it and bony union to the body of the sternum ensues.

THE THORACIC CAGE AND ITS MOVEMENTS

The ribs form a series of obliquely curving slats which incompletely enclose the conical thorax. They are supported by the axial pillar of the vertebral column posteriorly, and all but the last two articulate with the sternum anteriorly through their cartilages. The combined elements form the skeletal *thoracic cage* which not only protects the thoracic viscera but also acts as an expansible framework by which the capacity of the thorax can be altered. The superior aperture of the thorax is closed by tissues of the neck, by the apices of the lungs, and by the structures (vessels, nerves, trachea, esophagus) which pass through it. The broader inferior thoracic aperture is bounded by the vertebral column, the lower ribs, and the fused costal cartilages on their way to the sternum. This aperture, the thoracic outlet, is closed by the musculofibrous sheet of the diaphragm which attaches around its boundaries. The lungs, heart, and other thoracic organs com-

pletely fill the thoracic cavity within the thorax. The lungs, however, are expansible organs. They have no mechanism of their own to cause them to expand, a process that is necessary in order for air to be drawn into them in breathing. The lungs expand because of differences in pressure between the atmospheric environment and the interior of the thorax when the capacity of the thorax is increased. The increase in thoracic volume is accomplished by a bellows action of the diaphragm, which descends as it contracts, and by movements of the bony elements of the thoracic cage. Action of the diaphragm lengthens the thorax in a superior-inferior direction. Increases in the transverse and anteroposterior dimensions are produced by movements of the ribs and sternum.

The rib bows out laterally between its sternal and vertebral joints of suspension in a manner that is universally described as that of a "bucket handle." When the intercostal and levatores costarum muscles contract in inspiration, a gliding and pivoting movement of the head of the rib and the tubercle at their articulations allows the curved shaft of the rib to swing outward and upward as in swinging up of the bucket handle from its position of rest beside the bucket. The combined movement of the ribs results in an increase in the transverse diameter of the thorax.

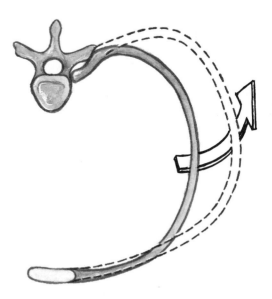

This movement is permitted anteriorly by a rotation of the costal cartilage in its joint with the sternum and by a twisting of the flexible cartilage. At the same time the movement of the ribs thrusts the sternum forward to increase the antero-posterior diameter of the thorax.

In ordinary breathing the expansion of the thorax is followed by expansion of the lungs into the space which is created and a consequent inflow of atmospheric air into the respiratory system. The termination of muscular contraction brings inspiration to an end. Expiration is accomplished by elevation of the diaphragm and the elastic recoil of the thoracic wall which compresses the lungs and thereby forces air from the respiratory system. In forced or labored breathing, muscles of the neck which attach to the sternum and first rib come into play to elevate these structures and deepen the thorax. Other muscles, such as the serratus anterior and the pectoralis minor, also pull upon the chest wall in an effort to increase the volume of the thorax to enhance respiration.

PECTORAL GIRDLE

The erect position of man has freed the upper limb from weight-bearing and locomotion. In their place, it has acquired free mobility, the strength to perform varied tasks, and a high degree of dexterity. These characteristics indicate that the attachment of the skeleton of the upper limb to the axial skeleton through the pectoral girdle will similarly possess the qualities of freedom and mobility. The pectoral girdle consists of the *scapula* and the *clavicle.* The head of the humerus, as the superior end of the skeleton of the upper extremity, articulates with the scapula.

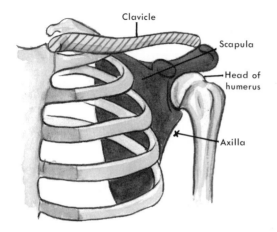

The scapula, however, is a flat bone, located on the posterolateral aspect of the thorax, which has no actual articulation with the axial skeleton. It is held in position mainly by the combined action of muscles which attach to its various surfaces and borders. In addition some of these muscles produce rotation of the bone which greatly increases the range of motion of the humerus, whereas others produce direct movements of the humerus and serve to hold it in its articulation with the scapula. Only the clavicle, extending between the manubrium of the sternum medially and the scapula laterally, provides a bony connection with the axial skeleton.

SCAPULA

The scapula is a flat bone with the general form of an inverted triangle which medially is molded to the posterolateral aspect of the thorax and laterally is thrust out on a tangent to it by the clavicle. Portions of the scapula produce the squared prominence of the shoulder above and behind the armpit or axilla; the flat remainder which is applied to the thorax has earned it the common name of the "shoulder blade." Since the thorax is cone-shaped with a narrow circumference superiorly and a larger girth inferiorly, the arm could not swing clear of the chest wall if the scapula were completely applied to the thoracic surface. The clavicle, therefore, serves a most valuable function in bracing the scapula, and thus the shoulder joint, out laterally from the thorax. The space formed between the inner surface of the arm and the chest wall is the axilla. The thin, platelike body of the scapula has an anterior (costal) surface, a posterior (dorsal) surface, three borders, and two major processes. Both surfaces are clothed by muscles attaching to them.

Borders and Angles

The long medial border of the scapula is vertically disposed and runs about 2 to 3 inches away from and parallel to the vertebral column, acquiring therefore the name *vertebral border.* At the rounded *inferior angle* the thin vertebral border meets the thicker lateral or *axillary border* which extends superolaterally to the *lateral angle.* The lateral angle flares out into a short *neck* which flattens into the shallow *glenoid fossa,* the point of articulation of the humerus with the scapula. The short *superior border* bears the coracoid process and is indented by the *suprascapular notch.*

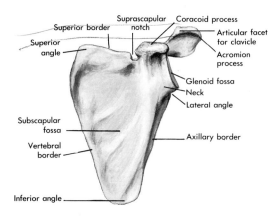

This border meets the vertebral border at the *superior angle.*

Surfaces

The *costal surface* is generally concave because of molding of this surface to the curved thoracic wall, but the concavity is deepened in the central area of the bone to form the *subscapular fossa.* The subscapularis muscle originates here and its thick mass is lodged in the fossa on its way to attach to the humerus. Superiorly the angulated coracoid process rises from the superior border with the suprascapular notch for transmission of the suprascapular nerve just medial to it. The superior angle, vertebral border, and inferior angle are flat and roughened for the attachment of the broad serratus anterior muscle which extends from the vertebral border to the thoracic wall.

The *dorsal surface* is generally convex. The *spine of the scapula* interrupts the upper part of this surface and divides it unequally into a narrower supraspinous fossa above the spine and a broader infraspinous fossa inferior to it. The *supraspinous fossa* is a shallow shelf between the spine and the superior border. The supraspinatus muscle takes origin here and passes laterally along the fossa toward the shoulder joint. The suprascapular nerve, noted passing through the notch in the superior border, innervates this muscle. The larger convex *infraspinous fossa* gives origin to the infraspinatus muscle which extends superolaterally along the undersurface of the spine on its way to the humerus. The lateral border of the dorsal

surface is thickened for the attachment of the teres minor muscle which extends to the humerus. Below this and spreading onto the dorsal surface of the inferior angle is the origin of the teres major muscle which also passes laterally to the humerus. The vertebral border of this surface is taken up, above the root of the spine, by an area for attachment of the levator scapulae muscle which descends from the neck. The border at the root of the spine affords attachment to the rhomboideus minor muscle and, between this point and the inferior angle, to the rhomboideus major muscle. These two muscles pass directly medially to the vertebral column.

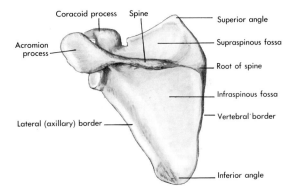

Processes

The *spine of the scapula* starts out from the vertebral border as an ever-widening ridge which extends the area of attachment for the muscles above and below it. It also strengthens the thin body of the bone as these muscles pull the head of the humerus to the glenoid fossa. The spine, in rising laterally, extends behind the neck of the scapula and the glenoid fossa to terminate in the broad flat *acromion process* which overhangs the shoulder joint. The inferior margin of the spine and adjacent surfaces of the acromion process give rise to the deltoid muscle which covers the shoulder and attaches to the humerus. The superior margin of the spine receives the insertion of part of the large trapezius muscle of the back. The anterior margin of the acromion process presents the *articular facet* for the lateral end of the clavicle.

The *coracoid process* projects superiorly from the superior border and then is twisted sharply laterally and anteriorly like a bent hook or beak. Its tip passes under the clavicle to curve over in front of the shoulder joint. This short bent lever is the point of attachment for several muscles which extend upward from the chest wall and the arm.

Glenoid Fossa

The neck of the scapula expands from the lateral angle of the scapula into a flat head which is largely composed of the shallow glenoid fossa. The fossa, which varies from a tear shape to an oval pear shape, is much smaller than the head of the humerus which turns and rotates upon it.

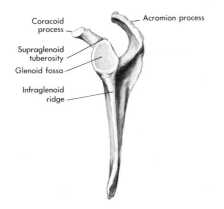

The ligaments, cartilage, and musculofibrous cuff which strengthen this shallow joint are described with the shoulder joint.

CLAVICLE

The familiar "collar bone" or *clavicle* is a doubly curved long bone which extends across the anterosuperior aspect of the thorax from the sternal manubrium across the first rib and coracoid process to the acromion of the scapula. The clavicle is the only part of the pectoral girdle to make connection with the axial skeleton; aside from providing attachment for muscles, it thrusts the scapula out from the thoracic wall to increase the mobility of the upper extremity. The clavicle is subcutaneous throughout its length. It is vulnerable to injury because of its exposed position and because of force transmitted to it from the shoulder. The medial part of the bone is convex in a forward direction, whereas the lateral portion is convex posteriorly.

Medial Portion

The knobby medial or *sternal end* of the clavicle bulges above and behind its articulation with

the manubrium of the sternum. This joint, like the temporomandibular joint, has an intermediate articular disk. The *articular facet* on the squared end of the clavicle extends a bit onto the inferior surface to contact the first costal cartilage which is immediately below the clavicle. The posterior surface receives an attachment of the sternocleidomastoid muscle, whereas the anterior surface is roughened for the clavicular head of the pectoralis major muscle. The central part of the shaft between the two curved ends is smooth and devoid of muscle attachments.

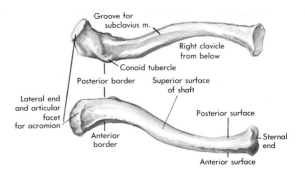

Lateral Portion

The lateral portion of the bone is flattened so that the anterior and posterior surfaces become borders between the superior and inferior surfaces. The broad superior surface is smooth and subcutaneous. A part of the deltoid muscle originates from the anterior border, whereas a portion of the trapezius muscle at the side of the neck attaches to the posterior border. The lateral end of the bone broadens to a blunt extremity which has an *articular facet* for the acromion on its lateral surface. The acromion is located lateral to the acromial end of the clavicle and forms the actual bony "point" of the shoulder.

Inferior Surface

The inferior surface is marked by a long groove for the subclavius muscle and by a *conoid tubercle* near the acromial end to which the strong coracoclavicular ligament is attached. This ligament binds the coracoid process, immediately below, to the clavicle and helps to position the scapula.

RELATIONS OF THE PECTORAL GIRDLE

This figure depicts the pectoral girdle as if one were to look at it from directly in front.

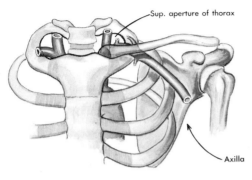

From the first rib the thoracic wall slants conically downward. The clavicle thrusts out against the acromion. Structures ascending through the superior aperture of the thorax can approach the upper limb by grazing over the first rib and over the slanting surface of the chest wall below. In doing so they pass behind the clavicle and in front of the scapula to enter the axilla, whose bony boundaries are made up medially by the uppermost ribs, posteriorly by the scapula, and laterally by the upper end of the humerus.

JOINTS OF THE PECTORAL GIRDLE

The joints of the pectoral girdle are those between the clavicle and the thorax and between the clavicle and scapula. A firm but movable connection between the clavicle and the sternum must be provided. The scapula must be braced by the clavicle, which will allow freedom for the scapula to increase the range of shoulder movement.

Sternoclavicular Joint

The synovial joint between the medial end of the clavicle and the manubrium of the sternum bears the brunt of the forces produced in propping the scapula outward. The squared knobby end of the clavicle does not make a good fit with the much smaller clavicular notch on the manubrium. The fit is improved by the *sternoclavicular articular disk* which divides the joint into two parts with separate synovial cavities. When the clavicle is elevated or depressed, as the shoulder joint moves up or down, the action takes place, as in a hinge joint, between the knobby end of the clavicle and the disk. In other motions at this joint the disk, carried upon the clavicle, glides upon the clavicular notch as the clavicle shifts forward or backward to follow similar movements of the scapula.

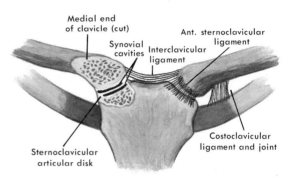

The poor fit of the articulating portions of the joint requires strengthening by accessory ligaments. *Anterior* and *posterior sternoclavicular ligaments* cross both surfaces of the joint from the end of the clavicle onto the sternum. These ligaments resist undue anterior or posterior movements of the clavicle which might result in dislocation in either direction. The medial slant of the sternoclavicular ligaments makes them effective in checking upward and lateral shifts of the clavicle. *Interclavicular ligament* fibers are stretched between the medial ends of the clavicles to resist any upward or lateral movement which would pull them apart from each other.

Costoclavicular Joint

A fibrous joint is formed between the first rib and the clavicle by the fibers of the *costoclavicular ligament*. These fibers ascend from the rib and its cartilage to the inferior surface of the clavicle. Undue upward and lateral movement of the end of the clavicle is prevented by this ligament.

Acromioclavicular Joint

A firm union between the lateral end of the clavicle and the acromion process is not desirable. A simple synovial plane joint permits gliding movements between the two bones. The acromioclavicular joint is strengthened superiorly by the *acromioclavicular ligament* which prevents the clavicle from overriding the acromion.

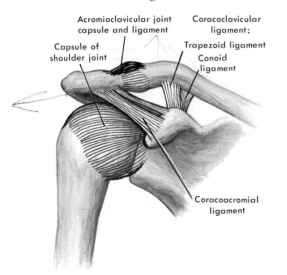

Acromioclavicular joint
capsule and ligament

Coracoclavicular
ligament:

Capsule of
shoulder joint

Trapezoid ligament

Conoid
ligament

Coracoacromial
ligament

When disruptive force shears apart the two bones at this joint, athletic trainers and sports physicians use the term "shoulder separation."

Coracoclavicular Joint

With motion permitted at the acromioclavicular joint, another point of union between the two bones must be found where the relationship is firm enough to permit the clavicle to thrust the scapula outward. A favorable location occurs where the coracoid process of the scapula bends under the lateral end of the clavicle. A fibrous joint is formed between the two bones by the two parts of the heavy *coracoclavicular ligament.* One part, the cone-shaped *conoid ligament,* extends upward from the coracoid process at the point of angulation to the conoid tubercle of the under surface of the clavicle. The other part, the *trapezoid ligament,* is broader and reaches upward to a more lateral attachment on the clavicle. The two parts of the coracoclavicular ligament strongly check more than slight movements at the acromioclavicular joint.

"SCAPULOTHORACIC JOINT"

This term is used functionally to refer to the movable apposition of the scapula to the thoracic wall where no true articulation exists. The scapula is held in place by muscles which extend between it and the axial skeleton. These muscles produce the following movements of the scapula which are concerned with and extend the range of motion of the shoulder joint.

1. *Elevation* occurs as the scapula is raised in the motion of "shrugging the shoulders," in which the shoulder joint is raised.

2. *Depression* involves the return of the scapula from the elevated position.

3. *Protraction* brings the scapula anterolaterally along the thoracic wall to a round shouldered position.

4. *Retraction* forces the scapula back along the thoracic wall in the position of military attention.

5. *Rotation* changes the direction of the glenoid fossa. The significant rotation is *lateral* or *upward rotation* in which the inferior angle swings laterally to turn the glenoid fossa upward. This greatly extends the degree to which the humerus, and therefore the arm, can be raised. *Medial* or *downward rotation* swings the inferior angle medially and turns the glenoid fossa downward.

LIVING ANATOMY OF THE PECTORAL GIRDLE

A finger can be inserted into the suprasternal fossa to palpate the suprasternal notch. By pushing laterally the crevice of the sternoclavicular joint and the knobby medial end of the clavicle can be felt. Another crevice separates these from the first costal cartilage immediately below. The fingers can now follow the superior surface of the clavicle laterally. A slight crevice can be felt separating the lateral end of the clavicle from the acromion process which forms the hard "point of the shoulder." If the fingers sweep medially and downward, the spine of the scapula can be traced through the overlying back muscles from the acromion to its root on the vertebral border of the scapula. The broad dorsal surface of the scapula can be distinguished above and below the spine through the muscles. The vertebral border can be followed inferiorly to the inferior angle. There the fingers can be hooked under the axillary border which can be followed superolaterally as it disappears amidst the muscles of the posterior wall of the axilla.

UPPER LIMB

The skeleton of the upper limb belongs to the appendicular skeleton. The skeletal elements offer rigidity, support, and muscular attachments for the limb. The upper limb is cylindrical although the hand flattens and splays out into the five digits. The various bones articulate at many joints. The muscles operating a particular bone will attach to the superior end of the bone, while its lower portion will afford origins for muscles which operate the succeeding bone or bones. The limb is always described as if it were in the anatomic position, which does not always coincide with natural positions of portions of the limb.

Particular terms are used in reference to the limb. *Proximal* refers to a portion of a bone or of the whole limb which is nearer the trunk. *Distal* indicates a part which is farther away from the trunk. *Flexor* or *volar* refers to a structure or a surface which is located anteriorly when the limb is in the anatomic position with the arms at the sides and the palms of the hands facing forward. Similarly *palmar* indicates the palm surface of the hand which is anterior in this position. *Extensor* or *dorsal* refers to a posterior structure or surface. In *pronation* the palm of the hand is turned posteriorly, in reference to the anatomic position, or downward if the arm is outstretched. *Supination* refers to the opposite position or action in which the hand is placed in the true anatomic position.

The joints of the upper limb are known by the rather familar names of shoulder joint, elbow joint, and wrist joint. In many of the joints, more than one articulation is encompassed by the common name. For example the elbow joint actually includes the humeroulnar, humeroradial, and superior radioulnar articulations within a common capsule.

HUMERUS

The humerus is the long bone of the brachium or anatomical arm, commonly referred to as the upper arm. This single bone articulates with the pectoral girdle proximally and with the two bones of the antebrachium (forearm, lower arm) distally. The humerus is moved by muscles which extend from the scapula and back posteriorly and from the clavicle and chest wall anteriorly. These muscles attach to the upper part of the bone while the distal part affords origin to muscles which cross the elbow joint to move the forearm.

Upper End

The *head* of the humerus is a smooth rounded articular surface which is tilted posteriorly and

medially to articulate with the glenoid fossa of the scapula. The head fits closely to the shaft with only the slightly depressed band of the *anatomical neck* intervening between it and two pillar-like projections for muscle attachments, the *tuberosities*. The *lesser tuberosity*, on the medial side of the anterior surface, receives the subscapularis muscle. The *greater tuberosity* bulges upward on the lateral surface. Three muscles insert upon it in the same order that they arise from the scapula. Highest is the supraspinatus, then the infraspinatus; lowest is the teres minor. Between the tuberosities on the anterior aspect is the long *bicipital groove* or intertubercular sulcus through which the tendon of the long head of the biceps brachii muscle passes. The upper end becomes cylindrical just below the tuberosities and is called the *surgical neck* because of the occurrence of fractures here in contrast to the anatomical neck.

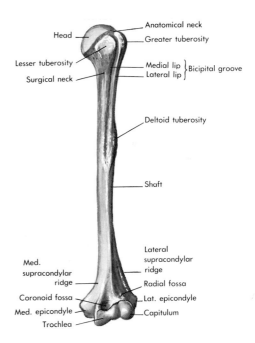

Shaft

The shaft of the humerus becomes cylindrical below the head. On the anterior aspect is the *bicipital groove*. Although a bicipital tendon occupies the groove, the tendon of the broad latissimus dorsi muscle of the back extends from the posterior aspect of the axilla to insert into the depths of the groove and along its medial wall.

The *medial lip* of the bicipital groove is roughly beaded by the attachment of the teres major muscle. The tendon of the pectoralis major muscle converges upon the groove from the anterior wall of the axilla to insert upon the rough *lateral lip* of the groove. The deltoid muscle tapers along the lateral aspect of the middle of the shaft to insert upon the rough raised *deltoid tuberosity*. At this level on the posterior surface the shallow *spiral groove* slants across the shaft downward and laterally, marking the course of the radial nerve and the deep brachial artery. The lower half of the shaft begins to flatten and widen.

Lower End

The lower end of the humerus, widened transversely and flattened anteroposteriorly, is sculptured by features relating to the elbow joint and to origins of muscles of the forearm. The lateral and medial borders of the shaft widen into the *lateral* and *medial supracondylar ridges,* which continue onto knobby expansions, the *lateral* and *medial epicondyles,* at the respective sides of the articular surface.

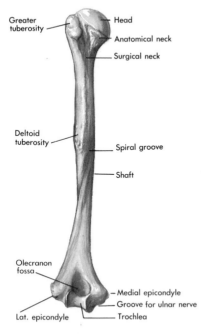

Greater tuberosity
Head
Anatomical neck
Surgical neck
Deltoid tuberosity
Spiral groove
Shaft
Olecranon fossa
Medial epicondyle
Groove for ulnar nerve
Lat. epicondyle
Trochlea

The lateral epicondyle and its supracondylar ridge provide the origin for a group of muscles which produce extension of the wrist and fingers of supination of the forearm and hand. The medial epicondyle and its supra-

condylar ridge give origin to another group of muscles with the opposite functions of flexing the wrist and fingers or pronating the forearm and hand.

The *articular surface* is divided into the rounded *capitulum* (small head) laterally and the scroll-like *trochlea* (pulley) medially. The depressed center of the head of one of the forearm bones, the radius, moves upon the rounded surface of the capitulum when the elbow is bent. The wrenchlike upper end of the other forearm bone, the ulna, surrounds the trochlea and moves upon it. The shallow *radial fossa* and the deeper *coronoid fossa* just superior to the capitulum and trochlea on the anterior surface receive the ends of the two forearm bones when the elbow is fully flexed.

The *posterior surface* is shaped much like the anterior. The medial epicondyle makes a prominent angle with the sharp edge of the trochlea. The ulnar nerve can be felt here as it winds behind the epicondyle. It is vulnerable to blows or pressure, with the familiar "crazy bone" tingling resulting. The posterior surface is deeply indented to form the *olecranon fossa* which receives the large jaw or olecranon process of the wrenchlike ulnar head when the elbow is extended.

Living Anatomy of the Humerus

Most of the humerus is overlaid with muscles which make palpation of its parts difficult. If one partially bends the right elbow, the tip of the left index finger can be placed on the prominent point of the elbow which is made by the olecranon of the ulna. The left thumb will naturally fall upon the medial epicondyle, which like the olecranon process is subcutaneous. The middle finger can be run across the olecranon and over the crevice between it and the posterior surface of the capitulum onto the subcutaneous surface of the lateral epicondyle. By deep pressure the thumb and middle finger can be brought superiorly along the epicondylar ridges. The shaft of the bone can be made out with deep pressure, but it will be noted that medially this is performed against ribbon-like nerves and vessels which are interposed between the surface and the bone. The greater tuberosity can be felt through the deltoid muscle just below the acromion process.

SCAPULOHUMERAL JOINT

The head of the humerus is an excellent example of a ball but the glenoid fossa of the scapula is a poor socket. The shallow glenoid fossa is more flat than concave and presents a much smaller articular surface than the larger head of the humerus.

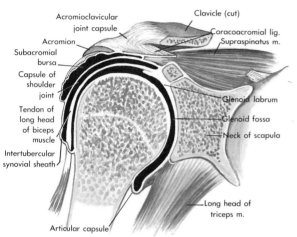

Although the fossa is widened and slightly deepened by a rim of fibrocartilage, the *glenoidal labrum,* the construction of the joint favors a wide range of movement at the expense of union and stability. The force of gravity is ever at work to separate the humeral head from the scapula. The articular capsule is lax in order to facilitate free movement. Neither the configuration of the articulating surfaces nor the disposition of the capsule furthers stability but other factors lessen the probability of dislocation.

Capsule and Capsular Ligaments

The articular capsule attaches to the margins of the glenoid fossa and around the anatomical neck of the humerus. Between these attachments the capsule broadens into a larger sac which loosely covers the head of the humerus. The synovial membrane follows the fibrous membrane closely. The internal surface of the fibrous membrane is thickened along the anterior aspect of the joint by three collagenous bands, the *glenohumeral ligaments.* The superior glenohumeral ligament crosses the capsule between the supraglenoid tubercle of the scapula and the lesser tuberosity of the humerus. Below it is the middle glenohumeral ligament. The inferior glenohumeral ligament stretches from the anterior

margin of the glenoidal labrum to the anatomical neck of the humerus.

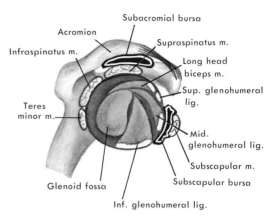

The capsule presents the openings of two extensions of its synovial membrane. One, the *intertubercular synovial sheath,* surrounds the tendon of the long head of the biceps brachii muscle. This synovial extension is strengthened, and the tendon kept in place, by the *transverse humeral ligament,* which crosses over the sheath between the greater and lesser tuberosities. The other synovial extension is a medial outpouching of the synovial membrane, the *subscapular bursa,* which reduces friction between the subscapularis muscle and the neck of the scapula. The bursa may be separate with no communication to the synovial cavity.

Accessory Ligaments

The *coracohumeral ligament,* extending over the shoulder joint from the coracoid process to the greater tuberosity, strengthens the joint superiorly. The *coracoacromial ligament,* extending in two bands between the coracoid process and the acromion, acts as a protective arch above the head of the humerus and the supraspinatus tendon.

Musculotendinous Cuff

It remains for some of the muscles which cross the shoulder joint and produce some of its movements to contribute the greatest stability. Four muscles approach the joint from the scapula. The subscapularis muscle passes in front of the joint. The supraspinatus crosses the joint superiorly. The infraspinatus muscle extends behind the joint. The teres minor muscle similarly is located posterior to the joint. These muscles and their tendons form a partly muscular, partly tendinous hood or cuff about the capsule that invests its anterior, superior, and posterior surfaces. The close investment of the capsule by these muscles and tendons has led to their designation as the *musculotendinous cuff* of the shoulder joint. These muscles hold the humerus to the glenoid fossa, strengthen the capsule, and flexibly resist undue movements of the humeral head in anterior, superior, and posterior directions. Since all these muscles produce rotations of the humerus, the functional structure is also referred to as the *rotator cuff.*

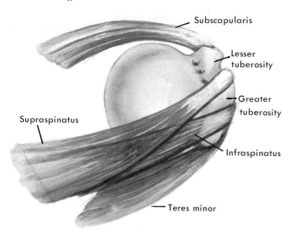

Other muscles contribute lesser degrees of stability. The tendon of the long head of the biceps brachii muscle presses against the upper end of the humerus as it passes along the intertubercular groove. The deltoid muscle assists in pulling the humerus toward the glenoid fossa as it produces various shoulder movements.

Bursae

The *subscapular bursa* has been mentioned. Another large bursa is placed between the acromion process and deltoid muscle externally and the musculotendinous cuff internally. This *subacromial bursa* facilitates movement between the internal and external structures. Instead of a single large bursa there may be separate smaller *subacromial* and *subdeltoid bursae.* The names are frequently interchanged in common usage. The subacromial bursa may acquire calcium deposits which painfully restrict shoulder movements. The subscapular muscle and its tendon are frequently irritated mechanically by the throwing motions of athletes.

Movements of the Shoulder Joint

The movements of the shoulder joint, their range increased by associated movements of the scapula, are movements of the upper limb. If, however, the upper limb is fixed in position (as in climbing or in chinning exercises), movements at the shoulder joint change the position of the trunk of the body. If the upper limb is rigidly braced and the muscles of the shoulder joint stabilize the articulation, the force generated by the muscles of the lower limbs and trunk may be transmitted effectively across the shoulder joint as in pushing an automobile.

The major movements which occur at the shoulder joint are:

1. *Flexion* in which the humerus is brought forward beside the thorax. Continued flexion carries the humerus upward as well as forward and, finally, upward and backward beside the head to a vertical position.

2. *Extension* in which the humerus is returned from any position of flexion to the anatomic position or is carried backward from the position beside the thorax.

3. *Abduction* in which the humerus moves laterally away from the body. Continued abduction carries the humerus upward as well as laterally and, finally, upward and medially to a vertical position beside the head.

4. *Adduction* in which the humerus is returned to the side of the body from any degree of abduction. The thorax prevents a further movement toward the midline of the body, but if adduction is combined with partial flexion the arms can be carried across the front of the chest and crossed. Since the glenoid fossa faces forward as well as laterally, a similar combination of adduction and extension is less free, but the same effect can be obtained by bending of the elbows as in clapping the hands behind the back.

5. *Rotations* in which the anterior aspect of the humerus turns medially (*medial* or internal *rotation*) or laterally (*lateral* or external *rotation*). Rotations combine with other movements. For instance, medial rotation occurs with and facilitates full flexion to the vertical position, and lateral rotation occurs increasingly with abduction to the vertical position.

6. *Combined movements* in which several movements occur simultaneously or progressively are the basis of most natural motions at the shoulder

joint. *Circumduction,* as a joint movement in general, was defined in terms of the shoulder joint. *Lateral rotation* of the scapula turns the glenoid fossa upward to make possible full abduction of the shoulder joint. Otherwise the humerus would impinge upon the acromion to stop the movement. *Retraction* of the scapula carries the glenoid fossa backward and turns it more laterally to facilitate extension of the shoulder joint. *Protraction* of the scapula occurs with shoulder flexion to turn the glenoid fossa forward.

BONES OF THE FOREARM

The *ulna* is the medial bone and the *radius* is the lateral bone of the forearm (antebrachium).

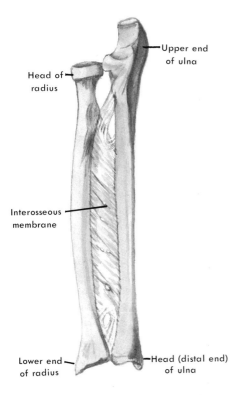

Head of radius

Upper end of ulna

Interosseous membrane

Lower end of radius

Head (distal end) of ulna

They lie side by side in the anatomical position, flexibly connected by the fibrous *interosseous membrane* which stretches between them. Both bones articulate with the humerus, but the more prominent upper end of the ulna is dominant in the formation of the elbow joint. The upper end of the radius is termed its head, whereas the head of the ulna is at its distal or lower end. The head of the radius articulates also with the upper end of the ulna and rotates against it in pronation and supination. The shafts of the bones are quite similar. They are triangular in cross section and have sharp interosseous borders which face each other across the interosseous membrane. The lower end of the radius is the larger and is the dominant member in the articulation with the wrist bones. The head of the ulna not only is the smaller but falls short of the actual wrist joint. Its articulation is by way of an intermediate fibrocartilage.

Pronation and Supination

Much of the dexterity and power of the hand would be lost if the upper limb were constructed to operate solely in the anatomical position. The natural position of the hands and wrists is midway between supination and pronation. Many movements require varying degrees of pronation. In this action the ulna remains almost stationary. The head of the radius rotates against the lateral side of the upper end of the ulna on its pivot against the capitulum. In full pronation the lower end of the radius swings in an arc anterior to the ulna so that the radial shaft crosses the ulna and the lower end comes to lie medial to the head of the ulna.

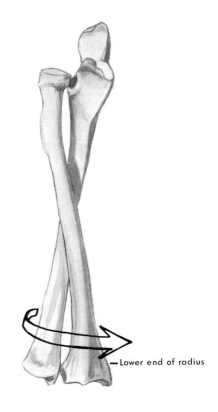

Lower end of radius

Ulna

The *upper end* resembles an open-end wrench. The upper and posterior jaw of the wrench is the proximal end of the shaft of the bone, the *olecranon process,* and the lower and anterior jaw is formed by a block of bone, the *coronoid process,* jutting anteriorly from the upper end of the shaft.

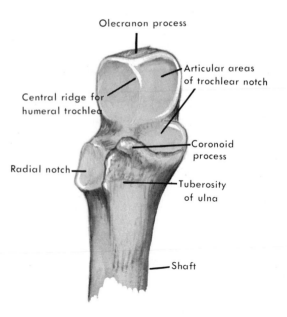

Olecranon process

Articular areas of trochlear notch

Central ridge for humeral trochlea

Coronoid process

Radial notch

Tuberosity of ulna

Shaft

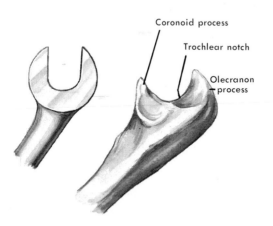

Coronoid process

Trochlear notch

Olecranon process

The two processes surround the trochlea of the humerus and are molded to its pulley-like shape to form the *trochlear notch* of the upper end of the ulna. Articular areas of the trochlear notch, therefore, are found on the anterior surface of the olecranon process and the superior surface of the coronoid process. Each is raised into a central ridge corresponding to the central depressed portion of the trochlea. When the forearm is fully extended, the sharp anterior margin of the olecranon slides into the olecranon fossa on the posterior aspect of the humerus. Full flexion brings the sharp anterior margin of the coronoid process into the coronoid fossa on the anterior aspect of the humerus. The superior aspect of the olecranon process receives the tendon of the triceps brachii muscle which extends the forearm. The posterior aspect is smooth. It lies just beneath the skin and forms the prominent point of the elbow. Below the coronoid process is a broad sloping surface, the *tuberosity of the ulna.* The tuberosity is greatly roughened for muscle attachments. The *radial notch* high on the lateral surface of the upper end of the ulna is the articular area upon which the head of the radius turns in pronation and supination.

The *shaft* of the ulna is triangular. The lateral or *interosseous border* provides attachment for the interosseous membrane. The rounded *anterior border* of the bone continues downward to end as a distal projection of the head, the *styloid process.* The *anterior surface* lies between these two borders. Its proximal two thirds is taken up by the origin of a large, deeply located muscle of the forearm which flexes (bends) the fingers. The distal third of the anterior surface gives origin to a pronator muscle which crosses the interosseous membrane to pull upon the adjacent radius. The *medial surface* of the shaft is along the medial and posterior aspects of the bone between the rounded anterior border and the *posterior border,* which is a sharp ridge running down the posterior aspect from the olecranon. The deep flexor muscle of the fingers spreads its origin from the anterior surface onto the upper two thirds of the medial surface. The *posterior surface* is found on the posterior and lateral aspects of the shaft between the posterior border and the sharp interosseous borders. The posterior surface is occupied by the successive origins of a group of muscles which produce varying movements of the thumb and wrist joint.

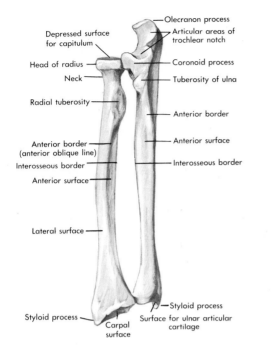

The *head* of the ulna is the rounded and knobby expansion of the anterior and posterior surfaces of the shaft. The medial surface continues onto the *styloid process*. The *ulnar articular cartilage* attaches to the distal surface of the head and intervenes between it and bones of the wrist joint. Along the lateral margin of the head is a rounded articular area on which the ulnar notch of the radius turns in pronation and supination.

Radius

The *upper end* bears the *head* which is a rounded platelike structure with a thick margin. The superior surface of the head is depressed centrally so that it can glide smoothly on the humeral capitulum in flexion and extension of the forearm and pivot on it in the rotatory movement of pronation and supination. The thick margin of the head also rotates against the radial notch of the ulna in the latter movement. Below this margin the upper end tapers into the cylindrical *neck* which is limited below by a prominent rough bulge, the bicipital or *radial tuberosity*. The tendon of the biceps brachii muscle of the arm attaches to this tuberosity.

The *shaft* of the radius becomes triangular below the tuberosity and shows a distinct lateral bowing which facilitates its crossing over the ulna in pronation and supination. The *anterior border* descends obliquely laterally from the radial tuberosity to become the smooth lateral edge of the bone when viewed from in front. The sharp *interosseous border* descends along the medial margin of the bone. The rather concave *anterior surface* extends distally between these borders from the tuberosity to the expanded lower end. The oblique upper course of the anterior border is also called the *anterior oblique line* of the radius. A large superficial flexor muscle of the fingers takes origin from this line. The supinator muscle comes across from the ulna to attach above the oblique line, while a flexor muscle of the thumb takes its origin from the upper two thirds of the anterior surface below the oblique line. The distal part of the anterior surface receives the attachment of a pronator muscle from the same area on the ulna. The *lateral surface* of the shaft merges with the rounded anterior border below the oblique line. In addition to the supinator attachment proximally, the lateral surface is roughened at the middle of the shaft for the attachment of another pronator muscle. The *posterior border* is only vaguely defined and the *posterior surface* is only vaguely delimited from the lateral surface. It is an origin of muscles to the thumb.

The *lower end* of the radius is a flared expansion of all the surfaces of the shaft. Their rolled margins surround a concave articular surface named the *carpal surface* where the radius articulates with the carpal (wrist) bones. The margin separates the carpal surface from the *ulnar notch* on the medial surface. The ulnar notch fits around the curved head of the ulna and moves upon it in pronation and supination. The lateral surface is prolonged into the radial *styloid process*.

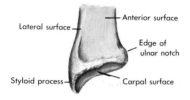

Lateral surface — Anterior surface

Edge of ulnar notch

Styloid process — Carpal surface

Living Anatomy of the Ulna and Radius

The posterior surface of the *ulna* can be felt subcutaneously from the olecranon at the point of the elbow along the entire forearm. The styloid process can be felt as a small knob at the back of the wrist. The head of the *radius* can be felt to move in pronation and supination distal to the lateral epicondyle of the humerus. The upper part of the radial shaft is overlaid by muscles, but all the surfaces of the lower shaft except the interosseous border can be felt. The styloid process can be noted to be at a lower level than its counterpart on the ulna.

HUMEROULNAR JOINT

The elbow joint consists of three articulations that are contained within a common capsule. All are synovial joints, the principal one being between the trochlea of the lower end of the humerus and the trochlear notch of the inner surfaces of the coronoid and olecranon processes of the ulna. The humeroulnar articulation is a hinge joint in which the coronoid process is carried into the coronoid fossa on the anterior aspect of the humerus in flexion and the olecranon process

moves into the olecranon fossa on the posterior aspect of the humerus in extension.

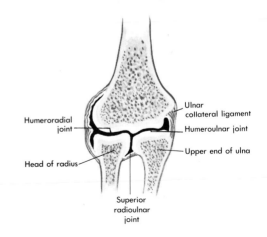

Humeroradial joint

Ulnar collateral ligament

Humeroulnar joint

Head of radius

Upper end of ulna

Superior radioulnar joint

Carrying Angle

The trochlea is tilted laterally so that the ulna articulates at an angle of about 15 degrees lateral to the long axis of the humerus. It can be readily seen that a similar lateral angulation of the forearm is produced in the anatomic position.

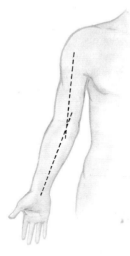

This *carrying angle* of the elbow permits the extended forearm to clear the side of the body in the movements of swinging the arms and carrying a load.

Joint Construction and Capsule

The scroll-like shape of the trochlea and the capitulum of the humerus make a good fit with the reciprocal forms of the articular surfaces of the trochlear notch of the ulna and the head of the radius. The construction of the ends of the articulating bones limits the movements to those of a hinge. Although the fit is good there is sufficient looseness, without loss of stability, to allow the ulna to make small sidewise adjustments during the movement of the radius in pronation and supination. The interlocking of the processes of the ulna with the fossae of the humerus lends stability to the humeroulnar articulation.

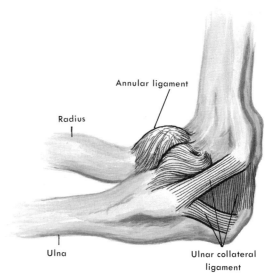

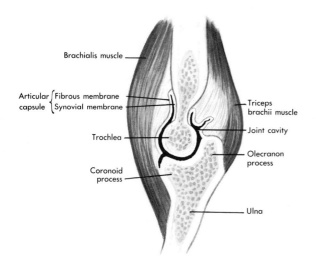

The *articular capsule* loosely covers the anterior and posterior aspects of all three joints of the elbow. The *fibrous membrane* attaches to the anterior and posterior surfaces of the humerus. Inferiorly, the fibrous membrane attaches to the anterior margin of the coronoid process. The *synovial membrane* follows the fibrous covering closely but at the superior attachments it turns back upon itself to line the fossae of the humerus.

Ligament

The capsule is thin and loose anteriorly and posteriorly to permit free hinge movements. The hinge is strengthened by the presence of collateral ligaments at each side of the joint. The triangular *ulnar collateral ligament* (medial ligament of the elbow) spreads out distally across the joint from the medial epicondyle to the medial sides of both the coronoid and olecranon processes. Its tightness prevents more than slight abduction at the hinge.

Movements

Flexion and extension are the principal movements at the humeroulnar joint. Flexion is arrested mainly by the apposition of the flexor surfaces of the forearm and arm but is finally stopped as the coronoid process engages the coronoid fossa. Extension is stopped at the fully straightened position of the elbow as the olecranon process impinges upon a fat pad of the olecranon fossa.

HUMERORADIAL JOINT

The radius accompanies the ulna in the hinge movements of the elbow joint. In these movements the concave superior surface of the head of the radius moves upon the round humeral capitulum. In pronation and supination it may also rotate upon the capitulum as a pivot at the same time that it is gliding around the capitulum in flexion and extension. The combination makes possible a variety of hand positions during flexion and extension of the forearm.

Capsule

The relations of the articular capsule and collateral ligament to the head of the radius are influenced by the presence of the annular ligament which loops around the radial head to hold it to the ulna and complete the sleeve in which it rotates. Firm attachment of the capsule to the radius would interfere with its rotation. Therefore the inferior attachment of the *fibrous membrane* is largely to the annular ligament instead of the bone. The *synovial membrane* lines the internal surface of the ligament and then turns back to the radial head to attach to the articular cartilage.

Ligament

The *radial collateral ligament* (lateral ligament of the elbow) is a strong, fan-shaped band which extends from the lateral epicondyle of the humerus to the lateral side of the annular ligament rather than to the radius, to avoid interference with the rotation of pronation and supination.

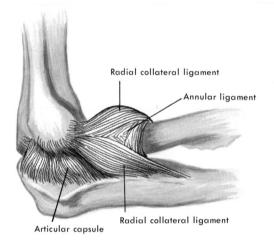

This collateral ligament prevents more than slight adduction at the hinge.

RADIOULNAR JOINTS

Three joints keep the radius in flexible alignment with the ulna but allow the radius to move around it with pronation and supination.

Superior Radioulnar Joint

The head of the radius articulates with the radial notch of the ulna which is in an anterior position on the lateral aspect of the coronoid process. The anterior location of the joint brings the upper portion of the shaft forward so that it can clear the ulna as it crosses over in pronation and supination. The thick margin of the radial head rotates in a fibro-osseous sleeve, the medial fifth of which is the cartilage-covered radial notch. The remainder of the sleeve is the syn-

ovium-lined *annular ligament*. In addition to guiding the rotation of the head of the radius, the annular ligament forms the superior point of union between the radius and the ulna.

Intermediate Radioulnar Joint

The radioulnar or antebrachial *interosseous membrane* forms a fibrous joint between the two bones. Collagenous fibers are arranged into a sheet which extends between the interosseous borders of the bones. The direction of the fibers is downward and medially. In addition to maintaining alignment of the shafts of the bones in the anatomic position, while going slack during pronation and supination, the interosseous membrane extends the area available for the attachment of muscles which arise from the surfaces of the bones contiguous to its margins.

Inferior Radioulnar Joint

Since the lower end of the radius swings in an arc about the head of the ulna, a freely movable joint is indicated between the rounded articular area on the ulnar head and the ulnar notch of the medial side of the lower end of the radius. The ulnar notch glides around the ulnar head as the radius swings about the ulna.

WRIST BONES

The movements of the arm and forearm at the shoulder and the elbow joints are like the coarse adjustments of a crane in bringing the hand into proximity with its point of actions. Pronation and supination bring the hand farther into a position for work. The delicate movements of the hand begin at the wrist or *carpus*. The hand, in the anatomical position, can be bent toward the anterior aspect of the forearm (flexed),

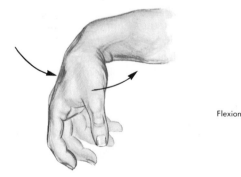

Flexion

straightened and bent toward the posterior aspect of the forearm (extension),

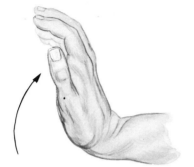

Extension

cocked medially toward the midline of the body (ulnar deviation or adduction),

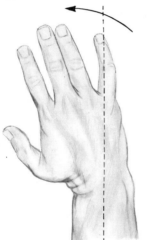

(Hand in pronation)

or cocked laterally away from the midline of the body (abduction or radial deviation).

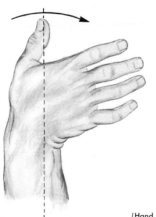

(Hand in pronation)

These movements may be performed singly or in combination. With pronation and supina-

tion added, the hand is given a remarkable range of motion upon the forearm.

Construction of the Wrist

The wrist is constructed of eight small, irregular *carpal bones* which fit together like pieces of a jigsaw puzzle. The bones are arranged in two rows of four each. The proximal row articulates with the radius and with the ulnar articular cartilage. In turn the bones of the first row articulate with the bones of the distal row which articulate with the bones of the hand. In addition each carpal bone articulates with the bones adjacent to it in its row. Since the articular surfaces presented by the radius and the ulnar cartilage are concave, the bones of the first row are arranged in a convex line. The bones of the distal row are longer to fit into the arch made by the proximal row and yet present a straighter distal articular border for the bones of the hand. The carpal bones as a whole are bound together so that the anterior surface of the carpus is concave forward and the posterior surface is slightly domed or rounded.

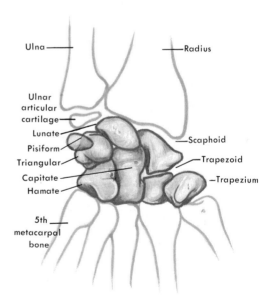

The Carpal Bones

A brief description of each carpal bone follows.
Proximal Row. The first bone from the lateral (thumb) side in the anatomic position is the *scaphoid* (navicular), named because of its fancied boat shape. Next medially is the *lunate* bone

which presents a semilunar appearance. Beside it medially is the *triangular* (triquetral) bone. The *pisiform* is a small bone, the approximate size and shape of a pea, which perches on the anterior and medial surfaces of the triangular bone.

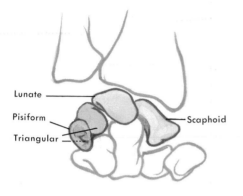

Distal Row. The first bone from the lateral side is the large irregular *trapezium* which, because of its many surfaces, was formerly called the greater multangular bone. It is placed farther laterally than the scaphoid and pushes forward a prominent beaklike ridge. The *trapezoid* bone next medially is a small many-sided bone, formerly called the lesser multangular, which fills in the odd crevices between the trapezium laterally, the scaphoid proximally, and the large *capitate* bone medially. The capitate, the third bone in this row, is named for its prominent rounded head which extends proximally to fill in the highest point of the convex arch of the first row. The fourth bone of the row is the *hamate,* only a small part of which reaches the medial side of the wrist because the triangular bone of the proximal row arches distally beside it. The hamate is named because of the prominent hook which it thrusts forward on the medial side of the wrist.

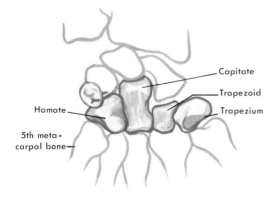

The general trough shape or forward concavity of the anterior surface of the carpus is accentuated by forward projections, two on each side of the wrist. On the lateral side is a tubercle on the lateral end of the scaphoid and the beak upon the trapezium. Medially the pisiform bone projects forward from the surface of the triangular bone and distal to it is the hook of the hamate. From these four points of suspension a fibrous band, the *flexor retinaculum,* stretches across the anterior surface of the carpus. It helps hold the concavity of this surface and, in addition, forms the fibrous roof of a tunnel which holds down the flexor tendons of the fingers as they cross the wrist.

Living Anatomy of the Carpal Bones

The anterior and posterior surfaces of the carpus are extensively overlaid by tendons and fibrous bands so that details of the bony structures beneath cannot be felt. On the anterior surface several skin creases, produced by folding of the skin during flexion, are apparent. By feeling at the medial (ulnar) side of the distal crease the small hump of the pisiform bone can be detected. The tubercle on the scaphoid bone can be felt at the lateral end of the same crease with deep pressure at the base of the thumb.

Wrist Joints

There are joints between adjacent carpal bones so that they are free to glide upon each other in adjusting to major movements of the wrist. The summation of these many small movements contributes to the motions of the wrist as a whole but the major movements of the wrist take place at the *radiocarpal* and *midcarpal joints.*

Radiocarpal Joints. At the radiocarpal joint movement takes place between the radius above and the scaphoid and lunate bones of the first row of carpal bones below. The ulna does not reach the wrist joint, but its articular cartilage extends the radiocarpal joint medially where the ulnar cartilage is opposed by the triangular bone and partly overlapped by the lunate. Therefore the radiocarpal joint can be said to be the wrist joint because the carpal bones carrying those of the hand and fingers can move upon the radius and ulnar disk as a unit. The articular capsule, lined internally by the synovial membrane, encloses the joint. The capsule is strengthened by the *ulnar* and *radial collateral ligaments* along the respective sides of the wrist and by *radiocarpal ligaments* which slant medially across both the palmar and dorsal surfaces of the joint from the radius to the proximal row of carpal bones.

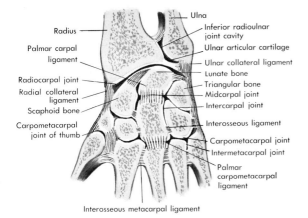

Interosseous metacarpal ligament

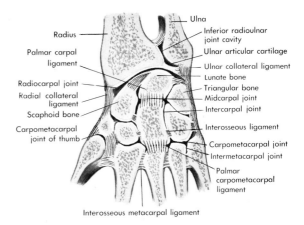

Interosseous metacarpal ligament

synovial cavity is reduced to a slit between the articulating bones. Movements between the two rows of carpal bones at the midcarpal joint provide greater flexibility of the wrist and supplement movements at the radiocarpal joint.

Small slitlike synovial cavities extend proximally and distally from the midcarpal joint between adjacent bones of both rows of carpal bones. These *intercarpal joints* permit gliding movements between adjacent bones. *Interosseous ligaments,* lined by synovial membrane, seal the intercarpal joints of the first row from the radiocarpal joint proximally and close off the intercarpal joints of the second row from the carpometacarpal joint distally. The interosseous ligaments maintain a flexible alignment of opposing bones. They are assisted along either side of the wrist by extensions of the collateral ligaments. A whole series of *intercarpal ligaments,* too numerous to name, crisscross between adjacent carpal bones on both palmar and dorsal surfaces to further assist the interosseous ligaments in maintaining flexible union and alignment.

Movements of the Wrist. The major movements occur together at the radiocarpal and midcarpal joints which supplement each other. In *flexion* the greatest bending motion is at the midcarpal joint with additional bending occurring at the radiocarpal articulation. In *extension,* straightening at the midcarpal joint supplements the greater radiocarpal movement. In *ulnar deviation,* the lunate bone slides fully into the concavity of the lower end of the radius from its anatomic position partly overlapping the ulnar articular disk. This leaves adequate room for the triangular bone to move freely upon the ulnar

These ligaments ensure that the hand will be carried with the radius in pronation and supination.

Carpal Joints. The *midcarpal joint* is between the first and second rows of carpal bones. Since the bones of the first row are arranged in a curving convex line to conform to the concavity of the radiocarpal joint, the midcarpal joint is made necessarily sinuous as the bones of the second row align themselves to those of the first row. The

disk so that ulnar deviation is completed at this joint. In *radial deviation,* however, the longer radial styloid process interferes with the free movement of the carpal bones. The two rows of carpal bones spread apart to widen the midcarpal joint in this movement and complete it, although the total movement is less than in ulnar deviation. Rotation at the radiocarpal joint cannot occur because of its condyloid nature which restricts movement to the two planes of flexion-extension and deviation. The combination of these movements results in *circumduction,* which in association with pronation and supination provides an extensive range of motion at the wrist.

BONES OF THE HAND

The five bones of the hand (metacarpus) are long bones in miniature. The *metacarpal bones* are remarkably alike and, except for the metacarpal of the thumb, differ mainly in length. The second to fifth metacarpals undergo little independent movement. They move *with* the carpal bones rather than upon them. In this way the metacarpals form a strong framework for the hand and a firm base for movement of the finger bones. However, the mobile first metacarpal bone gives the thumb a wide range of motions.

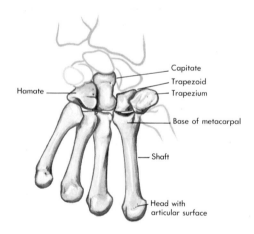

Typical Metacarpal

Each of the second to fifth metacarpal bones has a squared proximal end or *base* which articulates with bones of the distal row of the carpus at the carpometacarpal joints. The second meets the trapezoid and a medial portion of the trapezium, the third articulates with the capitate, the

fourth abuts the hamate and a medial bit of the capitate, while the fifth forms its joint on the lateral part of the hamate. The bases closely abut each other and are tied together by ligaments. The *shafts* taper to slender columns which are slightly curved toward the ventral surface. Rounded knobby *heads* form the prominent knuckles. Since there is very little extension of the fingers on the metacarpals past the plane of the hand, the *articular surfaces* of the heads are located on the distal and palmar surfaces only.

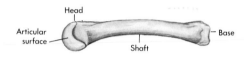

When the fingers fold in flexion their bones slide onto the palmar part of the articular surfaces. When the fingers extend to the plane of the hand from flexion the distal part of the articular surface is used.

First Metacarpal

The metacarpal bone of the thumb is shorter and broader than the others and its position is different from the rest. The thumb metacarpal has been carried over to the anterolateral side of the wrist and turned to a plane at right angles to the other metacarpals so that its anterior surface faces medially and its posterior surface laterally.

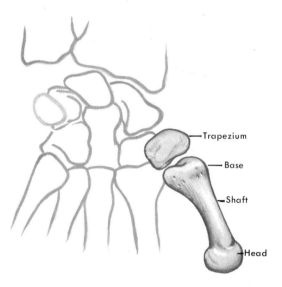

The concavity of the ventral surface of the carpus places the trapezium, with which the thumb metacarpal articulates, forward ahead of the other carpal bones. The articular surface of the trapezium is shaped like a saddle with sloping sides to receive the somewhat trough-shaped base of the first metacarpal. The turned plane of the bone, its forward position on the wrist, and the mobility of its joint confer upon the thumb its great range of motion. The movements will be described with the thumb muscles whose disposition must be understood to give meaning to the movements.

Living Anatomy of the Metacarpals

The bases of the metacarpals cannot be distinguished from the bones of the wrist. The shafts are felt easily along the posterior surface of the hand amidst the extensor tendons; the heads can be felt and seen as the "knuckles" when the fist is clenched.

CARPOMETACARPAL JOINTS

A common carpometacarpal joint exists between the second to fifth metacarpal bones and the distal row of carpal bones. The thumb has a separate joint.

Common Carpometacarpal Joint

The articular relation of each of the metacarpal bones to the carpal bones has been described. The long crevice of the synovial common carpometacarpal joint winds between the ends of the articulating bones. *Intermetacarpal joints* extend a short distance distally from the common cavity between the bases of the metacarpal bones. The bases of the metacarpal bones are bound closely together by firm *interosseous metacarpal ligaments.* The articular capsule of the common joint and its extensions is strengthened by numerous *dorsal* and *palmar carpometacarpal ligaments* and by *dorsal* and *palmar metacarpal ligaments* which cross the many joint crevices concerned. These small ligaments limit the movement of the metacarpal bones upon each other and ensure that these bones of the hand will move with the carpal bones as a unit.

Carpometacarpal Joint of the Thumb

This articulation is entirely different from the common joint because the thumb metacarpal has

been turned to meet its carpal bone at a plane at a right angle to the other metacarpals. This joint has its own synovial cavity, a separate articular capsule, and mobility not shared by the other metacarpal bones. The base of the thumb metacarpal faces the trapezium at the anterior and lateral extremity of the distal row of carpal bones. There is no intermetacarpal ligament between the thumb and second metacarpal, and the articular capsule of the thumb joint is loose as should be expected where free movement exists.

BONES OF THE FINGERS

The finger bones are also long bones, smaller than the metacarpals. An individual finger bone is called a *phalanx.* Collectively they are known as *phalanges.* The fingers have three bones, the *proximal, middle,* and *distal phalanges,* whereas the thumb has only two. The *base* of a proximal phalanx is concave for articulation with the rounded head of the metacarpal.

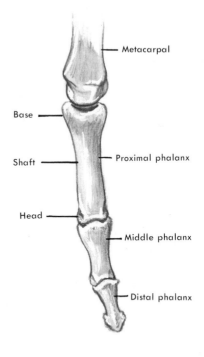

The base tapers to a slender shaft which presents a rounded, pulley-like *head.* The base of the shorter middle phalanx is raised at the center to conform to the head of the proximal phalanx with which it articulates. The nature of this joint permits flexion and extension but prohibits motion to the sides. Spreading of the fingers, therefore, must occur at the metacarpophalangeal joints. The short shaft of the middle phalanx terminates in a head similar to that of the proximal bone. The base of the distal phalanx is molded to the pulley surface of the middle bone.

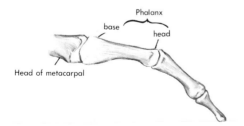

A short narrow shaft terminates in a flat, spade-shaped head which supports the tissues of the finger tip. The surfaces of the phalanges are covered by fibrous sheaths, hoods, and tendons by which muscles attach to operate the fingers.

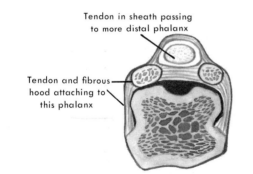

METACARPOPHALANGEAL JOINTS

Important movements of flexion and extension occur at the articulation of the knobby heads of the metacarpal bones with the bases of the proximal phalanges of the fingers. These are simple synovial joints in which the ends of the articulating bones are enclosed by a thin articular capsule. The concave bases of the proximal phalanges slide around onto the palmar surface of the metacarpal heads in *flexion*. In *extension* the phalangeal bases glide dorsally onto the distal surface of the metacarpal heads but not past the plane of the hand. The curving of the articular surface slightly onto the sides of the heads of the second, fourth, and fifth metacarpal heads allows the corresponding fingers to be moved away from or toward the line of the middle finger. Movement

away from this line is *abduction*. Movement toward it is *adduction*. *Deep transverse metacarpal ligaments* extend from the head of each of the four metacarpal heads (other than the thumb) to the adjacent heads.

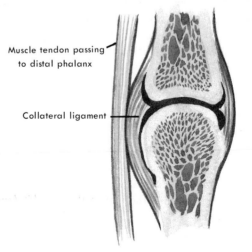

The tautness of these ligaments precludes adduction of the metacarpals and thus makes these bones a firm stable base for similar movements of the fingers. Similarly the ligaments, by holding the metacarpals steady during flexion of the metacarpophalangeal joints, greatly increase the strength of finger flexion. This fixation of the joints during flexion is aided by *collateral ligaments* along each side of the joints, which tighten during flexion and draw the ends of the bones closely together.

INTERPHALANGEAL JOINTS

The pulley-like head of a proximal phalanx moves upon the domed base of a distal phalanx at a simple synovial joint. The hinge character of the interphalangeal joints indicates that the movements are those of flexion and extension. Strong *collateral ligaments* on each side of the joint aid the conformation of the articulating surfaces in precluding lateral movements so that all spreading of the fingers is performed at the metacarpophalangeal joints.

PELVIC GIRDLE, HIP BONES, AND PELVIS

The upper limb is light, free, and mobile. It has been seen to have an equally free attachment to the axial skeleton through the pectoral girdle. At the pelvic girdle different factors are operating which require a sturdier attachment of the lower limb. The erect position places greater forces upon the lower elements of the axial skeleton. In weight-bearing and in locomotion these must be transmitted to the lower limb, yet movement must be possible between the limb and the trunk. In addition the abdominal and pelvic organs must be supported from below and protected. The pelvic girdle performs functions not only similar to those of the scapula and clavicle against greater forces but also similar to those of the sternum and ribs.

PELVIC GIRDLE

The pelvic girdle consists of the two *hip bones* which meet each other anteriorly and articulate with the sacrum of the axial skeleton posteriorly. Each hip bone is composed of three bones which fuse together—the *ilium, ischium,* and *pubis.* Because of its composite formation the hip bone was once called the *innominate bone.* The fusion of the individual bones greatly restricts movement within the girdle itself. Limited movement occurs anteriorly between the hip bones where they are joined by a fibrocartilaginous disk at the *symphysis pubis.* Otherwise rigidity of the girdle is characteristic.

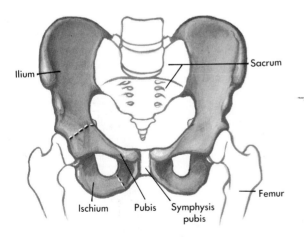

The *bony pelvis* is formed by the two hip bones with the sacrum as a keystone between them. This structure, a composite between the sacral part of the axial skeleton and the pelvic girdle, may be compared to the thoracic vertebrae, ribs, and sternum in that it encloses and protects vital organs. The pelvis in the female protects and supports the uterus during pregnancy and makes up the bony boundaries of the birth canal. It also provides bony points of attachment for the muscles and ligaments of the abdominal wall superiorly and for those of the lower limb inferiorly.

HIP BONE

The hip bone may be compared to an electric fan complete with a hub and two twisted blades.

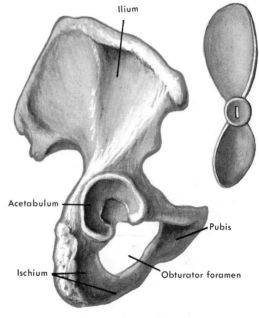

The hub is a cup-shaped block of bone, the *acetabulum.* It is the socket which receives the head of the thigh bone at the hip joint. The superior blade of the fan, the *ilium,* flares upward and outward from the acetabulum to its broad thick crest which one can feel above the hips in the familiar hands-on-hips position. The *pubis* is an irregular arch which is directed forward to meet its fellow of the opposite side at the symphysis pubis. One limb of the pubis originates from and contributes to the acetabulum. The other limb turns inferiorly at the midline to form its part of the symphysis and fuses with the third bone, the *ischium.* The U-shaped ischium arises from its part of the acetabulum to arch inferiorly and posteriorly into the depth of the pelvic region. It rises anteriorly as the other limb of the

"U" to meet and fuse with the pubic bone below the symphysis. The two pubic bones enclose the lower portion of the trunk anteriorly. The ischium with its fellow on the opposite side forms the prominent bony points upon which one sits. The limbs of the pubis and ischium enclose an aperture in the lower blade of the hip bone, the *obturator foramen,* which is closed in life by a fibrous membrane.

Acetabulum

The socket of the acetabulum faces laterally, anteriorly, and inferiorly. It presents a thin rounded margin except inferomedially where the rim of the socket is deficient at the *acetabular notch.* The acetabulum is formed by all three bones which have been molded around the head of the thigh bone. The anteromedial fifth faces the pubis and is formed by it. The inferior two fifths represent the acetabular portion of the ischium. The superior two fifths of the acetabulum are contributed by the ilium. Most of the acetabulum is smoothed by the articular surface which is shaped like a horseshoe. The nonarticular portion of the acetabulum is depressed between the limbs of the horseshoe and its open end to form the *acetabular fossa* which is continuous below with the acetabular notch in the rim.

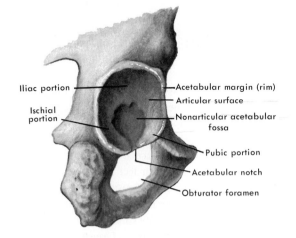

Iliac portion
Ischial portion

Acetabular margin (rim)
Articular surface
Nonarticular acetabular fossa
Pubic portion
Acetabular notch
Obturator foramen

The head of the thigh bone glides upon the articular area which presses heavily upon it to transmit the forces of weight-bearing from the ilium. A plate of cartilage which builds up the socket is affixed to the internal surface of the rim of the acetabulum, and ligaments strengthening the capsule

surrounding the joint attach to the external surface.

Ilium

The iliac portion of the hip bone expands from the acetabulum into a broad flaring plate of bone. The ilium presents a broadly convex lateral surface which faces the gluteal region or buttock and a mainly concave medial surface which faces the lower abdomen and pelvis. These surfaces are limited superiorly by the long *iliac crest.* The crest ends anteriorly in the *anterior superior iliac spine* to which is attached the strong *inguinal ligament* of the anterior abdominal wall. Below this spine the ilium presents a bulge known as the *anterior inferior iliac spine* which affords attachment to several muscles of the front of the thigh. Muscles of the anterior and lateral abdominal wall attach to the surface of the iliac crest. The iliac crest terminates posteriorly in the *posterior superior iliac spine.* From this point the posterior border of the bone descends in a forward direction to the *posterior inferior iliac spine.* The posterior border and its spines provide attachment for ligaments extending to both the sacrum and the ischium. The remainder of the posterior border passes sharply forward from the lower spine to the acetabulum. It is deeply indented on its way by the *greater sciatic notch.*

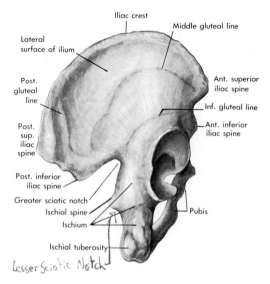

Iliac crest
Middle gluteal line
Lateral surface of ilium
Post. gluteal line
Ant. superior iliac spine
Inf. gluteal line
Post. sup. iliac spine
Ant. inferior iliac spine
Post. inferior iliac spine
Greater sciatic notch
Ischial spine
Pubis
Ischium
Ischial tuberosity
Lesser Sciatic Notch

The lateral surface of the ilium is marked by three lines which radiate outward in order from the acetabulum. These are the *inferior, middle,* and *posterior gluteal lines* which mark the attachments of muscles of the buttock. The *medial surface* of the ilium is divided into two areas. The anterior two thirds is the shallowly concave *iliac fossa*, the borders of which converge upon the superior part of the acetabulum where the junction of the ilium and pubis is marked by the *iliopectineal eminence*. The iliac fossa is largely taken up by the origin of the iliacus muscle which passes in front of the hip joint to the upper end of the thigh bone. The posterior third of the medial surface is roughened in its lower part into the *auricular surface* for the articulation of the ilium with the sacrum and in its upper part into the *iliac tuberosity* where ligaments of the sacroiliac joint attach. The *iliopectineal line* extends anteriorly from the auricular surface to the iliopectineal eminence. The ilium continues below this line to border the greater sciatic notch and become continuous with the ischium.

Ischium

The ischial portion of the hip bone can be followed inferiorly from its junction with the ilium on the medial surface. It is a broad column of bone made bulky on the opposite lateral surface by its contribution to the acetabulum. The medial border of the ischium forms the forward margin of the greater sciatic notch which is limited inferiorly by a medial projection, the *ischial spine*, to which a ligament from the sacrum attaches. The *lesser sciatic notch* indents the posterior margin below the ischial spine. The posterior surface of the bone is formed into the large *ischial tuberosity* to which the posterior thigh muscles and the heavy sacrotuberous ligament of the pelvis attach. Anterior to the tuberosity the ischium sends the slender flattened *ischial ramus* forward and superiorly to meet the *inferior ramus* of the pubic bone. The two rami are together termed

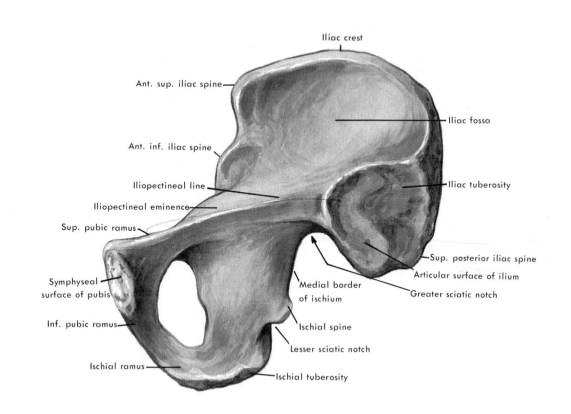

Iliac crest

Ant. sup. iliac spine

Iliac fossa

Ant. inf. iliac spine

Iliac tuberosity

Iliopectineal line

Iliopectineal eminence

Sup. pubic ramus

Sup. posterior iliac spine

Articular surface of ilium

Symphyseal surface of pubis

Medial border of ischium

Greater sciatic notch

Inf. pubic ramus

Ischial spine

Lesser sciatic notch

Ischial ramus

Ischial tuberosity

the *ischiopubic* or *conjoined ramus.* Muscles of the thigh and hip joint arise from lateral and medial surfaces of the ischium and its ramus.

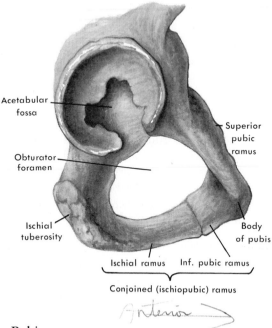

Acetabular fossa

Obturator foramen

Ischial tuberosity

Superior pubic ramus

Body of pubis

Ischial ramus   Inf. pubic ramus

Conjoined (ischiopubic) ramus

Anterior

Pubis

The hip bone is completed anteriorly by the pubis, which consists mainly of two bars of bone, the *superior* and *inferior pubic rami.* The superior ramus forms the pubic portion of the acetabulum and extends medially over the obturator foramen. The short inferior ramus rises from its junction with the ischial ramus. The two pubic rami meet at the midline in the *body* of the pubic bone. The medial surface of the body is roughened and is termed the *symphyseal surface,* for it is at this point that the pubic bone meets its fellow of the opposite side at the symphysis pubis.

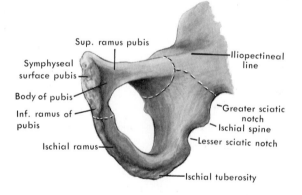

Sup. ramus pubis

Symphyseal surface pubis

Body of pubis

Inf. ramus of pubis

Ischial ramus

Iliopectineal line

Greater sciatic notch

Ischial spine

Lesser sciatic notch

Ischial tuberosity

JOINTS OF THE PELVIC GIRDLE

The joints of the pelvic girdle are the two sacroiliac joints and the symphysis pubis. These joints are constructed so that very little motion can occur.

SACROILIAC JOINT

The sacroiliac joint is one of the most important joints of the body because it must bear the tremendous loads which are placed upon it.

Construction of the Joint

The auricular surface of the ilium articulates with the auricular surface of the sacrum. The articulating surfaces are roughened into many uneven eminences and depressions which fit into reciprocal forms on the opposite bone. This dovetailing stabilizes the joint and interlocks the bones to reduce movement. The close apposition of the articular surfaces reduces the synovial membrane and fibrous capsule to a minimum.

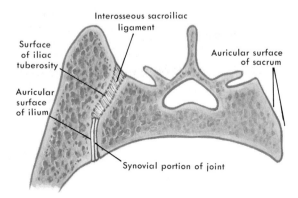

Interosseous sacroiliac ligament

Surface of iliac tuberosity

Auricular surface of ilium

Auricular surface of sacrum

Synovial portion of joint

The synovial portion is but one part of the sacroiliac joint. A fibrous union is also effected by the short heavy collagenous fibers of the *interosseous sacroiliac ligament* which crosses the narrow cleft that lies posterior and above the synovial auricular area between the sacrum and the iliac tuberosity. The interosseous ligament is both a fibrous joint and an intrinsic ligament. It counters any tendency of the sacrum to rotate upon the two hip bones.

Ligaments of the Joint

The load that is placed upon the sacrum tends to cause it to rotate between the hip bones. The tendency to rotation is always present despite the interlocking of the synovial auricular surfaces and the interosseous fibrous union. Intensely strong, extrinsic collateral and bracing ligaments extend between the sacrum and portions of the hipbone to reinforce the sacroiliac joint.

Sacroiliac Ligaments. The periosteum of the ala of the sacrum thickens into a collagenous band which fans out across the anterior aspect of the sacroiliac joint as the *anterior sacroiliac ligament.* Posteriorly a series of bands from the fused transverse tubercles of the sacrum collect into the *posterior sacroiliac ligament,* which bridges the joint line and attaches to the iliac tuberosity and posterior superior iliac spine. Above the level of the joint, the *iliolumbar ligament* connects the transverse process of the fifth lumbar vertebra with the iliac crest. All these collateral ligaments tighten the sacroiliac joint.

Functional Factors

The strong sacroiliac joint is not easily strained by the load placed upon it or by unusual physical activities. In pregnancy circulating hormones soften the interosseous and extrinsic ligaments. This process, associated with softening of the cartilage of the symphysis pubis, results in looseness and movement of the joints of the pelvic girdle.

SYMPHYSIS PUBIS

The pubic portions of the hip bones meet at the midline of the pelvis to form the anterior joint of the pelvic girdle, the *symphysis pubis.* The articulation is a true cartilaginous symphysis in which the symphyseal surface is united to its fellow on the opposite side by the firm fibrocartilage of the *interpubic disk.* The flexible union of the pubes is reinforced by the *superior pubic ligament,* which crosses the joint superiorly between the pubic tubercles, and by the *arcuate pubic ligament,* which arches under the inferior border of the disk between the pubic rami.

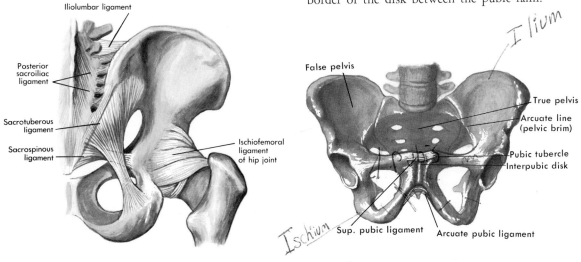

Iliolumbar ligament

Posterior sacroiliac ligament

Sacrotuberous ligament

Sacrospinous ligament

Ischiofemoral ligament of hip joint

False pelvis

I lium

True pelvis

Arcuate line (pelvic brim)

Pubic tubercle

Interpubic disk

Ischium — Sup. pubic ligament

Arcuate pubic ligament

Sacrotuberous Ligament. The lower end of the sacrum is braced, to prevent its backward rotation, by the *sacrotuberous ligament.* This is a thick band running from the iliac tuberosity and the posterior surface of the lower sacrum and coccyx to attach to the medial aspect of the ischial tuberosity.

Sacrospinous Ligament. A similar *sacrospinous ligament* extends from the lateral borders of the lower sacrum and coccyx to attach to the spine of the ischium.

PELVIS

The *pelvis* consists of the two hip bones and the sacrum which is wedged between the two bones posteriorly. Two different pathways through the pelvis are taken in the transmission of force and weight. When the body is erect the path is through the ala of the sacrum, the ilium, to the acetabulum and the head of the thigh bone.

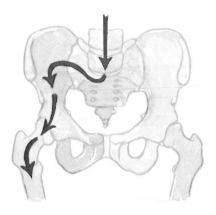

If the body is in the sitting position the pathway passes inferiorly through the ilium to the ischium and its tuberosity. The ischial tuberosity is cushioned by a bursa and a pad of fat.

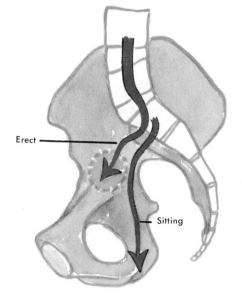

Construction of the Pelvis

The pelvis is divided by the *arcuate line* into the *false pelvis* and the *true pelvis*. The arcuate line can be followed forward from the sacral promontory onto the ala of the sacrum, across the medial surface of the ilium, and through the iliopectineal eminence onto the superior pubic ramus. This broad line of demarcation is also called the *pelvic brim* and is the inlet to the true pelvis. Superior to the arcuate line is the lower abdomen with its organs supported by the medial surface of the ilium and by abdominal muscles attaching to its crest and to the superior margin of the pubis. The *true pelvis* lies below the arcuate line. It surrounds the *pelvic cavity* which contains the lower organs of the alimentary and genitourinary systems. The bony boundaries of the true pelvis are the sacrum and coccyx posteriorly, the part of the ilium below the arcuate line and the ischium laterally, and the ischial and pubic rami anteriorly. Whereas the inlet to the pelvic cavity is bounded by the pelvic brim, the outlet of the bony pelvis is formed by a series of bony points. These are the coccyx posteriorly; the greater sciatic notches, ischial tuberosities, and lesser sciatic notches laterally; and the pubic arch anteriorly. The pubic arch is formed by the ischiopubic rami meeting at the symphysis pubis.

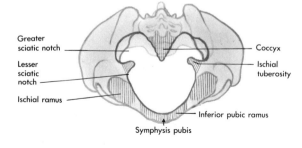

Gaps between bony points marking the outlet are filled in by ligaments. The *sacrotuberous* and the *sacrospinous ligaments* convert the greater and lesser sciatic notches into the corresponding *greater* and *lesser sciatic foramina* through which pass the muscles, vessels, and nerves of the gluteal region and thigh. In the living person the pelvic outlet is closed by a series of muscles and fibrous membranes which, with the soft tissues of the skin and genital organs, constitute the *perineum*.

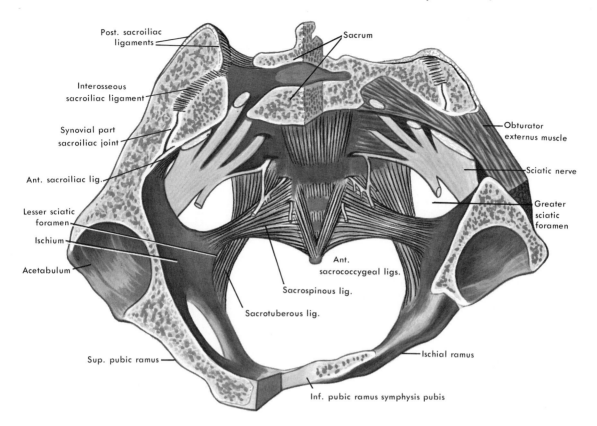

Post. sacroiliac ligaments

Sacrum

Interosseous sacroiliac ligament

Synovial part sacroiliac joint

Obturator externus muscle

Ant. sacroiliac lig.

Sciatic nerve

Lesser sciatic foramen

Greater sciatic foramen

Ischium

Acetabulum

Ant. sacrococcygeal ligs.

Sacrospinous lig.

Sacrotuberous lig.

Ischial ramus

Sup. pubic ramus

Inf. pubic ramus symphysis pubis

Sex Differences

The male and female pelves differ in general respects and in characteristics concerned with the relation of the female pelvis to the birth canal. The *male pelvis* tends to be larger, heavier, and more coarsely marked by ligamentous and muscle attachments. The pelvic cavity is deeper, its side walls straighter, and the sacrum and coccyx project more anteriorly than is the case in the female pelvis, to make a heart-shaped outline. The pelvic outlet is narrow with the limbs of the pubic arch closer together.

The *female pelvis,* lighter and more delicate in its bony framework, is more shallow. The outline of the pelvic cavity is round or oval.

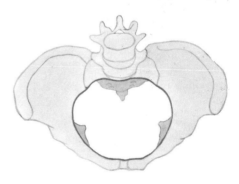

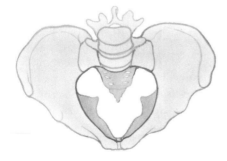

The sacrum and coccyx are flatter. They do not project as far forward as in the male and do not impinge on the birth canal. The side walls of the shallower female pelvic cavity are wider apart because of a turning out of the ischial tuberosities, longer pubic rami, and a wider angle with the pubic arch. The wider pubic arch associated with a wider sacrum produces a wider pelvis with hip bones set farther apart. The upper ends of the thigh bones are similarly farther apart. These factors are responsible for the wider curvature of the hips which is characteristic of the female.

Living Anatomy of the Pelvis

The iliac crest is easily located in the hands-on-hips position. The crest can be followed anteriorly to the anterior superior iliac spine. The ischial tuberosity is felt easily through the muscles of the buttock by upward pressure in the sitting position. A skin dimple just lateral to the sacrum and at the posterior end of the iliac crest marks the location of the palpable posterior superior iliac spine.

LOWER LIMB

Comparison of the bones of the lower limb with those of the upper limb provides a striking contrast. The upper limb and its pectoral girdle gain mobility and dexterity at the expense of stability; the lower limb and the pelvic girdle are characterized by sturdiness and stability. These qualities are necessary to receive and transmit the forces of weight-bearing at the same time that the lower limbs participate in the movements of the body. The movements are simpler but powerful, and the size of the muscles concerned contributes to the greater mass of the lower limb. The proximal portion of the lower limb is the "thigh." Only the portion between the knee joint and the ankle is referred to in anatomy as the "leg."

FEMUR

The *femur,* previously referred to as the thigh bone, is the longest and largest of the long bones. It possesses an upper end, a long shaft, and a lower end. Since the hip joints are set rather wide apart by the construction of the pelvis, the femur passes obliquely medially from the hip joint to the knee.

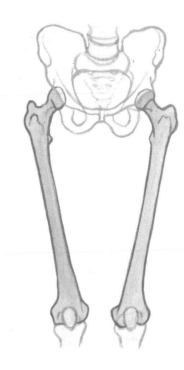

Upper End

The upper end of the femur is constructed for articulation with the hip bone and for the attachment of muscles which extend from the pelvis to the femur to operate the hip joint. Since the acetabulum faces laterally, downward, and inferiorly, the upper end of the femur is bent upward and medially into a long *neck* which brings the rounded *head* into a ball-and-socket joint with the acetabulum. The entire surface of the head, slightly greater than a hemisphere, is formed into the smooth globular *articular area.* The articular area is broken only by the *fovea,* a circular depression with slightly roughened edges, located on the medial surface of the head. (See illustration at bottom of page 130.) The fovea is for the attachment of the short ligament of the femoral head which helps hold the head into the acetabulum and carries blood vessels into the head. The head quickly tapers into the thick *neck.* The tapered part is grasped firmly by the cartilage that builds up the edges of the acetabulum.

A prominent irregular process, the *greater trochanter,* rises from the lateral aspect of the upper end of the shaft. A blunt projection, the *lesser trochanter,* extends posteromedially from the medial side of the bone below the neck. The trochanters form large areas for the attachment of many of the muscles which extend to the upper end of the femur from the pelvis to move the hip joint.

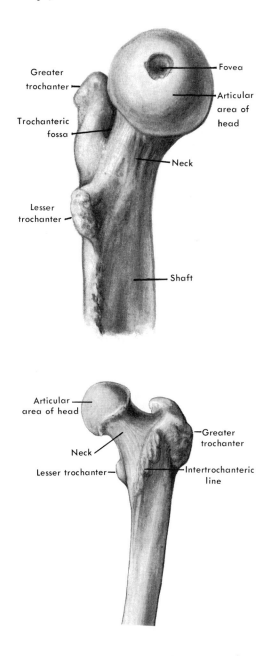

Greater trochanter

Trochanteric fossa

Lesser trochanter

Fovea

Articular area of head

Neck

Shaft

Articular area of head

Neck

Lesser trochanter

Greater trochanter

Intertrochanteric line

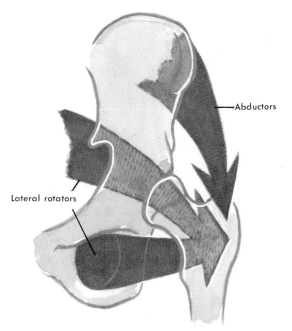

Abductors

Lateral rotators

A coarse *intertrochanteric line* on the anterior surface of the bone between the trochanters demarcates the neck of the bone from the shaft. The *intertrochanteric crest* is a ridge on the posterior surface between the two trochanters which delimits the neck from the shaft posteriorly. A deep excavation on the posteromedial aspect of the greater trochanter, the *trochanteric fossa,* receives one of the lateral rotator muscles.

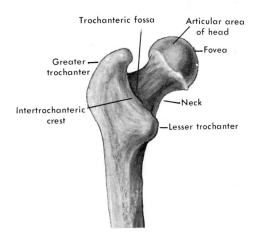

Trochanteric fossa

Articular area of head

Fovea

Greater trochanter

Intertrochanteric crest

Neck

Lesser trochanter

As a general rule the muscles attaching to the greater trochanter pull in such a way as to swing the thigh away from the midline (abduction) or turn the femur outward (lateral rotation). The lesser trochanter receives the attachment of a powerful compound muscle from the lumbar vertebral column and from the iliac fossa of the ilium which pulls the femur anteriorly.

Shaft

The long shaft is smooth, rounded, and almost without landmarks. It is bowed anteriorly and is marked posteriorly by a long, raised, and beaded ridge, the *linea aspera,* which forks superiorly to run to each trochanter. The entire anterior, medial, and lateral surfaces of the upper three-quarters of the shaft and the edges of the linea aspera are given over to the origin of a group of muscles which cross the knee joint to extend (straighten) the knee. The linea aspera provides a long line of attachment for the adductor group of thigh muscles which draw the femur toward the midline. Posteriorly another group of muscles, the hamstrings, extend across the back of the knee joint to flex (bend) the knee.

Lower End

At the lower end of the femur the shaft expands into two smooth curved surfaces, the femoral *condyles,* for articulation with the tibia. The condyles are best appreciated from the posterior aspect where the linea aspera divides into the *medial* and *lateral supracondylar lines* which enclose the posterior surface of the lower end of the bone. This triangular surface is also called the *popliteal surface* as it forms the bony floor of part of the popliteal fossa at the back of the knee joint. The condyles are raised like two large casters with the deep *intercondylar fossa* between them.

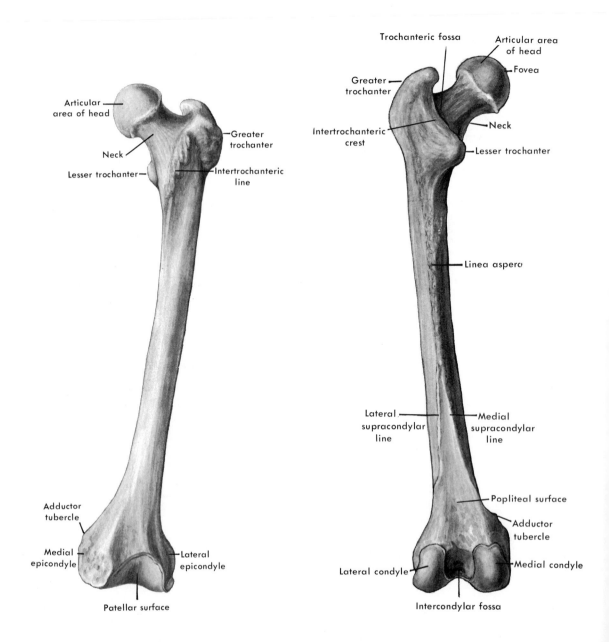

The condyles turn forward over the end of the bone and blend together to limit the fossa anteriorly. The confluent articular surface then turns superiorly as the *patellar surface*. This surface is concave to receive the sesamoid bone, the *patella*, which lies in front of the knee joint and forms the "knee cap." The medial and lateral surfaces of the lower end of the femur are formed into the *medial* and *lateral epicondyles* which are positioned like hubcaps on the casterlike condyles. The medial supracondylar line extends the line of attachment of the adductor muscles of the thigh downward from the linea aspera to an elevation at the superior margin of the medial epicondyle which is called the *adductor tubercle*.

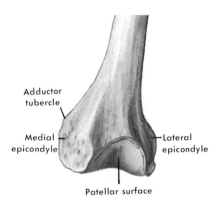

Living Anatomy of the Femur

The shaft and much of the upper end of the femur are masked by the heavy muscles of the thigh and hip. The greater trochanter can be felt through the muscles at the side of the hip. The lower end of the femur is much more palpable because few muscles arise from it and those of the thigh which cross the knee joint are becoming tendinous at this level. The patella can be felt in front of the knee joint and, on each side of it, the joint crevice between the femoral condyles and the tibia. The epicondyles form squared bony prominences on each side. Posteriorly the tendons of the hamstring muscles delimit the popliteal fossa.

HIP JOINT

The hip joint combines all the typical components of a synovial joint into an effective mechanism for support, transmission of weight, and movement.

Construction of the Joint

A perfect example of a synovial ball-and-socket joint makes possible the articulation of the head of the femur with the acetabulum of the hip bone. The femoral head moves upon the articular cartilage of the acetabulum and upon the *transverse ligament* of the acetabulum which bridges the nonarticular acetabular fossa. The acetabulum is deepened by the cartilaginous *acetabular labrum* which builds up the rim of the acetabulum and grasps the femoral head beyond the equator of its hemisphere. The femoral head is held into the acetabular socket by a rim that has a smaller diameter than that of the head. The rim must be disrupted before the head can be dislocated from its socket.

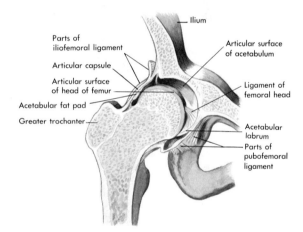

The Synovial Cavity and the Articular Capsule

Although the synovial cavity is reduced to a potential crevice between the lubricated articulating cartilages, elsewhere the entire synovial cavity is extensive because of the disposition of the articular capsule. The *fibrous membrane* of the articular capsule attaches to the margins of the acetabular labrum and encloses much of the neck of the femur. The *synovial membrane* is lax and follows the fibrous capsule. It is reflected at the distal attachments onto the neck of the femur. In the acetabular fossa the membrane is modified into the *acetabular fat pad*.

Ligaments

The ligaments of the hip joint check the movements of the joint rather than reinforce its already sturdy construction. They are thickenings of the capsule rather than true extrinsic bands but are among the largest and strongest ligaments of the body. One ligament runs to the femur from each of the three divisions of the hip bone.

Iliofemoral Ligament. This largest of the ligaments attaches superiorly to the anterior inferior iliac spine and the adjacent iliac portion of the acetabulum. It stretches across the front of the hip joint in two limbs of an inverted "Y" which attach inferiorly to the intertrochanteric line.

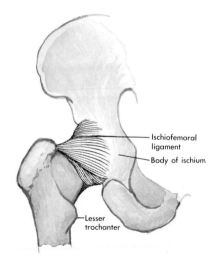

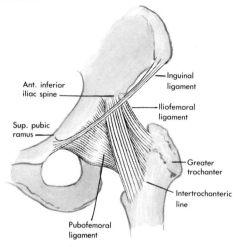

This ligament prevents the backward bending of the pelvis upon the head of the femur. The erect position of the body can be preserved by the iliofemoral ligaments without muscular actions to fix the hip joints, because the tightness of these ligaments checks backward rotation of the pelvis upon the femoral heads. In other words the pelvis and trunk of the body are balanced upon the heads of the femurs and are prevented from rolling backward upon them by the tautness of these strong ligaments.

Pubofemoral Ligament. The fibers of this ligament spiral from the superior ramus of the pubis along the anterior and inferior aspects of the joint to the undersurface of the femoral neck. These fibers grow tight with abduction to check this movement.

Ischiofemoral Ligament. From origins on the body of the ischium, this ligament spirals upward behind the hip joint to attach to the neck of the femur.

This ligament acts with the iliofemoral ligament in checking extension.

Movements

The conventional movements of a ball-and-socket joint occur at the hip. In *flexion* the thigh is moved forward as in taking a step. If the knee is bent, the thigh may be flexed until it meets the abdomen.

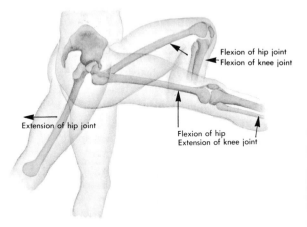

In *extension* the thigh is brought back from any position of flexion or is carried posteriorly as in taking a step backward. Extension is much more limited than flexion. In *abduction* the lower limb can be carried away from the side of the body and the opposite limb. This movement is soon checked by the pubofemoral ligament and by the antagonistic action of the muscles which adduct the limb. In *adduction*

the limb is returned from the abducted position until it strikes the opposite limb. By combining some flexion and lateral rotation, the movement of adduction can be continued across and in front of the other limb.

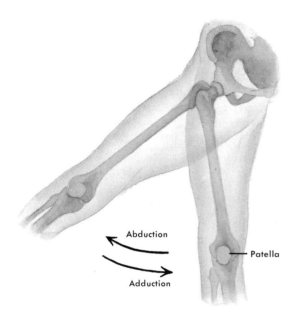

Abduction

Adduction

Patella

Medial and *lateral rotation* have the same meanings as at the shoulder joint. None of the movements at the hip joint have the range of motion of the shoulder joint assisted by scapular movements, but this is compensated for by the stability and sturdiness that are provided. If the lower limbs are fixed, the pelvis moves upon the femoral heads instead of the femur upon the acetabulum. In such a changed situation the movement of flexion of the trunk of the body, as in bending to touch the toes, takes place mainly at the hip joints.

PATELLA

The patella is a triangular bone with rounded edges and angles which lies in front of the knee joint in a position to protect it. This, however, is not the main function of the patella, which, as a sesamoid bone, acts to increase the mechanical advantage of the muscle which extends the leg and straightens the knee. The bone is thickened on its posterior surface to fit into the concave trough between the anterior portions of the femoral condyles. The patella is held in position

by the attachment of the tendon of the thigh muscle to the anterior aspect of the tibia. The patella does not move in a longitudinal direction but rather the tibia rolls against its posterior surface.

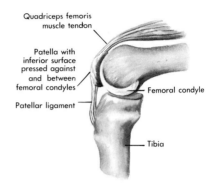

Quadriceps femoris muscle tendon

Patella with inferior surface pressed against and between femoral condyles

Patellar ligament

Femoral condyle

Tibia

As the leg is flexed the bending of the knee puts the tendon under tension and the patella is pressed into the trough between the front of the femoral condyles.

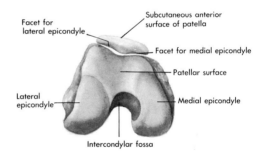

Facet for lateral epicondyle

Subcutaneous anterior surface of patella

Facet for medial epicondyle

Patellar surface

Lateral epicondyle

Medial epicondyle

Intercondylar fossa

With extension of the leg the patella rises from the trough and becomes more prominent as the knee straightens. Although the patella is in an exposed position and is in danger of fracture with direct anterior trauma to the knee, the patella does not take direct force in kneeling. In the kneeling position the patella is on the front of the bent limb. It is the upper end of the tibia that contacts the surface knelt upon.

BONES OF THE LEG

The anatomical leg is roughly comparable to the forearm in that it is the second component of a limb which articulates with the proximal component by a hinge joint and it is supported by two long bones, the *tibia* and the *fibula*.

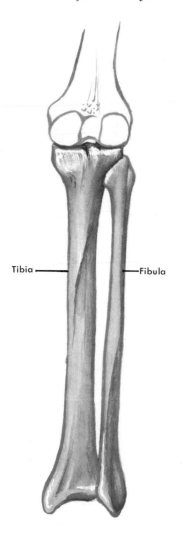

Tibia — — Fibula

and supination do not occur in the lower limb, there is no movement of the tibia and fibula around each other.

Tibia

The tibia or "shinbone" lies close to the anterior aspect of the leg where its entire extent can be felt along the medial side of the leg from the knee joint to the ankle. The tibia is the bone of the leg which bears the weight of the body, receiving it from the femur at the knee joint and transferring it to the foot at the ankle joint.

Upper End. The upper end of the tibia expands superiorly to form a broad thick table top for articulation with the large femoral condyles.

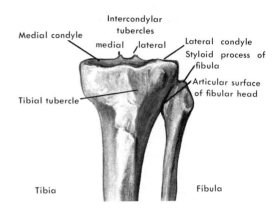

Intercondylar tubercles
Medial condyle   medial lateral   Lateral condyle
Styloid process of fibula
Articular surface of fibular head
Tibial tubercle
Tibia   Fibula

Whereas the elbow joint is hinged anteriorly, the knee joint is hinged posteriorly. In the leg, there is one dominant bone, the tibia. Since pronation

The expansion of the upper end is carried posteriorly so that most of the articular surface is set posterior to the long axis of the shaft. The *articular surface* is divided into an elongated oval *medial condyle* and a more circular *lateral condyle*.

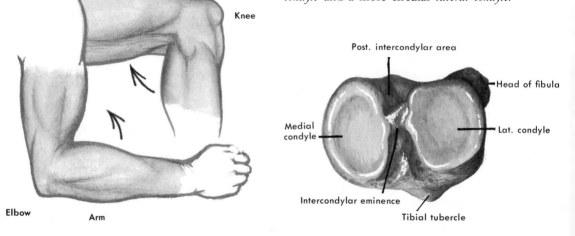

Knee

Elbow   Arm

Post. intercondylar area
Head of fibula
Medial condyle   Lat. condyle
Intercondylar eminence
Tibial tubercle

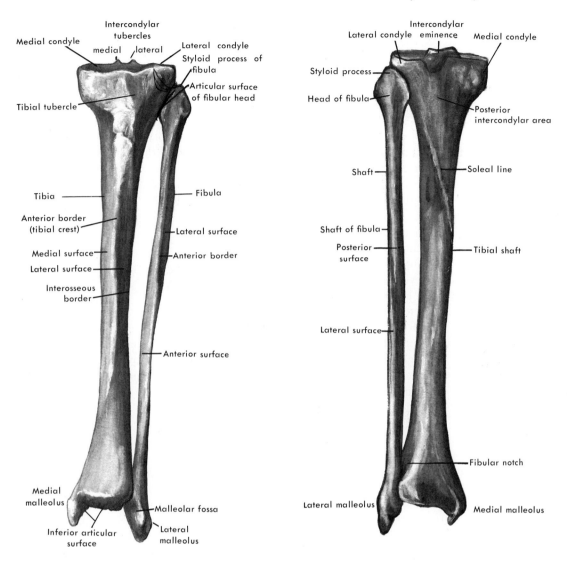

Medial condyle

Intercondylar tubercles
medial — lateral

Lateral condyle

Styloid process of fibula

Articular surface of fibular head

Tibial tubercle

Tibia

Fibula

Anterior border (tibial crest)

Lateral surface

Medial surface

Anterior border

Lateral surface

Interosseous border

Anterior surface

Medial malleolus

Malleolar fossa

Inferior articular surface

Lateral malleolus

Intercondylar eminence

Lateral condyle

Medial condyle

Styloid process

Head of fibula

Posterior intercondylar area

Shaft

Soleal line

Shaft of fibula

Posterior surface

Tibial shaft

Lateral surface

Fibular notch

Lateral malleolus

Medial malleolus

These two articular areas are separated by the roughened *intercondylar eminence* which is raised toward each condyle into two summits, the *medial* and *lateral intercondylar tubercles*. The articular tibial condyles, although slightly concave, provide more of a gliding surface for the rounded femoral condyles than sockets to hold them. Flat semicircular wedges of fibrocartilage, the medial

and lateral *semilunar cartilages,* placed on the edges of the tibial condyles, deepen the articular surfaces.

The lateral condyle projects over the head of the fibula which fits against a *fibular facet* on the posterolateral aspect of the side of the condyle. The sides of the condyles receive the attachment of many thigh muscles. The anterior surface of the upper end of the bone is drawn out into the prominent *tibial tubercle* for attachment of the ligament from the patella.

Shaft. The shaft of the tibia, slightly bowed anteriorly, presents a sharp anterior border, broad medial and lateral surfaces, and a narrower rounded posterior surface. The sharp *anterior border,* commonly known as the "shin," makes several gradual curves downward as the *tibial crest*

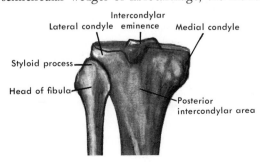

Intercondylar eminence

Lateral condyle

Medial condyle

Styloid process

Head of fibula

Posterior intercondylar area

before it curves medially to become the medial border of the lower end of the bone. (See illustration at top of page 137). The broad *medial surface* is smooth and, lying directly under the skin, has no muscle attachments. The *lateral surface* is enclosed by the sharp anterior border and a long raised line on the lateral aspect of the bone. The latter line, or *interosseous border,* marks the attachment of the interosseous membrane between the tibia and the fibula. The lateral surface swings forward onto the anterior aspect of the lower end of the shaft as the anterior border curves medially. The upper two-thirds of the lateral surface is taken up by the origin of a muscle which sends its tendons across the front of the ankle to lift the foot. The lower part of the lateral surface, now having swung onto the anterior aspect of the bone, has no muscle attachments but is covered by tendons, vessels, and nerves crossing the front of the ankle. The *posterior surface* is generally rounded below the expanded upper end. A rough beaded *soleal line* slants medially and downward across the upper third. One of the calf muscles arises from a fibrous band which is anchored to this line. (See illustration at top of page 137.) Another muscle, which acts to shift the femur and tibia upon each other, attaches to the posterior surface above the soleal line. Two muscles take origin from the middle of the shaft below the line. On the medial side is a muscle which bends the toes, whereas on the lateral side is the muscle which bends the foot backward at the ankle. The lower portion of the posterior surface has no muscle attachments but is crossed by tendons, vessels, and nerves passing behind the ankle joint.

Lower End. The lower end of the tibia expands into a block of bone which is shaped inferiorly for articulation with the talus bone at the ankle joint. The medial surface is prolonged as a distal projection, the *medial malleolus,* which grasps the medial side of the talus and forms the prominent hump at the medial side of the ankle.

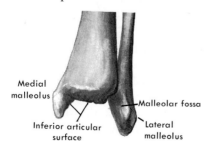

Medial malleolus

Malleolar fossa

Inferior articular surface

Lateral malleolus

The inferior or distal surface faces the talus at the hinged ankle joint. The inferior end of the tibia is hollowed out to form a concave socket for the talus. The *inferior articular surface* extends onto the internal aspect of the medial malleolus which reaches distally beyond the level of the end of the tibia to grasp the side of the talus. The lower end of the fibula fits into the *fibular notch* on the lateral side of the lower end of the tibia.

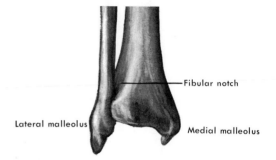

Fibular notch

Lateral malleolus

Medial malleolus

Ligaments of the ankle joint attach to the anterior, medial, and posterior borders of the lower end, and ligaments binding the fibula to the tibia attach to the lateral border about the fibular notch.

Fibula

The fibula is a slender twisted rod, triangular in cross section, which is placed laterally and somewhat posteriorly to the tibia. The fibula has little to do with weight-bearing inasmuch as it does not articulate superiorly with the femur. The fibula extends the area available for muscle attachments in the leg, a role it shares with the interosseous membrane between it and the tibia.

Upper End. The fibula rises under the overhanging lateral condyle of the tibia to expand into a small, knoblike *head.* Most of the superior surface of the head is formed into an *articular facet* which fits against the fibular facet on the condyle of the tibia. The remainder of the head extends superiorly as the *styloid process* beside the condyle.

Shaft. The *shaft* begins immediately below the head and continues distally as a twisted rod which comes slightly forward beside the tibia until it ends in an expanded lower end at the lateral side of the ankle joint. The twisting of the shaft makes it difficult to follow the changing direction of its borders and surfaces. The lateral surface gives attachment to muscles whose tendons pass around the lower end of the bone to the outer surface of the foot. These muscles lift the outer edge of the foot to turn the sole outward. The anterior surface provides the origin of a group of muscles whose tendons cross the front or dorsum of the foot to lift (extend) the toes and bend the foot at the ankle joint. A posterior tibial muscle extends its origin from the tibia across the interosseous membrane to the posterior surface of the midshaft of the fibula. The more distal portion of the posterior surface is given over to one of the flexor muscles of the toes.

Lower End. The lower end of the shaft fits into the fibular notch of the tibia. The lower end turns out in a downward and lateral direction in a prolongation beside the talus, which is the *lateral malleolus.* The fibula grasps the lateral surface of the talus. Tendons uniting the fibula to the tibia and to the talus attach into the malleolar fossa on the distal surface, whereas other ligaments of the ankle joint are affixed to the borders of the tip of the malleolus.

Living Anatomy of the Leg Bones

The tibial condyles can be felt on each side of the patella. The medial condyle can be traced from its anterior to its medial aspect. As the lateral condyle is followed around toward the lateral aspect of the knee the flattened prominence of the head of the fibula is felt. The tibial tubercle is palpable just below the tendon extending to it from the patella. The tibial crest and the subcutaneous medial surface of the tibia can be followed the whole length of the shin and out onto the medial malleolus. As the foot is bent upward the tendons crossing the front of the ankle joint come under tension and lateral to them the lateral malleolus of the fibula is found. The lateral malleolus projects farther distally than its counterpart.

KNEE JOINT

Both the femur and the tibia articulate with intermediate disks of fibrocartilage which partly subdivide the joint, and the patella forms a gliding joint with the femur within a common capsule.

Construction of the Joint

The femur approaches the tibia on a medially directed course which produces an angular junc-

tion in which much of the weight falls upon the lateral femoral condyle.

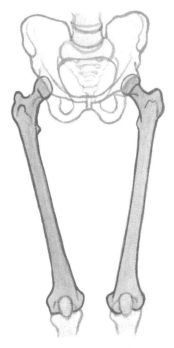

The knees should be separated by approximately the thickness of a hand. If there is less of a medial convergence of the two femurs, the distance between the knees is increased. This is the condition of "bow legs" or *genu varum.* In the opposite condition, the medial direction of the femurs is increased, the knees are brought together, and the tibia are turned relatively outward in the "knock kneed" or *genu valgum* position.

The movements of the articular surfaces are a combination of the gliding of the tibial head around the femoral condyles and a part rolling, part skidding movement of the condyles upon the tibia. When no weight is borne the tibial movement is more pronounced. When weight is borne and the knee moves in walking, the tibia is fixed and the femoral condylar movement predominates. The total length of a femoral condyle exceeds that of the tibial condyle on which it moves. Furthermore, the length of articular surface of the lateral femoral condyle is less than that of its medial counterpart. Therefore there is not enough tibial surface to accommodate the rolling femoral condyle, and the femoral condyles cannot just turn upon the articular surface but must roll and skid upon it simultaneously. By the time the shorter lateral femoral condyle has completed its movement, the medial one still has some of its circuit to turn. This is completed by a lateral rotation of the tibia which turns the oval medial tibial condyle to present a longer axis at the end of the movement.

The knee joint is inherently unstable. The multitude of reinforcing structures are best described before attempting to explain the complicated synovial cavity.

Capsule and Ligaments of the Knee Joint

The fibrous membrane of the articular capsule is replaced anteriorly by the quadriceps tendon and the patella. The capsule is thin posteriorly where it faces the popliteal fossa, and is greatly strengthened laterally and medially by collateral ligaments.

Collateral Ligaments. Side-to-side movements are prevented by collateral ligaments which grow tight as the joint is straightened into full extension, where the greatest stability is needed to support the weight of the body. The lateral or *fibular collateral ligament* tightens between the side of the lateral femoral condyle and the head of the fibula. The medial or *tibial collateral ligament* extends between the sides of the medial femoral and tibial condyles.

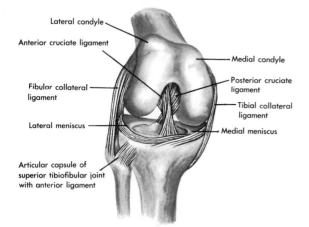

The *oblique popliteal ligament* is a thickening of the posterior part of the capsule which grows tight as the joint straightens to prevent overextension.

Cruciate Ligaments. Two intra-articular ligaments arise from the central nonarticular portion of the tibia to cross as they rise upward in the intercondylar fossa between the femoral condyles.

The *anterior cruciate ligament* originates anterior to the intercondylar eminence of the tibia and extends upward and backward to the internal surface of the lateral femoral condyle. The *posterior cruciate ligament* rises upward from a position posterior to the intercondylar eminence and passes forward to cross the anterior ligament and attach internal to the medial femoral condyle. The cruciate ligaments prevent anteroposterior displacement of the tibia and femur upon each other. The anterior ligament prevents displacement of the tibia anteriorly or dislocation of the femur backward. The posterior ligament prevents displacement of the tibia posteriorly or dislocation of the femur forward. Since the ligaments cross, some of their fibers are tense in any direction of movement and the cruciate ligaments assist in preventing side-to-side motions and in keeping the articulating bones together in flexion and extension. They are the general stabilizing tendons of the knee joint.

Intra-articular Cartilages

Wedge-shaped crescentic rims of fibrocartilage are placed at the margins of the medial and lateral tibial condyles. These *menisci* build up the edges of the tibial surface and deepen it for articulation with the femoral condyles. Anterior and posterior tips or horns face the intercondylar eminence. The *medial meniscus,* oval shaped to fit its tibial condyle, is firmly attached to the intercondylar eminence by both horns and its rim is attached to the tibial collateral ligament. The *lateral meniscus,* more circular in shape, is only weakly attached to the eminence and barely at all to the fibular collateral ligament.

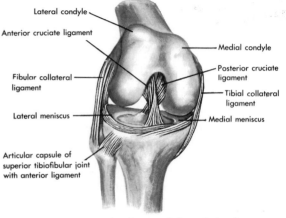

As the condyles of the femur

roll upon the cartilaginous rims, the lateral meniscus moves slightly to keep out of the way of the crushing force of articulation. The more firmly attached medial meniscus can shift only very slightly and is sometimes torn in violent wrenching movements of the knee.

Synovial Cavity and Membrane

The synovial cavity is capacious because of the large intercondylar fossa and many crevices that occur because of the incongruous fit of the bones. The synovial membrane can be traced from the margin of the articular cartilage on the anterior aspect of the femoral condyles and the patellar area. The synovial membrane here passes superiorly for several inches under the quadriceps tendon and muscle before it turns inferiorly to be applied to them. This upward reflection forms the *suprapatellar pouch* or bursa. This extension of the synovial cavity acts as a true bursa in facilitating free movement of the extensor mechanism upon the femur. The synovial membrane passes distally to surround the articular surface of the patella. Below the patella the membrane specializes into the *infrapatellar fat pad* from which synovial *alar folds* continue the membrane to the intercondylar area of the tibia. This nonarticular area receives a synovial investment which also continues upward around the cruciate ligaments into the intercondylar fossa.

Articular capsule

Synovial pouch

Synovial fluid

Articulating bones

Fat pad

Synovial fringe

Articular cartilage

The actual bearing surfaces of the bones are uncovered as are the menisci, but the membrane lines the fibrous capsule and collateral ligaments distally

from the inferior surface of the menisci to the articular margins of the tibial condyles. Posteriorly the synovial membrane extends almost uninterruptedly, with the fibrous capsule from the tibia behind the femoral condyles.

Movements of the Knee Joint

The major movements of the knee joint are the hinge movements of *flexion* and *extension* with either tibial or femoral components predominating. Abduction and adduction are prohibited by the collateral and cruciate ligaments as well as by the nature of the articulation. Slight medial and lateral rotation of the tibia occurs in adjusting the articulating surfaces to each other.

Tibiofibular Joints

Certain movements of the ankle require that the fibula spring a bit in its union to the tibia, but for the most part it remains firmly attached so that the leg bones can provide a strong base for ankle movement.

Superior Tibiofibular Joint. An articular facet on the fibular head makes a plane synovial joint with the fibular facet on the side of the tibial condyle. The articular capsule is strengthened by fibrous bands extending from the side of the tibial condyle to the fibular head, anterior and posterior to the slanting joint. These are appropriately called the *anterior* and *posterior ligaments of the fibular head.*

Interosseous Membrane. An intermediate, fibrous tibiofibular joint is formed by the interosseous membrane which extends between the interosseous borders of the two shafts. Since the

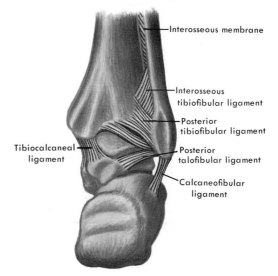

Interosseous membrane

Interosseous tibiofibular ligament

Posterior tibiofibular ligament

Posterior talofibular ligament

Calcaneofibular ligament

Tibiocalcaneal ligament

bones of the leg do not move about each other in pronation or supination, this fibrous articulation is tighter than the radioulnar interosseous membrane. The attachments of muscles to the adjacent surfaces of the tibia and fibula are extended onto the membrane.

Inferior Tibiofibular Joint. The lower end of the fibula is bound to the long concave fibular notch of the tibia by fibers of the *interosseous tibiofibular ligament* in a typical fibrous joint. *Anterior* and *posterior tibiofibular ligaments* strengthen the respective surfaces of the joint.

BONES OF THE FOOT

The foot is turned forward at right angles to the rest of the limb to provide a stable base for weight-bearing and to form a lever for springing off the ground in locomotion. The arrangement of bones and joints in the foot must be strong in order to support the weight of the body and to resist the force which gravity presses down upon it. At the same time the joints must be resilient and elastic in order that the bony components may move smoothly and progressively in locomotion and take up the shocks and jars which come as the foot strikes the ground. Even when the body is still, in the erect position, opposing muscle groups are at work in the leg using the bones of the foot as an anchor in order to maintain balance and posture.

The Foot in General

The broad flat *heel* projects posteriorly below the ankle joint. The upper surface of the foot is arched and there are few soft structures of tendons or vessels between the skin and the bones. Commonly referred to as the "instep," this arched upper surface is referred to in anatomy as

the *dorsum* of the foot. The opposite surface is the sole of the foot or the *plantar surface.* It extends from the heel, under the *arch* of the foot, to the ball of the foot from which the toes extend forward.

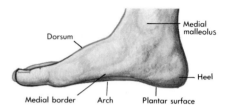

Each foot has *lateral* and *medial borders.* The foot is arched both longitudinally and transversely. The arches aid in weight-bearing, in twisting movements of the foot, and in providing a passageway for nerves, vessels, and tendons to the sole of the foot. Since the foot is turned forward at right angles to the longitudinal axis of the body in the anatomic position, the movements of the foot at the ankle joint cannot be described by the usual terms of "flexion" and "extension." When the foot is bent upward so that the dorsum of the foot approaches the anterior aspect of the leg, the movement is called *dorsiflexion.* When the foot is bent downward to move the plantar surface of the foot posteriorly and to increase the angle between the foot and the front of the leg, the movement is called *plantar flexion.* The bones of the foot also permit the sole of the foot to be twisted inward or outward. The sole faces inward or medially in *inversion,* whereas it faces outward or laterally in *eversion.*

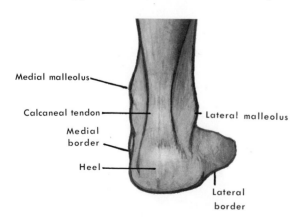

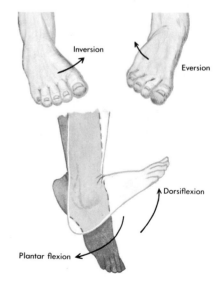

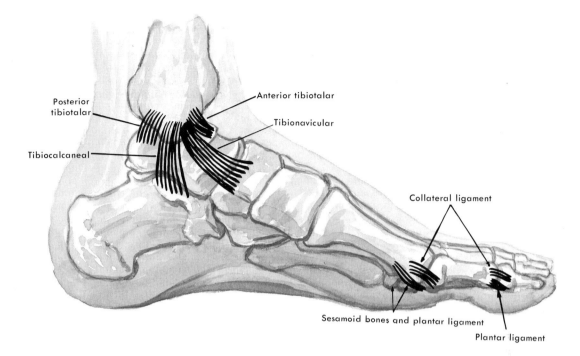

Bony Construction of the Foot

The skeleton of the foot consists of three groups of bones. The *tarsal bones* are somewhat equivalent to the carpal bones of the wrist, and the *metatarsal bones* are similar to the metacarpal bones of the hand. The *phalanges* are like the bones of the fingers.

calcaneus. The *talus,* in older terminology referred to as the astragalus, is the only bone of the foot to articulate with the tibia and fibula. It acts as a rocker by which the foot as a unit can be dorsiflexed or plantar flexed at the hinge of the ankle joint. The talus receives the entire weight borne by its limb. Half of this weight is transmitted forward to the bones forming the arch of the foot and half downward to the heel. The *calcaneus,* or os calcis, is the bone of the heel. It supports the talus, withstands shock as the heel strikes the ground, and transfers forward the portion of body weight it receives from the talus.

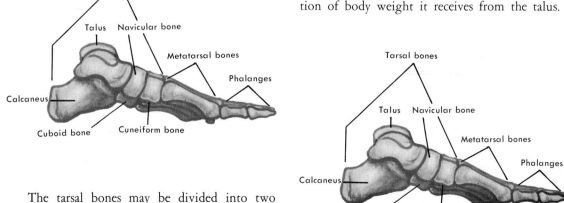

The tarsal bones may be divided into two groups. The first group includes the *talus* and the

Five other tarsal bones lie anterior to the talus and the calcaneus. These increase the flexibility of the foot, particularly in its twisting movements, and form the keystones of the longitudinal arch of the foot. The *navicular bone* lies in front of the talus on the medial side of the foot and articulates with the three *cuneiform bones.* The *cuboid bone* is anterior to the calcaneus but, being longer, also comes to lie beside the third cuneiform bone on the lateral side of the foot. The five metatarsal bones extend forward from the cuboid and the cuneiforms and articulate with the proximal phalanges.

Individual Bones of the Foot

The major features of each of the bones of the foot follow.

Talus. The *talus* is a right-angled bone which rests upon the calcaneus below and is grasped by the two malleoli above. The *body* of the talus bears upon its superior surface the rounded *trochlea,* the articular surface for the tibia and the malleoli. The inferior surface is irregularly concave and bears posterior and medial *calcaneal facets* for the calcaneus. The neck projects forward from the medial side, bearing the rounded *head* which articulates with the navicular bone.

Calcaneus. The long thick *calcaneus* lies under the talus and projects posteriorly beyond it as the bone of the heel. The calcaneus, as the

posterior pillar of the longitudinal arch of the foot, is set obliquely upward and forward along the lateral side of the foot. The posterior part strikes the ground at the enlarged *medial* and *lateral tubercles.* The anterior surface projects forward under the talus to articulate with the cuboid bone. The superior surface of the calcaneus bears laterally a large convex *posterior facet for the talus* upon which the latter bone can rock. A ledge of bone, the *sustentaculum tali,* projects from the medial surface to support the talus. This ledge bears a *medial facet* which articulates only with the posterior surface of the head of the talus, leaving the anterior surface free for movements with the navicular bone.

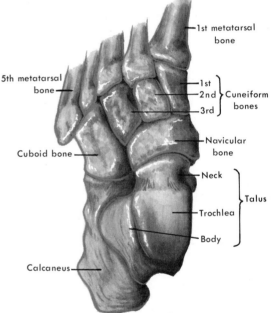

Navicular. The *navicular bone* has a concave posterior surface for the rounded head of the talus and an irregularly convex anterior surface with three articular facets for the cuneiform bones.

Cuneiform Bones. The three *cuneiform bones* are numbered from the first on the medial side to the third on the lateral side. Each is wedge shaped with a broader base toward the dorsum of the foot. As they interlock toward the plantar surface these bones create the transverse arch of the undersurface of the foot. Each cuneiform bone provides a base for a metatarsal bone of the same number. The third or lateral cuneiform bone also articulates laterally with the cuboid bone.

Cuboid. The longer *cuboid bone* lies along the lateral surface of the foot beside both the navicu-

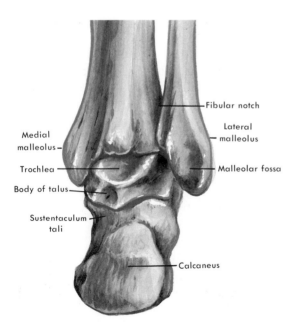

lar and the third cuneiform bones. Posteriorly it articulates with the calcaneus and anteriorly it provides a base for the fourth and fifth metatarsals.

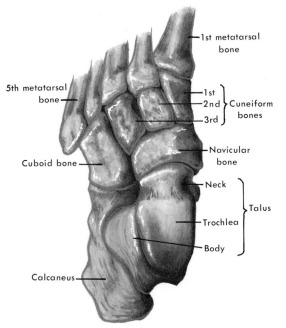

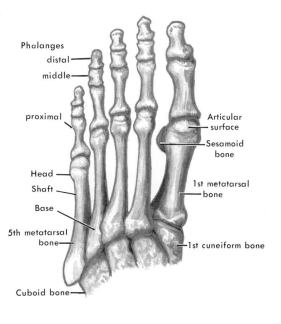

Metatarsals. The slender *metatarsal bones* form the anterior pillar of the longitudinal arch of the foot. Their *bases* articulate with the tarsal bones and their *heads* underlie the ball of the foot. The shaft of the first metatarsal bone is thicker and shorter than the rest. All the metatarsals are closely bound together by ligaments so that stability is gained at the expense of mobility. Two sesamoid bones are affixed to the plantar surface of the head of the first metatarsal bone. Since the toes can bend upward in an extreme degree of extension (hyperextension), a characteristic not shared by the fingers, the articular surface on the heads of the metatarsal bones continues from the plantar surface, over the anterior aspect, to cover the upper surface.

Phalanges. The great toe has two phalanges; the others have three. The proximal phalanx of each toe is made up of a squared base, a more slender shaft, and a slightly expanded head. The heads, as in the fingers, are pulley shaped but with more of the articular surface on the dorsal aspect. The proximal phalanges become progressively shorter and smaller from the great toe toward the fifth. The distal phalanges, except for that of the great toe, are short, with expanded ends but virtually no shaft, and rudimentary.

TALOCRURAL (ANKLE) JOINT

The ankle joint is an excellent example of a synovial hinge articulation.

Construction and Movements

The domed upper surface of the trochlea of the talus is received into the concave articular surface of the tibia. The lateral and medial malleoli grasp the sides of the trochlear articular surface. Movement of the ankle joint is limited to *plantar flexion* and *dorsiflexion.* The talus is broader anteriorly than posteriorly. As the foot is dorsiflexed the broader anterior portion glides between the malleoli and tends to force them apart. The inferior tibiofibular joint gives only slightly and the force is transmitted upward to the superior synovial joint, which can move. As the foot is plantar flexed, the narrower posterior portion of the talus moves between the malleoli and some lateral rocking motion is possible. The articular capsule requires strengthening against such lateral movement and especially against unusual shocks or strains that are encountered on rough uneven ground.

Collateral Ligaments

Strong collagenous bands cross from the tibial and fibular malleoli to the talus, and from the malleoli to the bones of the foot below and behind the talus. These collateral ligaments strengthen the sides of the joint and prevent the forward displacement of the tibia upon the talus by the pull of muscles whose tendons curve under

of the talus is the *posterior tibiotalar ligament*. The deltoid ligament is injured when the ankle is sprained with the foot turned out in eversion.

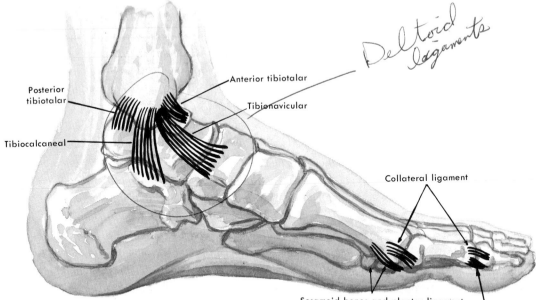

Deltoid ligaments

the malleoli from the back of the leg. The ligaments are named for the bony connections.

Lateral Ligaments. Two ligaments reinforce the joint posteriorly and resist forward displacement. The *posterior talofibular ligament* runs posteriorly from the lateral malleolus to the posterior extremity of the talus. The *calcaneofibular ligament* extends posteriorly and downward from the malleolus to the calcaneal bone. The *anterior talofibular ligament* reinforces in an anterior direction from the malleolus to the neck of the talus. These lateral ligaments are injured in the ankle sprain that occurs when the foot is violently twisted inward in inversion.

Medial Ligaments. Four ligamentous bands cluster so closely together on the medial side of the joint that their fanlike spread gives them the common name of the *deltoid ligament*. From the medial malleolus two bands extend forward. One to the medial side of the talus below the articular surface is the *anterior tibiotalar ligament*. The other to the anteroinferiorly located sustentaculum tali part of the calcaneal bone is the *tibiocalcaneal ligament*. The *tibionavicular ligament* drops downward beside the joint to the tuberosity of the navicular bone. A posteriorly directed band between the malleolus and the posterior portion

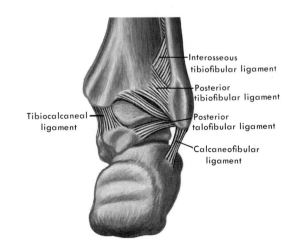

Arches of the Foot

The *longitudinal arch* is lower along the lateral than along the medial side of the foot. It is divided into medial and lateral segments. The *medial segment* rises through the calcaneus through its sustentaculum to the head of the talus. From this high point the arch descends forward through the anterior pillar formed by the navicular, the three cuneiforms, and the first three metatarsals whose heads form three of the six points of contact of the arched foot. The lower *lateral segment* begins, as does the medial, at the posterior point of contact, the calcaneus. Its line rises through the lateral part of the calcaneus by passing the talus to reach its high point in the cuboid. The anterior pillar continues downward through the fourth and fifth metatarsals to their heads which complete the six points of contact of the longitudinal arch. The arch of the foot is also domed by a *transverse arch* which rises across the width of the foot between the medial and lateral borders. Concavities on the inferior surfaces of the navicular and cuboid bones begin the transverse arch proximally. The arch is continued by the interlocking wedges of the cuneiform bones which hold the bases of the metatarsals into the same domed disposition. The arches of the foot could not be maintained against the force of weight-bearing by bony congruity alone. Ligaments and the interplay of opposing muscles have prominent roles in the maintenance of the arches.

soft tissues of the heel. The ledge of the sustentaculum tali is palpable below the medial malleolus. The other tarsal bones can be felt only vaguely through the tendons and ligaments on the dorsum and scarcely at all through the soft tissues of the sole. The fifth metatarsal bone is felt along the lateral side of the foot, and its head forms a prominent tubercle. A small tuberosity on the navicular bone can be felt halfway along the medial border of the longitudinal arch. The heads of the first and fifth metatarsals are palpable on their respective sides of the ball of the foot and those of the others can be vaguely felt by deep pressure on the ball of the foot. Most of the surfaces of the phalanges are palpable through the tissues covering them.

JOINTS OF THE FOOT

Articulations occur at many places in the foot, wherever its many bones move upon each other. Many movements of the tarsal bones are in the

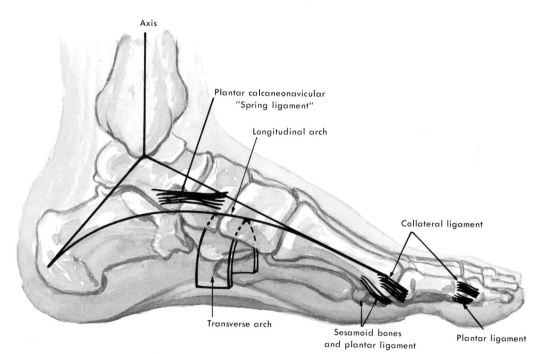

Axis

Plantar calcaneonavicular "Spring ligament"

Longitudinal arch

Collateral ligament

Transverse arch

Sesamoid bones and plantar ligament

Plantar ligament

Living Anatomy of the Bones of the Foot

The calcaneal tendon can be pinched posteriorly between the two malleoli and followed inferiorly onto the posterior surface of the calcaneus whose lateral surface can be felt above the

nature of short shifting adjustments that are associated with the movement of the foot as a unit. At other joints more important movements occur which are significant in the maintenance of balance, in weight-bearing, and in walking. These are emphasized in the following section.

Subtalar Joint

The talus has no muscles attached to it. Its movements are governed by its relationships to the tibia above, the calcaneus below and behind, and the navicular in front. The talus performs the function of a rocker between the leg above and the other bones of the foot below. The *subtalar joint,* also known as the talocalcanean joint, is a synovial articulation between the posterior calcaneal facet of the inferior surface of the talus and the posterior talar facet on the superior surface of the calcaneus. Half of the weight-bearing load of a limb passes across this joint to the heel. The lateral and medial ligaments of the ankle strengthen this joint also, and in addition there are short *talocalcaneal ligaments* on each side. The fibrous capsule is converted to the *interosseous talocalcanean ligament* which unites the bones and closes off the synovial cavity from the complex joint in front of it. Part of the support of the talus by the calcaneus is provided through the subtalar joint which also permits the heel to share in inversion and eversion of the foot.

Transverse Tarsal Joint

The transverse tarsal joint is not an anatomic entity but is an important functional grouping of two joints which occur anterior to the talus. The two joints lie almost side by side and their synovial cavities frequently communicate to form the common joint described as the transverse tarsal joint. It is along the transverse line of this joint that much of the inversion and eversion of the foot occurs. Essentials of the components of the combined joints follow.

Talocalcaneonavicular Joint.   A short, longitudinally oriented synovial cavity between the medial facet of the talus and the medial facet on the sustentaculum tali of the calcaneus exists anterior to the interosseous talocalcanean ligament. This cavity is continuous with the transversely oriented synovial cavity between the rounded head of the talus and the concave posterior surface of the navicular bone. At the talonavicular part of the articulation, the talus transfers the other half of the weight-bearing load to the medial side of the foot.

Calcaneocuboid Joint.   The anterior extremity of the calcaneus meets the posterior surface of the cuboid bone at a synovial joint whose synovial cavity is continuous with the talonavicular joint to form the lateral part of the midtarsal joint.

Other Tarsal Joints

Plane synovial joints occur between the adjacent surfaces of all the other tarsal bones. The anterior surface of the navicular bone articulates with the posterior surfaces of all three cuneiform bones at the *cuneonavicular joint. Intercuneiform joints* are found between the first and second and between the second and third cuneiform bones. The *cuneocuboid* joint is located between the third cuneiform bone and the medial side of the cuboid bone. A common synovial cavity intervenes between the bones at all these joints, and it might be said that the large cuneonavicular synovial cavity extends also between the cuneiforms and between the third cuneiform and the cuboid. The synovial cavity is shut off from the midtarsal joint behind by a strong cubonavicular interosseous ligament. It is closed off anteriorly by tarsal interosseous ligaments between the cuneiforms and between the third cuneiform and the cuboid.

Tarsometatarsal Joints

A separate medial *tarsometatarsal joint* cavity is found between the first metatarsal bone of the great toe and the first cuneiform bone. A common *intermediate tarsometatarsal joint* cavity occurs between the bases of the second and third metatarsal bones and the second and third cuneiforms. A common *lateral tarsometatarsal joint* cavity intervenes between the fourth and fifth metatarsal bases and the cuboid bone. These separate joints are closed off from each other by *interosseous cuneometatarsal ligaments.* Short extensions of the synovial cavities pass distally between the bases of adjacent metatarsal bones as the *intermetatarsal joints.*

Tarsal Ligaments

Short ligamentous bands connect all the tarsal bones. These cross the many joint crevices on both dorsal and plantar surfaces as *intertarsal ligaments* which should be distinguished from the interosseous ligaments. Both are specially named for the bones which are connected, but whereas the interosseous ligaments cross the articular cavities or seal them from each other, the intertarsal ligaments extend between surfaces and strengthen the articular capsules. Both types maintain the alignment of the bones and bind them into a flexible tarsal unit. There are certain other ligaments that deserve special mention.

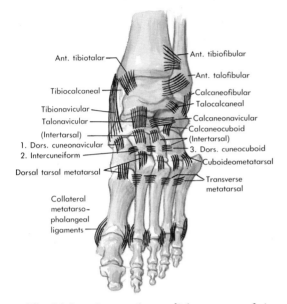

The high point on the medial segment of the longitudinal arch is at the articulation of the head of the talus, supported by the sustentaculum tali of the calcaneus, with the navicular bone. An intensely strong and elastic ligament runs longitudinally under this articulation. It is coated with hyaline cartilage so that the bones may move smoothly upon it. This *plantar calcaneonavicular ligament* is often called the "spring ligament" because of its elasticity in preserving the longitu-

dinal arch and in resisting its flattening through depression of the head of the talus by the weight of the body.

The high joint on the much lower lateral segment of the longitudinal arch is at the calcaneocuboid portion of the transverse tarsal joint. *Long plantar* and *plantar calcaneocuboid ligaments* support this articulation and strengthen this segment of the longitudinal arch.

Distal Joints of the Foot

The heads of the metatarsal bones form synovial joints of the condyloid type with the proximal phalanges of the toes at the *metatarsophalangeal joints.* All are interconnected by transverse metatarsal ligaments so that spreading of the toes in adduction and abduction is limited. Extension (dorsiflexion) and plantar flexion occur but less freely than in the hand. Strong collateral and plantar ligaments reinforce the articular capsules. Synovial hinge joints are found at the *interphalangeal articulations* where the articular capsules are strengthened by collateral ligaments to confine the movements to limited extension and flexion and by plantar ligaments which strengthen the weight-bearing surface.

Coordinated Movements

After the muscles of the leg and foot have been studied it will be appropriate to consider the movements of units of the foot as they are coordinated in balance, weight-bearing, and locomotion.

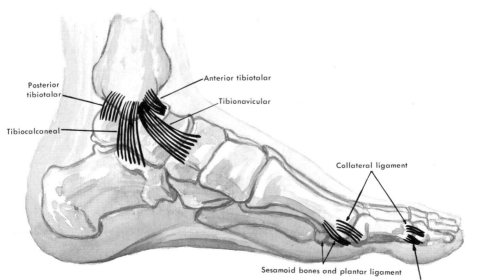

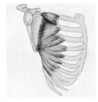

4

The

Muscular

System

Muscles are the contractile units of body structure. They are found wherever one part of the body moves upon another whether it is in the rapid flickering of the eyes or in the long, sustained pull or lift of the arms in doing heavy work. The study of this system involves a large number of muscles and classification is necessary in order to acquire knowledge about muscles in a purposeful manner. Muscles are usually described by anatomists in groups according to their regional location or their relationships in a common stratum of a body part. Since the objective of your study is as much to learn what muscles do as what they are, it will be found in this book that the function of a group of muscles is the basis of organization. The manner in which the muscles are grouped, the wording of section headings, and the titles of the tables are designed as specific guides to learning about muscles from a realistic, functional approach.

MUSCLES IN GENERAL

Before individual muscles are studied it is necessary to consider the nature of muscular tissue and the manner in which it is combined with other tissues to form muscles.

MUSCLE TISSUE AND MUSCLE

Muscular tissue is the fundamental tissue of the body which has become specialized to possess in the living man the attribute of contractility. Its metabolism is directed, beyond the maintenance of life within it, to transforming the energy derived from chemical interactions in its protoplasm into contraction of the living muscle substance. The cells are dominant and the intracellular medium is greatly reduced in amount by the close packing of the contractile cells. A delicate network of reticular fibers invests the cells. Nerve fibers and their specialized end organs wind between the cells, and a rich capillary network extends through the tissue mass.

Muscle Cells and Muscle Fibers

The cytoplasm of muscle cells has developed into intercellular organelles called muscle *fibrils* or *myofibrils*. These complicated linkages of protein molecules as part of the living substance of the muscle cell possess the property of contractility. A linear cluster of myofibrils surrounds the nucleus and fills the cell. When these minute threads contract, the entire cell shortens. The ribbon-like form of the muscle cell, created by the myofibrils within it, accounts for the synonym *muscle fiber* used in microscopic anatomy. It will be seen shortly that the gross anatomist uses the term muscle fiber in a slightly different manner. Only in the type of muscle known as smooth muscle is the cell clearly demarcated. In the other types the cell membranes are not clearly seen.

Types of Muscle

Three types of muscle can be identified, according to their structural, functional, and positional differences.

Smooth Muscle. Slender tapered cells formed into thin sheets are arranged into spirals and interlacing networks in the walls of organs of the circulatory, digestive, respiratory, and genitourinary systems. Historically known as *smooth muscle,* this type of muscle contracts in response to nerve impulses from a part of the nervous system that is not directly under voluntary control.

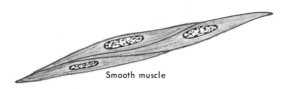

Smooth muscle

Moreover, an awareness of the activities of this type of muscle does not ordinarily rise into the realm of consciousness. It therefore is also known as *involuntary muscle.* The myofibrils present a homogeneous appearance under the microscope, giving origin to another synonym, *nonstriated muscle,* because of its contrast to other types.

Delicate longitudinal striations seen by light microscopy in the smooth muscle cell have been shown by electron microscopy to be *myofilaments.* When bundled together the myofilaments form *myofibrils.* The myofilament is believed to be the unit of contraction in the muscle cell. Chemical studies reveal that myofilaments are composed of two protein macromolecules, *actin* and *myosin.* In the presence of adenosine triphosphate (ATP) a complex protein, *actomyosin,* forms. This complex, in a long chainlike fibrous form, most probably is the basis of the contractile myofilament.

Mitochondria and the Golgi complex are present in muscle cells between myofilaments and especially near the nucleus. These organelles

are intimately concerned with the energy reaction wherein the ATP can be said to provide the fuel for muscular contraction.

Cardiac Muscle. The muscle tissue of the heart is similar to smooth muscle in that its contractions are regulated by the portion of the nervous system that is concerned with the more automatic and involuntary functions. In addition, cardiac muscle has the inherent ability to initiate its own impulse to contraction, independent of the nervous system. Its myofibrils present a series of bands or cross striations under the microscope. Thus, cardiac muscle can be described as *striated involuntary muscle* to distinguish it from smooth muscle. Cell boundaries in cardiac muscle are not distinct, and myofibrils seem to pass from the area of one nucleus to that of another without interruption. Other characteristics of cardiac muscle are described in the discussion on the circulatory system.

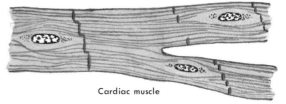

Cardiac muscle

Voluntary or Skeletal Muscle. The gross units of body structure that are called the "muscles" are under the control of the will even though their function may become semiautomatic through repetition and training. This type of muscle tissue is therefore called *voluntary muscle.* Since these muscles are for the most part attached to and move portions of the skeletal system, the tissue of which they are composed is also called *skeletal muscle.* The striped appearance of the myofibrils under the microscope has led to the further designation, *striated voluntary muscle.*

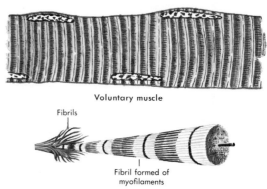

Voluntary muscle

Fibrils

Fibril formed of myofilaments

The voluntary muscle fiber is larger and broader microscopically than the other types and has no clear cell membranes separating adjacent nuclei. It is customary, therefore, to regard this tissue arrangement as a multinucleated, voluntary muscle fiber rather than as a true assemblage of cells. This is the type of muscle that will be considered throughout the study of the muscular system.

The striated, voluntary muscle cell is surrounded by a typical *unit membrane* which encloses the cytoplasmic components and multiple *nuclei.* During its development, a syncytial protoplasmic arrangement forms, in which the continuity of protoplasm and the contractile element of a number of primitive cells combine to make the multinucleated, larger, and broader adult muscle cell.

A striking identifying feature of voluntary muscle is *cross striation* of the cell, which is evident also in the fibrils and myofilaments. The cross striations are described as a dark *A band,* which remains constant in size during contraction, and a shorter, lighter *I band,* which becomes much shorter during contraction of the myofilaments. Also recognized is a narrow, very dark *Z band,* which crosses the center of the light I band.

Electron microscopy confirms these findings and, in addition, reveals the presence of any mitochondria near the nucleus and wedged in narrow crevices between the myofilaments. They are believed to be important in the ATP energy reaction which is necessary for contraction. A tubular, smooth endoplasmic reticulum (*sarcoplasmic reticulum*) is a tight, delicate meshwork which invests each myofibril.

It is now known that there are two sets of myofilaments which are grouped to form the myofibril. There are *thick myofilaments,* which are composed of the complex protein, *myosin,* and *thin myofilaments* which are formed of *actin.* The two forms are grouped so that the thick filaments form the A band, and the thin filaments make the I band. The *sliding theory* of filament contraction explains muscular contraction on the basis that the thin myofilament shortens and slides past the thick myofilament. The overall effect within a muscle fiber, and within a whole muscle, is shortening and contraction.

Voluntary muscle cannot contract unless it is stimulated by nerve impulses. The sarcoplasmic reticulum is believed to provide a tubular network for the spread of the activating impulse. The mechanism for nerve stimulation is described later in the discussion of the nerve supply of muscle.

STRUCTURE OF VOLUNTARY MUSCLES

Muscles are not composed only of muscle tissue. The large muscles that make up the muscular system are organs in their own right in the sense that they are combinations of a number of tissues that work together in the common function of producing effective movement. That function could not be carried out by muscle tissue alone. Various forms of connective tissue, nerve fibers, and end organs as well as components of the vascular and lymphatic systems are associated with muscle tissue to make their function possible.

Fibers and Fascicles

Individual muscle fibers, each 0.01 to 0.1 mm. in thickness and 5 to 10 cm. in length are surrounded by a delicate network of reticular connective tissue fibers. These fibers, the *endomysium,* cloak the muscle fiber, attach it to adjacent fibers, and hold capillaries and nerve fibers in relation to it. The endomysial connective tissue fibers are the most delicate representatives of the connective tissue skeleton that permeates, supports, and subdivides the muscular contractile elements of the organ. The muscle fibers are bundled together in groups of 12 or more by a slightly denser sleeve of collagenous connective tissue fibers to form a larger unit, the muscle *fascicle.* This connective tissue is called the *perimysium.* In addition to bundling individual muscle fibers into groups, the perimysium provides a pathway for nerve fibers and blood vessels to reach these units of muscle structure.

Amidst the perimysial connective tissue are found special nervous system receptor organs, which detect the strength of contraction or the amount of stretch placed upon the muscle. These organs are really underdeveloped contractile units which have connections with the nervous system. They are termed *intrafusal muscle fibers,* in contrast to the ordinary contractile units of voluntary muscle which are the *extrafusal muscle fibers.*

The muscle fascicle is the smallest unit of muscle structure that is large enough to be seen by the unaided eye. It is for this reason that another use of the term "muscle fiber" arises in

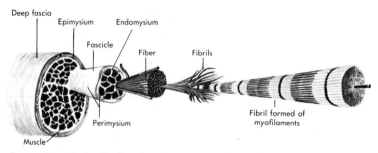

the description of muscles of the body. An important part of such descriptions is to note the "course and direction of muscle fibers" from which the direction of muscle pull and consequent direction of skeletal movement are deduced. The muscle fiber in this usage is actually a fascicular bundle of individual fibers surrounded by perimysium, which is located at the surface of the whole muscle where it can be seen and its linear direction observed. (See also color plate 4.)

The Whole Muscle Assembled

Fascicles are assembled into groups of fascicles and these into larger groups by coarser and coarser sleeves of perimysial connective tissue. As the units become larger the connective tissue takes the form of plates and coarse sheets which visibly separate parts of the muscle into functional subdivisions. The heaviest are called *muscle septa* because of their appearance as partitions when a muscle is cut across. The largest blood vessels, lymphatics, and nerves travel in the septa. The septa continue to the surface to become continuous with and fasten down an external covering of connective tissue, the *epimysium,* which binds the whole muscle into a functioning entity and gives each one its characteristic shape and form.

The deep fascia also invests each muscle, separates it from its neighbors, and permits the muscles to move smoothly upon each other.

Length and Attachment of Muscle Fibers

It is very rare for a single muscle fiber to extend the complete length of a muscle. The length of a muscle depends on a combination of the bundling of muscle fibers longitudinally into a series in the fascicles and the use of connective tissue fibers to complete the "run" to the bone to be moved. Within a fascicle some fibers start near the more fixed attachment of the bone. These fibers terminate in long, tapering points which are overlapped by new muscle fibers which are firmly bound to the original fibers by the reticular endomysium between them. The intermediate fibers extend a short distance to tapered ends to be overlapped and succeeded by other fibers.

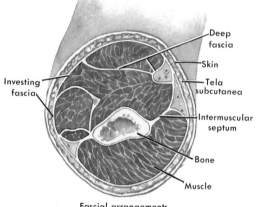

Fascial arrangements

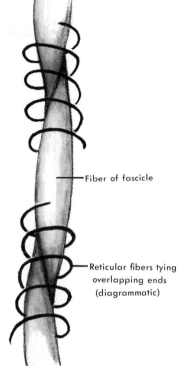

Fiber of fascicle

Reticular fibers tying overlapping ends (diagrammatic)

Ultimately a final fiber in the fascicle extends to the end of the fleshy part of the muscle. A tendon of noncontractile collagenous tissue extends to the attachment on the bone to be moved. A fascicle built up from such a series of fibers tied together in linear succession acts exactly as a single fiber of the same total length. When an individual muscle fiber contracts, it shortens to approximately half its length. The summation of the half shortening of a linear series of small short fibers gives the whole fascicle the shortening effect of a single long fiber of the same length.

Strength of a Muscle. The length of a muscle has little to do with its strength. A more important factor is the actual number of fascicles in the muscle which are available to contract. A muscle with all its fibers or fascicles running the entire length of the muscle assumes a narrow straplike form or the shape of a broad thin sheet. Such muscles are usually weak. The strongest muscles have many fascicles of varying lengths packed into the mass of tissue and tend to assume a bulkier, fusiform shape. If the fascicles vary in length, the subdivisions of the tendon must extend within the central portion of the muscle mass; this enables the fascicles to attach to the connective tissue that actually transmits their pull to the skeletal part that is to be moved. It is for this reason that the actual point of transition on the surface of a muscle, where the reddish-brown fleshy mass of the muscle gives way to the silvery collagenous tendon, is not the true muscle-tendon junction. All it means is that beyond this point there are no contractile units.

Muscle-Tendon Relations. From the foregoing account it should be apparent that there is no abrupt transition between the gross muscle and its tendon. A muscle really has a very extensive tendinous "core" or skeleton which permeates the whole muscle and is associated with and continuous with the perimysial connective tissue. Wherever the contractile fascicle comes to an end, there is a collagenous ribbon, thread, or cable to attach it deep within the muscle mass. The actual junction of a muscle fiber with a tendon, or of the terminal fiber in a fascicle with a tendon, is the union of two dissimilar tissues. There is no continuity of myofibrillae with collagenous fibrils. Instead a tight delicate network of reticular connective tissue fibers, which are continuous with the tendinous fibers, makes a cup-shaped

receptacle into which the blunt terminal end of the fiber fits. Loops of reticular fibers are thrown around the muscle fiber to bind it into the receptacle and to ensure that shortening results in pull on the tendon.

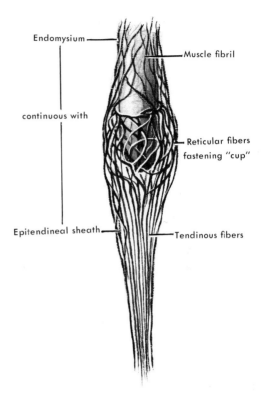

In many muscles the tendinous core runs up through one side of the muscle and all the fascicles approach the tendon from one side on an angled course. Their appearance, similar to that of one half of a feather, causes such muscles to be called *unipennate muscles.*

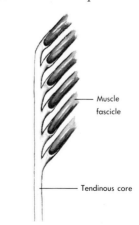

If the tendinous core runs up through the center of a

relatively flat muscle, the fascicles can gain attachment from both sides; this is the *bipennate* arrangement.

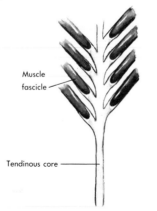

A bulkier muscle has a number of subdivisions of the tendinous framework which permit many fascicles to approach their tendinous attachments multidimensionally; this is termed a *multipennate muscle*.

A muscle with all fascicles running parallel to the direction of its pull has the fewest fascicles and is the weakest kind of muscle. More fascicles are accommodated in a unipennate muscle so that it is stronger. A bipennate muscle has an even greater mass of fascicles and is even stronger. A multipennate muscle has the greatest cross-sectional area of fascicles and, therefore, is the strongest of all.

The Tendon and Its Attachment.   All the collagenous tendon fibers coming from their fascicles assemble together to form a ribbon, cable, or sheet. These unyielding structures transmit the pull of muscle contraction to bone. They must be anchored securely so that the bone moves and they do not pull out when the muscle shortens. Anchorage is achieved in several ways. The collagenous fibers of the tendon may fan out to intermingle with the collagenous fibers of the fibrous periosteal envelope surrounding the bone.

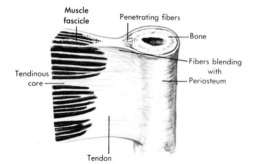

Or the collagenous fibers of the tendon may, in the development of the bone, have penetrated into the osseous structure of the bone to be surrounded and enmeshed by the bone matrix. Most tendons gain anchorage by a combination of the two methods. (See also color plate 4.)

Tendon Protection.   As the tendon extends from the muscle toward its point of attachment on the bone to be moved, it runs amidst the loose arrangement of deep fascia that separates any adjacent structures in the body. The loose fascia, frequently containing small amounts of fat, does not impede the pull of the tendon in response to muscle shortening. When a tendon courses through a congested area where there may be many tendons, arteries, and nerves on parallel courses, friction and close packing may reduce the effectiveness of the tendon movement. Furthermore, as a joint is crossed or a bone passed over, the friction against a bony surface may actually damage the tendon. Two modifications of connective tissue spaces provide protection.

1. *Bursae* are flattened sacs which form a cushion between a muscle or tendon and a bony surface. These delicate structures arise by the coalescence of many small tissue spaces in connective tissue to form a larger cavity. The cells and intercellular fibers of the tissue mat together to form a surrounding membrane, which, though not an epithelial sheet, functions like one to retain the tissue fluid. A thin collection of fluid in the sac allows free movement of the walls of the bursa. As the tendon or muscle presses against one wall, the membrane moves upon the opposite wall. In this way the tendon is cushioned from the bony surface and movement is facilitated.

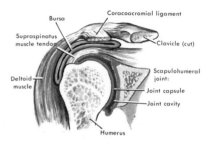

2. *Tendon sheaths* are an extension of this concept in which the double-walled sac is wrapped almost completely around the tendon and is stretched out along more of the linear course of the cable. One wall fits against the tendon while the other abuts adjacent structures. The tendon

moves through a fluid-filled sleeve that almost surrounds it.

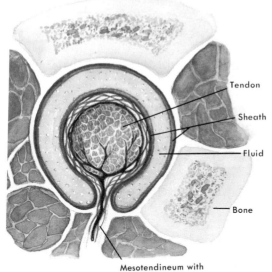

Mesotendineum with
blood vessels

Where the double walls reduplicate with each other, a web of connective tissue, the *mesotendineum,* is continuous with the tendon itself to position it and to convey the tendon's arterial supply.

THE MUSCLE
AND ITS ATTACHMENT

It has been mentioned that muscle tissue cannot be successfully attached to bone and that connective tissue unites these dissimilar body materials. The tendon of a muscle has been described in its role of connecting the muscle to the skeletal member of the part to be moved. But it must be remembered that the muscle must be anchored at its other end to a more fixed part or parts of the skeletal system. As the muscle contracts, it pulls against this more fixed end just as it pulls through the tendon upon the part to be moved. Since the fixed end is a more stable attachment, the bulk of the power of muscle contraction is transferred through the tendon to the part to be moved.

Origin of the Muscle

The end of the muscle that is ordinarily the fixed end is called the *origin.* Usually, but not always, this is the end that is more centrally located in terms of nearness to the axial skeleton

or more proximally located in terms of position in a limb.

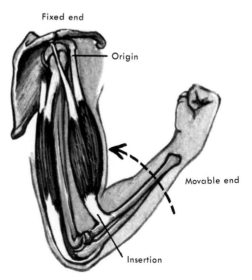

(See also color plate 4.) The muscle fascicles nearest this fixed end are connected by collagenous fibers to the skeletal system just as described for a muscle-tendon junction. Such *fibers of origin* are frequently very short so that the muscle appears to have a "fleshy" origin from the bone. Some muscles encompass an extensive area in gaining their fixed attachment so that they may have a very broad origin from numerous skeletal points.

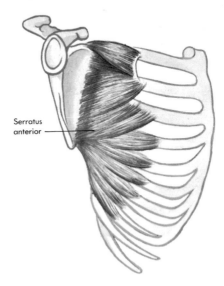

They may have to utilize aponeuroses, intermuscular septa, or interosseous membranes to secure a sufficient area for attachment.

Insertion of the Muscle

Regardless of the arrangement of fascicles and tendons within the muscle, most skeletal muscles run a straight course between the fixed and movable ends. The only exceptions are sphincter muscles in which the fascicles surround a body orifice and attach to themselves. The *insertion* of a muscle is its point of attachment to the skeletal member of the part of the body to be moved. It is, therefore, considered to be the movable end of the muscle. The *tendon of insertion* is the final assembly of all the strands of the tendinous core into a cable or ribbon which then completes the course of the muscle and actually forms the attachment of the movable end of the muscle. The point of attachment of some muscles is at a distance from the fleshy part of the muscle so that the tendon may be longer than the contractile part of the muscle. Matters of leverage and direction of pull fairly precisely determine the point of insertion on a bone. This point is usually smaller than the area of origin. The fleshy part of a muscle usually converges upon the tendon of insertion from broader origins. Although the tendon may be cylindrical or platelike, the point of insertion of the muscle is circumscribed or limited linearly to a small area of the bone. It may even compete for space on a tubercle, ridge, process, or tuberosity with several related muscle tendons.

Functional Reversal of Origin and Insertion

An example of origins and insertions may be seen in the latissimus dorsi muscle. This superficial muscle of the back has a broad origin from the lower thoracic vertebrae, the crest of the ilium, and from aponeurotic fibers of origin from a thick band of fascia connecting these areas. It sweeps upward and laterally along the back and grazes the inferior angle of the scapula; its fibers converge upon a tendon which inserts upon the humerus. The humeral end is the movable end in the usual work of this muscle. The humerus, however, may be fixed and unable to move freely, as when one clings to a pole or tree with the arms above the head. Gravity and the weight of the body, in this instance, have made the humeral end or insertion now the fixed end. In this situation, contraction of the muscle results in the trunk of the body being lifted as in climbing a pole. The fixed end has become movable. In this sense the origin and insertion have become functionally reversed. Such functional reversals of origin and insertion occur, but the description of the muscles concerned lists the fixed origin and movable insertion according to the usually expected action in the anatomic position.

NERVE SUPPLY OF MUSCLE

An electrical impulse, generated in cells of the central nervous system and conducted over nerve fibers, is necessary to induce the contraction of the myofibrillae within the fibers of a voluntary muscle.

Essentials of Nervous System Organization

A brief look at the organization of the nervous system is necessary to understand its role in the function of the muscular system. The *central nervous system* comprises a long mass of nerve cells and their fibers which is enclosed and protected by the axial skeleton. A highly developed portion, the *brain,* is located within the skull. The brain contains the higher centers that exert voluntary control over the activities of the body, including the voluntary muscles. The central nervous system extends inferiorly below the skull as the *spinal cord,* which is enclosed and protected by the vertebral column. The spinal cord contains the actual centers from which nerve impulses are sent out to voluntary muscle fibers. The *peripheral nervous system* consists of the nerve fibers, bundled into cables called *nerves,* which leave and enter the central nervous system. These fibers are integral protoplasmic parts of nerve cells located within the central nervous system but which, in the process of reaching the structures they innervate, are spread throughout the whole body.

Nerve Supply to Voluntary Muscle

The nerve supply to a muscle is referred to as its *innervation.*

Nerve-Muscle Relations.   There are many nerve fibers in the nerve cable that extends peripherally to a particular muscle. As the nerve reaches the muscle, it breaks up into branches which course through the connective tissue core

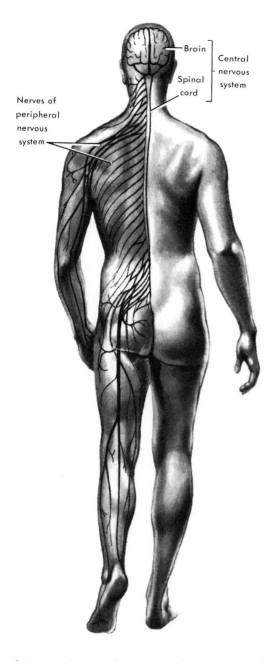

Nerves of peripheral nervous system

Brain

Spinal cord

Central nervous system

nerve cell. The number of muscle fibers in a motor unit varies in voluntary muscles. A large coarse muscle which, though powerful, does not carry out delicate movements may have several hundred voluntary muscle fibers per motor unit. Muscles which perform fine precise movements have fewer muscle fibers per motor unit, approaching a 10:1 to even a 3:1 ratio. As the motor nerve impulse reaches the point of branching of the nerve fiber, it passes over all branches so that if an impulse is sent to a motor unit, all the muscle fibers in that unit will contract. This does not mean that all the muscle fibers in an entire muscle are contracting at the same time. Only as many motor units as are needed for the performance of a specific function receive the impulse to contract. This is regulated within the central nervous system so that some motor units may be resting as others are stimulated to contract. As the need for greater strength of contraction appears, other units are brought into action.

The actual connection between the muscle fiber and the terminal branch of the nerve fiber is made by a specialized nerve ending called the *motor end plate.*

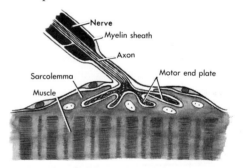

Nerve

Myelin sheath

Axon

Motor end plate

Sarcolemma

Muscle

(See also color plate 12.) Tiny filaments of the nerve ending spread over the surface of the muscle fiber under which is a collection of muscle nuclei. It is believed that the nerve impulse releases a chemical substance, acetylcholine, at the end plate and that this substance initiates the contraction of the myofibrillae within the muscle fiber.

There are two unit membranes which are involved with the transfer of the activating impulse from the nerve ending to the sarcoplasm of the muscle fiber. The first unit membrane is that of the nerve ending, the second that of the muscle fiber. They are related through the actual motor end plate. At this nerve-muscle junction, the unit membrane of the branched nerve ending is folded to lie in

of the muscle. Each branching of the nerve sends a smaller group of nerve fibers into a smaller portion of the muscle. It is rare for one nerve fiber, representing a single motor nerve cell, to be responsible for the innervation of a single muscle fiber. As a single nerve fiber approaches a fascicular group of muscle fibers, or even a group of fascicles, the motor nerve fiber branches to supply them.

The Motor Unit. A *motor unit* consists of all the muscle fibers which are innervated by the branching of the nerve fiber from a single motor

clefts, called *synaptic troughs,* on the surface of the muscle fiber. Here the nerve ending and the muscle fiber are separated only by their protein polysaccharide unit membranes. Mitochondria cluster in the ending of the nerve associated with a number of submicroscopic sacs or *synaptic vesicles.* It is believed that the chemical substance, acetylcholine, is released by the nerve impulse, and that this substance crosses the synaptic trough and passes through the unit membrane of the muscle fiber. Although the acetylcholine is shortly destroyed by the enzyme acetyl cholinesterase in the muscle fiber, this does not occur until the flow of ions across the membranes has set up an electrical potential. This potential spreads through the sarcoplasmic reticulum to induce shortening of the myofilaments, with subsequent contraction of the muscle fiber.

Sensory Mechanisms. Information sent to the central nervous system from any part of the body is termed a *sensation.* Specialized sensory nerve endings are found among muscle fibers or in the tendon of the muscle. These detect the amount of contraction that is taking place, the degree of tension or stretch being placed upon the tendon, and the range of motion occurring in the body part that is being moved. Sensory impulses generated by spindle-shaped nerve endings pass into the central nervous system to be acted upon by other nerve cells. Much of this action is automatic and the sensory information only rarely rises to conscious levels.

Reflex Arcs. A great deal of the action of the nervous system affecting voluntary muscle takes place in the spinal cord. Although the higher centers of the brain can superimpose the influence of the will upon the motor nerve cells of the spinal cord, much of the control of muscle action is automatic in response to sensory information. An involuntary automatic action of the nervous system governing muscle action is called a *reflex.*

The simplest reflex is seen in the case of a threatened burning of the hand when a person reaches to pick up a supposedly cool iron. The heat coming from the iron is a stimulus to sensory nerve endings in the skin of the fingers. These generate a sensory impulse which travels to the spinal cord. (See color plate 12.) The impulse is transferred to motor nerve cells which generate impulses to motor units of muscles which contract to produce a withdrawal movement of the hand. This action is almost instantaneous and requires no action of the will. It occurs even before the sensory information can be conveyed to higher centers in the brain whereby one actually perceives that the iron is hot. The short course through the spinal cord is known as a *reflex arc.* (See color plate 12.)

Another example is found in the case of voluntary muscle which is already contracting. A nerve ending in the tendon generates a sensory impulse that automatically apprises the central nervous system of the degree of stretch or pull upon the tendon. More motor nerve cells can be stimulated to bring further motor units into play if needed. It must be remembered that, although there are many reflex activities which are simple arcs and are handled at a local level in the nervous system, many are very complex and require several nerve pathways to and from higher centers or to multiple motor areas to elicit the coordinated response of a group of muscles.

Gamma Motor Fibers and Muscle Spindles. A variation of the simple reflex arc makes possible the detection and modification of the degree of voluntary muscle contraction. Two structures are involved in understanding this concept, the *gamma motor nerve fiber* and the *muscle spindle.*

Most motor nerve fibers (also called alpha motor fibers) are large, myelinated nerve fibers which extend from the ventral horn cells of the gray matter in the spinal cord to the voluntary muscle fibers by way of the motor end plate. The nerve impulse conducted along these fibers stimulates the contraction of striated voluntary muscle.

The *gamma motor nerve fibers* are much smaller nerve axons which come from different motor nerve cells in the ventral horn of the spinal cord. These fibers travel in the spinal nerve in company with the alpha fibers. The gamma nerve fibers terminate by innervating certain somewhat underdeveloped muscle fibers, which are really proprioceptive sense receptor organs within the muscle. These *muscle spindles* have rudimentary contractile elements within their sarcoplasm which respond to the nerve impulse carried by the gamma nerve fiber. Their special role, however, is to provide sensation, as they contract, of the degree of contraction of the muscle, or of the amount of stretch placed upon the muscle fascicles by postural changes. A similar structure in the tendon of the muscle, the *Golgi tendon organ,* detects the degree of stretch placed upon the muscle tendon when the bone to which it is attached is moved.

Sensory nerve impulses are initiated by the muscle spindles (and tendon organs). These impulses return to the spinal cord through a spinal nerve, ending upon ventral horn motor nerve cells directly or via short association neurons. This completes the mechanism whereby the contraction impulse going out over alpha motor fibers can be modified. In this way the degree and strength of contraction can be controlled in response to information received from the muscle itself.

The gamma motor and muscle spindle concepts are developed further in the section on proprioceptive pathways in Chapter 5.

FUNCTIONAL MECHANICS OF MUSCLE ACTION

A number of mechanical principles of muscle construction have already been presented. Certain functional principles should be emphasized.

Leverage

Muscles, bones, and joints participate in typical lever systems in which the contraction of muscle is the force applied, the bone attached to and moved is the arm of a lever, and the joint between two bones is the fulcrum. Two facts relating to the leverage of muscles are important in assessing their functions.

1. Muscles that attach close to the joint produce a greater total range of motion than muscles attaching farther distally.

2. Muscles that attach close to the joint have less leverage than muscles attaching farther distally and, therefore, produce a less powerful movement.

It can be seen, in summary, that the size of a muscle alone is a poor indicator of its strength. It is necessary to appraise the number of fascicles, the direction of fascicular pull, fascicular-tendon patterns, the number of motor units, and leverage in order to assess the strength and range of muscle action.

Functional Classification of Muscle Actions

The effect that is produced on the part of the body that is moved is the *action* of a muscle. This action is determined not so much by the control of the nervous system which induces contraction as by the course and direction of the muscle pull and the place of attachment of the tendon upon the bone to be moved. Muscles are related to the movement of body parts in the following ways.

Prime Movers. The muscle or muscles whose contraction is directly responsible for a given movement. More than one muscle may be the prime mover or *agonist* in a particular movement. For example, the biceps brachii of the arm produces flexion of the elbow but so does the brachialis muscle which lies beneath it. A given muscle may be a prime mover in more than one type of movement. For example, the biceps muscle also produces supination of the forearm. Frequently a muscle may be more effective as a prime mover in one of its actions. Its other actions are secondary actions in which the muscle may be less effective but of important assistance to another prime mover in a group action.

Antagonists. The muscle or muscles which produce an action opposite to that of prime movers. For example, muscles which extend the elbow are antagonists of those which flex it. The integrating action of the nervous system is of greatest importance in this regard. If both prime movers and antagonists are in a full state of contraction, the joint will be held motionless, and this may be important in stabilizing the joint to bear weight. Usually the nervous system gradually inhibits the action of an antagonist as the prime mover contracts. In this way the action of the prime mover is steadied and smoothed as the antagonist relaxes. (See also color plate 4.)

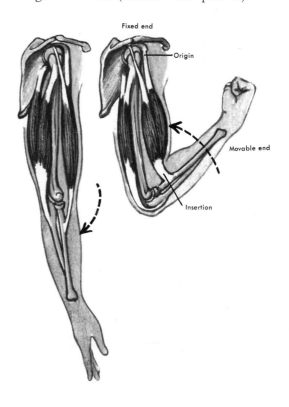

Fixators. Muscles whose contractions stabilize one part of the body so that it will be a firm base for the action of muscles which move another part. For example, muscles that attach to the scapula steady it to make a stable anchorage for muscles which move the arm.

Synergists. Muscles which by their own action facilitate the action of other muscles. For example, flexion of the fingers is weak if the wrist is also flexed. Muscles that extend the wrist are synergists of the finger flexors because extension of the wrist increases the power of finger flexion.

Frequently muscles will fulfill more than one of these functions. Fixator muscles which stabilize a more proximal joint may in so doing facilitate the action of muscles which move a more distal joint and, thereby, act simultaneously as synergists.

triangular like the Greek letter delta, whereas the trapezius or rhomboid muscles have geometric forms. A teres muscle is round, a latissimus muscle broad, and a quadratus muscle quadrangular.

Divisions. For example, a biceps muscle has two heads of origin, a triceps three, and a quadriceps four.

Attachments. Many muscles are named for their skeletal points of origin and insertion. The sternocleidomastoid muscle indicates by its name its attachments to the sternum, clavicle, and mastoid process. Frequently several of these factors are combined in the name of a muscle. The flexor digitorum superficialis is a muscle which flexes the fingers and lies in a stratum of the forearm that is superficial to the flexor digitorum profundus.

THE STUDY OF MUSCLES

The many muscles of the body will be described briefly as individual structures. Their multitude will not be so bewildering if attention is paid to how the muscles are grouped according to their relationship to common skeletal parts or the movements they produce in common.

Naming of Muscles

The name of a muscle is frequently a clue to some important morphological or functional characteristic of the muscle. Muscles are named according to:

Location. For example, the pectoralis major is the great muscle of the breast or chest. The supraspinatus is located above the spine of the scapula.

Direction. For example, the rectus abdominis runs a straight course along the longitudinal axis of the anterior abdominal wall. The direction of the inferior oblique muscle in relation to the orbit and that of the transversus abdominis in relation to the abdominal wall are indicated in the names.

Action. For example, the addition of functional modifiers such as flexor, extensor, or pronator to the name of a muscle indicates its major action.

Shape. For example, the deltoid muscle is

MUSCLES OF THE HUMAN BODY

The important product of the study of muscles is not what muscles are but what muscles do. In this book the muscles will be considered as component units of motive power in the movement of parts of the body. The headings of sections and descriptive tables will indicate the general function of a group of related muscles. Frequently these muscles will also be related to each other in terms of position or similarities of attachment to the skeletal system, but the prime reason for their association together will be that of function. At times it may be necessary to include muscles under a main heading even though they have different functions, mainly because they are located in proximity to the major muscles under consideration. Such muscles will be clearly indicated as "associated" muscles.

MUSCLES THAT MOVE THE HEAD AND NECK

The head is balanced atop the axial skeleton in the erect anatomic position. It is held in this balanced position partly by its own weight pressing down upon the atlanto-occipital joint and partly by the coordinated action of the muscles that approach the base of the skull from all directions. The action of antagonists is important in movements of the head and neck because frequently a movement, once started, is largely completed by gravity. The antagonists then have the role of regulating the gravity movement and checking it at the appropriate point. For example, when the head and neck are bent forward in flexion, the muscles that extend the neck become quite active in slowing and controlling the forward fall of the head produced by gravity. If the body is in the supine position, however, gravity tends to resist flexion of the head, which then can be performed only by active contraction of muscles which can produce flexion.

The mobility of the cervical part of the vertebral column greatly extends the range of motion of the head. It will be found that many of the muscles which produce movements of the neck also move the head either by direct attachments to the skull or because the head is carried along as the neck is inclined in various directions.

Muscles Moving the Head Alone

The head alone can be moved without accompanying movements of the neck. Several small short muscles under the base of the skull are grouped together as the *suboccipital muscles*. There are also several other muscles that run longer independent courses to move the head.

Suboccipital Muscles. These muscles are clustered together in a ring formation immediately below the base of the skull. They run a short course from attachments on the atlas or axis to the occipital bone of the skull or, in one case, from the axis to the atlas. These muscles usually work together to extend or flex the skull by a rocking motion of the occipital condyles on the atlas, or to rotate the skull and atlas together around the pivot joint with the axis. The short, quick, almost automatic movements of the head in following movements of the eyes are largely the function of these muscles.

Sternocleidomastoid Muscle. A long strap-like muscle crosses the neck obliquely from origins on both the anterior surface of the manubrium and the medial third of the clavicle as it ascends to insert upon the mastoid process of the skull. The *sternocleidomastoid* muscle has the effect

of dividing the neck into two geographical areas, a superior and medial *anterior triangle* and a posterior and lateral *posterior triangle*. The *triangles of the neck* are convenient for designating the location of structures in the neck.

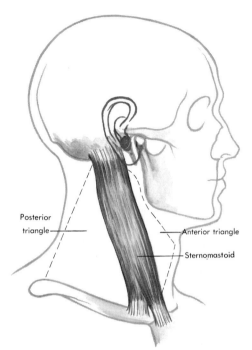

The action of the sternocleidomastoid muscle is to pull the skull downward on the same side and to draw it forward. The result is that the head is tilted to the same side and, at the same time, the face is rotated toward the opposite side. The tendinous origins of the two muscles can be felt diverging at the base of the neck above the sternum where they outline the suprasternal fossa. When one's head is turned to the right, the taut borders of the muscle on the left can be followed obliquely upward to its insertion. If both muscles act at the same time, there can be no turning of the head because the muscles oppose each other. Flexion of the head will result instead, particularly when the body is supine and it is necessary to raise the head, as from a pillow, against the resistance of gravity.

Associated Muscles. A long series of deep muscles, closely related to the vertebral column, exist to produce movements of the axial skeleton. The most superior of these deep muscles extend between the cervical vertebrae and the skull.

SUBOCCIPITAL MUSCLES

MUSCLE	Rectus capitis anterior	Rectus capitis lateralis	Obliquus capitis superior	Rectus capitis posterior minor	Rectus capitis posterior major	Obliquus capitis inferior
LOCATION AT BASE OF SKULL	Anterior	Lateral	Posterior	Posterior	Posterior	Posterior
ORIGIN	Lateral mass of atlas	Transverse process of atlas	Transverse process of atlas	Posterior tubercle on posterior arch of atlas	Spine of axis	Spine of axis
FIBER DIRECTION	Upward, medially	Upward, laterally	Upward, medially	Upward	Upward, laterally	Laterally, slightly upward
INSERTION	Occipital bone anterior to occipital condyle	Occipital bone lateral to occipital condyle	Lateral end of inferior nuchal line of occipital bone	Medial end of inferior nuchal line	Inferior nuchal line between obliquus capitis superior and rectus capitis posterior minor	Transverse process of atlas
HEAD MOVEMENT PRODUCED	Flexion (forward bending)	Lateral flexion to its own side	Extension (backward bending)	Extension	Extension, lateral flexion, rotation to its own side	Extension, lateral flexion, rotation to its own side

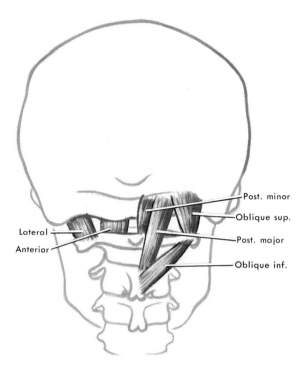

While their parent muscle masses are concerned with movements of the vertebral column, the following components move the head in company with those movements. It should be remembered that when they act, other muscles are producing movements of the neck or of the entire back.

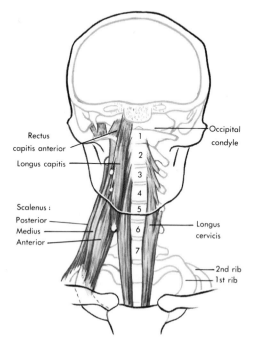

ANTERIOR MUSCLE. The *longus capitis* is a thin muscle which arises from the transverse processes of the third to sixth cervical vertebrae. It runs upward along the front and sides of the upper cervical vertebrae to insert upon the base of the skull in front of the rectus capitis anterior. Its action, similar to that of its suboccipital companion, is that of flexion of the head.

POSTERIOR MUSCLES. Posteriorly located along the vertebral column are the superior portions of the deep back muscles. The *longissimus capitis* and the *semispinalis capitis* continue the long columns of muscles upward to the skull where they attach to the posterior aspect of the base of the skull, overlapping the posterior suboccipital muscles. The *splenius* muscle, which runs from the thoracic spinous processes to the cervical transverse processes, has a part, the *splenius capitis*, which extends from the ligamentum nuchae of the cervical vertebrae upward and laterally to attach behind the sternocleidomastoid muscle on the mastoid process. All these muscles control the forward fall of the head with gravity in flexion

and, as head extensors, bring it back from the flexed position to its normal attitude.

Muscles That Move the Neck

The neck is moved by muscles which are located anterior or posterior to the bodies of the cervical vertebrae. These muscles have their lower attachments to the axial skeleton below the neck.

Scalene Muscles. Three slender fusiform muscles descend obliquely laterally from attachments to the anterior surface of the transverse processes of the cervical vertebrae to the first or second rib. The *scalene muscles* have an anteroposterior relation to each other and thus are named the *scalenus anterior, medius,* and *posterior.*

SCALENE MUSCLES

MUSCLE	ORIGIN	INSERTION
Scalenus anterior	Third to sixth cervical transverse processes	Scalene tubercle of first rib
Scalenus medius	All cervical transverse processes	First rib posterior to subclavian groove
Scalenus posterior	Fourth to sixth cervical transverse processes	Second rib

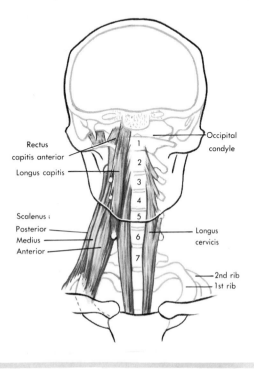

The scalene muscles are active in lateral flexion of the neck but, coming forward in their descent, can contribute to forward bending. It should be noted, however, that the sternocleidomastoid muscle, while pulling on the head, is powerful in lateral flexion of the neck. The two sternocleidomastoid muscles acting together are stronger in flexion of the head and neck than the scalenes. The most effective action of the scalene muscles is to elevate the first and second ribs and, therefore, to suspend the superior thoracic aperture. They become accessory respiratory muscles by virtue of their effort to lift the thorax and thereby increase breathing capacity in forced or labored respiration.

Longissimus Cervicis.   Along the sides and front of the cervical vertebrae is a flat muscle, the *longus cervicis,* which arises from the bodies of the first to third thoracic vertebrae. Other fibers are added to the muscle from the lower cervical bodies as it ascends to insert upon the upper cervical bodies. As it is inserting, the longus capitis muscle is beginning from transverse processes to run upward to the skull. The longissimus cervicis muscle is primarily a flexor muscle of the neck.

Associated Muscles.   The long posterior muscle masses of the deep back muscles have attachments to the back of the neck. Cervical portions of the *longissimus, semispinalis,* and *splenius* muscles described with the head muscles form a group of posterior cervical muscles. These, to be more fully described with muscles of the vertebral column, act with the muscles of the back to control flexion of the neck and to return the flexed neck and head to the upright position.

Hyoid Muscles

Two sets of muscles are related to the hyoid bone which is unarticulated to the rest of the skeleton, and is located in the neck between the mandible and the larynx.

Hyoid Bone.   The *hyoid bone* is a slender, U-shaped bone which is suspended from the styloid processes of the base of the skull by the stylohyoid ligaments. The muscles attaching to it serve to fix the bone in position by their contraction and also use it for their own support and leverage. The hyoid bone consists of the curved cylindrical *body,* which faces forward, and two

arms, projecting backward, each of which is termed the *greater cornu* (horn). A small upward tip, the *lesser cornu,* is located at the junction of each greater cornu with the body. The muscles which attach to the hyoid bone are called the *suprahyoid* and *infrahyoid muscles* because of their positional relationship to the bone.

Muscles Anchoring the Hyoid Bone.   The *infrahyoid muscles* run upward from the axial skeleton and pectoral girdle to anchor the hyoid bone against the pull of the suprahyoid muscles, which use the bone for their lower attachment. The infrahyoid muscles are described in the following table.

Muscles Controlling the Floor of the Mouth. The *suprahyoid muscles* form or shape the floor of the mouth. All utilize the hyoid bone for their lower attachment and, therefore, depend on the contraction of the infrahyoid muscles to provide a fixator action on the hyoid bone. The suprahyoid muscles are described in the table on the following page.

MUSCLES OF MASTICATION

The muscles of mastication are the muscles which act upon the temporomandibular joint to close the jaws. The masticatory muscles are intensely strong. They produce the motive power to close the jaws forcibly, to clench the teeth, and to grind the food. These muscles protrude and retract the mandible, and move it from side to side. In contrast to the action of the masticatory muscles, opening of the jaws is largely the result of gravity during inhibition of masticatory muscle action, assisted by the contraction of the suprahyoid and platysma muscles. The muscles of mastication are tabulated on Page 169.

MUSCLES THAT MOVE THE VERTEBRAL COLUMN

Just as the head is balanced atop the vertebral column, so the axial skeleton and pelvic girdle are balanced upon the heads of the femurs at the ball and socket hip joints. The line of the center of gravity of the body falls in front of the vertebral column, through the hip joints, and just in front of the knee joints. The vertebral column supporting the trunk of the body is kept erect (*Text continued on page 170*)

MUSCLES ANCHORING THE HYOID BONE

MUSCLE	Sternohyoid	Omohyoid		Sternothyroid	Thyrohyoid
SHAPE	Strap	Two bellies united by tendon		Strap, under sternohyoid	Strap
ORIGIN	Back of manubrium of the sternum	*Inferior belly* from upper border of the scapula	*Superior belly* from the central tendon	Back of manubrium of the sternum	Thyroid cartilage of the larynx
FIBER COURSE	Upward in the neck	Forward and upward under sternocleidomastoid to end in tendon	Upward from central tendon toward hyoid bone	Upward toward larynx	Continues plane of sternothyroid upward toward larynx
INSERTION	Lower border of body of hyoid bone	In central tendon	Lower border of body of hyoid bone	Thyroid cartilage of larynx	Greater cornu of hyoid bone
ACTION		All muscles anchor hyoid bone by drawing it downward			

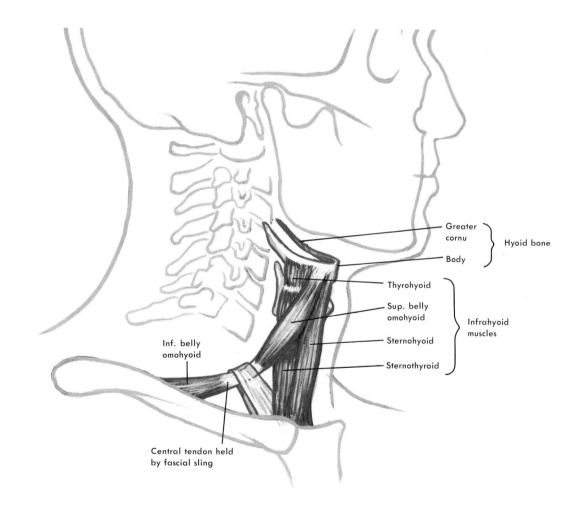

MUSCLES CONTROLLING THE FLOOR OF THE MOUTH

MUSCLE	Digastric		Stylohyoid	Mylohyoid	Geniohyoid
SHAPE	Fusiform, two bellies united by a tendon		Slender cylinder	Flat sheet	Thin band
ORIGIN	*Posterior belly* from mastoid notch of temporal bone	*Anterior belly* from digastric fossa of mandible near symphysis	Styloid process	Mylohyoid line on internal surface of mandible	Genial tubercle of symphysis of mandible
FIBER COURSE	Forward and downward toward hyoid bone	Backward and downward under floor of mouth	Along posterior border of the *digastric,* splitting to allow passage of its tendon	Sheet of fibers forms floor of mouth as it crosses to meet fellow of other side	Backward above mylohyoid muscle
INSERTION	The two bellies unite to a central tendon which is attached to the body and greater cornu of the hyoid bone by a fascial loop		Hyoid bone at the junction of the body and greater cornu	A central fibrous line in floor of mouth formed as the two fellows fuse; posterior fibers to hyoid	Front of body of hyoid bone
ACTION	Pulls chin backward and downward in opening mouth		Draws hyoid bone up and backward but resisted by infrahyoid muscles; total effect is elongation of floor of mouth	Forms floor of mouth; Elevates floor of mouth in swallowing; Supports tongue	Pulls hyoid bone forward, shortening floor of mouth

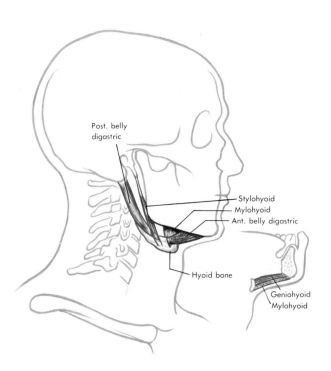

Post. belly digastric

Stylohyoid
Mylohyoid
Ant. belly digastric

Hyoid bone

Geniohyoid
Mylohyoid

MUSCLES OF MASTICATION

MUSCLE	Masseter		Temporalis	Medial pterygoid	Lateral pterygoid
SHAPE	Rhomboid		Fan	Quadrilateral	Two-headed
ORIGIN	Zygomatic arch: *superficial part* from anterior two thirds of inferior border	*Deep part* from the medial surface	Inferior temporal line of skull bordering temporal fossa	Medial surface of lateral pterygoid plate of base of skull	*Upper head* from greater wing of sphenoid bone; *lower head* from lateral surface of pterygoid plate
FIBER COURSE	Downward and backward	Downward vertically	Converge along floor of temporal fossa to pass under zygomatic arch	Downward and backward along medial surface of ramus of mandible	Backward through infratemporal fossa, heads converging
INSERTION	Lateral surface of ramus of mandible		Coronoid process of mandible	Medial surface of angle of mandible	Front of neck of mandible; capsule and disk of temporomandibular joint
ACTION	Draws mandible upward		Draws mandible upward; posterior fibers retract mandible	Draws mandible upward; protrudes mandible with lateral muscle	Protrudes mandible; aids in side-to-side movement of mandible

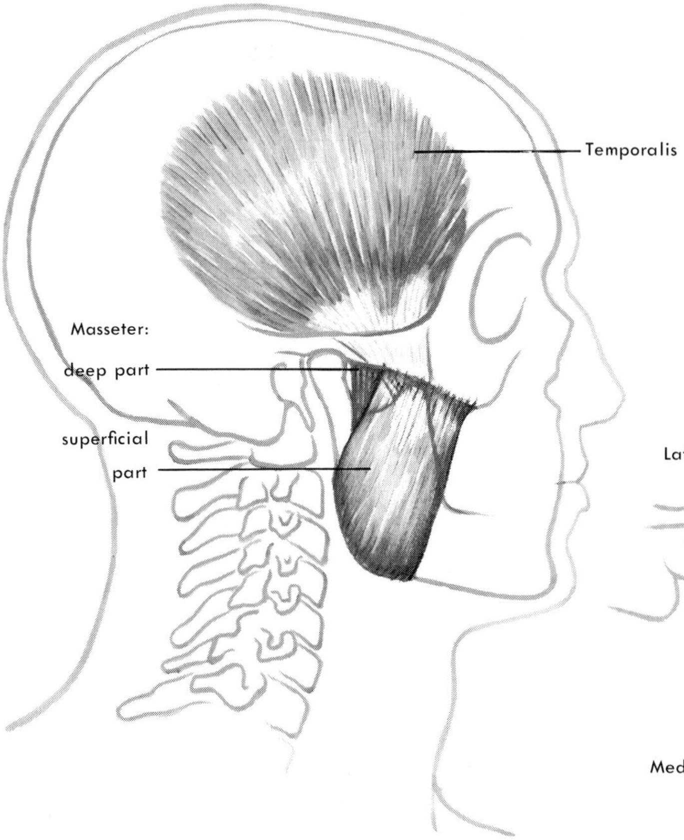

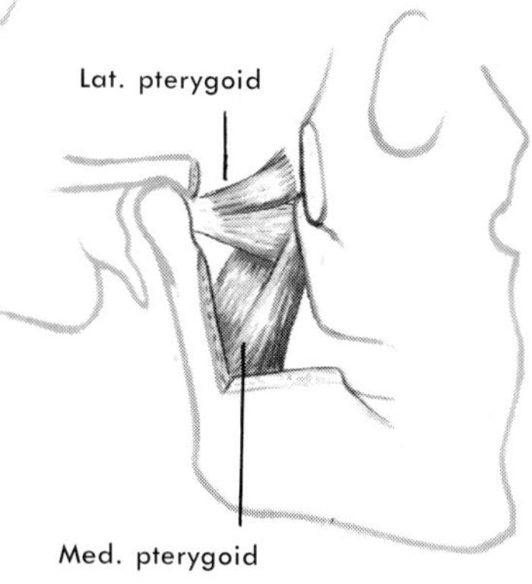

more by muscles of the back of the thighs, which prevent the pelvis (and, therefore, the vertebral column) from falling forward, than by the muscles of the vertebral column itself. The muscles which move the vertebral column act to change the position of the trunk of the body, but are usually not required to maintain the erect position.

The muscles which are most closely related to the vertebral column are located posterior to it and are called the *postvertebral* muscles. Other muscles acting in front of the vertebral column and consisting largely of abdominal muscles are the *prevertebral* muscles.

Postvertebral Muscles

If one's hand is placed across the midline of the back, a long prominent mass of muscle can be felt on each side. These longitudinal muscle masses are the result of the fusion of many groups of small muscles which, in development, spanned an interval from only one vertebral segment to the next. Three layers of postvertebral muscles make up the longitudinal mass on each side.

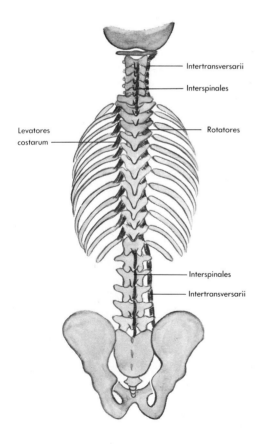

Intertransversarii

Interspinales

Levatores costarum

Rotatores

Interspinales

Intertransversarii

Deep Postvertebral Muscles. In the cervical and lumbar regions, small bands of muscle extend between parts of adjacent vertebrae. The *interspinales* form a series passing from one spinous process to another, whereas the *intertransversarii* similarly extend between adjacent transverse processes. In the thoracic region, short *rotatores* ascend medially to each vertebral lamina from the transverse process below, and *levatores costarum* pass to the rib below. All these short muscles can affect the entire vertebral column only in the summation of their movements. They are more effective in the short, shifting movements of aligning adjacent vertebrae. The suboccipital muscles moving the head are developmentally of this group. (See illustration in right column.)

Intermediate Postvertebral Muscles. More fusion is present in the longitudinal mass of muscle covering the deep muscles. Muscle fascicles arise from the transverse processes of all of the vertebrae and the posterior surface of the sacrum to pass upward and medially to vertebral spines above. Many of the fascicles, through fusion, span several vertebral segments, and all mingle into a complicated mass which is collectively called the *transversospinalis muscle complex.* Certain component parts can be identified:

1. The *multifidus* is a thick block of fascicles of the muscle located in the lumbosacral region.

2. *Semispinalis thoracis* is the group of fascicles attaching up to the midthoracic region of the vertebral column.

3. *Semispinalis cervicis* is a portion of the fused muscle mass which originates from the upper thoracic transverse processes and extends to cervical spinous processes. As an entity it has been mentioned as a posterior neck muscle. (See illustration in left column of following page.)

4. *Semispinalis capitis* is the most superior part of this long muscle mass in which the fibers extend from cervical transverse processes to the posterior aspect of the skull. It has been described previously as if it were a muscle moving the head alone, as it does in company with muscles moving the vertebral column as a whole.

Superficial Postvertebral Muscles. The more external and more lateral portions of the long back muscles are collectively called the *erector spinae* or *sacrospinalis complex.* In this complex the muscle fibers show even greater fusion into well-defined, vertically coursing muscle groups. The lowermost portion is a bulky mass of fibers aris-

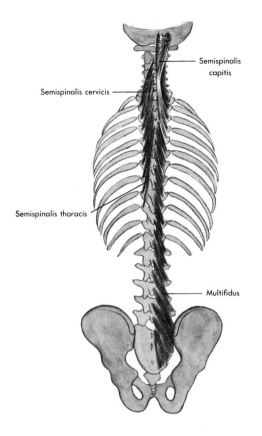

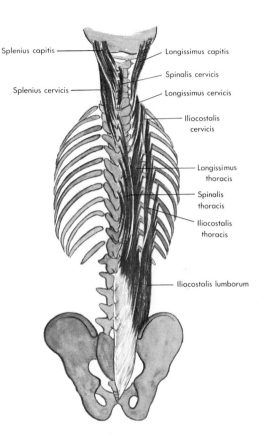

located laterally to the rest of the complex. The *iliocostalis lumborum* continues the erector spinae

ing from the ilium and lumbar vertebrae and from a strong fascial sheet stretched across the interval between them. These fibers sweep upward and laterally as the true *erector spinae* or *sacrospinalis* to end upon the lower ribs. Succeeding fused masses of muscle fibers continue the long chain of superficial fibers upward as three well-defined columns.

1. The medial column is an insignificant narrow band known as the *spinalis thoracis,* which is continued in the neck as the *spinalis cervicis.*

2. The intermediate column continues the thick erector spinae or sacrospinalis upward by a new supply of fused fibers that bridge the ribs as the *longissimus thoracis.* This long mass sends inserting fibers at segmental intervals to the ribs and to the thoracic transverse processes. New fibers join the mass to extend the column upward to the cervical vertebrae as the *longissimus cervicis* and to the base of the skull as the *longissimus capitis.* The latter two portions have been referred to as muscles moving the neck and head in company with movements of the vertebral column.

3. The lateral column, given the name of the *iliocostalis,* represents successive groups of fibers

upward as a lateral band attaching to the lower ribs. It is succeeded by the *iliocostalis thoracis,* which spans the lower and upper ribs. The *iliocostalis cervicis* completes this column between the upper ribs and the transverse processes of the fourth to sixth cervical vertebrae.

Splenius Muscle. The previously mentioned *splenius* muscle is applied like a restraining band over the other postvertebral muscles in the upper thoracic and cervical regions. Its *splenius cervicis* portion extends upward and laterally from the upper thoracic spinous processes to the lower cervical transverse processes, whereas its *splenius capitis* portion takes a similar course from the lower cervical spinous processes and the ligamentum nuchae to the mastoid process of the skull. These work in harmony with the other postvertebral muscles although their effect is upon the head and neck.

Actions of the Postvertebral Muscles. All the postvertebral muscles work together to extend the vertebral column as a whole through their effect upon the parts of the column, to which they attach. Complete arching of the back with backward bending of the head is an example of their coordinated actions. Similar movement of the neck or head alone can be performed by individual components of the muscle complexes. Equally important actions of the postvertebral muscles are to control or regulate the forward flexion movements of the trunk, neck, and head which occur largely by gravity and to restore these parts of the body from the flexed position. If muscles on one side act alone, lateral flexion of the neck or trunk will be produced. Rotation of the trunk is accomplished by the action of the postvertebral muscles, which course obliquely on the side of rotation, and by the contraction of the prevertebral abdominal muscles.

Prevertebral Muscles

The prevertebral muscles oppose the actions of the postvertebral muscles. The main prevertebral muscles are not located close to the vertebral column but enclose the trunk laterally and anteriorly. These are collectively known as the *abdominal muscles.* They are described separately. Another muscle, the *iliopsoas,* extends from the sides of the lumbar vertebrae and the iliac fossa to the femur. It is more concerned with movements of the femur. Although these prevertebral muscles do have other primary functions, they are the natural antagonists of the postvertebral group and are brought into action to control extension of the vertebral column. Furthermore, when the body is in the supine position, gravity resists rather than assists flexion of the vertebral column. Then the abdominal muscles contract powerfully to flex the vertebral column, as when one sits up in bed.

MUSCLES PROVIDING FACIAL EXPRESSION

A group of thin muscles that are closely applied to the scalp and to the face have developed primarily to move the scalp or to guard the openings of the skull. The contraction of these muscles also produces characteristic movements of the face in company with the display of emotion by the individual. The group of scalp and facial muscles, therefore, has come to be described as the *muscles of facial expressions.*

Muscles Moving the Scalp

The scalp is a thick mobile membrane of skin and fibrous tissue which invests the cranium. The mobility of the scalp is a factor in warding off and absorbing much of the force of blows to the skull. Movement of the scalp with the force of a blow also minimizes injury to the scalp itself. The skin and subcutaneous tissue are tightly bound down to a smooth collagenous sheet which invests the cranium. This membrane is the *epicranial aponeurosis* (galea aponeurotica). Only a loose connective tissue unites the epicranial aponeurosis to the periosteum of the cranial vault. The scalp can move upon the cranium because of the loose connective tissue under the aponeurosis.

The collagenous membrane of the epicranial aponeurosis is replaced posteriorly and anteriorly by voluntary muscle fibers. The combination of the muscle fibers with the aponeurosis is termed the *occipitofrontalis.* The *occipital* fibers of the muscle spread posterolaterally over the skull to insert upon the occipital bone along the superior nuchal line, above the attachments of posterior neck muscles. The *frontal* fibers of the muscle extend over the forehead to insert by blending with those facial muscles which encircle the orbits or cover the root of the nose. The occipital fibers draw the scalp backward; the frontal fibers draw it forward. The frontalis portion also wrinkles the forehead and raises the eyebrows.

Small *auricular muscles* arise from the epicranial aponeurosis anterior, superior and posterior to the external ear. The *anterior auricular* slants posteroinferiorly, the *superior auricular* passes inferiorly, and the *posterior auricular* extends anteriorly to the medial surface of the framework of the external ear. These small muscles contribute very little to movements of the scalp and, except in occasional individuals, in the human have lost their power to move the ears.

Muscles Producing Facial Movements

Facial movements and the variety of facial expressions are largely the by-products of the action of muscles which exist primarily to guard the orifices of the orbits, nose, and mouth. The muscles responsible for the protection of these openings act as sphincters or dilators to control the degree of opening or closure of the eyelids, nostrils, or lips.

Eyelid Closure. In the skull the orbit is bounded by the bony orbital margins, but in the living person the apparent aperture is the opening between the eyelids known as the *palpebral fissure.* When the eyes are fully opened, the palpebral fissure is an oval aperture, rounded laterally and drawn to a point medially, through which the white and colored portions of the eye can be seen. The fissure is obliterated as the eyelids are closed. Gentle closure of the eyelids is possible by gravity alone if the muscle responsible for lifting the upper eyelid is relaxed or inhibited from acting. Then, as in fatigue or in composing oneself for sleep, the upper lid is permitted to drift gently downward over the eyeball to meet the lower lid.

Active closure of the eyelid requires the action of the *orbicularis oculi.* Part of the fibers of this muscle are contained in the substance of the eyelids. These muscle fibers, comprising the *palpebral portion* of the orbicularis oculi, produce gentle voluntary closure of the eyelids. The *orbital portion* of the muscle consists of a wider concentric group of fibers which swing around the eyelids onto the frontal bone above, the zygoma laterally, and over other facial muscles below. When the orbital fibers contract, usually accompanied by contraction of those of the palpebral portion, the eyelids are closed forcibly and quickly, the eyelashes are buried between the lid margins, and tears are forced into the lacrimal drainage ducts. This is the type of closure that protects the eyes from a threatening object or occurs in the grimacing expressions that accompany pain, disgust, or weeping. Blinking, a reflex phenomenon in which quick movements of the eyelids keep the eyeball moist by the spreading

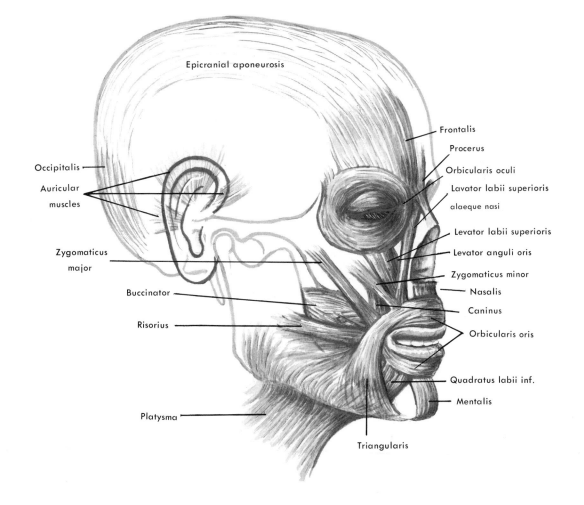

Epicranial aponeurosis

Occipitalis

Auricular muscles

Zygomaticus major

Buccinator

Risorius

Platysma

Frontalis

Procerus

Orbicularis oculi

Lavator labii superioris alaeque nasi

Levator labii superioris

Levator anguli oris

Zygomaticus minor

Nasalis

Caninus

Orbicularis oris

Quadratus labii inf.

Mentalis

Triangularis

of tears across its surface, is accomplished mainly by the palpebral portion of the muscle alone.

Eyelid Opening.   There is no muscle of facial expression which, applied flatly to the surface of the skull, can oppose the orbicularis oculi and act to open the eyelids. The *levator palpebrae superioris* arises deep in the interior of the orbit, but instead of attaching to the eyeball, this muscle passes forward to enter into the upper lid. It will be described more fully with the muscles of the eye. Contraction of the muscle produces elevation of the upper lid. Frequently in conscious efforts to keep the eyes opened, the frontalis muscle attempts to assist by elevating the eyebrows and skin of the forehead.

Nasal Movements.   The nose and nostrils of man do not have the great range of motion found in animals. The most that can be expected is some dilation and constriction of the nostrils accompanying the emotions of fear, anger, and disdain.

When breathing is labored, the nostrils dilate in an attempt to admit more air to the respiratory system.

Dilation of the nostrils is produced by two muscles, the *depressor septi* and the *dilator naris*, small muscles which extend upward from the maxilla to the septum and wing of the nose respectively. They draw these structures downward to enlarge the nostrils.

Constriction of the nostrils is produced by the *compressor naris* of each side which extends from the maxilla over the cartilages of the wing of the nose to meet its fellow over the bridge of the nose. As these muscles contract, they press the nasal cartilages inward to reduce the size of the nostrils.

Wrinkling of the nose is produced by a number of small muscles, including the *compressor naris*. Situated on the forehead above the root of the nose is the *procerus*, which is developmentally part

MUSCLES OPENING THE MOUTH

MUSCLE	Zygomaticus major	Zygomaticus minor	Levator labii superioris	Levator labii superioris alaeque nasi	Levator anguli oris
ORIGIN	Zygoma	Zygoma under orbicularis oculi	Maxilla at infra-orbital foramen	Frontal process of maxilla	Body of maxilla Deep to other muscles
FIBER DIRECTION	Downward, medially	Downward	Downward	Downward, laterally	Downward, laterally
INSERTION	Angle of mouth	Lateral part of upper lip	Medial part of upper lip	Medial part of upper lip and ala of nose	Lateral part of upper lip and angle of mouth
ACTION	Draws corner of mouth upward	Draws upper lip upward	Draws upper lip upward	Draws upper lip upward; pulls ala of nose up	Draws upper lip upward

MUSCLE	Risorius	Depressor anguli oris	Depressor labii inferioris	Mentalis
ORIGIN	Fascia over parotid gland	Mandible below corner of mouth	Same	Front of mandible
FIBER DIRECTION	Horizontally, medially	Upward	Upward, medially	Downward, medially
INSERTION	Angle of mouth	Angle of mouth	Its side of lower lip	Skin of chin
ACTION	Draws corner of mouth laterally	Pulls corner of mouth downward	Draws lower lip downward	Wrinkles skin of chin; pulls lower lip forward

of the frontalis muscles on each side. These fibers insert into the deep layers of the skin between the eyebrows. Their contraction wrinkles the skin of the root of the nose upward. Under the medial encircling fibers of the orbital portion of the orbicularis oculi are short muscular bands of the *corrugator supercilli,* which run transversely from the frontal bone at the root of the nose to insert into the skin of the eyebrows. When these fibers contract, the eyebrows are drawn together and vertical wrinkles occur in the forehead down to the root of the nose, as in deep thought or frowning. In addition, muscles of the upper lip which insert into the skin under and about the nose will wrinkle the nose when the lip is drawn sharply upward.

Opening of the Mouth. Opening or closing the mouth can be done passively as an accompaniment to opening or closing the jaws, which involves a different set of muscles from those which control the lips. Opening of the lips is performed by a number of small facial muscles which extend from origins on the facial bones to insertions into the deep layers of the skin of the lips or at the angles of the mouth.

The insertion of all these muscles into the deep layers of the skin of the lips or the area about the lips facilitates the movement of the very mobile facial skin, in addition to separating the lips or drawing portions of them upward or downward. The mouth is opened, but in addition a variety of facial expressions may be produced. All these muscles work together, although certain ones will make the major contribution to a particular expression. The zygomaticus major is particularly active in drawing up the corners of the mouth in smiling. The depressor anguli oris is predominant in the opposite movement in expressions of sadness and despair. When the upper lip is wrinkled upward in disbelief, negativity, or

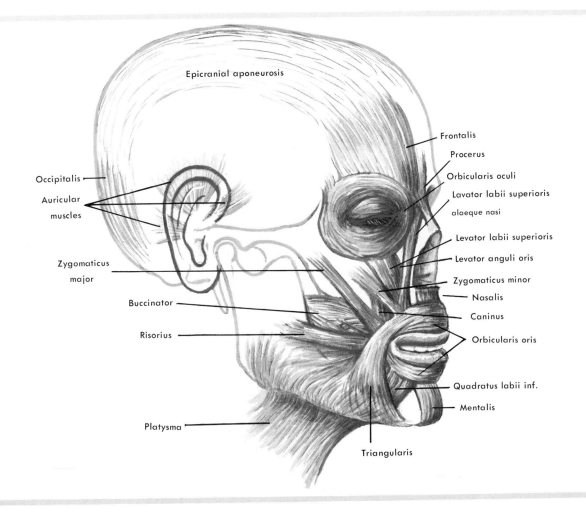

Epicranial aponeurosis

Frontalis

Procerus

Orbicularis oculi

Lavator labii superioris

alaeque nasi

Occipitalis

Auricular muscles

Levator labii superioris

Levator anguli oris

Zygomaticus major

Zygomaticus minor

Nasalis

Buccinator

Caninus

Risorius

Orbicularis oris

Quadratus labii inf.

Mentalis

Platysma

Triangularis

disdain, the zygomaticus minor and the levators of the upper lip are active. The risorius is in full play when the teeth are bared or the corners of the mouth pulled sideways in grimaces. Note that the levator labii superioris alaeque nasi is a muscle of the upper lip that also produces wrinkling of the nose. All the muscles which insert into the skin of the angle of the mouth also blend their fibers together in an interlacing common muscle bundle. Many rapidly changing movements of the corners of the mouth in speaking, chewing, smiling, and grinning may be the result of the contraction of groups of muscle fibers which change from instant to instant. Although the mentalis does not open the mouth, it is a most expressive muscle as it wrinkles the chin or causes the lower lip to protrude during thought, doubt, or disagreement.

If all the lip-opening muscles contract at once, the lips are opened to their maximum extent, as in a wide separation to show a dentist the bite of one's teeth. If the mouth must be opened wider, the masticatory muscles are called upon to open the jaws. Such independent action is rare, for a certain amount of jaw opening accompanies any separation of the lips, whether in speaking, smiling, yawning, or eating. (See table on preceding page.)

Closing of the Mouth. Closure of the oral aperture combines closure of the jaws and approximation of the lips. Tight sealing of the aperture is performed by the lip-closing muscles alone. The primary muscle in closing of the mouth is the *orbicularis oris,* a sphincter muscle surrounding the mouth and making up much of the substance of the lips. Only a few of its fibers swing completely around the mouth. Most of them insert into the common muscle mass at the corners of the mouth or cross at the corners into the lateral part of the opposite lip. Many of the vertically disposed muscles of the upper and lower lips run into the substance of the orbicularis oris laterally and at the corners of the mouth. Partial contraction of the orbicularis oris closes the lips gently. Complete contraction seals the lips and sets them tightly in line.

Another muscle, the *buccinator,* comes forward from origins about the pharynx and from the maxilla and mandible near the molar teeth. As this muscle extends forward toward the corner of the mouth it makes up most of the fleshy mass of the cheek. Superior and inferior fibers pass into the lateral parts of the upper and lower lips respectively, while the central fibers crisscross (decussate) into both. The buccinator compresses the cheeks against the jaws to hold food between the teeth and tongue for chewing. The muscle also acts with the lip muscles which insert into the lateral part or corners of the mouth to pull the lips close upon the teeth and to set them firmly in determined or angry expressions.

Despite the antagonism of the muscles which open and close the lips, many of their fibers act simultaneously to align the lips in the many varieties of expression, particularly in "pursing," whistling, singing, and as the result of changing moods.

Nerve Supply. All the muscles of the scalp and of facial expression are innervated by the seventh cranial nerve, the facial nerve, except for the levator palpebrae superioris. Since the facial nerve is the sole motor supply of the many facial muscles, disease or injury affecting it will have widespread effects which are disastrous to the individual. *Facial paralysis* results in a flat expressionless face on the side affected. The angle of the mouth droops and saliva cannot be controlled. Food cannot be well kept between the teeth. Smiling and pursing of the lips are impossible. The person cannot wrinkle his forehead or elevate his eyebrow on the paralyzed side. His eyes cannot be closed, and in the absence of blinking, tears roll over the margins of the eyelids. This condition is distressing not only because of the mechanical problems but also because, by interference with facial expression, the personality of the person suffers both outwardly and inwardly.

Platysma Muscle. A sheet of voluntary muscle runs superficially in the tela subcutanea of the anterior and lateral aspects of the neck. This sheet, the *platysma muscle,* is much like the muscles of facial expression in that it is related to the skin and is innervated by the facial nerve. The platysmal fibers take origin from the skin and tela subcutanea over the clavicle and the deltoid muscle on each side. The fibers run upward through the tela subcutanea of the neck to insert upon the lower border of the mandible and into the tissues of the corner of the mouth. This muscle functions mainly to smooth the skin of the neck and to prevent the formation of a sharp concavity between the jaw and the neck which might result in pressure upon the neck veins. The muscle can be demonstrated by tensing the jaw and drawing out the corners of the mouth. Its strands then stand out as cords in the neck.

MUSCLES ACTING UPON THE TRUNK

The trunk is the cylindrical mass of the body, exclusive of the head, neck, and extremities.

The Trunk and the Body Wall

The trunk, when viewed externally, is more familiarly divided into the back posteriorly and the thoracic and abdominopelvic regions anterolaterally. These regions are supported by the thoracic and lumbosacral vertebrae. The anterior and lateral aspects of the abdominal region have no skeletal support as do the thoracic and upper abdominal organs which are protected and supported by the sternocostal portions of the axial skeleton. The muscles and fasciae of the trunk close the abdominal wall and support the organs within it.

Trunk Muscles in General

The trunk muscles spiral about the trunk from the first rib superiorly to the iliac crest and the inguinal ligament inferiorly. Fusion, migration, and interweaving of the muscles make the abdominal wall intact. Many abdominal muscles have attachments or courses related to the thoracic "cage." Furthermore, it will be seen that many of the abdominal muscles, and the subcutaneous tissues overlying them, are innervated by the lower thoracic spinal nerves (intercostal and subcostal) which follow the paths taken by the migration of embryonic musculofascial and dermal plates to form the abdominal wall. These interrelations of the thoracic and abdominal portions of the trunk should indicate their common enclosure of these regions of the body.

Muscles Moving the Ribs

Movement of the ribs is an important part of the increase and decrease in the volume of the thorax during breathing. Rib movements take place about the pivots formed by the articulations of the ribs anteriorly with the sternum and posteriorly with the vertebrae. The muscles responsible for these movements also are aligned to form curtains which close the thoracic interspaces between the ribs. (See table on following page.) The focusing of attention upon the regular curving bars of the ribs detracts from a proper perspective of the thoracoabdominal trunk musculature. Even though the thoracic trunk muscles are separately named or numbered, they are but ribbonlike, segmentally located portions of sheets of voluntary musculature which are interrupted by the ribs. The true nature of the trunk musculature is seen inferior to the thoracic skeleton in the continuous sheets of the abdominal muscles.

SERRATUS POSTERIOR. Two thin fibromuscular sheets are found spreading from each side of the vertebral column across the deep back muscles and the posterior thoracic wall under the more superficial muscles. The *serratus posterior superior,* originating from the ligamentum nuchae and the spinous processes of the lower cervical and first few thoracic vertebrae, passes under the rhomboid muscles to the posterior surface of the second to fifth ribs. The *serratus posterior inferior* crosses from the lower thoracic and upper few lumbar vertebral spinous processes under the latissimus dorsi to attach to the lower four ribs. These muscles are inconstant, variable, and so thin that no justifiable function can be ascribed to them other than that of slightly contractile deep fascia.

QUADRATUS LUMBORUM. A four-sided muscle, the *quadratus lumborum,* forms part of the posterior wall of the abdomen in front of the erector spinae portion of the long back muscles (p. 171). The quadratus lumborum ascends from its origin on the posterior part of the iliac crest to attach to the transverse processes of the upper four lumbar vertebrae and to the twelfth rib. This muscle holds the last rib down against the pull of the diaphragm in inspiration and also functions to assist in lateral bending of the vertebral column.

Action of the Rib Musculature. The actual function of each of the sets of muscles just tabulated is still the subject of controversy. Some texts indicate functions of elevation or depression of the ribs for one or another muscle that are at variance with the statements of other authors. Electromyographic studies show that the rib muscles stabilize the thoracic interspaces and regulate their size rather than act as prime movers in breathing.

It is suggested that the main changes in thoracic volume during breathing should be ascribed

MUSCLES ACTING UPON THE RIBS

Muscle	External intercostal	Internal intercostal *Innermost intercostal* †Subcostal	Transversus thoracis	Levatores costarum
Location	Outer plane of muscles between ribs, from tubercle to costochondral junction	Inner plane of muscles between ribs, from sternum to angle of rib	Inner surface of sternum and costal cartilages	On internal surfaces of ribs near their posterior articulations
Origin	Lower margin of each rib except twelfth	Lower margin of costal cartilage and rib	Posterior surface of lower half of sternum and xiphoid process	Transverse processes of thoracic vertebrae
Fiber Direction	Downward and forward	Downward and backward	Thin flat bands course superolaterally	Inferolaterally
Insertion	Upper margin of rib below	Upper margin of rib below	Costal cartilages of ribs 2 to 6	Rib below between tubercle and angle

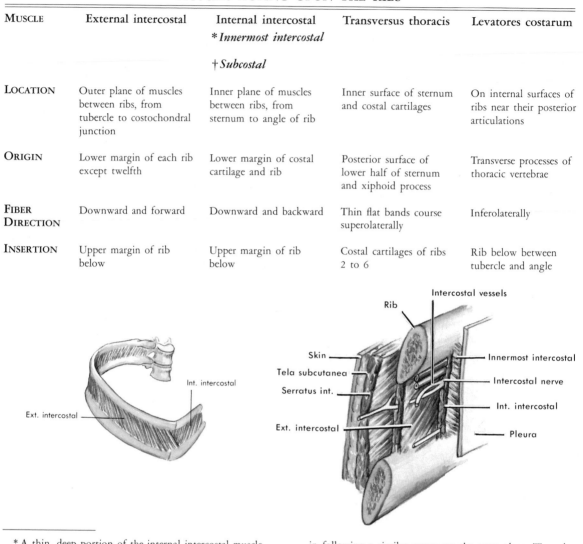

*A thin, deep portion of the internal intercostal muscle sometimes described separately because, in the lower intercostal spaces, it is split away by the passage of the intercostal nerve.

†Morphologically similar to internal intercostal muscles

in following a similar course on the same plane. The subcostal muscles are flat bands near the angles of the ribs. These muscles extend from a rib over several interspaces to insert on the internal surface of the second or third rib below.

to the bellows action of the diaphragm and the lifting of the upper thorax by the scalene and sternocleidomastoid muscles of the neck. The external and internal intercostal muscles and the levatores costarum certainly act to assist in the elevation of the upper ribs especially, but are not prime movers in the action. All the muscles tabulated above are active in stabilizing the intercostal spaces against the change of intrathoracic pressures during breathing. These muscles preserve the intervals between the ribs so that the ribs are not hauled violently upward, like the louvers of a shutter, by the pull of the neck muscles or downward by the pull of abdominal muscles. The various rib muscles help in inspiration by transmitting an even pull from rib to rib and, importantly, prevent the intercostal spaces from being pushed outward. During expiration, as the intrathoracic pressure decreases, the rib muscles act to prevent inward collapse of the tissues of the interspace. The transversus thoracis stabilizes the lower thoracic margins by drawing the sternum and costal cartilages inferiorly against the upward pull of other muscles. It thereby assists the diaphragm by a fixator action as does the quadratus lumborum.

The rib muscles of one side acting together assist the back and abdominal muscles of that side in producing lateral bending of the trunk.

Abdominal Muscles Acting Upon the Trunk

The primary functions of the abdominal trunk muscles are to:

1. Work in concert with the back muscles to produce trunk movements and to maintain the erect position.

2. Compress the abdomen into its roughly cylindrical form to maintain body form.

3. Change the intraabdominal pressure in visceral functions such as breathing, defecation, urination, coughing, and vomiting.

4. Confine and protect the abdominal organs by forming a contractile body wall in this region.

5. Stabilize the pelvis when movements of the lower extremities are made from the supine or prone position.

General Plan of the Abdominal Muscles. The abdominal muscles fall into two groups. The *first group* is composed of three platelike muscles which encircle the abdominal region, arising from extensive origins between the lower thoracic skeleton superiorly, the vertebral column and associated aponeuroses posteriorly, and the iliac crest of the pelvic girdle inferiorly. The muscles of the first group are the *external oblique, internal oblique,* and *transversus abdominis.* These muscles follow the path of embryonic muscle masses, which migrated around the abdominal viscera toward the anterior midline of the trunk to close the broad defect between the thoracic skeleton above and the pelvic girdle below. The muscles of this group are associated closely together in a platelike formation much like the thin laminated sheets in a plywood panel. The analogy can be carried further in that the fibers of the abdominal muscles of this group do not run in the same plane. The external and internal oblique muscles surround the abdomen in such a way that the direction of muscle fibers of the external oblique on one side is continued by the course of the fibers of the internal oblique of the opposite side. Since the thoracic and abdominal muscles are regional components of the spiraling sheets of trunk musculature, it should be noted in the Table of Abdominal Muscles (p. 181) that a given abdominal muscle continues the depth plane of a counterpart muscle in the thorax.

Each of the three flat abdominal muscles, in encircling the abdomen in development, falls short of the anterior midline. The midline defect which would otherwise exist is filled by fascial extensions of the muscles which become related to the *second group* of abdominal muscles. This group consists mainly of the longitudinally disposed *rectus abdominis* muscles and the midline fascial band between them, the *linea alba.* The

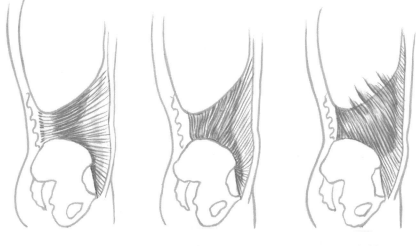

Transversus abdominis   Internal oblique   External oblique

pyramidalis belongs to this group, but it is small and insignificant and functions only to tighten the linea alba fascial band. The rectus abdominis is actually the fusion product of several midline, segmental, embryonic muscle masses. The marks of fusion are shown in the adult by horizontal *tendinous inscriptions* on the surface of the muscle which mark the functional, as well as developmental, separation of this long muscle into segmental groups of muscle units.

Table of the Abdominal Muscles. The abdominal muscles are tabulated in the table on page 181. Several morphological tissue arrangements must be understood before studying this table. These are the thoracolumbar aponeurosis and the relationship, near the anterior midline, of the fascial extensions of three flat abdominal muscles to the fascial sheath of the rectus abdominis.

1. THORACOLUMBAR APONEUROSIS. Many of

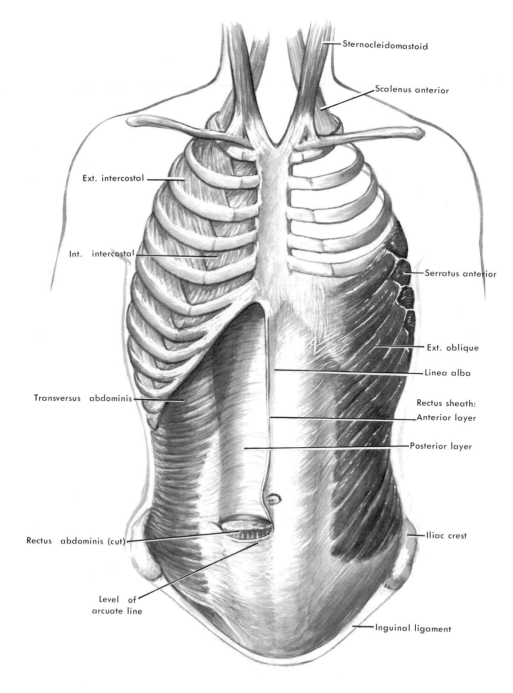

- Sternocleidomastoid
- Scalenus anterior
- Ext. intercostal
- Int. intercostal
- Serratus anterior
- Ext. oblique
- Linea alba
- Transversus abdominis
- Rectus sheath:
- Anterior layer
- Posterior layer
- Rectus abdominis (cut)
- Iliac crest
- Level of arcuate line
- Inguinal ligament

TABLE OF ABDOMINAL MUSCLES

MUSCLE	External oblique		Internal oblique	
THORACIC COUNTERPART	External intercostal		Internal intercostal	
ORIGIN	Slips from lower eight ribs interdigitating with serratus anterior and latissimus dorsi*	Thoracolumbar aponeurosis		
FIBER DIRECTION	Downward and medially	Upward and medially	Upward and medially and transversely	Downward and medially
INSERTION	Posterior and lower fibers to iliac crest; anterior and upper fibers into the fibrous anterior rectus sheath and linea alba*	Lower three ribs and costal margin	Fibrous anterior rectus sheath	Crest of pubic bone

Labels (left figure): Linea alba in midline; Ext. oblique: Muscle; Aponeurosis; Aponeurosis of transversus abdominis joins below arcuate line; Rectus abdominis m.; Int. oblique: Muscle; Aponeurosis; Arcuate line; Int. oblique m.; Inguinal ligament

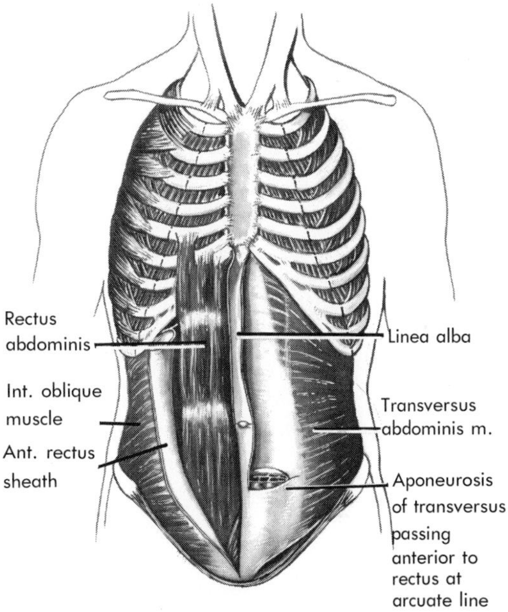

Labels (right figure): Rectus abdominis; Int. oblique muscle; Ant. rectus sheath; Linea alba; Transversus abdominis m.; Aponeurosis of transversus passing anterior to rectus at arcuate line

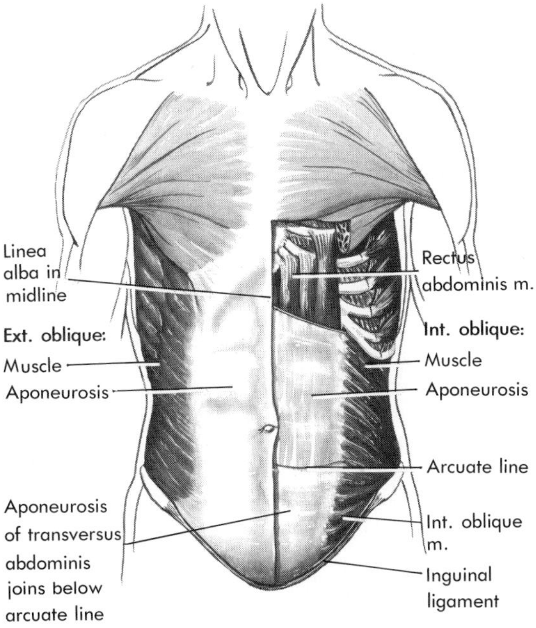

MUSCLE	Transversus abdominis		Rectus abdominis	Pyramidalis
THORACIC COUNTERPART	Innermost intercostal, transversus thoracis		Pectoral muscles	Pectoral muscles
ORIGIN	Inner surface of lower six costal cartilages, thoracolumbar aponeurosis, iliac crest, and iliac fascia		Xiphoid process and fifth to sixth costal cartilages *	Body of pubis
FIBER DIRECTION	Transversely in upper part	Downward and medially in lower part	Vertically downward *	Vertically upward
INSERTION	Fibrous posterior rectus sheath	Crest of pubic bone	Pubic symphysis and pubic crest*	Lower part of linea alba

*See comment under action of the abdominal muscles, pp. 182.

the muscles which migrate from the vertebral region toward the scapula and humerus, or around the abdominal wall, are broad and sheet-like. Neither the spinous and transverse processes of the thoracic and lumbar vertebrae nor the margins of the pelvic girdle can provide sufficient surface area for the attachment of all of the muscle fibers. A strong aponeurotic membrane, the *thoracolumbar aponeurosis* (lumbodorsal fascia in older terminology), attaches to the vertebral column, especially in the lumbar region, and to the posterior margin of the iliac bone. It provides an extremely strong base which extends the surface area for muscle attachments beyond what the bones can provide. (See illustration on page 180.)

2. Anterior Fascial Relationships. The fibrous sheath surrounding the rectus abdominis is principally the product of the fascial extensions of the three flat abdominal muscles at the lateral border of the rectus abdominis. These extensions interweave with each other, surround the rectus abdominis to form its sheath, and attach into the tough midline raphe of interlaced collagenous fibers which form the linea alba. There is much variation in the fascial patterns of this area: at different vertical levels in the same person, and between individuals.

1. The rectus abdominis muscles are vertically disposed on each side of the midline with the linea alba extending from the xiphoid process to the symphysis pubis as a tough midline fascial band between them.

2. In reaching toward the midline for attachment, the fascial extensions of the three flat abdominal muscles fuse with each other in varying manners, ensheath the rectus abdominis, and interlace into the linea alba. (See table on page 181.)

3. The aponeuroses of the external oblique and internal oblique muscles pass anterior (external) to the rectus abdominis to form the *anterior layer of the rectus sheath*. Between the xiphoid process and a variable point inferior to the umbilicus (*arcuate line*), the aponeurosis of the transversus abdominis passes posterior (deep) to the rectus abdominis to form the *posterior layer of the rectus sheath*.

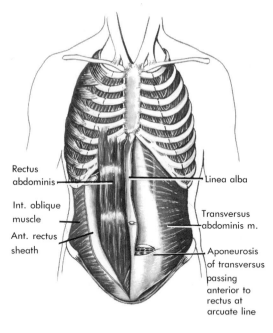

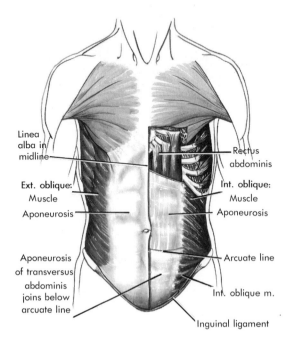

The following facts should suffice for an understanding of the anterior attachments and relations of the abdominal muscles near the midline.

4. Below the arcuate line the fascial extension of the transversus abdominis joins that of the internal oblique to pass anterior to the rectus abdominis. The posterior layer of the rectus sheath from here to the symphysis pubis is formed only by an internal abdominal fascial layer related to the deep surface of the transversus abdominis muscle (*transversalis fascia*). The transversalis fas-

cia surrounds the whole abdomen just external to the peritoneum, much as does the endothoracic fascia of the thorax with which it is continuous.

Inguinal Ligament and Inguinal Rings.   At this point a review of the pubic and iliac bones would be helpful. The lower part of the aponeurosis of the external oblique muscle thickens and turns under between attachments to the pubic tubercle and the anterior superior iliac spine. The rolled-under portion forms the *inguinal ligament,* which stretches across the boundary zone between the front of the abdomen and the anterior aspect of the upper thigh. A somewhat triangular aperture appears in the aponeurosis of the external oblique muscle just before it turns under. This opening, the *superficial inguinal ring,* extending between the medial end of the inguinal ligament and the pubic tubercle, transmits the spermatic cord from the abdomen into the scrotum in the male and transmits a ligament of the uterus into vulvar tissues in the female.

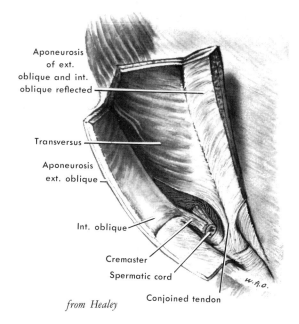

from Healey

At about the midpoint of the inguinal ligament externally, another opening exists in the transversalis fascia internally. This aperture is the *deep inguinal ring.*

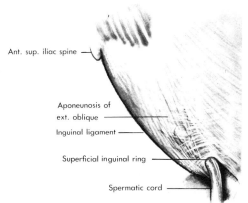

from Healey

Posterior to the external oblique aponeurosis, the lowest fibers of the internal oblique muscle take origin from the inguinal ligament. These fibers pass over the spermatic cord to join with the lower fibers of the transversus abdominis. The joined muscle fibers arch medially and downward to attach to the pubic tubercle as the *falx inguinalis* or *conjoined tendon.* Posterior to these structures is the transversalis fascia.

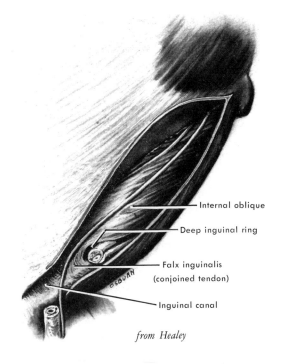

from Healey

The spermatic cord in the male and the round ligament of the uterus in the female gain egress from the abdominopelvic cavity into the tissues of the abdominal wall through the deep inguinal ring. A cleftlike passageway passing medially between the two inguinal rings is termed the *inguinal canal.*

Inguinal Canal. The inguinal canal is more of a cleft between the musculofibrous tissues of the lower abdominal wall than it is a canal. The cleft is only about an inch and a half long and follows a medial-inferior course which inclines superficially.

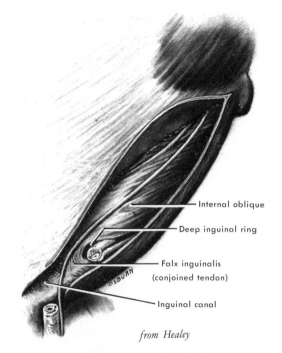

from Healey

The aponeurosis of the external oblique muscle forms the anterior boundary; the rolled-under inguinal ligament forms the floor; and in various parts, the internal oblique muscle, transversalis muscle and transversalis fascia, and the conjoined tendon form the posterior wall of the inguinal canal. The significance of the inguinal canal lies not so much in its contents or boundaries, but in the fact that abdominal contents may come to protrude through the canal and superficial inguinal ring to form an *inguinal hernia.*

Cremaster Muscle. A few of the lowest fibers of the internal oblique muscle separate from the main muscular sheet to become draped in a series of loops over the spermatic cord. These fibers are designated separately as the *cremaster muscle.*

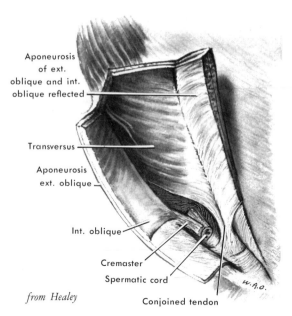

from Healey

Visceral and Parietal Actions of the Abdominal Muscles. The general functions of the abdominal muscles were discussed at the beginning of this section. A host of subsidiary functions can be summarized as follows:

1. Protection of the abdominal viscera by tensing the abdominal wall.

2. Maintenance of the cylindrical form of the trunk by compression of the body wall.

3. Increase of intraabdominal pressure as in coughing, vomiting, and straining.

4. Indirect aid to visceral functions by assisting, through compression and increase of intraabdominal pressure, in the moving or evacuation of the contents of abdominal organs.

Actions upon the Trunk. The main skeletal function of the abdominal muscles is in movements of the trunk.

1. FLEXION OF THE TRUNK. The rectus abdominis muscle is the prime mover. In a simple bending forward of the body from the erect position, gravity is the main influence, with the movement being regulated by the antagonistic action of muscles which extend the back. The rectus abdominis comes into play in the erect position when the trunk must be flexed against resistance. If the body is in the supine recumbent position, contraction of the rectus abdominis lifts the thorax toward an upright sitting position. If neck flexor muscles contract to prevent the head from falling backward because of gravity, the neck and head will follow the flexion of the trunk. Simultaneous contraction of the external

oblique muscles assists the rectus abdominis in flexing the trunk.

The rectus abdominis action and the conventional tabulation of its origin and insertion illustrate a problem in describing muscles functionally. The origin of the rectus abdominis should be its less movable end, whereupon it should be expected that anatomists would list the pelvic attachment as the origin. Yet most anatomists do not. The more movable end is the attachment to the thorax which, therefore, should be described as the insertion. This paradox in classifying muscle structure does not affect the function of the muscle at all.

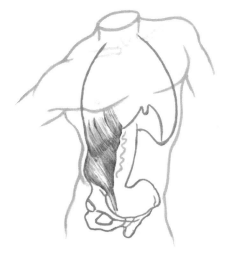

Lat. motion
External oblique

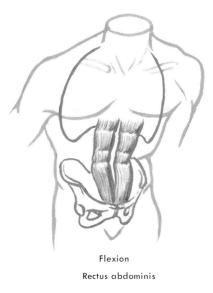

Flexion
Rectus abdominis

2. ROTATION OF THE TRUNK. The oblique muscles of the abdomen work together with the back muscles to produce rotation of the trunk by rotating the vertebral column and pelvis. Remember the "counterpart direction" of fibers of one external oblique and the opposite internal oblique muscles. The external oblique muscle rotates the trunk to the same side, assisted by the internal oblique muscle of the opposite side.

3. LATERAL BENDING OF THE TRUNK. Both oblique muscles and the transversus abdominis, assisted by the rectus abdominis of the same side and the intercostal muscles, work together with the back muscles to bend the trunk laterally.

The aforementioned descriptive problem of structural origin and insertion versus function applies equally well to the external oblique muscle.

If you place your hands above your hips while standing, walking, or making short shifts of position, it will be evident that the abdominal muscles are continually active in assisting the back muscles in maintaining the erect position of the trunk. If then you stand or hop on only one leg, it will be appreciated that the abdominal muscles of the same side contract actively to preserve body balance. The combined actions of the abdominal muscles and the back muscles are excellent examples of both cooperation and regulation of muscle actions by antagonistic groups of muscles.

THE DIAPHRAGM: A MUSCLE ACTING WITHIN THE TRUNK

The *diaphragm* (more appropriately named the thoracoabdominal diaphragm) does not fit into other muscle classifications. It is a musculofibrous sheet which divides the interior of the trunk into the thoracic cavity superiorly and the abdominopelvic cavity inferiorly. The diaphragm serves as a base of support for the thoracic viscera in the erect position and participates in the increase and decrease of the volume of the thorax which make breathing occur.

Structure of the Diaphragm

The diaphragm is a broad sheet, rounded by the shape of the cylindrical thorax, which drapes over the liver at the bases of the lungs and heart.

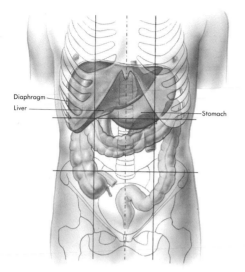

It is composed of many muscular bands which take origin from points around the inferior thoracic aperture. These are the:

1. Posterior surface of xiphoid process anteriorly.

2. Internal surface of the fused seventh to tenth costal cartilages and the eleventh and twelfth ribs.

3. Transverse processes of the twelfth thoracic and first lumbar vertebrae.

4. Bodies of the first and second lumbar vertebrae by long ligamentous bands known as the *right* and *left crura of the diaphragm.*

All the muscular bands of the diaphragm are directed medially, anteriorly, and posteriorly, depending upon their origins about the body wall, to insert into a strong fibrous plate, the *central tendon of the diaphragm.* Contraction of the muscular elements flattens the dome-shaped diaphragm and causes it to descend. Abdominal viscera, notably the liver and stomach, are pushed downward as the volume of the thorax is increased for inspiration.

Openings of the Diaphragm

During development the muscular elements of the diaphragm grew around structures already coursing longitudinally within the trunk. In the

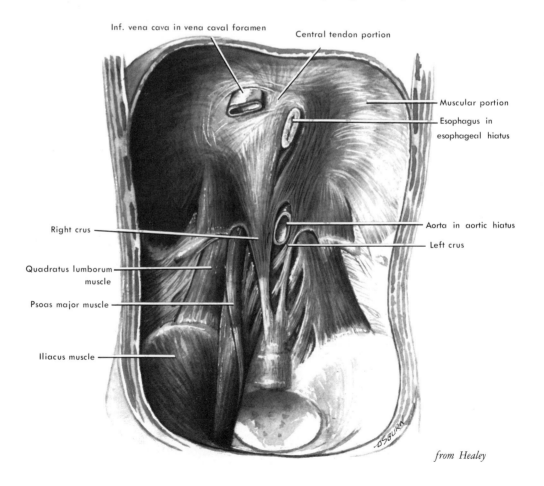

from Healey

adult, however, these structures are said to pass through openings of the diaphragm. These are the:

1. *Aortic hiatus,* between the two crura posteriorly, for the aorta, azygos vein, and thoracic duct. (See illustration on opposite page.)

2. *Esophageal hiatus,* located anteriorly and to the right of the aortic opening, for the esophagus, the arteries and veins passing to it from abdominal vessels, and the vagus nerves enroute to the abdomen.

3. *Vena caval foramen,* in the right part of the central tendon, for the inferior vena cava and the right phrenic nerve.

Other structures, such as vessels of the trunk musculature, left phrenic nerve, thoracoabdominal nerves, and autonomic nerves to abdominal organs, pierce the diaphragm separately or pass between muscular bands.

MUSCLES CLOSING THE PELVIC OUTLET

The bony pelvis is open anteroinferiorly between the boundaries formed by the coccyx, the inferior margins of the sacrum and iliac bones, the ischial tuberosities, and the pubic arch. The openings of the alimentary tract (rectum and anus), the urinary system (urethra), and the reproductive system (external genitalia) are associated with the pelvic outlet. It is through this aperture in the pelvis that the developed fetus must be expelled at the end of pregnancy. Although an outlet to the pelvis must be provided, the pelvic organs must be supported against the downward pressure of gravity in the erect position and the weight or distention of abdominal organs above them. Closure of the pelvic outlet is effected by a combination of muscles and fasciae which form the *pelvic diaphragm* and the *urogenital diaphragm.* These musculofibrous membranes are associated with the tela subcutanea, skin, and external genitalia to form the *perineum.* This term is applied to the region of the body extending externally between the medial surfaces of the thighs from the coccyx posteriorly to the symphysis pubis anteriorly.

Since the muscles concerned in the formation of the perineum are principally muscles associated with the genitourinary system, the description of the musculature closing the pelvic outlet would require the introduction here of a host of new terms. To avoid duplication, the muscles are described with the genitourinary system. Perineal muscles associated with the rectum and anus are discussed with the digestive system.

MUSCLES MOVING AND POSITIONING THE SCAPULA

The scapula, positioned by its associated muscles, performs two functions. It anchors and makes possible the articulation of the upper limb with the trunk, but it also extends the range of mobility of the arm. To anchor the upper limb the scapula must be fixed in a number of positions relative to the actual position of the limb. However, the scapula must be mobile in order to extend the range of mobility of the arm. It will be recalled that the scapula is only applied to the posterolateral thoracic wall and that it articulates with trunk only through the clavicle. The dual functions of the scapula, therefore, are made possible only by the action of a series of muscles. It would be wise to review the structure of the scapula and its movements before proceeding.

Muscles Moving the Scapula

The movements of the scapula are elevation, depression, protraction, and retraction. Although many muscles have attachments to the scapula, not all of them produce scapular movements unless the limb is fixed so that functional reversal of the muscle attachments results in movement of the trunk (as in rope climbing) through the medium of scapular attachments. The muscles tabulated act primarily upon the scapula; other muscles which have scapular attachments act mainly on the upper limb. (See table on page 188.)

Muscles Positioning the Scapula

The scapula virtually floats on the posterolateral surface of the thorax. Only the acromioclavicular and coracoclavicular ligaments bind it into the pectoral girdle and tend to suspend the bone against its own weight in the erect position. The action of muscles using the scapula for attachment would continually displace this mobile bone if it were not for the coordination and natural antagonistic actions to be found among

MUSCLES ACTING PRIMARILY UPON THE SCAPULA

MUSCLE	Trapezius	Levator scapulae	Rhomboideus minor	Rhomboideus major	Serratus anterior
SHAPE	One is triangular; both make a trapezoid	Thick band	Narrow band	Flat band	Broad, flat with fingerlike origins
ORIGIN	Medial third superior nuchal line, external occipital protuberance, ligamentum nuchae, all thoracic spinous processes	Posterior tubercles of transverse processes of cervical vertebrae 1 to 4	Lower ligamentum nuchae; C7 and T1 spinous processes	Spinous processes of T2 to T5	Digitations from lateral surfaces of upper eight ribs
FIBER DIRECTION	Cranial and upper cervical fibers pass inferolaterally; lower cervical and upper thoracic pass laterally; lower thoracic pass superolaterally	Downward and slightly laterally	Obliquely downward and laterally	Obliquely downward and laterally	Posteriorly along medial wall of axilla to scapula
INSERTION	Cranial and upper cervical fibers to posterior surface of lateral third of clavicle; lower cervical and upper thoracic to acromion and crest of spine of scapula; lower thoracic to crest of scapular spine	Vertebral border of scapula from superior angle to root of spine	Vertebral border of scapula at root of spine	Vertebral border of scapula from root of spine to inferior angle	Vertebral border of scapula from superior angle to inferior angle
ACTION ON SCAPULA	Elevate scapula and suspend pectoral girdle; middle and lower fibers together retract scapula to "square shoulders"	Elevation; if scapula is fixed, rotates neck to same side; turns glenoid fossa downward	Elevation; draw scapula toward vertebral column; help fix scapula to provide a firm base for movement of upper extremity; turn glenoid fossa downward		Protraction; turns glenoid fossa upward

Note: Depression of the scapula is produced largely by gravity aided by the pectoralis minor and the rhomboids.

them. The elevators of the scapula (levator scapulae, rhomboids) counter the effects of gravity, the downward pull of the pectoralis minor muscle, and the outward or downward pull upon the scapula when the muscles of the upper extremity are acting against resistance, as in lifting or carrying a heavy weight. The trapezius joins in all these functions, the upper and middle fibers being most effective. It will be seen shortly that the teres major, attaching to the axillary border of the scapula, is a powerful muscle in moving the humerus. If the rhomboids are not functioning to position the scapula, the unopposed pull of the teres major will draw the inferior angle of the scapula right into the axilla. Similarly, other muscles (subscapularis, supraspinatus, and infraspinatus) which act across the shoulder joint are counterbalanced in their own actions by the restraining effects upon the scapula of the serratus anterior and trapezius which hold the bone in place and keep it closely applied to the chest wall.

Study carefully the illustrations and descriptions of the muscles attaching to the scapula. Feel your own scapula while putting your arm through a full range of motion. Attention to these factors is a rewarding study of the mutual relationships and coordination to be found in the antagonism, fixation, and synergism of groups of muscles. A return to this section after completing the next section on muscles acting upon the shoulder joint will enhance your understanding of the scapular muscles.

MUSCLES MOVING THE ARM AT THE SCAPULOHUMERAL JOINT

The muscles which extend to the upper end of the humerus from the thorax and the pectoral girdle are individually capable of producing more than a single movement of the shoulder joint and collectively are involved in many coordinated movements. To avoid duplication of description, the muscles are described first and then their complex interrelationships are discussed. A review at this time of earlier descriptions of the scapulohumeral articulation is indicated.

Muscles Acting Across the Shoulder Joint

The muscles which produce movements of the shoulder joint are described in the table on pages 192, 193.

Relation to the Shoulder and Axilla

Muscles acting across the shoulder joint contribute to the formation of both the shoulder and the axilla. The bony landmarks of the shoulder are made by the pectoral girdle, but these prominences are rounded and softened by the deltoid muscle. This may be illustrated by clasping one's right upper arm and shoulder just below the bony point of the shoulder. The thumb of the left hand will lie along the anterior deltoid, the thumb and base of the index finger will enclose the middle part of this muscle, and the tips of the index amd middle fingers will lie upon the posterior deltoid. The fourth and fifth fingers will cover the posterior axillary fold within which the teres major, latissimus dorsi, and subscapularis approach the humerus. In their course these muscles form the fleshy posterior wall of the axilla. Anteriorly, the pectoralis major forms the substance of the anterior axillary fold and the anterior wall of the axilla as it courses laterally across the front of the axilla to insert upon the humerus.

The function of the muscles extending across the scapulohumeral joint in strengthening and stabilizing the joint cannot be emphasized too strongly. All these muscles, including the long head of the biceps brachii of the arm, help hold the head of the humerus in apposition to the glenoid fossa and reinforce the weak capsular ligaments. The muscles of particular importance in this regard are the subscapularis, supraspinatus, infraspinatus, and teres minor. The intimate association of these muscles with the capsule of the shoulder joint has been previously described under the name of the *musculotendinous* or *rotator cuff*. This will be found in the description of the shoulder joint, which should be reviewed before proceeding. (See illustration at top of following page.)

Muscle Actions at the Shoulder Joint

The movements which take place at the scapulohumeral joint are seldom simple ones confined to but one direction or to one plane. They may be analyzed, however, to consist of one or more components of flexion, extension, abduction, adduction, medial rotation, and lateral rotation. The movement of circumduction is a combination of all these components. There are many gradations and combinations of the simple categories of movement which contribute to the

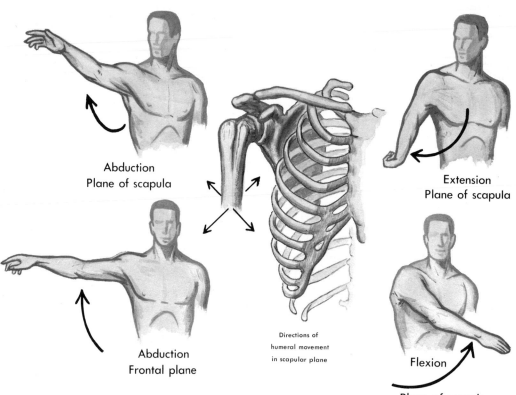

Abduction
Plane of scapula

Extension
Plane of scapula

Abduction
Frontal plane

Directions of
humeral movement
in scapular plane

Flexion
Plane of scapula

extremely free mobility of the upper limb at the shoulder joint. In the tables on pp. 194 and 195 are listings of the muscles which produce given movements. It is felt that this approach is more useful than memorizing all the possible actions of one muscle before proceeding to another. The muscles listed first are prime movers in the particular movement. Those appearing lower on a list, and marked by an asterisk, may have other more direct and primary actions but are known to assist, even if weakly, in the action under consideration.

Planes of Movement at the Shoulder. The position of the scapula influences the direction of movement of the humerus. The scapula is placed obliquely along the posterolateral aspect of the thorax so that the articulating surface of the glenoid fossa does not lie in either of the conventional planes of the body. The surface of the glenoid fossa faces anteriorly as well as laterally from the shoulder joint. Reference to an articulated skeleton will show, therefore, that movement of the humerus is not in the coronal (frontal) plane, but at an oblique angle to it. The humerus does not swing exactly forward or backward in flexion or extension, but forward and medially, or backward and laterally. It follows that movements of abduction or adduction are not in the frontal plane at a right angle away from or toward the thorax, but anterolaterally in abduction and posteromedially in adduction. Because of the oblique position of the scapula, movements of the humerus can be thought of as occurring either *in the plane of the scapula* or as movements *in the conventional planes of the body.*

The necessary introduction of the concept of

scapulohumeral movements occurring in the plane of the scapula need not be a problem of comprehension if the skeleton is studied again. When the humerus moves outward and upward in abduction *in the direct frontal plane of the body,* it can be seen that the head of the humerus cannot move very far without the greater tuberosity striking the acromion process to limit the movement. In order to continue the movement of abduction, lateral humeral rotation is necessary to move the greater tuberosity out of the way of the acromion. When the humerus is considered to move obliquely outward and upward *in the plane of the scapula,* no rotation of the humerus is necessary.

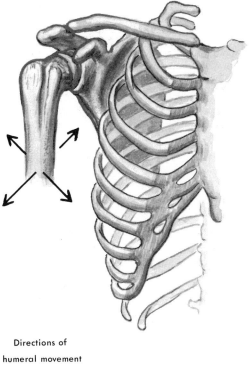

Directions of
humeral movement
in scapular plane

One can counter by saying that it is quite possible to abduct the arm at a right angle to the surface of the thorax exactly in the frontal plane. This is correct. All one needs to remember is that when the humeral movement parallels the frontal plane, abduction must be accompanied by lateral rotation in order to complete the movement.

Special Notes on Muscle Actions. Points to consider in relation to muscles or movements marked with a dagger in the tables are:

1. DELTOID MUSCLE. The actions of this tripartite muscle have been the subject of long debate and of much recent electromyographic research. The three parts of this muscle may act separately, in combination with one or both other parts, or in functional association with different muscles as indicated in the tables. If all three parts of the muscle act together, the arm will be abducted in the *scapular* plane. If only the posterior and middle parts are activated by the central nervous system, the arm will be abducted in the *frontal* plane.

Despite its morphological disposition as an abductor, the deltoid muscle would waste much of its contractile force in pulling the head of the humerus directly upward and out of the glenoid fossa. The efficiency of the deltoid muscle is helped by the musculotendinous cuff muscles, which hold the humeral head in its approximation to the glenoid fossa in normal circumstances. The work of the deltoid may then be applied to abduction and elevation of the arm. Much discussion has centered about the ability of the deltoid to initiate abduction because its tendon of insertion parallels the shaft of the humerus. It has been believed that the supraspinatus is situated better to initiate abduction. The more prevalent belief today is that both muscles work together.

The deltoid is an important stabilizer (fixator) of the shoulder joint when the arm has been abducted (elevated) to a horizontal position. By holding the arm at this level, other movements in a horizontal plane are facilitated. Examples are drawing a line across a blackboard or positioning the arm at the horizontal level for forearm and hand movements or exercise in isometric procedures or weight-lifting. When the deltoid and supraspinatus have moved the humerus into this position (approximately 90 degrees of elevation), movements of the humerus toward or away from the thorax are termed *horizontal adduction* or *horizontal abduction.* As an antagonist to adductor muscles, the deltoid regulates and smoothes their action.

2. SUPRASPINATUS MUSCLE. Debate has also centered upon the role of the supraspinatus muscle in the production of shoulder movements. The supraspinatus muscle, through its rotator cuff role, acts as much to hold the head of the humerus in place during deltoid activity as to participate in abduction of the arm. It is commonly

(*Text continued on page 195*)

MUSCLES ACTING ACROSS THE SHOULDER JOINT

MUSCLE	Pectoralis major	Pectoralis minor	Deltoid	Supra-spinatus	Infra-spinatus
SHAPE	Broad fan	Blunt triangle	Thick inverted triangle	Rounded thick band	Thick band
LOCATION	Anterior aspect of thorax and axilla	Anterior aspect of thorax deep to pectoralis major	Caps and surrounds shoulder	Crosses shoulder joint superiorly	Crosses shoulder joint posterosuperiorly
ORIGIN	Anterior surface of medial half of clavicle; lateral margin of sternum; and cartilages of upper six ribs	External surface of second to fifth ribs	Anterior surface of lateral third of clavicle; lateral margin of acromion and crest of scapular spine	Medial two thirds of supraspinous fossa	Medial three fourths of infraspinous fossa
FIBER COURSE	Clavicular fibers pass inferolaterally; upper sternocostal fibers laterally; lower sternocostal fibers superolaterally anterior to axilla	Superolaterally anterior to axilla	Clavicular (anterior) part inferolaterally and posteriorly; acromial (lateral) part inferiorly; and spinous (posterior) part inferolaterally and anteriorly	Superolaterally over capsule of shoulder joint	Superolaterally across posterior surface of capsule of shoulder joint
INSERTION	Fibers converge to insert by folded, U-shaped tendon into crest of greater tuberosity of humerus	Medial border of coracoid process	All parts converge to a tapered insertion upon deltoid tuberosity of humerus	Highest of three attachments to greater tuberosity of humerus	Middle of three attachments to greater tuberosity of humerus

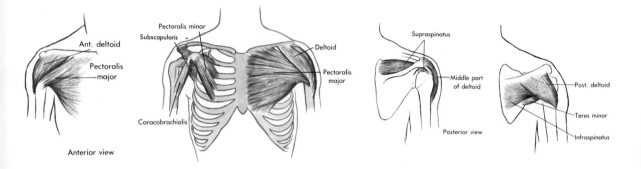

MUSCLES ACTING ACROSS THE SHOULDER JOINT (Continued)

Teres minor	Teres major	Subscapularis	Latissimus dorsi	Coracobrachialis	MUSCLE
Narrow band	Thick band	Flat fleshy band	Broad flat triangle	Short band	SHAPE
Close association with above	Along axillary border of scapula and posterior wall of axilla	Posterior wall of axilla	Main part superficial on back; terminal part along posterior wall of axilla	Lateral wall of axilla and proximal part of arm	LOCATION
Lateral part of infraspinous fossa	Oval area on dorsal surface of inferior angle of scapula	Medial two thirds of subscapular fossa	Spinous processes of lower six thoracic vertebrae; lumbar and sacral vertebrae via thoracolumbar fascia; and iliac crest	Tip and lateral surface of coracoid process of scapula	ORIGIN
As with infraspinatus; inferior to infraspinatus	Superolaterally along axillary border of scapula and posterior axillary fold to front of humerus	Laterally in front of capsule of shoulder joint	Superolaterally across back, across inferior angle of scapula, along posterior wall of axilla in posterior axillary fold	Along lateral wall of axilla	FIBER COURSE
Lowest of three attachments to greater tuberosity of humerus	Crest of lesser tuberosity of humerus	Lesser tuberosity of humerus	Floor of intertubercular groove and its medial lip	Middle third of humerus along its medial surface	INSERTION

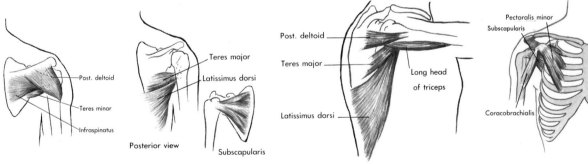

Posterior view

Note: To complete the description of muscles acting across the shoulder joint, refer to the descriptions of the triceps brachii and biceps brachii of the arm.

MUSCLE ACTIONS AT THE SHOULDER JOINT

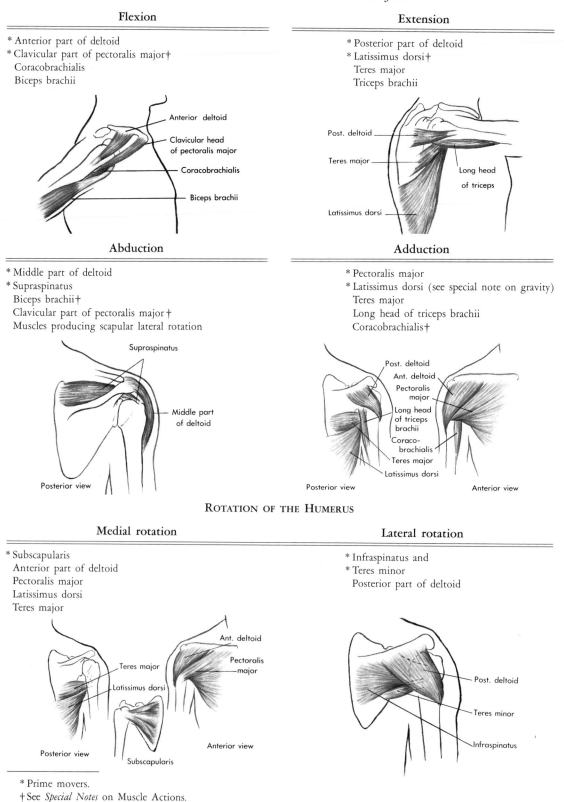

Flexion

* Anterior part of deltoid
* Clavicular part of pectoralis major†
  Coracobrachialis
  Biceps brachii

Anterior deltoid

Clavicular head
of pectoralis major

Coracobrachialis

Biceps brachii

Extension

* Posterior part of deltoid
* Latissimus dorsi†
  Teres major
  Triceps brachii

Post. deltoid

Teres major

Long head
of triceps

Latissimus dorsi

Abduction

* Middle part of deltoid
* Supraspinatus
  Biceps brachii†
  Clavicular part of pectoralis major†
  Muscles producing scapular lateral rotation

Supraspinatus

Middle part
of deltoid

Posterior view

Adduction

* Pectoralis major
* Latissimus dorsi (see special note on gravity)
  Teres major
  Long head of triceps brachii
  Coracobrachialis†

Post. deltoid
Ant. deltoid
Pectoralis
major
Long head
of triceps
brachii
Coraco-
brachialis
Teres major
Latissimus dorsi

Posterior view   Anterior view

ROTATION OF THE HUMERUS

Medial rotation

* Subscapularis
  Anterior part of deltoid
  Pectoralis major
  Latissimus dorsi
  Teres major

Ant. deltoid

Pectoralis
major

Teres major

Latissimus dorsi

Posterior view

Subscapularis

Anterior view

Lateral rotation

* Infraspinatus and
* Teres minor
  Posterior part of deltoid

Post. deltoid

Teres minor

Infraspinatus

  * Prime movers.
  † See *Special Notes* on Muscle Actions.

MUSCLE ACTIONS AT THE SHOULDER JOINT (Continued)

RETENTION OF HUMERAL HEAD IN GLENOID FOSSA

Facilitate	Tends to disrupt

* Musculotendinous cuff muscles:
  Supraspinatus
  Infraspinatus
  Subscapularis
  Teres minor
Biceps brachii
	Deltoid†

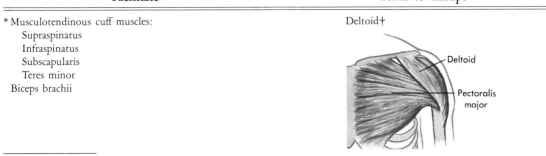

* Prime movers.
† See *Special Notes* on Muscle Actions.

believed that, although the deltoid can perform abduction without the assistance of a paralyzed supraspinatus and with other musculotendinous cuff muscles holding the humeral head in place, the supraspinatus is a weak abductor if it must function alone in a case of deltoid paralysis and cannot accomplish, even weakly, a full range of abduction.

3. FLEXION. This movement occurs not only from the anatomic position but is also applied to the return of the arm from a position of extension.

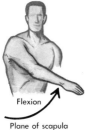

Flexion
Plane of scapula

If the arm is already in full extension, the clavicular part of the pectoralis major is in a position of stretch and mechanical advantage to help initiate flexion, although it would not be a flexor from the anatomic position.

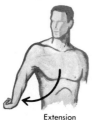

Extension
Plane of scapula

4. EXTENSION. Extension also implies a return of the arm from any degree of flexion. The latissimus dorsi is especially effective in extending the arm from such positions and also against resistance.

5. ADDUCTION. The thorax restricts this movement to the return of the abducted arm to the side of the body. To continue adduction of the arm across the front of the thorax toward the midline requires simultaneous flexion so that the humerus can clear the thorax. The same holds true for adduction posteriorly (as in clapping the hands behind the back) in which partial extension of the humerus is necessary to evade the posterior thoracic wall.

Gravity is the main force accomplishing adduction to the side of the body in the erect position. Then abductors, such as the middle part of the deltoid and the supraspinatus, smooth and regulate the movement as antagonists. The coracobrachialis departs only slightly from the axis of the humerus and, therefore, can be only a weak adductor, and that only when other muscles already have the arm in motion.

6. ABDUCTION. Other points to bear in mind about this movement are that neither the biceps brachii nor the clavicular part of the pectoralis major can initiate the movement effectively but can, by the nature of their course and attachments, contribute to abduction once other prime movers have initiated the movement.

Abduction includes elevation of the arm past the horizontal level to a position beside the head.

Lateral scapular rotation (under the influence of the serratus anterior and trapezius) prevents the acromion from blocking the upward movement of the humerus. Unless these muscles act to direct the scapula further upward, the full range of elevation cannot be accomplished.

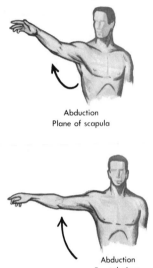

Abduction
Plane of scapula

Abduction
Frontal plane

7. ROTATIONS.   Rotation of the humerus medially or laterally facilitates movements at the shoulder joint and brings the brachium into positions which enable it to function as the base from which muscles of the forearm, wrist, and hand operate.

Regional Correlations.   Now that considerable information has been acquired about muscles it is pertinent to remember that as a muscle extends from one point of skeletal attachment to another it is not alone: it needs neighboring structures for its own function. A muscle wrapped in its fascial envelope is often one of several which share a common location and reasonably similar points of origin and insertion. These facts indicate that such muscles may have a common group function, such as lateral rotation of the upper extremity. The rotator cuff muscles lying above and behind the shoulder joint and axilla do have many common morphological and functional relationships. They are loosely encased in a regional deep fascia which separates them from other muscles of the region which have different planes and different actions. For instance, the point of the shoulder is capped by the middle deltoid and supraspinatus muscle fibers which together produce abduction of the shoulder joint.

Muscles lying in a similar location, and having a common function, generally share the same nerve and arterial blood supply. A book arranged for the study of separate systems cannot incorporate much regional information, but it can be gained by constant cross reference to Chapters 5 and 6. For example, in Chapter 5 it is noted that the abductor muscles of the shoulder joint are induced to contract by nerve impulses traveling in the suprascapular nerve and in the axillary nerve which course in the same region. Similarly, muscles lying in the same general location will usually receive arterial blood from the artery which courses in the same area, or from several if they extend a considerable distance. In Chapter 6 it will be found that the suprascapular artery is the vessel which supplies the muscles of the shoulder joint but that the humeral attachments of abductor muscles are nourished by the posterior humeral circumflex artery.

If one correlates a common location with functional muscle-nerve-artery relations, comprehension of living anatomy is easier.

MUSCLES MOVING THE FOREARM AT THE HUMEROULNAR AND HUMERORADIAL JOINTS

The muscles which produce the hinge actions of flexion and extension at the elbow joint extend along the humerus from origins on the pectoral girdle or on the humerus itself and cross the joint to attach proximally on either the ulna or the radius. The brachium is divided transversely into *anterior* and *posterior compartments* within which the muscles lie. Compartmenting is the product of the heavy platelike *medial* and *lateral intermuscular septa* which form shelves between the humerus and the tubelike brachial investing fascia. In general, muscles which flex the elbow are located anterior to the humerus (preaxially) and are found in the anterior or preaxial compartment. Muscles which extend the elbow hinge are in the posterior or postaxial compartment.

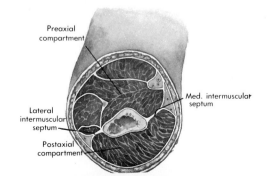

Preaxial
compartment

Med. intermuscular
septum

Lateral
intermuscular
septum

Postaxial
compartment

Muscles Acting Across the Elbow Joint

These muscles are described morphologically in the table on pages 198 and 199.

Special Notes on Muscle Actions

1. The double roles of the biceps brachii and the pronator teres should be noted. Whereas both can flex the forearm (the pronator teres being notably weaker in this role), they have opposite effects upon the upper radioulnar joint in that the biceps supinates while the other pronates. These two muscles have different innervations: the biceps (along with brachialis and coracobrachialis) from the musculocutaneous nerve and the pronator teres from the median nerve. The significance of the differing innervation on the hinge joint of the elbow is that if the action of the main flexors is impaired through injury to the musculocutaneous nerve, flexion is still possible although not powerful.

2. The biceps brachii and the triceps brachii are natural antagonists by virtue of both position and function. (See color plate 4.) Each regulates and smooths the action of the other. Their antagonism is advantageous when both contract powerfully at the same time, for then the elbow joint is held in full extension, particularly when acting against resistance. A stable, straight elbow is provided for pushing movements and to form a firm base for pronation-supination or wrist-hand movements.

3. Muscle actions about the elbow joint are frequently combined or take place at hinge angles of varying degrees. Most frequent in upper extremity positions of action are flexion-pronation and extension-supination. The writing-drawing position is an example of the first combination; lifting of objects held in the palms typifies the second.

Muscle Actions at Humeroulnar and Humeroradial Joints

In the hinge movements at the elbow, the radius and ulna move together upon the scroll-like articulating surfaces of, respectively, the capitulum and the trochlea of the humerus. The *carrying angle* described in the skeletal-articular chapter appears as the elbow is straightened but disappears with flexion. Tables of muscle action follow with prime movers marked by asterisks.

FLEXION AT ELBOW	EXTENSION AT ELBOW
* Brachialis	* Triceps brachii†
* Biceps brachii	Anconeus
* Brachioradialis	
Pronator teres	

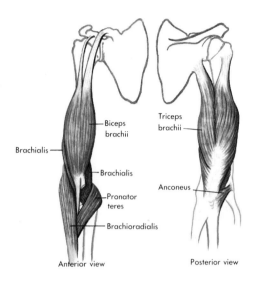

Brachialis — Biceps brachii — Brachialis — Pronator teres — Brachioradialis — Anterior view

Triceps brachii — Anconeus — Posterior view

---

* Prime movers in order of strength.
† See *Special Notes.*

OTHER ACTIONS OF MUSCLES CROSSING THE ELBOW JOINT

Pronator teres	Pronation,* weak flexion of forearm
Supinator	Supination*
Biceps brachii	Supination,* flexion of forearm*

---

* Prime mover action of the particular muscle.

MUSCLES MOVING THE FOREARM AT THE RADIOULNAR JOINTS

A review of the superior and inferior radioulnar joints and the movements of the radius about the ulna is indicated at this point. The mechanisms of pronation and supination were described in Chapter 3. It remains to determine the muscles which produce these movements of the forearm in which, by the nature of the articulations of the wrist, the hand is carried along.

MUSCLES ACTING ACROSS THE ELBOW JOINT

MUSCLE	Biceps brachii	Brachialis	Brachioradialis	Pronator teres
COM-PARTMENT	Anterior	Anterior	Not primarily in the brachium	Not primarily in the brachium
LOCATION	Superficially on anterior aspect of brachium	Deep to biceps, anterior to humerus	Posterior and lateral surfaces of forearm	Antecubital fossa
ORIGIN	Short head from medial aspect of tip of coracoid process; long head from supraglenoid tubercle of scapula	Distal two thirds of anterior, anteromedial, and anterolateral surface of humerus	Lateral supracondylar ridge of humerus	Superficial head from medial epicondyle and medial epicondylar ridge; deep head from coronoid process of ulna
FIBER COURSE	Form two bellies uniting to form a fusiform muscle directed distally in axis of brachium	Distally along humerus	Distally along lateral side of humerus	Distally and laterally into depths of antecubital fossa
INSERTION	Bicipital tuberosity of radius and via bicipital aponeurosis into fascia of forearm	Anterior surface of coronoid process and tuberosity of ulna	Lateral side of distal end of radius above styloid process	Middle part of lateral surface of radius

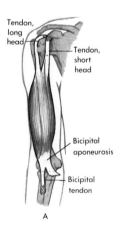

Tendon, long head
Tendon, short head
Bicipital aponeurosis
Bicipital tendon

A

Biceps brachii

B

Brachialis

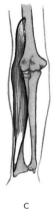

C

Brachioradialis

D

Pronator teres

MUSCLES ACTING ACROSS THE ELBOW JOINT (Continued)

Triceps brachii	Anconeus	Supinator		MUSCLE
Posterior	Posterior	Not primarily in the brachium		COM-PARTMENT
Posterior aspect of brachium	Triangular continuation of the triceps	Deep, around the lateral side of the elbow joint		LOCATION
Long head from infra-glenoid tubercle of scapula; lateral head from posterior surface of humerus above radial groove; medial head below groove	Posterior surface of lateral epicondyle	Superficial head from lat-eral epicondyle of humerus	Deep head from supinator fossa and crest, and oblique line of ulna	ORIGIN
Distally, the three heads joining lower on the humerus	Distally	Distally and medially obliquely along lateral side of elbow joint	Encircle head of radius, wrapping it posteriorly and laterally	FIBER COURSE
Via triceps tendon onto upper part of posterior surface of olecranon process of ulna	Lateral surface of olecranon process	Oblique line on radius between bicipital tuber-osity above and pronator teres insertion below	Upper third of shaft of radius opposite bicipital tuberosity	INSERTION

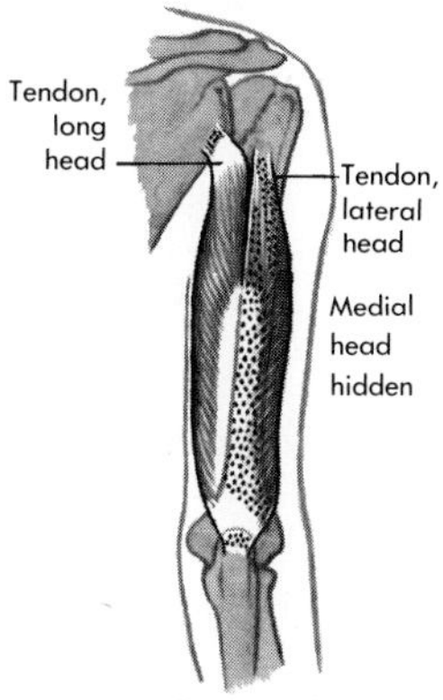

Tendon, long head
Tendon, lateral head
Medial head hidden
Posterior view
E
Triceps brachii

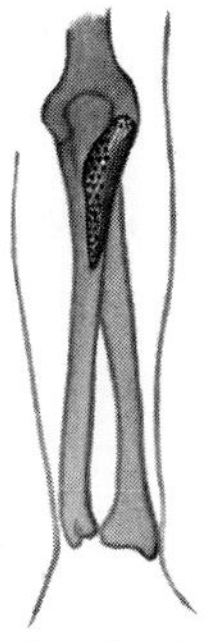

Posterior view
F
Anconeus

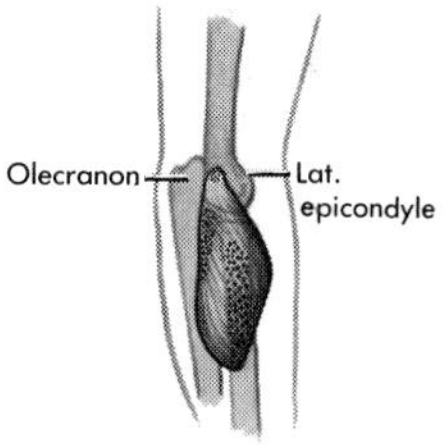

Olecranon — Lat. epicondyle
Lateral view
G
Supinator

In addition to the brachioradialis, pronator teres, and supinator, other forearm muscles have origins low on the humerus. Such muscles are not listed here because they do not produce movements of the forearm even though all or part of their fibers arise above or cross the elbow joint.

MUSCLES ROTATING THE HEAD OF THE RADIUS

LATERALLY IN SUPINATION	MEDIALLY IN PRONATION
* Supinator	* Pronator teres
Biceps brachii†	Pronator quadratus†

* Prime mover.
† See *Special Notes*.

Muscles Acting Upon the Superior Radioulnar Joint

No new muscles are involved in movement of the head of the radius upon the upper end of the ulna. All are either related to hinge movements or to origins on the lower end of the humerus which necessitate their crossing of the elbow joint. The muscles are tabulated above.

Notes on Muscles

1. The biceps brachii is an effective supinator, but the morphological arrangement of the supinator muscle gives it a more direct, prime-mover function. Part of the power of the biceps brachii is devoted to flexing the forearm. The difference in nerve supply (biceps from the musculocutaneous nerve, supinator from radial nerve) preserves supination by the biceps if injury to the radial nerve or its deep branch to the supinator causes paralysis of the supinator muscle.

2. The pronator quadratus muscle is listed even though its contraction more directly influences the arclike movement of the lower end of

the radius. Naturally the radial head must continue to rotate in keeping with the movement of the whole bone. It should be noted that the transverse disposition of the pronator quadratus in the fully supinated anatomic position precludes its effective institution of the pronation movement.

Muscles Acting Upon the Inferior Radioulnar Joint

The muscles which move the superior radioulnar joint also contribute to the movement of the lower end of the radius about the ulna. This is particularly true for the muscles producing supination which have no counterparts at the distal part of the forearm. A strong and effective pronator, the *pronator quadratus,* however, is located distally.

Pronator Quadratus. This flat, quadrilateral band of muscle is situated deeply in the distal forearm under the tendons of muscles passing to the hand. The pronator quadratus takes origin from the anterior surface of the distal part of the ulna. Its fibers pass laterally to insert upon the anterior surface of the radius above the ulnar notch. The directness of pull of this muscle makes it a powerful pronator once pronation has been started by the pronator teres so that the quadratus fibers can come into full play.

PRONATING AND SUPINATING MUSCLES

PRONATION		SUPINATION
Instituting	Completing	Full Range of Movement
Pronator teres	Pronator quadratus	Supinator
	Pronator teres	Biceps brachii

MUSCLES MOVING THE WRIST AND HAND AS A WHOLE

Muscles of the forearm, other than those presented previously, have the function of moving the hand as a unit upon the forearm at the wrist joint or of moving the fingers and thumb. Some of the forearm muscles perform both types of function when a primary role of movement of the digits also results in movement of the wrist joint because their tendons cross the joint. First it is proper to study the muscles which cause the hand as a unit to move upon the forearm, either at the radiocarpal and ulnar carpal joints or through associated movements of the carpal joints themselves.

A review of the construction of these joints is indicated at this point along with the types of movements which may occur. It is possible for the hand and wrist to be either flexed or extended and to be either radially deviated (abduction) or ulnarly deviated (adduction).

Muscles Acting Upon the Wrist and Hand

These muscles are tabularly described in two groups: those acting primarily on the wrist and hand, and those whose movement of the digits secondarily produce wrist movements.

Tables of Muscles. Both types of muscles are grouped in the forearm according to general function. This is seen well at the proximal part of the forearm where two well-defined masses of muscles form the medial and lateral boundaries of the antecubital fossa.

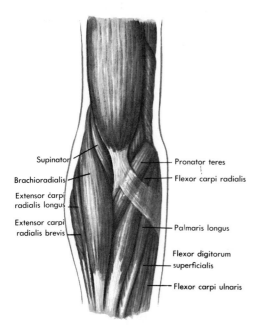

Supinator

Brachioradialis

Extensor carpi radialis longus

Extensor carpi radialis brevis

Pronator teres

Flexor carpi radialis

Palmaris longus

Flexor digitorum superficialis

Flexor carpi ulnaris

Muscles which generally have a flexor and pronator function are grouped along the medial aspect of the fossa and have origins extending medially from the lower end of the humerus or from the ulna. Muscles which generally have an extensor and supinator function are grouped along the lateral aspect of the elbow and have origins extending laterally from the lower end of the humerus or from the radius. Below the antecubital fossa the flexor group shifts to the anterior aspect of the forearm and the extensor group moves around the lateral side of the radius to a posterolateral position in the forearm. (See table on page 202.)

The muscles tabulated below exert their effect primarily upon the fingers and thumb but, by pulling upon the digits, secondarily produce movements of the hand and wrist upon the forearm. They are not described in detail here but are listed for completeness.

Muscles Secondary to Hand and Wrist Movements

Flexor digitorum superficialis
Flexor digitorum profundus
Extensor digitorum
Extensor digiti minimi

Muscle Actions Upon the Wrist and Hand

The movements produced by the muscles primarily acting upon the wrist and hand depend upon whether a muscle acts alone, in concert with either another muscle of its flexor or extensor group, or simultaneously with a muscle of the opposite, antagonistic group. The movement(s) produced are more important than the possible function(s) of an individual muscle. (See table on page 203.)

Synergistic Actions. Finger movements are facilitated by the muscles which flex and extend the wrist. Flexion of the fingers and the grasp of the hand are much weaker when the wrist is flexed. One can easily prove with his own hand that finger flexion is stronger if the wrist joint is extended. Since the pull of the muscles which flex the fingers also secondarily flexes the wrist, the extensor muscles of the wrist act together to maintain the hand in a position to form a stable base for finger movements.

MUSCLES PRODUCING FINGER MOVEMENTS

The hand could not accommodate within itself the fleshy mass of muscle fibers necessary to pro-

MUSCLES ACTING PRIMARILY UPON THE WRIST AND HAND

Muscle	Flexor carpi ulnaris	Palmaris longus	Flexor carpi radialis	Extensor carpi ulnaris	Extensor carpi radialis longus	Extensor carpi radialis brevis
Location	Medial side of forearm	Anterior aspect of forearm; may be small or absent	Lateral side of forearm	Posterolateral side of forearm	Posterolateral side of forearm	Posterolateral side of forearm
Origin	Common tendon from medial epicondyle of humerus and from medial side of olecranon of ulna	Common tendon from medial epicondyle of humerus	Common tendon from medial epicondyle of humerus	Common tendon from lateral epicondyle of humerus and posterior border of ulna	Lower part of lateral supracondylar ridge of humerus	Common tendon from lateral epicondyle of humerus
Fiber Direction	Distally along medial side of forearm	Distally, superficially over median nerve and deeper muscles	Distally along medial side of forearm medial to radial artery	Distally along posteromedial surface of forearm under extensor retinacular band of restraining fascia	Distally along lateral side of wrist deep to extensor retinaculum	Distally along lateral side of wrist deep to extensor retinaculum
Insertion	Into pisiform bone of the wrist and via interconnecting ligaments to hook of hamate bone and base of fifth metacarpal	Into flexor retinaculum (restraining fascia at the wrist) and into palmar fascia	Crosses scaphoid and trapezium bones under flexor retinaculum into bases of second and third metacarpal bones	Base of fifth metacarpal bone	Base of second metacarpal bone	Bases of second and third metacarpal bones

    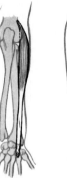

Flex. carpi ulnaris   Palmaris longus   Flex. carpi radialis   Ext. carpi ulnaris   Ext. carpi rad. longus   Ext. carpi radialis brevis

MUSCLE ACTION TABLES

MUSCLES PRIMARY IN FLEXION	MUSCLES PRIMARY IN EXTENSION
Flexor carpi ulnaris	Extensor carpi ulnaris
Flexor carpi radialis	Extensor carpi radialis longus
Palmaris longus	Extensor carpi radialis brevis

MUSCLES PRODUCING ULNAR DEVIATION (ADDUCTION)	MUSCLES PRODUCING RADIAL DEVIATION (ABDUCTION)
Extensor carpi ulnaris as part of extension, or pure deviation when acting with Flexor carpi ulnaris	Extensor carpi radialis longus and brevis as part of extension, or pure deviation when acting with Flexor carpi radialis

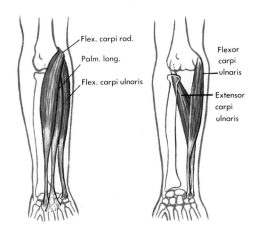
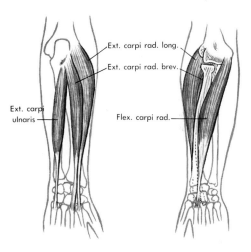

duce its many dexterous movements and still preserve its dexterity. The very nature of the hand and fingers as the prehensile mobile appendages of the distal part of the upper extremity requires that their bony framework be overlaid with a minimum of soft tissues. These soft tissues are blood vessels, nerves, and tendons. The main structures are the free-running tendons of muscles which are located in the forearm. The only exceptions are the small intrinsic muscles of the fingers and the short muscles of the thumb. Since the opposable thumb is set at a right angle to the plane of the hand, its movements involve a different terminology. Therefore, the muscles of the thumb are described separately.

Muscles Acting Upon the Fingers

Since very little movement occurs at the carpometacarpal and intermetacarpal joints, the muscles described in the following tables may be thought of as producing movement of the fingers. For tabular purposes the finger muscles are divided into (1) muscles located in the forearm whose tendons extend across the wrist and hand to attach to the fingers, and (2) intrinsic muscles of the hand which move the fingers.

SPECIAL NOTE ON TENDON SHEATHS AND INSERTIONS. Both the flexor and the extensor muscles just tabulated insert upon the phalanges by unusual tendon and fascial arrangements.

1. *Flexor tendon insertions:* The finger tendons of the flexor digitorum superficialis and profundus pass under the restraining fascial band of the flexor retinaculum at the wrist together. At this point they are enveloped within a lubricating *common synovial flexor sheath,* which facilitates their gliding in response to the contraction of their muscle fibers. The two sets of tendons pass across the palm together with the superficialis

MUSCLES OF THE FOREARM WHICH MOVE THE FINGERS

MUSCLE	Flexor digitorum superficialis	Flexor digitorum profundus	Extensor digitorum	Extensor digiti minimi	Extensor indicis
LOCATION	Middle stratum of anterior aspect of forearm	Deep stratum of ulnar aspect of forearm	Superficially on posterior aspect of forearm but under extensor carpi radialis muscles	On posterior aspect of forearm to ulnar side of extensor digitorum	Deep on posterior aspect of lower forearm
ORIGIN	Humeroulnar head from common tendon from medial epicondyle of humerus and medial surface of ulnar coronoid process; radial head from upper anterior border of radius	Extensive origins from upper two thirds of anterior surface of ulna, corresponding medial surface, medial side of coronoid process, and interosseous membrane	Common tendon from lateral epicondyle of humerus	Common tendon from lateral epicondyle of humerus	Distal part of posterior surface of ulna and adjacent interosseous membrane
FIBER DIRECTION	Muscle fibers of the two heads combine only to separate into superficial and deep, distally coursing portions	Distally to separate into four tendons passing under flexor retinaculum deep to superficialis tendons	Distally under extensor retinaculum, dividing into four tendons which cross the back of the hand interconnected by fibrous bands	Distally under extensor retinaculum sending one tendon to fifth finger	Distally into back of hand under extensor retinaculum medial to second finger tendon of extensor digitorum
INSERTION	Superficial portion sends two tendons under flexor retinaculum, one to middle phalanx of third and fourth fingers; deep portion sends two tendons, one to the middle phalanx of second and fifth fingers	One tendon passes to each of the second to fifth fingers to insert upon the base of its distal phalanx	One tendon passes to each of second to fifth fingers, interconnected and sending slips to adjacent tendons, to insert by special "extensor hood"* arrangement upon middle and distal phalanges	Joins fifth finger tendon of extensor digitorum to insert via "extensor hood"* arrangement	Joins "extensor hood"* arrangement on fifth finger to insert in common with fifth finger tendon of extensor digitorum

Flexor digitorum superficialis   Flexor digit. profundus   Extensor digitorum   Extensor digiti minimi   Extensor indicis

* Described on page 206.

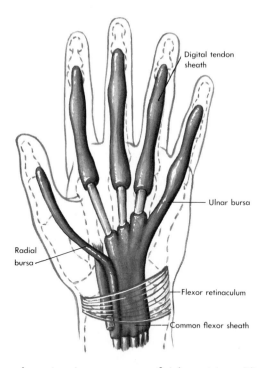

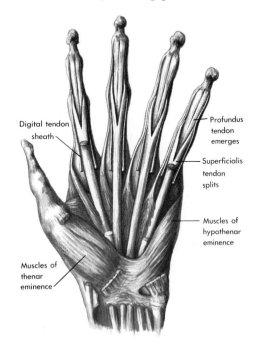

reunited into one tendon, which, at its middle phalanx insertion, splits again. Each half spreads to its own margin of the anterior phalangeal surface to insert without impeding the movement of the more distally coursing profundus tendon.

tendons in the more superficial position. The synovial sheath stops at mid palm, except for an extension about the fifth finger tendons, thereby earning its older name of the *ulnar bursa*. Near the metacarpophalangeal joints the tendons to the second, third, and fourth fingers acquire separate *digital tendon sheaths*.

Reference to the base of one's own finger between the webs to adjacent fingers demonstrates that in this area the morphological situation becomes crowded. The bulk of the tissue mass is taken up by the bone of the proximal phalanx. In addition to blood vessels, nerves, joint structures, and tela subcutanea, two flexor tendons must glide freely toward insertions upon the phalanges. The arrangement of structures is complicated by the fact that the deeper-lying profundus tendon destined for the distal phalanx must get past the superficialis tendon while the latter is inserting upon the anterior surface of the middle phalanx. A morphological impasse is avoided in development by the splitting of the superficialis tendon over the proximal phalanx. (See illustration at top of right column.) At the point of splitting, the profundus tendon emerges to be embraced by the split parts of the superficialis tendon. The profundus tendon is seen passing distally to its distal phalanx insertion, unimpeded by the superficialis tendon. The superficialis tendon is now the deeper and has

2. *Extensor tendon insertions:* A different mechanism of tendon sheaths and insertion exists on the opposite extensor surface of the wrist and hand. The *extensor retinaculum* is a restraining band of fascia extending across the back of wrist. All the tendons extending from the posterolateral surface of the forearm to the wrist, thumb, and fingers pass under the extensor retinaculum. A

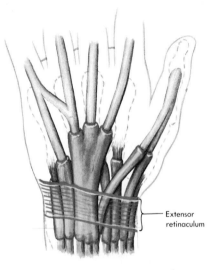

number of separate *extensor synovial sheaths* envelop these tendons, but only those related to the fingers are of concern at this point. The four tendons of the extensor digitorum and the single extensor indicis tendon are enclosed within a short synovial sheath at the back of the wrist. The extensor digiti minimi has its own sheath to the medial side of the other.

Extensor tendons insert into their phalanges in association with a hood of connective tissue fibers which are disposed transversely over the posterior surface of the bones. This "extensor hood" is also termed the *extensor expansion* or *dorsal aponeurosis* of the finger. The central thickened core of an extensor hood is made by the extensor digitorum tendon, even in the case of the second and fifth fingers, which also have their separate "indicis" and "minimi" muscles. The extensor digitorum tendon splits over the proximal phalanx to form a central and two lateral slips or bands. The central slip proceeds to insert into the posterior surface of the middle phalanx while the lateral slips, considerably enmeshed with the transverse fibers of the extensor hood and with the tendons of the intrinsic hand muscles, unite distally to attach to the distal phalanx. The picture will become complete with the description of the intrinsic muscles.

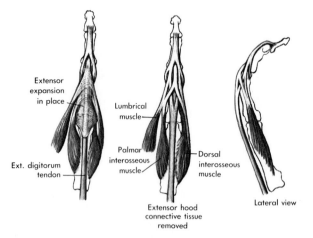

Extensor expansion in place

Lumbrical muscle

Ext. digitorum tendon

Palmar interosseous muscle

Dorsal interosseous muscle

Extensor hood connective tissue removed

Lateral view

Intrinsic Hand Muscles Moving the Fingers. A number of short muscles are localized in the hand. These are described as *intrinsic muscles of the hand.* They may be allotted to three groups: muscles of the thumb, muscles of the fifth finger, and muscles of the deep palm related primarily to moving the fingers to either side of the axis of the hand.

The intrinsic muscles of the thumb are de-

scribed with the muscles that move it. They make a rounded eminence on the lateral side of the palm at the base of the thumb, the *thenar eminence.* The intrinsic muscles to the fifth finger climb along the medial side of the palm to form the more elongated *hypothenar eminence.* The muscles of third group are located deep in the palm (lumbricals) under the long flexor tendons, or fill the crevices between the metacarpal bones (the interossei).

Intrinsic Muscles of the Fifth Finger. The little finger juts distally from the medial side of the hand, making for a poorer mechanical advantage of the long forearm muscles. In addition the little finger and thumb enter into special movements in which they may be approximated to add to the grasp and dexterity of the hand.

Deep Muscles of the Palm. The deep muscles of the palm function primarily to move the fingers to either side of the axis of the hand, although because of the nature of their insertions, they enter into certain flexor-extensor actions. The *axis of the hand* runs through the center of the shaft of the third metacarpal bone, which may be extended distally along the center line of the third finger. Movements of abduction and adduction of the fingers are referred to this axis, not to the axis of the body. A tabulation of the deep palm muscles is on page 208.

The tendons of the deep palm muscles to the fingers enter into a complex of interweaving and splitting within the substance of the extensor expansion. As a general rule it is sufficient to state that the dorsal interossei insert both into an extensor hood and into the base of a proximal phalanx, while the palmar interossei enter an extensor hood just distal to the base of the proximal phalanx. The lumbricals enter into the extensor expansion slightly more distally along this phalanx. The extension of the hood distally over the middle and distal phalanges indicates that the pull of the deep palm muscles is transferred along to these phalanges also.

Muscle Actions in Finger Movements

All the muscles which have been described or tabulated contribute to the total range of motion of the fingers. Since the hand is capable of many coordinated movements, the analysis of these many dexterous motions into individual muscle actions can be very complex. In the tabulations

INTRINSIC MUSCLES OF THE FIFTH FINGER

MUSCLE	Abductor digiti minimi brevis	Flexor digiti minimi brevis	Opponens digiti minimi	Fourth palmar inter-osseous
LOCATION	Hypothenar eminence along medial side of palm	Hypothenar eminence to lateral side of abductor	Deep to abductor digiti minimi brevis and flexor digiti minimi brevis	
ORIGIN	Pisiform bone distal to insertion of flexor carpi ulnaris	Hook of hamate bone	Hook of hamate bone	See tabulation of deep palm muscles.
FIBER COURSE	Distally	Incline medially in distal course	Laterally and distally, wrapping front of fifth metacarpal bone	
INSERTION	Medial side of proximal phalanx of fifth finger	In common with abductor	Medial side of shaft of fifth metacarpal bone	

Abductor digiti minimi brevis — Flexor digiti minimi brevis

Opponens digit. minimi

that follow the major muscles which produce or contribute to a particular pure motion are listed. It must be remembered that there are many shades and degrees of movement, that a finger motion may be carried out from a position already established by other muscles, and that the movement (circumduction of a finger, for example) may be the result of many combined movements. If a further example of complexity is needed, the crooking of the little finger in holding a teacup or playing a violin might be pondered. Major essential movements required in such a case include flexing the elbow, placing the forearm in a position between pronation and supination, flexion and ulnar deviation of the wrist, and flexion-adduction of the little finger, which is partially opposed toward the thumb.

MUSCLES MOVING THE THUMB

The thumb is described separately because it is set at an angle to the plane of the palm. Its position, although adding greatly to the dexterity of the hand, complicates study because a different usage of familiar terminology is required to explain the many movements of this digit. A review should be made of the position and construction of the thumb in Chapter Three. It should be noted again that the first metacarpal is turned so that its anterior surface faces medially across the palm and its posterior surface faces laterally. The turning of the metacarpal bone of the thumb combines with a more anterior location at the higher lateral side of the concave palm to place the thumb in a position of greater mobility.

DEEP PALM MUSCLES OF FINGERS

MUSCLE	Lumbricals	Palmar interossei	Dorsal interossei
	1, 2, 3, 4 (none for thumb; numbered in order from lateral to medial starting with index finger)	1, 2, 4, 5 (none for third [axis] finger; numbered for finger concerned)	1, 2, 3, 4 (numbered in order from lateral to medial; thumb and fifth finger have separate abductors)
LOCATION	Associated with flexor profundus tendons 1. Most lateral 4. Most medial	Between metacarpal bones on palmar side; positioned to adduct finger of same number toward third finger axis	Between metacarpal bones on dorsal side; positioned to abduct second to fourth fingers away from third finger axis; third has two sets and can be abducted both medially and laterally
ORIGIN	From flexor digitorum profundus tendons 1. Lateral side of second finger tendon 2. Lateral side of third finger tendon 3. Adjacent sides of third and fourth finger tendons 4. Adjacent sides of fourth and fifth finger tendons	1. Medial side of shaft of first metacarpal bone 2. Medial side of shaft of second metacarpal bone 4. Lateral side of shaft of fourth metacarpal bone 5. Lateral side of shaft of fifth metacarpal bone	1. By heads from medial side of first and lateral side of second metacarpal; acts on second finger 2. By heads from medial side of second and lateral side of third metacarpal; acts on third finger to abduct toward thumb 3. By heads from medial side of third and lateral side of fourth metacarpal; acts on third finger to abduct toward fifth finger 4. By heads from medial side of fourth and lateral side of fifth metacarpal; acts on fourth finger to abduct toward fifth finger
FIBER COURSE	Deep in palm and distally across respective metacarpophalangeal joint	Distally between heads of proximal phalanges beside metacarpophalangeal joints, keeping to side named in origin	Distally between heads of proximal phalanges beside metacarpophalangeal joints, keeping to side named in origin
INSERTION	Into lateral (thumb or radial) side of extensor expansions of second to fifth fingers distal to metacarpophalangeal joint	Into extensor expansion of finger of same number as metacarpal bone of origin, and on side toward axis of third finger as named in origin	Into extensor expansion of finger number 1 into second finger 2 into third finger 3 into third finger 4 into fourth finger

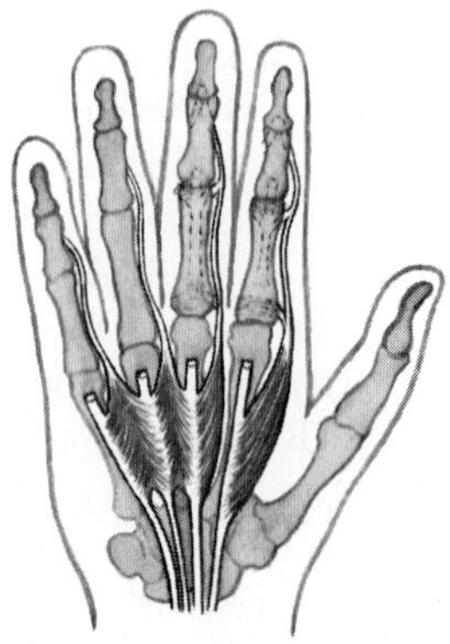

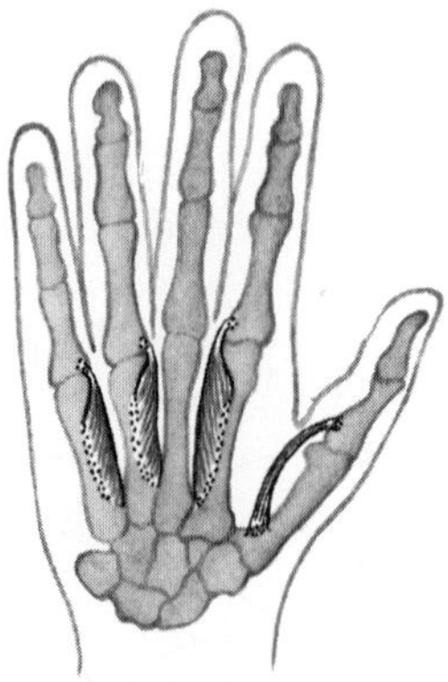

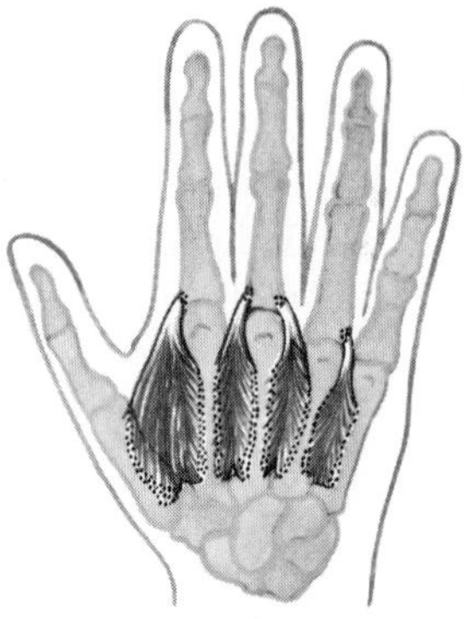

ACTIONS OF MUSCLES IN FINGER MOVEMENTS

FINGER FLEXION			FINGER EXTENSION		
Proximal Phalanges on Metacarpals	Middle Phalanx on Proximal	Distal Phalanx on Middle	Proximal Phalanges on Metacarpals	Middle Phalanx on Proximal	Distal Phalanx on Middle
Interossei Lumbricals Flexor digiti minimi brevis (fifth only)	Flexor digitorum superficialis (including fifth)	Flexor digitorum profundus (including fifth)	Extensor digitorum (including fifth) Extensor indicis (second only) Extensor digiti minimi (fifth only)	Interossei Lumbricals	Interossei Lumbricals

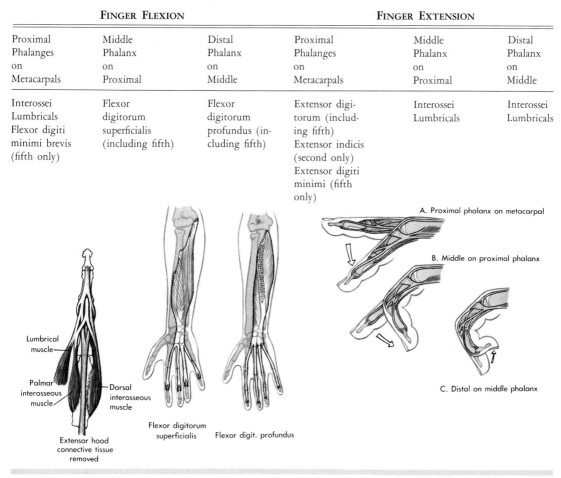

Lumbrical muscle

Palmar interosseous muscle — Dorsal interosseous muscle

Extensor hood connective tissue removed

Flexor digitorum superficialis   Flexor digit. profundus

A. Proximal phalanx on metacarpal

B. Middle on proximal phalanx

C. Distal on middle phalanx

FINGER ADDUCTION	SPECIAL ACTION	FINGER ABDUCTION
Toward third finger axis Palmar interossei (except third finger)	Opponens digiti minimi pulls fifth metacarpal forward, deepening the hollow of the hand, to place fifth finger in position to be opposed by thumb	Away from third finger axis Dorsal interossei (including third finger, including thumb, excluding fifth finger) Abductor digiti minimi (fifth only)

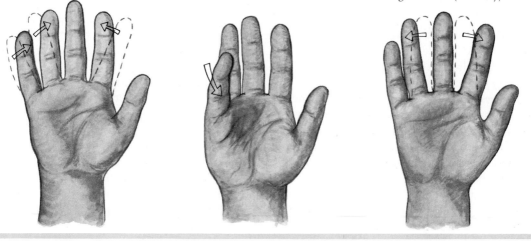

Although the range of motion of the thumb is better explained after a discussion of the course and insertion of the muscles which produce the movements, an introduction to the appearance and movements of the thumb follows so that the grouping of muscles can be better appreciated.

Appearance and Movements of the Thumb

Look at your own right hand, palm up, in a position of rest. The third to fifth fingers curl naturally in flexion and are in close approximation. The index finger is slightly apart from the other three and is less flexed because of the tone in relaxation of two extensor muscles. The thenar eminence (ball of the thumb) makes a rounded oblong on the lateral side of the hand distal to the wrist creases.

The thumb seems to spring out of this eminence, and it does if one discounts that the thumb metacarpal is amidst the fleshy thenar mass and thinks of the thumb only in terms of its two phalanges. The latter is a common misconception; the thenar eminence is more a part of the thumb as a finger than it is a part of the palm from which its muscles and bone have been turned and set apart. In the rest position, the first phalanx of the thumb comes out of the thenar eminence at the metacarpophalangeal joint at an angle toward the center of the palm. Since it is bent toward another body part, it is flexed even though this term has a different meaning of direction in respect to the other fingers. The distal phalanx is slightly bent across the palm so that its tip may be touching the side of the tip of the index finger.

The rest position of the thumb, therefore, is that of partial flexion, and *thumb flexion* is the sweeping of the entire thumb across the palm medialward. Be careful in performing this motion to keep the thumb close to the anterior surface of the other fingers. If you allow the thumb to come toward you in this movement, you are combining it with another movement (abduction). *Thumb extension* is the bringing of the thumb laterally, back across the palm again close

to the finger surfaces, to the rest position. Extension is continued by moving the thumb away from the hand laterally but keeping it still in the plane of the palm.

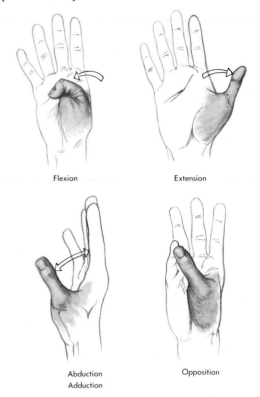

Flexion   Extension

Abduction
Adduction   Opposition

In *thumb abduction* the thumb moves as a unit toward you in the same plane in which the index finger flexes. This movement alone is not a natural one, but is frequently combined with other movements which are facilitated if the thumb is brought forward from the rest of the hand. A combination of thumb abduction and partial flexion is essential in grasping a ball or other object to be held in the hand. *Thumb adduction* returns the thumb from the abducted position.

Opposition of the thumb is a new term which is applied only to the thumb. If one places the pad of the thumb against the anterior (palmar) surface of each of the finger tips in succession, the thumb is being carried in opposition to each and toward the little finger. Returning the thumb to the rest position gains the little used description of *reposition*. Although one rarely does oppose the thumb to the little finger, the ability to perform this movement is essential to grasping any object and, in its pure form, is a diagnostic test of motor function and nerve integrity in the hand.

The main movements of the thumb have been described. Others, such as rotation, are essential to the combination of the pure movements into the finely adjusted motions of the thumb, which have many gradations including the employment of all in *thumb circumduction.*

Muscles Acting Upon the Thumb

The muscles responsible for movements of the thumb can be divided into two groups: the forearm muscles to the thumb and the intrinsic thumb muscles of the hand.

FOREARM MUSCLES TO THE THUMB

MUSCLE	Flexor pollicis longus	Abductor pollicis longus	Extensor pollicis brevis	Extensor pollicis longus
LOCATION	Deep anterolateral aspect of forearm over radius	Deep in middle of forearm becoming superficial on posterolateral aspect in lower part	Distal to abductor pollicis longus in same relationships	Medial side of abductor and extensor brevis muscles in same relationships
ORIGIN	Upper three fourths of anterior surface of radius and adjacent interosseous membrane	Upper posterior surface of interosseous membrane and adjacent surfaces of radius and ulna	Posterior surface of *radius* and adjacent interosseous membrane just distal to the abductor	Middle of posterior surface of *ulna* and adjacent interosseous membrane
FIBER COURSE	Distally over radius beneath flexor retinaculum with tendon in separate synovial sheath also called the "radial bursa" along medial aspect of thenar eminence	Distally and laterally becoming superficial to cross obliquely over extensor carpi radialis and brevis tendons enclosed in a synovial sheath	Close association with the abductor muscle; enclosed within same or separate synovial sheath; same oblique crossing relationships	Swings medial to the abductor and extensor brevis to reach base of thumb in own synovial sheath; same oblique crossing relationships
INSERTION	Palmar (medial) surface of distal phalanx	Reaching base of thumb, attaches to lateral side of first metacarpal bone	Passes along lateral surface of thenar eminence to insert upon posterior surface of *proximal* phalanx	Base of *distal* phalanx

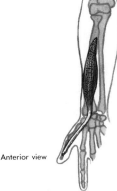

Anterior view

Flex. pollicus long.

Abduct. poll. long.

Posterior views

Ext. pollicus brevis.

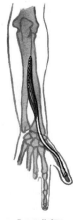

Ext. poll. long.

Forearm Muscles to the Thumb

The forearm muscles to the thumb, by traveling to the radial side of the forearm from deeper aspects, wrap over the lower parts of the extensor radialis longus and brevis to reach the base of the thumb. The tendons of the thumb muscles of the forearm form a topographical landmark or hollow as they cross the radial side of the wrist obliquely. This *anatomical snuff box* has as its lateral border the associated abductor pollicis longus and extensor pollicis brevis tendons. Bordering the hollow medially is the tendon of the

INTRINSIC THUMB MUSCLES OF THE HAND

MUSCLE	Abductor pollicis brevis	Flexor pollicis brevis	Opponens pollicis	Adductor pollicis
LOCATION	Superficially on lateral side of thenar eminence	Medial part of thenar eminence	Deeper in thenar eminence under abductor muscle	Deep in palm
ORIGIN	Anterior surface of flexor retinaculum, tubercle of trapezium, tubercle of scaphoid bone	Anterior surface of flexor retinaculum in common with abductor pollicis brevis	Flexor retinaculum in common with abductor pollicis brevis and flexor pollicis brevis and arising from trapezium bone	By two heads: *oblique* head from capitate and trapezoid bones and from base of second metacarpal bone; *transverse* head from the anterior surface of third metacarpal bone
FIBER COURSE	Obliquely distally and laterally in thenar eminence			The two heads fuse to form a triangular muscle, which courses obliquely, laterally, and distally into thenar eminence, being joined by first palmar interosseous muscle
INSERTION	By common tendon into lateral side of base of *proximal* phalanx; sometimes into lateral sesamoid bone of thumb		Lateral border of first *metacarpal* bone	By common tendon into medial side of base of *proximal* phalanx and medial sesamoid bone of thumb in common with first palmar interosseous muscle

Abductor pollicis brevis   Flexor pollicis brevis 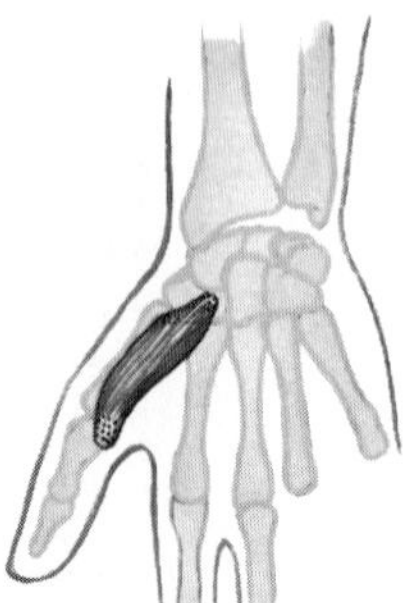   Opponens pollicis 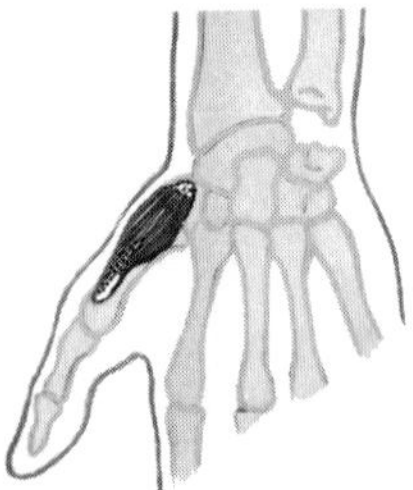   Adductor pollicis

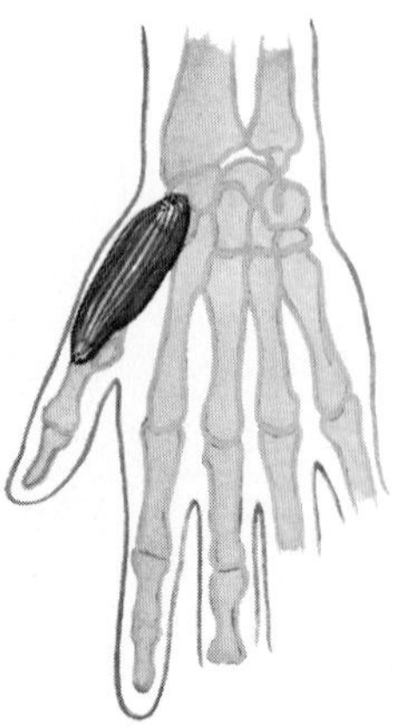

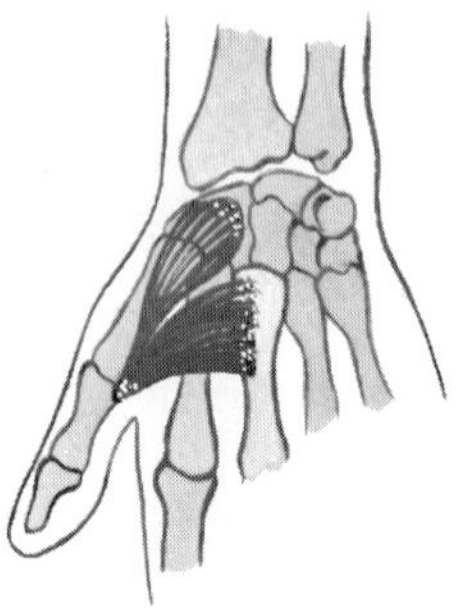

The first palmar and dorsal interossei are associated in function with these muscles.

extensor pollicis longus. The anatomical snuff box is demonstrated easily by cocking the thumb out laterally in extension. Its importance today is not as a receptacle for the taking of snuff as in earlier times, but as a means for locating the superficial branch of the radial nerve and the radial artery, which cross the area medially and obliquely on their way to the hand.

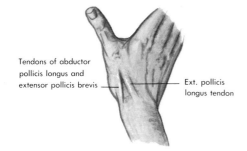

Tendons of abductor pollicis longus and extensor pollicis brevis

Ext. pollicis longus tendon

Intrinsic Thumb Muscles of the Hand. The forearm muscles to the thumb have given way to their tendons when the base of the thumb is reached. The fleshy bulk of the thenar eminence is made up of the "short muscles of the thumb," which, as intrinsic muscles of the hand, gain mechanical advantage from coursing obliquely or transversely across the hand from origins within the hand or upon the wrist. These muscles are described in the table on page 212.

Actions of Thumb Muscles

Despite differing origins and courses, the forearm and intrinsic hand muscles to the thumb function together to produce the movement of the thumb. Therefore the muscles responsible for the pure movements of the thumb are tabulated together in the second column of this page.

MUSCLES PRODUCING MOVEMENTS AT THE HIP JOINT

One of the largest assemblages of muscles in the body is concerned with moving or stabilizing the hip joint. The musculature is not limited to producing a range of movement of a free extremity such as has been learned of the scapulohumeral joint. With the body in the erect position the muscles acting upon the hip joint are called into play also to:

1. Strengthen and stabilize the joint to bear and transmit the weight of the trunk, upper limb, neck, and head.

ACTIONS OF THUMB MUSCLES

THUMB FLEXION

Proximal phalanx	Distal phalanx
Flexor pollicis brevis	Flexor pollicis longus

THUMB EXTENSION

Proximal phalanx	Distal phalanx
Extensor pollicis brevis Indirectly by continued pull of extensor pollicis longus	Extensor pollicis longus Aided by most other short muscles secondary to their main actions

Stabilization of base of thumb, provided by abductor pollicis brevis, is essential to extension.

THUMB ABDUCTION

Abductor pollicis longus aided by abductor pollicis brevis and extensor pollicis brevis

THUMB ADDUCTION

Adductor pollicis aided by first palmar interosseous

THUMB OPPOSITION

Opponens pollicis assisted by abductor pollicis brevis, flexor pollicis brevis, and adductor pollicis, which aid opposition by rotating thumb, stabilizing it, and guiding its movement across the palm

THUMB ROTATION

Occurs medially mainly during opposition by opponens pollicis, abductor pollicis brevis, and flexor pollicis brevis

2. Help preserve the erect position while supporting this weight.

3. Move the body through space (walking, running, jumping) while helping to maintain balance and support the weight of the body.

4. Provide a stable platform, if the lower extremities are fixed, so that the pelvis and trunk can move upon the legs at the hip joint (trunk flexion, for example).

Only when the body is supported in a recumbent or a sitting position can the lower limb be moved freely without consideration of these additional factors. But even the weight of the heavy bulk of bone and muscle of the lower limb forces

GLUTEAL MUSCLES

MUSCLE	Gluteus maximus	Gluteus medius	Gluteus minimus	Piriformis	Obturator internus
SHAPE	Thick quadrilateral mass	Thinner triangular mass	Thin band	Small slender triangle	Small slender triangle
LOCATION	Superficially, forming buttock	Underlies gluteus maximus	Underlies gluteus medius	Deep in buttock under lower part of gluteus maximus	Lateral wall of deep pelvis
ORIGIN	Ilium, posterior to posterior gluteal line; posterior surface of sacrum and coccyx; sacrotuberous ligament	Ilium between anterior and posterior gluteal lines	Ilium between anterior and inferior gluteal lines	Pelvic surface of sacrum; lowest part of gluteal area of ilium; sacrotuberous ligament	Pelvic surface of hip bone and obturator membrane
FIBER COURSE	Laterally and slightly downward behind hip joint	Medial fibers laterally and downward; lateral fibers almost vertically downward behind hip joint	As with gluteus medius, with fibers merging with medius	Laterally and slightly downward along lower border of gluteus medius to pass through greater sciatic foramen	Leaves pelvis through lesser sciatic foramen to turn sharply forward toward hip joint
INSERTION	Lower fibers into gluteal tuberosity of femur; upper fibers into iliotibial tract of fascia lata of thigh*	Lateral surface of greater trochanter of femur	Anterolateral border of greater trochanter femur	Posterior surface of greater trochanter of femur	Medial surface of greater trochanter of femur
ACTION	Extension of thigh at hip joint; also, if leg is fixed, extends pelvis and trunk in trunk movements; lateral rotator of thigh at hip joint	Both muscles abduct and medially rotate thigh at hip joint, especially in walking, to clear the leg from the ground; also when opposite leg is swinging, hold pelvis up by pulling against fixed leg of own side		Stabilizes hip joint; abducts thigh Lateral rotation of thigh at hip joint	Stabilizes hip joint; abducts thigh Lateral rotation of thigh at hip joint

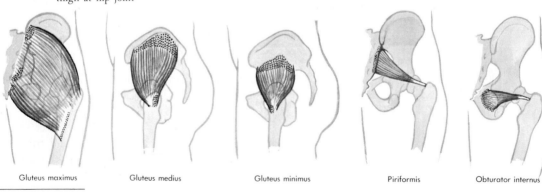

Gluteus maximus   Gluteus medius   Gluteus minimus   Piriformis   Obturator internus

*See description of iliotibial tract which follows.

GLUTEAL MUSCLES (Continued)

Gemellus superior	Gemellus inferior	Quadratus femoris	Obturator externus	Tensor fasciae latae	MUSCLE
Small slender triangle	Small slender triangle	Short quadrangular mass	Flat plate	Long band	SHAPE
Superior to obturator internus	Inferior to obturator internus	Inferior to gemellus inferior	External surface of obturator membrane	Anterolateral aspect of upper thigh	LOCATION
Ischial spine	Ischial tuberosity	Ischial tuberosity	External surfaces of pubis and ischium surrounding obturator foramen and external surface of obturator membrane	Outer lip of iliac crest and vicinity of anterosuperior iliac spine	ORIGIN
Follows obturator internus, upper border	Follows obturator internus lower border	Laterally behind hip joint	Crosses laterally behind hip joint	Downward to blend into tough fascia lata on anterolateral aspect of thigh*	FIBER COURSE
Converges to insert on obturator internus tendon behind hip joint	Converges to insert on obturator internus tendon	Intertrochanteric crest of femur	Trochanteric fossa of femur	Iliotibial tract of fascia lata of thigh*	INSERTION
		Adduct thigh	Adduct thigh		ACTION
Lateral rotation of thigh at hip joint	Lateral rotation of thigh at hip joint	Lateral rotation of thigh at hip joint	Lateral rotation of thigh at hip joint	Flexes and medially rotates thigh	

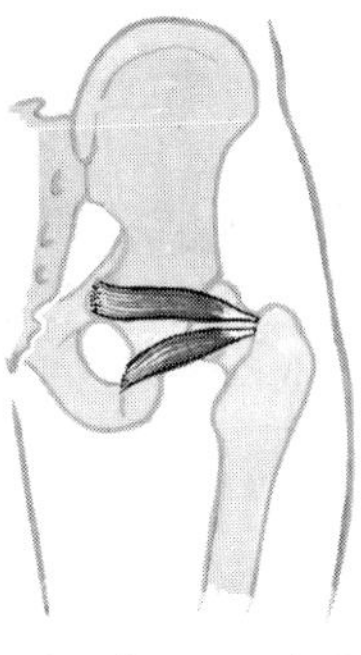

Gemellus sup. and inf.

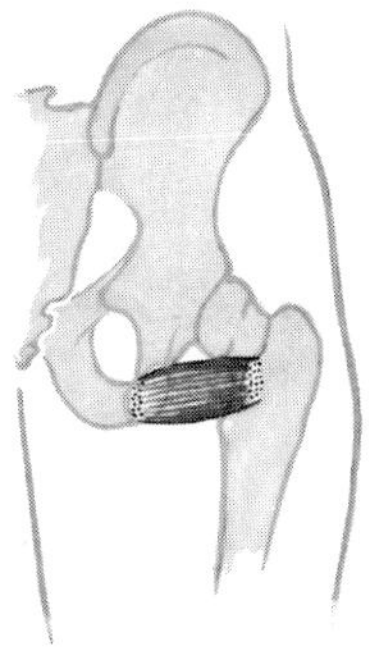

Quadratus femoris

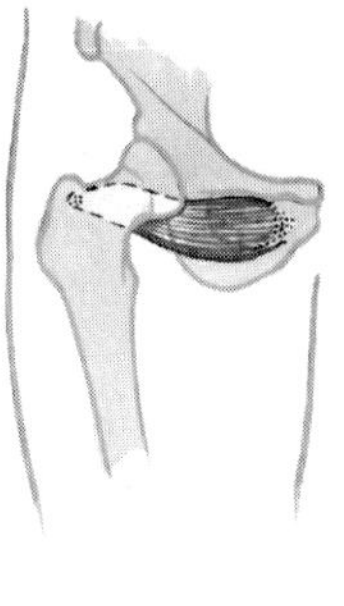

Obturator externus

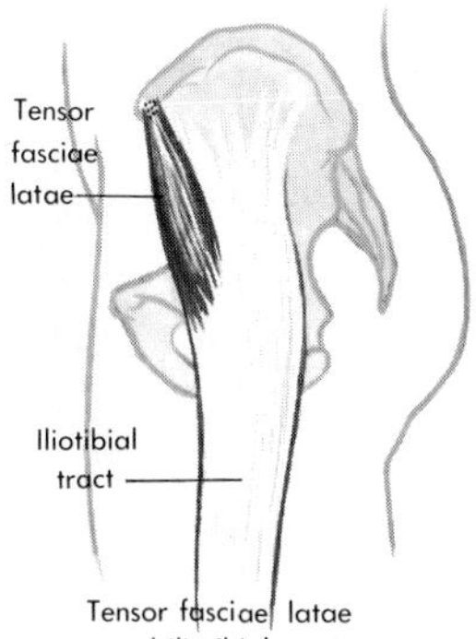

Tensor fasciae latae and iliotibial tract

* See description of iliotibial tract which follows.

the hip joint muscles to act against gravity, although the effect of gravity can be mitigated somewhat, as in physiotherapeutic exercising with the body immersed in water.

The muscles of the hip joint will be described according to their morphological characteristics, first by groups according to their positions about the hip joint. Brief accounts of their actions will be included. At the end of this section a composite tabulation of all the muscles acting upon the hip joint will be found.

Since the muscles have attachments to the

MUSCLES OF THE BACK OF THE THIGH

MUSCLE	Biceps femoris	Semitendinosus	Semimembranosus
SHAPE	Two heads; long flat muscle	Long flattened cylinder	Long flat sheet
LOCATION	Posterolateral aspect of thigh and lateral aspect of popliteal fossa	Posteromedial aspect of thigh and medial aspect of popliteal fossa	Deep to semitendinosus
ORIGIN	*Long head* from medial aspect of ischial tuberosity; *short head* from linea aspera of femur, lateral supracondylar line, lateral intermuscular septum	Medial aspect of ischial tuberosity in common with long head of biceps femoris	Superomedial aspect of ischial tuberosity by a long tendon
FIBER COURSE	Two heads combine in vertical course at lower third of thigh; tendon swings to lateral side of popliteal fossa	Vertically, forming a long tendon at midthigh which veers to medial side of popliteal fossa	Muscle fibers begin in upper thigh, passing vertically to become tendinous below midthigh
INSERTION	Head of fibula and into fascia of leg	Upper medial surface of tibia and fascia of leg	Posteromedial aspect of medial condyle of tibia
ACTION	Extends thigh at hip joint if knee is not fully flexed; flexes leg at knee joint if hip is not fully extended; extends pelvis if legs are fixed		

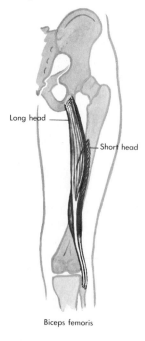

Long head

Short head

Biceps femoris

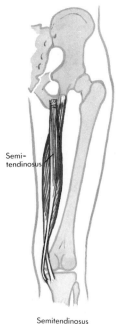

Semi-tendinosus

Semitendinosus
Semimembranosus

pelvic girdle and to the femur, a review of the bones as well as of the hip joint will be helpful at this time.

The groups of muscles to be considered by their locations are: muscles of the gluteal region, muscles of the back of the thigh, muscles of the front of the thigh, and muscles located medially on the thigh.

Gluteal Muscles

The gluteal region of the body is an area of morphological transition over the posterior surface of the hip bone, between the posterolateral aspect of the trunk and the back of the thigh. It is given the characteristic rounded form, termed the *buttock*(s), because of a mass of muscles which course laterally and downward over the posterior aspect of the hip joint to attachments on the upper part of the femur. One of the secondary sexual characteristics of the female is the deposition of fat into the tela subcutanea of the buttock to add to rounding prominence of the gluteal region. The muscles of this region are presented on pp. 214 and 215.

Iliotibial Tract and Its Muscles. The investing fascia of the thigh is called the *fascia lata*. It is thick and extremely strong over the anterolateral and lateral aspects of the thigh, in part because it restrains the heavy musculature of the thigh and maintains the conical shape of the thigh. This fascia also receives some of the fibers of the gluteus maximus muscle as well as the fibers of the tensor fasciae latae. The interweaving of the connective tissue fibers of the fascia with the ligamentous fibers of insertion from the muscles forms a strong tendinous band on the lateral aspect of the thigh. This *iliotibial tract* continues downward along the lateral side of the knee to the lateral tibial condyle where it blends with the deep fascia of the leg. The strong musculofibrous complex is important in flexing and medially rotating the thigh at the hip joint. It is also an effective contributor to the maintenance of the erect posture by extending the pull of the two muscles along the thigh to the upper part of the leg.

Muscles of the Back of the Thigh

A group of muscles, familiarly known as the "*hamstring* muscles," shape the posterior aspect of the thigh from below the buttock to the knee. These long fleshy muscles course vertically from the posteroinferior part of the pelvic girdle to cross the posterior aspects of both the hip and knee joints. They have the unusual functions of extending one joint (hip) and flexing another (knee). The muscles are described in the table on page 216.

Anterior Thigh Muscles

Some of the muscles at the front of the thigh act mainly upon the knee joint. In the table on page 218, only the muscles which act primarily upon the hip joint are included.

Medial Muscles of the Thigh

A group of muscles cross from the pubic and ischial parts of the hip bone to the medial side of the femur. These muscles, which are mainly adductors of the thigh, make up the fleshy mass of the medial aspect of the thigh.

Special Notes on Medial Thigh Muscles. The obturator externus muscle, although deeply related to the gluteal muscles, acts, in addition to lateral rotation, as a subsidiary adductor muscle, as may be seen by reviewing the course of its fibers.

The *femoral triangle* is a topographic area of

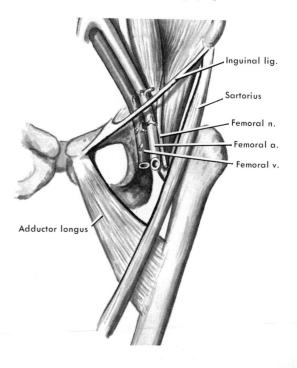

Inguinal lig.

Sartorius

Femoral n.

Femoral a.

Femoral v.

Adductor longus

| MUSCLE | Iliopsoas, two parts: | | Rectus femoris | | Sartorius | Tensor fasciae latae |
	Iliacus	Psoas major and minor*				
SHAPE	Broad, tapering	Thick long mass	Long fleshy mass		Long strap	
LOCATION	Iliac fossa of hip bone	Lateral to lumbar verte-brae	Part of quadriceps femoris mass on front of thigh		Across front of thigh	
ORIGIN	Upper part of iliac fossa and ala of sacrum	Slips from twelfth thoracic and all lumbar intervertebral disks and adja-cent margins of vertebrae	*Anterior head* from antero-inferior iliac spine	*Posterior head* from postero-superior margin of acetabulum	Anterosuperior iliac spine	
FIBER COURSE	Converges in medial and inferior direc-tion across iliac fossa toward hip joint	Laterally and downward beside vertebral column, inclin-ing along brim of pelvis and under inguinal ligament toward front of hip joint	Straight course inferiorly toward patella	Comes forward along lateral aspect of hip joint in down-ward course to merge with anterior head	Spirals across thigh infero-medially	Previously described with gluteal muscles; acts to flex thigh at knee joint in addition to effecting medial rotation
INSERTION	Mainly into tendon of psoas major muscle with few fibers directly into lesser trochanter of femur	Passes in front of hip joint to attach to lesser trochanter of femur	Long insertion tendon divides: part attaches to base of patella and re-mainder continues as ligamentum patellae to insert into tibial tuber-osity		Superior part of medial aspect of tibia	
ACTION	Strongest flexor of thigh at hip joint; if thigh is fixed, flexes trunk at hip joint Psoas major portion assists in lateral flexion of trunk		Assists iliopsoas in flexing thigh; also acts as part of quadriceps femo-ris muscle to extend knee		Flexes thigh at hip joint; assists in flexing leg at knee joint	

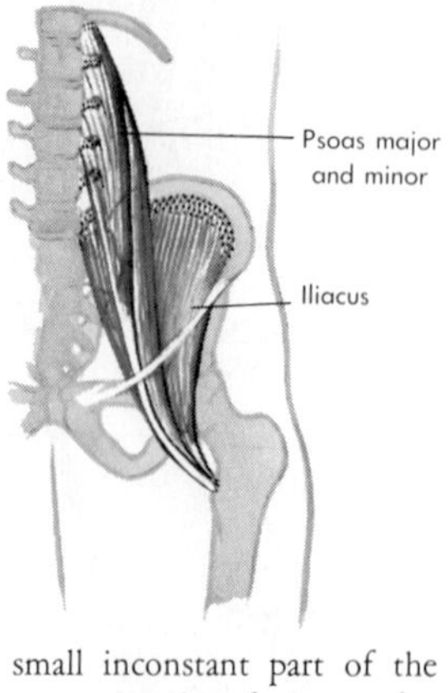

Psoas major and minor

Iliacus

Post. head

Ant. head

Rectus femoris

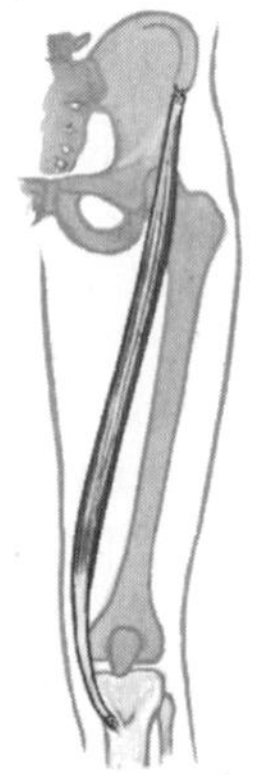

*Psoas minor is a small inconstant part of the psoas muscle mass. When present, it arises from vertebrae T12 and L1 and attaches to the arcuate line of the hip bone, thus having no action in flexing thigh, but possible action in assisting the psoas major muscle to flex the trunk.

the upper anteromedial aspect of the thigh. This triangle has as its base the inguinal ligament. The lateral border is formed by the medial border of the sartorius, and the medial border is formed by the adductor longus. The pectineus crossing obliquely at a deeper level forms the floor of a triangular compartment which is roofed over by the investing and superficial fasciae. The femoral triangle is significant as a compartment through which pass the femoral artery, femoral vein, and femoral nerve.

The combination of movements produced by the muscles of the upper part of the adductor group is important to the adductor function. It will be recalled that adduction of an upper extremity is stopped by the chest wall, and that to continue adduction of the arm across the chest it is necessary to combine this movement with flexion at the shoulder joint. A similar situation exists at the hip joint. One thigh stops the adduction of the other. In order to swing a thigh medially across the other, it is necessary for the moving thigh to be flexed also at the hip joint. (See tables on pages 218, 220 and 221.)

Muscle Actions Upon the Hip Joint

The actions of individual muscles have been listed in preceding tables from which an understanding should have been gained of the morphological grouping of muscles that act upon thigh and trunk in this important area of weight-bearing, posture, and locomotion. It will now be more realistic to tabulate the motions of the hip joint in terms of the muscles which enter into each, remembering that pure motions are rarely found in life. Flexion-adduction-medial rotation of the hip joint are frequently combined as are abduction-lateral rotation.

MUSCLES PRODUCING MOVEMENTS AT THE KNEE JOINT

Despite its common designation as a hinge joint, the knee joint undergoes more than flexion and extension. The manner in which the articulating bones move was described in Chapter 3. It was also indicated that there is a component of rotation at the knee, particularly in the swinging of the lateral femoral condyle about the axis

of rotation of this joint, which extends through the femoral head to the medial tibial condyle. Knee joint movements are further complicated in that if the leg is firmly fixed by planting of the feet, lateral rotation of the thigh usually accompanies knee flexion, and medial thigh rotation is an accompaniment of knee extension. If the leg is free but the hip joint is fixed (as in sitting in a chair), knee flexion is accompanied by medial rotation of the tibia upon the femur, whereas extension is associated with lateral rotation of the tibia. Since the knee joint separates the thigh and the leg, much of the rotation component with knee movements depends on which body part (thigh or leg) is free to move.

Muscles Acting upon the Knee Joint

Most of the muscles which act upon the knee joint have already been described in connection with the hip joint because they also cross the hip joint and produce thigh movements. These muscles are the biceps femoris, semimembranosus, semitendinosus, sartorius, gracilis, and rectus femoris. Other muscles which produce movements of the knee joint are described in the table on page 223.

Actions of Muscles upon the Knee Joint

The actions of muscles upon the knee joint can be summarized, the main consideration being the movements of the leg upon the thigh. This is done in the table on page 224. The knee joint, however, is concerned with weight-bearing and the preservation of balance. Functional antagonisms between flexor and extensor mechanisms are brought into play to ensure a stable, fixed knee in various stages of weight-bearing and balance preservation. Other correlated actions are necessary in locomotion, the main mechanisms of which are presented at the end of the chapter.

MUSCLES PRODUCING MOVEMENTS OF THE ANKLE AND FOOT

Movements of the distal parts of the lower extremity occur at three main points. The foot
(*Text continued on page 225*)

MUSCLE	Pectineus	Adductor longus	Adductor brevis	Adductor magnus (two parts)	Gracilis
SHAPE	Strap	Short narrow fan	Short broader fan	Long broad fan	Long thin band
LOCATION	Floor of femoral triangle;* most superior of adductor group	Medial border of femoral triangle* below pectineus	Lies under pectineus and adductor longus along inferior margin of obturator externus	Two heads remain separate as two-part muscle lying deep to pectineus, adductor longus, and adductor brevis	Most medial of adductor group
ORIGIN	Pectineal line of pubic bone	Body of pubic bone facing femur	Body and inferior ramus of pubic bone	*Adductor part:* from ischiopubic ramus *Extensor part:* from ischial tuberosity	Lower margin of body and ramus of pubic bone
FIBER COURSE	Inferolaterally	From narrow origin fans out laterally and with more inferior declination than pectineus	From narrow origin fans out laterally and inferiorly	*Adductor part:* horizontally to posteromedial aspect of femur *Extensor part:* almost vertically to medial side of lower two thirds of femur; lower part interrupted to allow femoral vessels passage to popliteal fossa	Vertically along medial thigh across knee joint
INSERTION	Upper part of pectineal line of femur	Medial lip of linea aspera of femur	Lower part of pectineal line of femur and upper part of linea aspera	*Adductor part:* linea aspera and gluteal tuberosity *Extensor part:* linea aspera, medial supracondylar ridge, and adductor tubercle	Upper part of medial aspect of tibial shaft
ACTION	Adducts thigh but, positioned anterior to the head of femur, also assists in flexion†	Adducts thigh and assists in flexion†	Adducts thigh and assists in flexion†	*Adductor part:* adducts thigh *Extensor part:* resembles hamstring muscles, primarily extending thigh	Weakly adducts and weakly flexes thigh

Pectineus   Adductor longus   Adductor brevis   Adductor magnus   Gracilis

* and †: See *Special Notes on Medial Thigh Muscles* on page 217.

ACTIONS OF MUSCLES UPON THE HIP JOINT

Hip Flexion	Hip Extension
Iliopsoas	Gluteus maximus
Tensor fasciae latae	Biceps femoris
Rectus femoris	Semitendinosus
Pectineus (assists)	Semimembranosus
Adductor longus (assists)	Extensor part of adductor magnus
Gracilis (weak)	

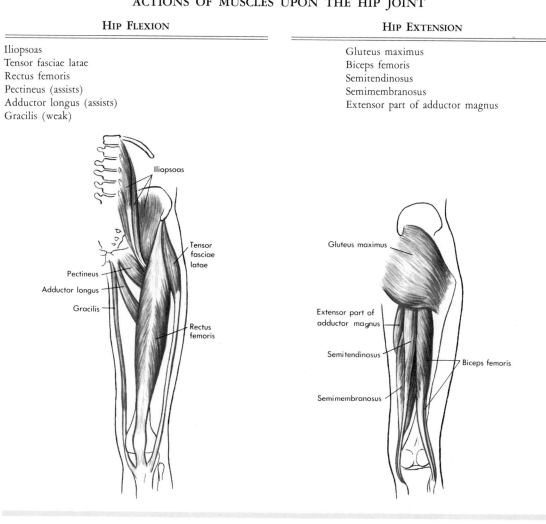

ACTIONS OF MUSCLES UPON THE HIP JOINT (Continued)

MEDIAL ROTATION OF HIP

Gluteus medius
Gluteus minimus
Tensor fasciae latae

LATERAL ROTATION OF HIP

Obturator internus
Obturator externus
Gemellus superior
Gemellus inferior
Quadratus femoris
Piriformis

HIP ABDUCTION

Gluteus medius
Gluteus minimus
Obturator internus
Piriformis

HIP ADDUCTION

Pectineus
Adductor longus
Adductor brevis
Adductor part of adductor magnus
Obturator externus
Quadratus femoris
Gracilis (weak)

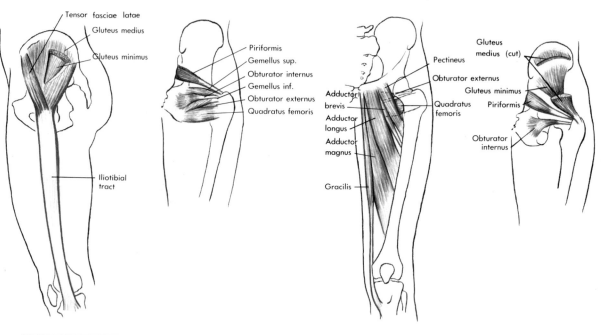

NEW MUSCLES PRODUCING MOVEMENTS OF THE KNEE JOINT

Quadriceps femoris—a four-part muscle including:

Muscle	Rectus femoris	Vastus lateralis	Vastus medialis	Vastus intermedius
Shape	Long fleshy mass	Long thick band	Thick fleshy mass	Flat fleshy sheet
Location	Front of thigh	Anterolateral aspect of thigh	Anteromedial aspect of thigh	Deep to rectus femoris, vastus lateralis, and vastus medialis over anterior aspect of femur
Origin		Narrow line downward from intertrochanteric line of femur, greater trochanter, gluteal tuberosity, linea aspera, and lateral intermuscular septum	Intertrochanteric line, spiral line, and medial intermuscular septum	Anterolateral surface of upper two thirds of femoral shaft and distal lateral intermuscular septum
Fiber Course	Refer to the table of anterior thigh muscles	In much of upper course is mixed with attaching fibers of gluteus maximus; fibers course downward to blend into lateral side of rectus femoris tendon	In upper course mixed with adductor muscle attachments; downward course covering medial aspect of shaft of femur	Fused with vastus lateralis fibers; passing deep to quadriceps tendon
Insertion		Aponeurotically into lateral side of rectus femoris (quadriceps) tendon	Aponeurotically into medial side of rectus femoris (quadriceps) tendon	Aponeurotically into deep surface of rectus femoris (quadriceps) tendon

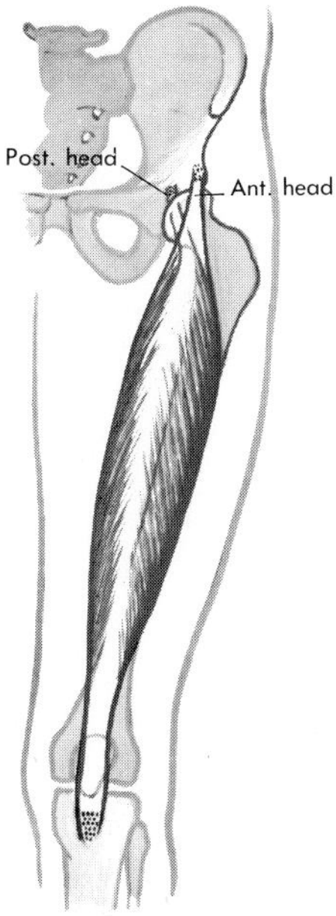

Post. head
Ant. head

Rectus femoris

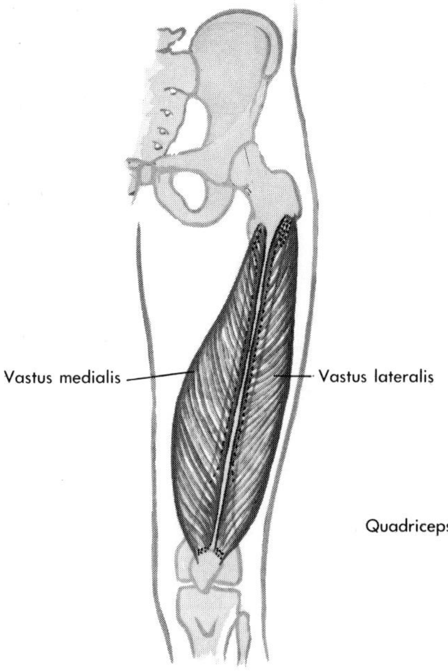

Vastus medialis — Vastus lateralis

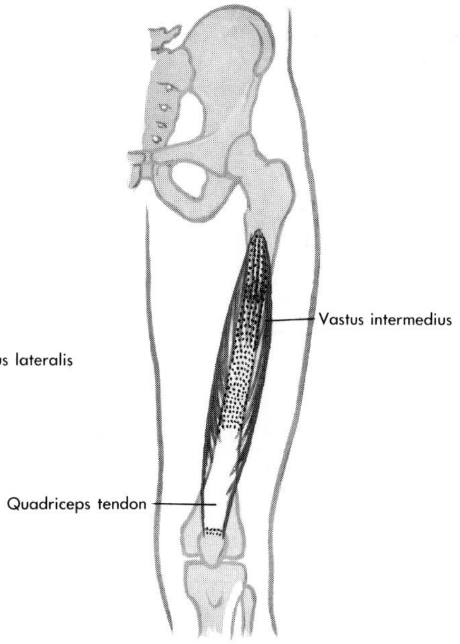

Quadriceps tendon

Vastus intermedius

Other Muscles.

Articularis genu An insignificant series of muscle slips extending from lower anterior aspect of femur into capsule of knee joint and suprapatellar bursa.

Gastrocnemius A powerful two-headed muscle of calf of leg which acts primarily on foot. The muscular heads take origin from lower end of femur and can act upon knee joint if foot is fixed.

Popliteus Small muscle of popliteal fossa, with origins on lateral condyle of femur and lateral knee cartilage; crosses knee joint medially to posterior surface of tibia. Primarily muscle of leg, it can act as knee joint stabilizer.

ACTIONS OF MUSCLES UPON THE KNEE JOINT

KNEE FLEXION

Biceps femoris
Semimembranosus
Semitendinosus
Sartorius (assists)
Gracilis (assists)
Gastrocnemius (assists)

KNEE EXTENSION

Quadriceps femoris:
  Rectus femoris
  Vastus lateralis
  Vastus medialis†
  Vastus intermedius

KNEE LATERAL ROTATION

Biceps femoris
Popliteus (if tibia is fixed)
Semimembranosus

KNEE MEDIAL ROTATION

Semitendinosus
Popliteus (if femur is fixed)

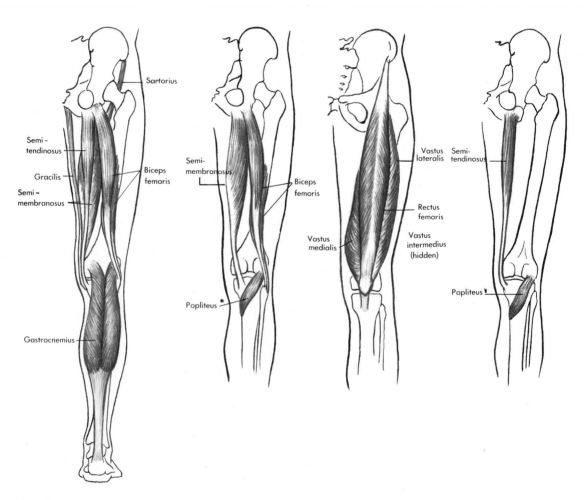

* If femur is in a fixed position.
† Some kinesiologists claim the vastus medialis to be the muscle which causes the last 10 to 15 degrees of extension.

Electromyographic studies do not confirm the claim. The vastus medialis is more effective in countering lateral displacement of the patella at the end of knee extension.

POSTERIOR MUSCLES OF THE LEG (Continued)

Flexor digitorum longus	Flexor hallucis longus	Tibialis posterior	MUSCLE
Flat sheet	Thicker sheet	Flat sheet	SHAPE
Medial side of leg deep to triceps surae	Lateral side of leg beside flexor digitorum longus	Deeply under flexor muscles	LOCATION
Posteromedial tibial surface below soleal line	Broad origin from lower two thirds of posterior fibular surface and adjacent intermuscular septum	Broad origin from interosseous membrane and adjacent surfaces of tibia and fibula	ORIGIN
Fibers enter a long tendon on medial side of muscle which descends to turn forward behind medial malleolus in its synovial sheath in company with flexor hallucis longus tendon; in sole divides into four tendons	Tendon passes with that of flexor digitorum longus behind medial malleolus in its own synovial sheath to extend forward through sole of foot	Converge upon large tendon at medial side of muscle which grooves back of medial malleolus medial to deep flexor tendons; turns forward deep in sole	FIBER COURSE
A tendon passes to each of lateral four toes, inserting into plantar surface of distal phalanx	Plantar surface of distal phalanx of large toe	Tendon spreads out to insert upon tuberosity of navicular bone, cuneiform bones, and bases of second, third, and fourth metatarsal bones	INSERTION
Flexion of distal phalanges of lateral four toes	Flexion of distal phalanx of large toe	Strongest muscle in inversion of foot	ACTION

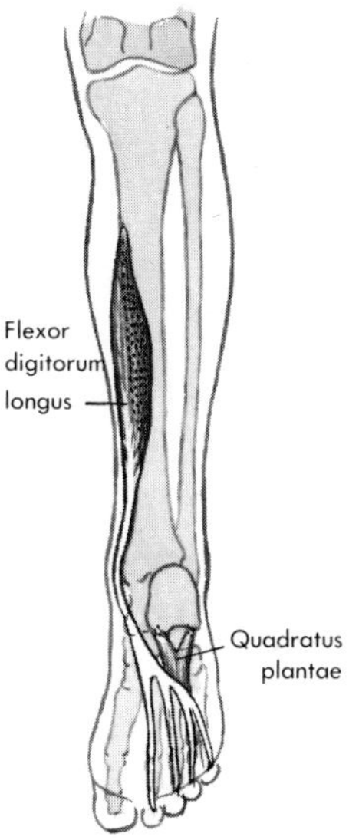

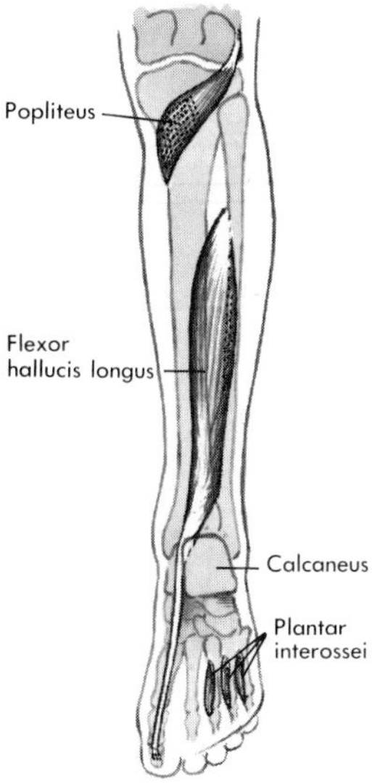

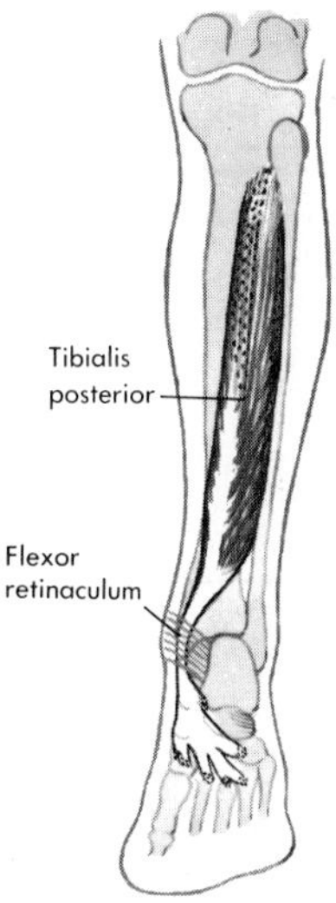

MUSCLE	Tibialis anterior	Extensor digitorum longus and peroneus tertius		Extensor hallucis longus
SHAPE	Long fusiform	Long flat sheet		Narrow fusiform
LOCATION	Parallels lateral border of tibia	Lateral to and partly over-lapped by tibialis anterior	Lowest part of extensor muscle is called peroneus tertius, which is frequently absent. If present, has own tendon which attaches distally to base of fifth metatarsal bone	Deep to tibialis anterior, extensor digitorum longus, and peroneus tertius
ORIGIN	Lateral tibial condyle, upper two thirds of lateral surface of tibial shaft and adjacent interosseous membrane	Lateral tibial condyle, upper three fourths of anterior surface of fibular shaft and adjacent interosseous membrane		Small area at middle of anterior surface of fibula and adjacent interosseous membrane
FIBER COURSE	Vertically into a long tendon inclining medially in front of ankle, acquiring a synovial sheath in passing under superior extensor retinaculum*	Vertically into long tendon passing in front of ankle to divide into four tendons. These with peroneus tertius tendon, if present, pass under superior and inferior extensor retinacula* in common synovial sheath		Tendon as with extensor digitorum longus and peroneus tertius, becoming superficial between diverging tibialis and extensor longus tendons. Passes under extensor retinacula* in synovial sheath medial to extensor longus tendon
INSERTION	Medial side of medial cuneiform bone and base of first metatarsal	The four tendons cross dorsum of foot to each of lateral four toes, forming fibrous expansion over each metatarsophalangeal joint: central part inserts into base of middle phalanx and lateral parts of tendon pass distally to base of distal phalanx		Crosses medial aspect of dorsum of foot to attach to upper surface of distal phalanx of large toe
ACTION	Dorsiflexion at ankle Inversion of foot by twisting up medial side of foot	Extends toes at metatarsophalangeal joint primarily, but aids in extending distal phalanges. Eversion of foot by twisting up lateral side of foot		Extends large toe; aids in dorsiflexion of ankle

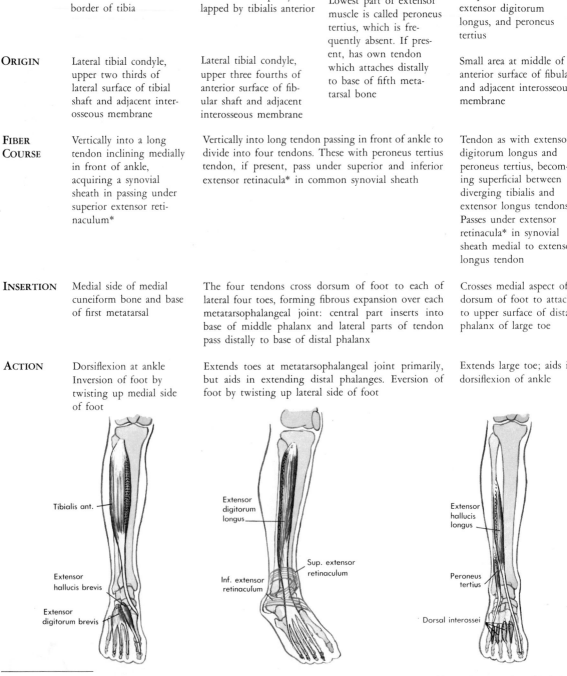

Tibialis ant.

Extensor hallucis brevis

Extensor digitorum brevis

Extensor digitorum longus

Inf. extensor retinaculum

Sup. extensor retinaculum

Extensor hallucis longus

Peroneus tertius

Dorsal interossei

---

* The tendons at front of the leg would become bowed out in front of the ankle and over the dorsum of the foot when their muscles contract if restraining bands did not confine them. The superior extensor retinaculum crosses the lower part of leg above the ankle, while the Y-shaped inferior extensor retinaculum presents one arm between the ankle malleoli and the other curving over the dorsum and around the medial surface of the foot.

THE HUMAN BODY
HIGHLIGHTS of STRUCTURE and FUNCTION

SKELETAL SYSTEM

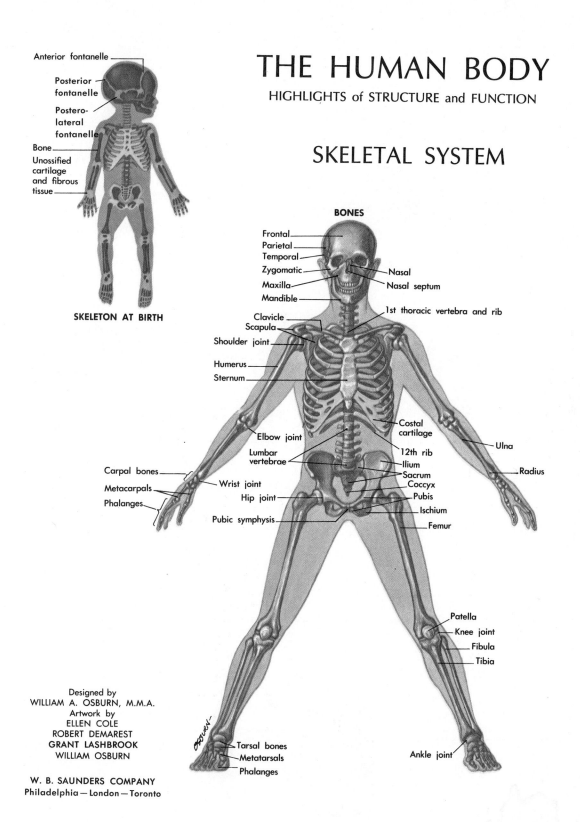

SKELETON AT BIRTH

Anterior fontanelle
Posterior fontanelle
Postero-lateral fontanelle
Bone
Unossified cartilage and fibrous tissue

BONES

Frontal
Parietal
Temporal
Zygomatic
Maxilla
Mandible
Nasal
Nasal septum
1st thoracic vertebra and rib
Clavicle
Scapula
Shoulder joint
Humerus
Sternum
Elbow joint
Lumbar vertebrae
Carpal bones
Metacarpals
Phalanges
Wrist joint
Hip joint
Pubic symphysis
Costal cartilage
12th rib
Ilium
Sacrum
Coccyx
Pubis
Ischium
Femur
Ulna
Radius
Patella
Knee joint
Fibula
Tibia
Tarsal bones
Metatarsals
Phalanges
Ankle joint

Designed by
WILLIAM A. OSBURN, M.M.A.
Artwork by
ELLEN COLE
ROBERT DEMAREST
GRANT LASHBROOK
WILLIAM OSBURN

W. B. SAUNDERS COMPANY
Philadelphia — London — Toronto

Plate 1

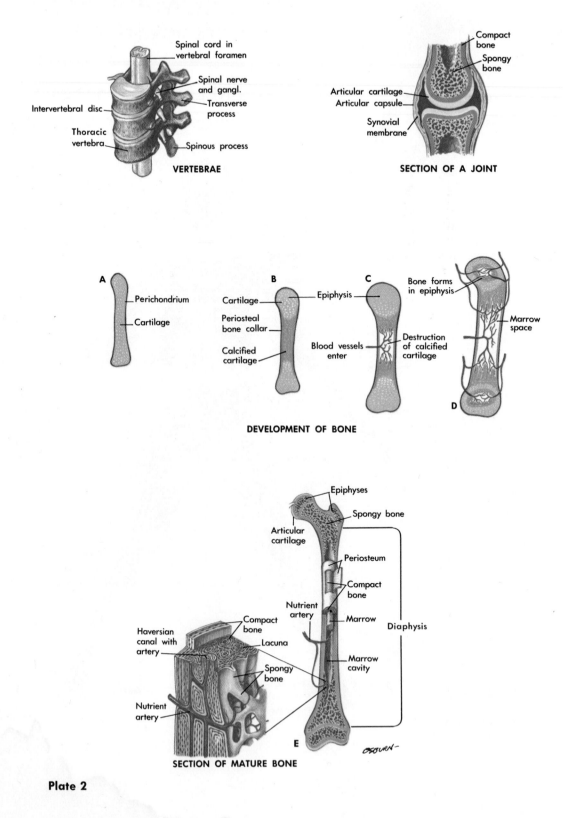

VERTEBRAE

Spinal cord in vertebral foramen
Spinal nerve and gangl.
Transverse process
Intervertebral disc
Thoracic vertebra
Spinous process

SECTION OF A JOINT

Compact bone
Spongy bone
Articular cartilage
Articular capsule
Synovial membrane

DEVELOPMENT OF BONE

A
Perichondrium
Cartilage

B
Cartilage
Periosteal bone collar
Calcified cartilage
Epiphysis

C
Blood vessels enter
Destruction of calcified cartilage

D
Bone forms in epiphysis
Marrow space

SECTION OF MATURE BONE

Epiphyses
Spongy bone
Articular cartilage
Periosteum
Compact bone
Nutrient artery
Marrow
Marrow cavity
Diaphysis

Haversian canal with artery
Compact bone
Lacuna
Spongy bone
Nutrient artery

E

OSBURN—

Plate 2

SKELETAL MUSCLES

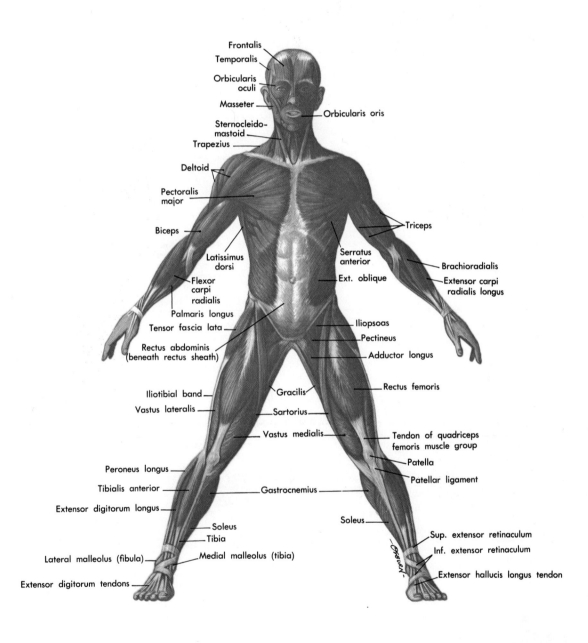

Frontalis
Temporalis
Orbicularis oculi
Masseter
Orbicularis oris
Sternocleido-mastoid
Trapezius
Deltoid
Pectoralis major
Biceps
Triceps
Latissimus dorsi
Serratus anterior
Brachioradialis
Flexor carpi radialis
Ext. oblique
Extensor carpi radialis longus
Palmaris longus
Tensor fascia lata
Iliopsoas
Pectineus
Rectus abdominis (beneath rectus sheath)
Adductor longus
Iliotibial band
Gracilis
Rectus femoris
Vastus lateralis
Sartorius
Vastus medialis
Tendon of quadriceps femoris muscle group
Peroneus longus
Patella
Tibialis anterior
Gastrocnemius
Patellar ligament
Extensor digitorum longus
Soleus
Soleus
Tibia
Sup. extensor retinaculum
Inf. extensor retinaculum
Lateral malleolus (fibula)
Medial malleolus (tibia)
Extensor digitorum tendons
Extensor hallucis longus tendon

Plate 3

HOW A MUSCLE PRODUCES MOVEMENT

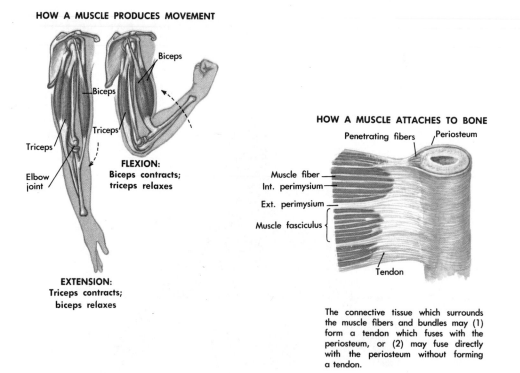

Biceps

Biceps

Triceps

Triceps

Elbow
joint

FLEXION:
Biceps contracts;
triceps relaxes

EXTENSION:
Triceps contracts;
biceps relaxes

HOW A MUSCLE ATTACHES TO BONE

Penetrating fibers   Periosteum

Muscle fiber

Int. perimysium

Ext. perimysium

Muscle fasciculus

Tendon

The connective tissue which surrounds
the muscle fibers and bundles may (1)
form a tendon which fuses with the
periosteum, or (2) may fuse directly
with the periosteum without forming
a tendon.

HOW A MUSCLE CONTRACTS

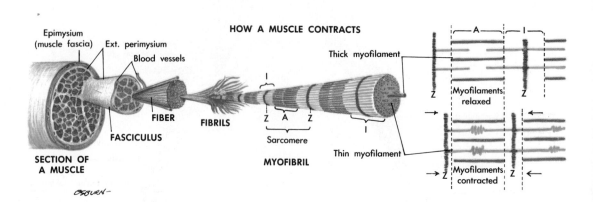

Epimysium
(muscle fascia)   Ext. perimysium

Blood vessels

Thick myofilament

FIBER

FIBRILS

FASCICULUS

Z   A   Z

Sarcomere

MYOFIBRIL

Thin myofilament

SECTION OF
A MUSCLE

A   I

Z   Z

Myofilaments
relaxed

Myofilaments
contracted

OSBURN-

Plate 4

RESPIRATION AND THE HEART

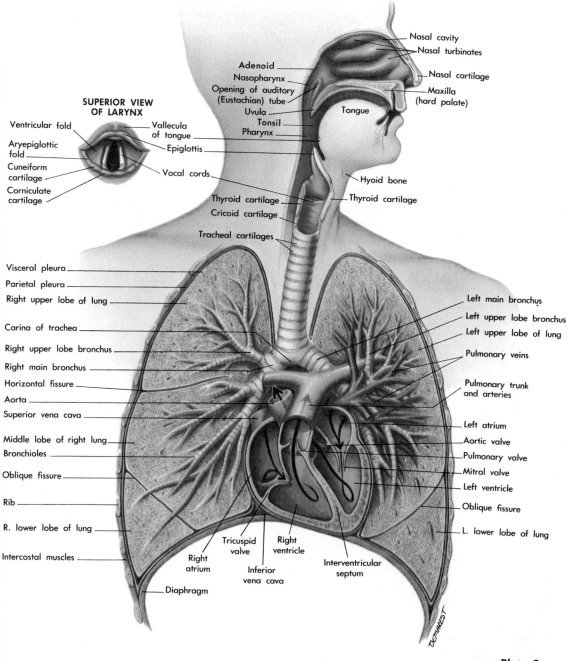

SUPERIOR VIEW OF LARYNX

Ventricular fold
Aryepiglottic fold
Cuneiform cartilage
Corniculate cartilage

Vallecula of tongue
Epiglottis
Vocal cords

Nasal cavity
Nasal turbinates
Adenoid
Nasopharynx
Nasal cartilage
Opening of auditory (Eustachian) tube
Maxilla (hard palate)
Uvula
Tongue
Tonsil
Pharynx
Hyoid bone
Thyroid cartilage
Thyroid cartilage
Cricoid cartilage
Tracheal cartilages

Visceral pleura
Parietal pleura
Right upper lobe of lung
Carina of trachea
Right upper lobe bronchus
Right main bronchus
Horizontal fissure
Aorta
Superior vena cava
Middle lobe of right lung
Bronchioles
Oblique fissure
Rib
R. lower lobe of lung
Intercostal muscles

Left main bronchus
Left upper lobe bronchus
Left upper lobe of lung
Pulmonary veins
Pulmonary trunk and arteries
Left atrium
Aortic valve
Pulmonary valve
Mitral valve
Left ventricle
Oblique fissure
L. lower lobe of lung

Tricuspid valve
Right ventricle
Right atrium
Inferior vena cava
Interventricular septum
Diaphragm

Plate 5

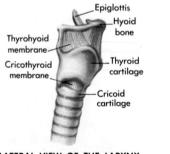

LATERAL VIEW OF THE LARYNX

Epiglottis

Hyoid bone

Thyrohyoid membrane

Cricothyroid membrane

Thyroid cartilage

Cricoid cartilage

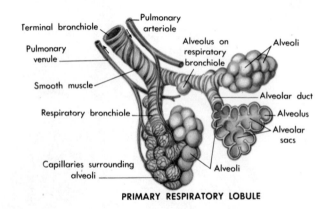

Terminal bronchiole

Pulmonary arteriole

Pulmonary venule

Alveolus on respiratory bronchiole

Alveoli

Smooth muscle

Respiratory bronchiole

Alveolar duct

Alveolus

Alveolar sacs

Capillaries surrounding alveoli

Alveoli

PRIMARY RESPIRATORY LOBULE

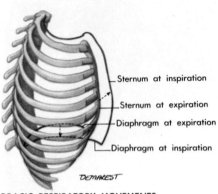

Sternum at inspiration

Sternum at expiration

Diaphragm at expiration

Diaphragm at inspiration

DEMAREST

THORACIC RESPIRATORY MOVEMENTS

Plate 6

GENITOURINARY SYSTEM

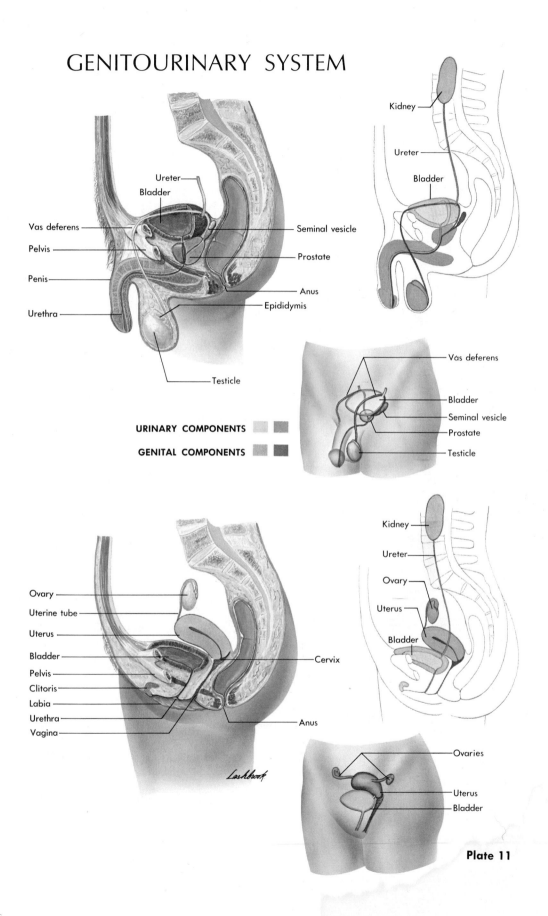

Ureter
Bladder
Vas deferens
Pelvis
Penis
Urethra
Seminal vesicle
Prostate
Anus
Epididymis
Testicle

Kidney
Ureter
Bladder

Vas deferens
Bladder
Seminal vesicle
Prostate
Testicle

URINARY COMPONENTS

GENITAL COMPONENTS

Ovary
Uterine tube
Uterus
Bladder
Pelvis
Clitoris
Labia
Urethra
Vagina
Cervix
Anus

Kidney
Ureter
Ovary
Uterus
Bladder

Ovaries
Uterus
Bladder

Lashbook

Plate 11

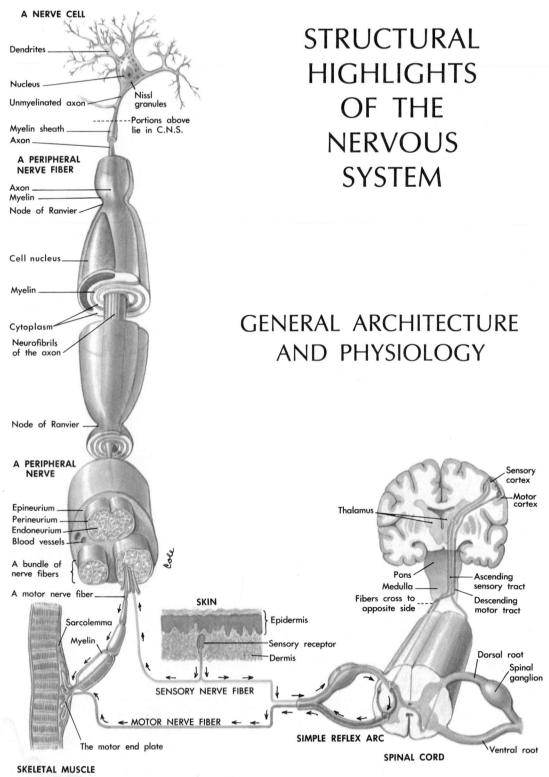

A NERVE CELL

Dendrites

Nucleus

Unmyelinated axon

Nissl granules

----Portions above lie in C.N.S.

Myelin sheath

Axon

A PERIPHERAL NERVE FIBER

Axon

Myelin

Node of Ranvier

Cell nucleus

Myelin

Cytoplasm

Neurofibrils of the axon

Node of Ranvier

A PERIPHERAL NERVE

Epineurium

Perineurium

Endoneurium

Blood vessels

A bundle of nerve fibers

A motor nerve fiber

Sarcolemma

Myelin

The motor end plate

SKELETAL MUSCLE

Plate 12

STRUCTURAL
HIGHLIGHTS
OF THE
NERVOUS
SYSTEM

GENERAL ARCHITECTURE
AND PHYSIOLOGY

Thalamus

Sensory cortex

Motor cortex

Pons

Medulla

Fibers cross to opposite side

Ascending sensory tract

Descending motor tract

Dorsal root

Spinal ganglion

SKIN

Epidermis

Sensory receptor

Dermis

SENSORY NERVE FIBER

← MOTOR NERVE FIBER ←

SIMPLE REFLEX ARC

Ventral root

SPINAL CORD

BRAIN AND SPINAL NERVES

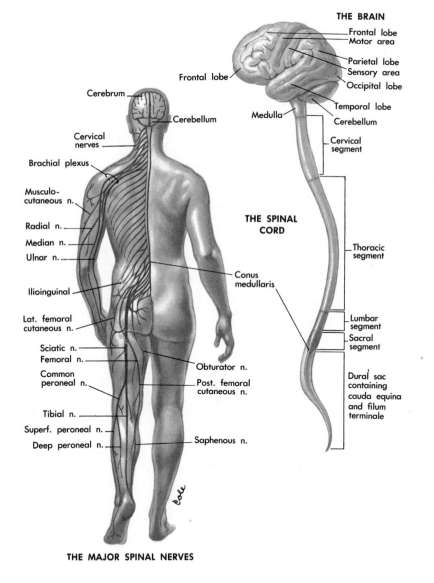

THE BRAIN

Frontal lobe
Motor area
Parietal lobe
Sensory area
Occipital lobe
Temporal lobe
Cerebellum

Frontal lobe

Cerebrum

Cerebellum

Medulla

Cervical
nerves

Cervical
segment

Brachial plexus

Musculo-
cutaneous n.

**THE SPINAL
CORD**

Radial n.

Median n.

Ulnar n.

Thoracic
segment

Ilioinguinal

Conus
medullaris

Lat. femoral
cutaneous n.

Sciatic n.
Femoral n.

Lumbar
segment

Sacral
segment

Common
peroneal n.

Obturator n.

Post. femoral
cutaneous n.

Dural sac
containing
cauda equina
and filum
terminale

Tibial n.

Superf. peroneal n.

Deep peroneal n.

Saphenous n.

THE MAJOR SPINAL NERVES

Plate 13

AUTONOMIC NERVES

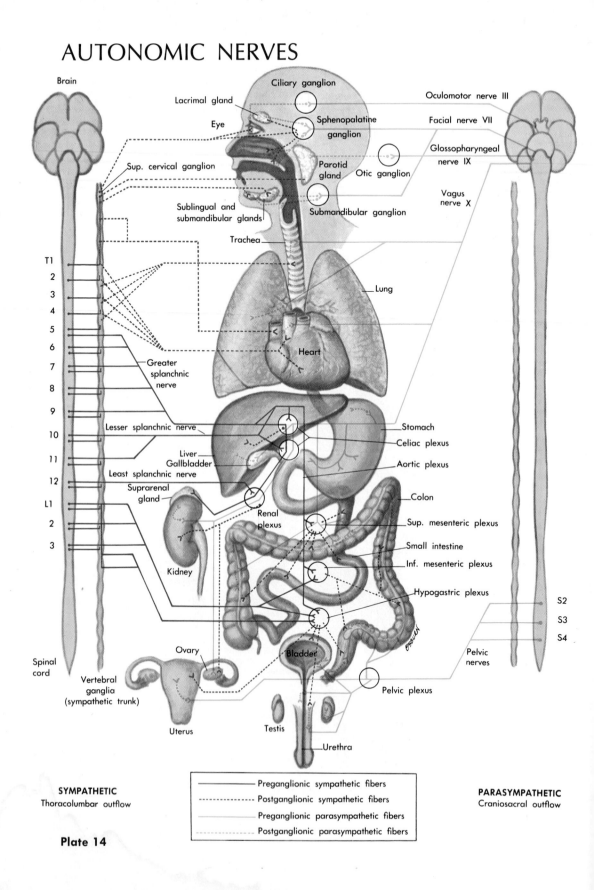

Brain

Ciliary ganglion

Lacrimal gland

Oculomotor nerve III

Eye

Sphenopalatine ganglion

Facial nerve VII

Glossopharyngeal nerve IX

Sup. cervical ganglion

Parotid gland

Otic ganglion

Vagus nerve X

Sublingual and submandibular glands

Submandibular ganglion

Trachea

T1
2
3
4
5
6
7
8
9
10
11
12
L1
2
3

Lung

Heart

Greater splanchnic nerve

Lesser splanchnic nerve

Stomach

Celiac plexus

Liver

Aortic plexus

Gallbladder

Least splanchnic nerve

Suprarenal gland

Colon

Renal plexus

Sup. mesenteric plexus

Small intestine

Inf. mesenteric plexus

Kidney

Hypogastric plexus

S2
S3
S4

Ovary

Bladder

Pelvic nerves

Spinal cord

Vertebral ganglia (sympathetic trunk)

Pelvic plexus

Uterus

Testis

Urethra

SYMPATHETIC
Thoracolumbar outflow

Preganglionic sympathetic fibers
Postganglionic sympathetic fibers
Preganglionic parasympathetic fibers
Postganglionic parasympathetic fibers

PARASYMPATHETIC
Craniosacral outflow

Plate 14

CENTRAL PLANTAR MUSCLES (2ND TO 5TH TOES)

MUSCLE	Flexor digitorum brevis	Quadratus plantae (flexor accessorius)	Lumbricals	Plantar interossei	Dorsal interossei
DEPTH IN SOLE	Superficial in central location with flexor longus tendon deep to it	Deep to flexor digitorum brevis	Deeper than quadratus plantae in central location	Deepest, against skeleton of foot in central location	As with plantar interossei but between tarsal bones
ORIGIN	Medial process of calcaneus and deep surface of plantar aponeurosis	Medial side and front of calcaneus by two heads	Adjacent sides of long flexor tendons to third to fifth toes; medial side of tendon to second toe	Medial side of base of third, fourth, and fifth metatarsals	Adjacent surfaces of all five metatarsal bones
FIBER COURSE	Divides into four tendons, each in a common synovial sheath with long flexor tendon; tendons pass forward in central part of foot diverging toward lateral four toes	Heads join and pass directly forward toward obliquely and laterally coursing flexor longus tendon	Adjacent heads join to pass toward medial side of third to fifth toes; single head passes to medial side of second toe	Tendons pass forward toward toes	Heads join to pass forward in intermetatarsal space toward webs between toes
INSERTION	Tendons divide, each permitting passage of a long flexor tendon; then divided parts attach to sides of middle phalanx of second to fifth toes	Into deep surface of flexor digitorum longus tendon	Medial side of proximal phalanx of toe concerned	Medial side of proximal phalanx for toe of same number	Into base of second, third, and fourth toes; second toe receives insertions on both sides; third and fourth toes on medial side
ACTION	Flexion of the four lateral toes at proximal interphalangeal joint; supports longitudinal arch of foot by its pull	Debatable but possibly pulls on flexor longus tendon to flex the four lateral toes when latter muscle is inhibited to prevent antagonism to dorsiflexion	Aid in flexion of four lateral toes at metatarsophalangeal joint; possible abduction-adduction role*	Flexion of third, fourth, and fifth toes at metatarsophalangeal joints; possible abduction-adduction role	As with plantar interossei

Flexor digitorum brevis

Quadratus plantae

Lumbricales

Lumbricales

Plantar interossei

Dorsal interossei

---

*The toes do not have the range of abduction-adduction of the fingers although they spread passively as weight is borne. A slight degree of active abduction and adduction can occur with the line of reference in the foot passing through the second toe. Theoretically, at least, the abductor hallucis abducts the large toe; the dorsal interossei can possibly abduct the second, third, and fourth toes, while the abductor digiti minimi could slightly abduct the little toe. The adductor hallucis is more likely to flex the great toe. The dorsal interosseous between the large and second toes theoretically can draw the second toe medially; the plantar interossei perform a similar adductor function for the third, fourth, and fifth toes.

ACTIONS OF MUSCLES UPON THE ANKLE, FOOT, AND TOES

MOVEMENTS OCCURRING AT THE ANKLE

Dorsiflexion of foot	Plantar flexion of foot
Tibialis anterior	Triceps surae
Extensor digitorum longus	Gastrocnemius
Extensor hallucis longus (assists)	Soleus
Peroneus tertius (weak)	Peroneus longus

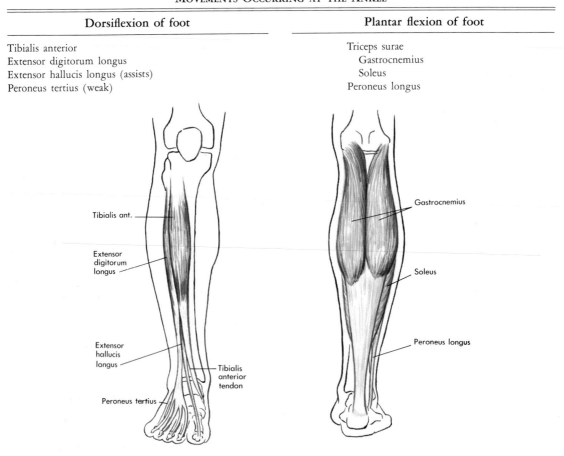

MOVEMENTS OCCURRING AT INTERTARSAL JOINTS

Inversion of foot	Eversion of foot
Tibialis posterior	Peroneus longus
Tibialis anterior (assists)	Extensor digitorum longus
Triceps surae (assists)	Peroneus tertius (assists)
Gastrocnemius	Peroneus brevis (weak)
Soleus	

SUPPORT OR PRESERVATION OF ARCHES OF THE FOOT

Longitudinal arch	Metatarsal (Transverse) arch
Flexor digitorum brevis	Flexor hallucis brevis (medial part)
	Abductor hallucis (medial part)
	Abductor digiti minimi (lateral part)
	Transverse head of adductor pollicis
	Interossei

MOVEMENTS OF THE TOES

Flexion

At metatarsophalangeal joints:
Lumbricales
Interossei (dorsal and plantar)
Flexor hallucis brevis

At interphalangeal joints:
Flexor digitorum longus (lateral four)
Flexor hallucis longus (large toe)
Flexor hallucis brevis (large toe)
Flexor digitorum brevis (lateral four)
Flexor digiti minimi (little toe)
Abductor hallucis (large toe)
Abductor digiti minimi (little toe, weak)

Extension

Extensor digitorum longus
Extensor hallucis longus
Extensor digitorum brevis (assists)

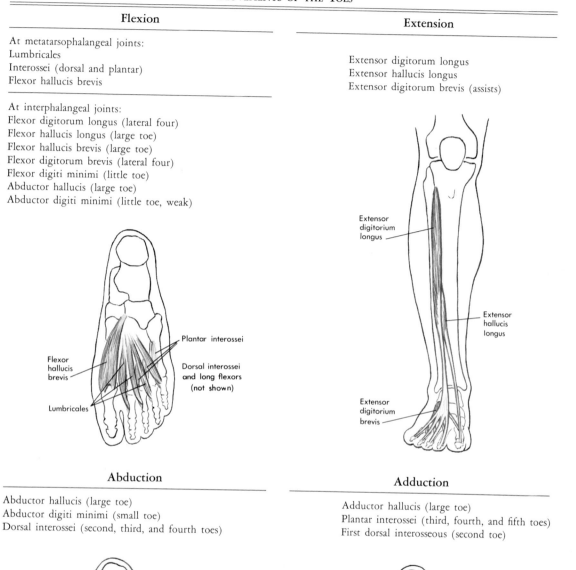

Abduction

Abductor hallucis (large toe)
Abductor digiti minimi (small toe)
Dorsal interossei (second, third, and fourth toes)

Adduction

Adductor hallucis (large toe)
Plantar interossei (third, fourth, and fifth toes)
First dorsal interosseous (second toe)

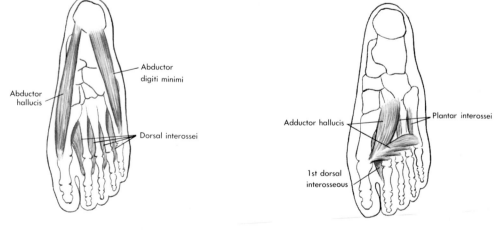

that, since the feet bear the weight of all the body above them, the force divided between them is transmitted equally to the calcaneus of the heel and to the heads of the metatarsal bones. The shapes of the bones of the foot and their alignment are insufficient to maintain the arches of the foot even with ligamentous assistance. The role of certain muscles of the foot, flexors particularly, is a very significant factor in functionally reinforcing the arches.

When a person stands on both feet in a quiescent, easy position, the trunk muscles are not particularly active, for the configuration of the axial skeleton with its compensatory vertebral curvatures and ligamentous support ensures the integrity of position of the trunk and head. From one's knowledge of the movements of the hip and knee joints it should be apparent that the erect position puts both these joints in extension in which they are strongest and best articulated for weight-bearing. With the line of gravity falling behind the hip joint, the weight of the trunk aids in the extension of the hip joint but is countered by the tightening of its capsule and the iliofemoral ligament. With greater weight pressing upon the knee joint and with the line of gravity passing in front of it, the knee would tend to hyperextend except for its strong liga-

ments and the muscular pull upon the hamstring muscles (knee flexors).

It is at this point in maintaining the erect position that muscular action becomes more evident. If one teeters to and fro slightly, the contraction and tightening of the hamstring muscles can be noted if the feet are firmly placed and not moved. The forward sway is resisted by actively-felt contractions of the calf muscles and flexors of the toes, which would, respectively, raise the heel and more firmly plant the toes. These actions are more pronounced when the weight of the body is supported by one limb only, as in standing upon one foot. Other factors enter the latter picture, however, because balance is more precarious. Tilting of the pelvis and lateral flexion or rotation of the trunk may become necessary in this circumstance.

Muscles and Regaining of Balance

When balance is not only threatened but is being lost, powerful reflex actions of many groups of muscles are brought into play to prevent falling. These vary according to the direction of the potential fall. Among the actions taken would be abduction-adduction movements of the hip, hip rotation, and flexion at the knee joints to produce quick changes of position of the feet. Dorsiflexion or plantar flexion at the ankle would occur to aid in moving the feet, to counter swaying, or to plant the feet more firmly after a change in position. If one were to make a series of movements designed to restore balance, and then analyze the movements and the muscle groups that produced them, good practice and use would be made of the knowledge gained of the muscular system.

THE MUSCLES AND LOCOMOTION

The science of kinesiology is concerned with the integration of muscular actions in the performance of many functions. One of the most complex is walking or its variants, such as running, leaping, and jumping and the many facets of sports and dancing. These are beyond the scope of this book but, as an example of the analyses that can be made by a person who understands the actions of muscles, consider the series of movements necessary in the seemingly very ordinary act of walking. Even in watching a child learn to walk one can hardly appreciate the laborious, conscious effort that must be made until spatial progression becomes, by repetition, a smoothly coordinated reflex activity, coordinated by the nervous system below the level of conscious activity.

Walking

Walking involves progression of the body by the lifting and swinging forward of one leg while the other supports an inclined body and provides a push-off which carries the body to the position of the swung leg, now planted upon the ground. Many joint movements and muscle actions are necessary to accomplish this rhythmic alternating process. These are summarized below, assuming that from the erect position a step forward is made with the right foot.

1. Notice in starting to walk that first there is a forward tilting of the trunk to move the center of gravity of the body forward. It might be said that the shift of the center of gravity is akin to starting a falling movement, so that the legs must move to bring the lower part of the body under the trunk again to preserve the line of gravity. The shift of gravity is accomplished largely by a partial relaxation of the hamstring and calf muscles, which act to maintain the erect position.

2. The right limb swings because of flexion of the hip and knee and dorsiflexion of the foot to clear the ground. There is an immediate sag of the pelvis to the right because of the weight of the unsupported, swinging right leg. This tilt is immediately countered by contraction of the *right* sacrospinalis muscle complex and the *left* gluteus medius and minimus muscles to pull the sagging pelvis back into position.

3. As the right limb swings forward off the ground, the *left* leg provides a push-off impetus to carry the body forward. Plantar flexion occurs at the ankle, and the toes flex to secure a good grip upon the ground (or the intervening sole of the shoe). The *left* hip and knee joints are extended, while the calf muscles and long flexor muscles contract to lift the *left* heel from the ground. The actual push-off occurs as the weight on the extended, forward-inclined *left* leg shifts to the front of the foot. The plantar flexors of the foot provide the final push-off impetus.

4. The next phase of walking occurs as the swinging *right* leg undergoes extension at the knee to straighten the extremity for impact, with dorsiflexion of the foot so that the heel strikes the ground first. The *right* foot at impact is in front of the body, but with the forward motion of the body and with the hip joint extending, the body comes over the foot to restore balance. The weight is transmitted forward to be distributed normally in the right foot.

5. Whether walking ensues or only one step is taken, the cycle then alternates to the left foot, which is brought forward by the same progression of movements just described. In taking but one step, the swing of the *left* foot would be lower and it would halt beside the other with the weight distributed between the two extremities.

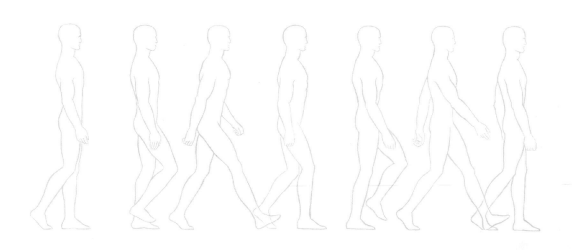

In walking, the *left* foot would continue its forward motion off the ground.

A NOTE IN CONCLUSION

All the rote learning and memorization that one is capable of will not bring a true understanding of the muscular system. Until the student puts his knowledge to use, he cannot be certain that he has learned. Close observation of many different movements while performing varied acts, accompanied by analysis of the components of muscle and joint activity, is a most useful way to further understanding of muscle function.

5

The

Nervous

System

THE NERVOUS SYSTEM IN GENERAL

The nervous system is formed by a continuous series of organs which are constructed of nervous tissue. The qualities of irritability and conductivity enable nerve cells to initiate and transmit electrical impulses through the organs which nervous tissue forms. The organs which constitute the nervous system are the *brain* and *spinal cord* from which *nerves* issue as cables which convey the electrical *nerve impulses* to and from every part of the body. The nervous system regulates the function of other systems of the body and provides for the awareness of changes which occur within the body or in its environment. The nervous system is also the structural and functional framework for thought, memory, and emotion. The higher planes of mental activity which are characteristic of man, such as reasoning, analysis of problems, and aesthetic, creative or idealistic thinking, are all the product of function in the nervous system.

GENERAL FUNCTIONS OF THE NERVOUS SYSTEM

Major functions of the nervous system are:
1. To detect changes in situations which affect the body. Because these changes must be sensed before they can be understood, this function is termed a *sensory* function or *sensation*. The changes detected are:

a. Changes in phenomena in the external environment to which the individual must adjust or against which he must defend. *Special senses* (sensation), such as vision, hearing and taste, and *general sensations,* such as touch, pain, temperature and pressure, are examples.

b. Changes in the spatial orientation of the body or its parts.

c. Changes in body function whether of visceral or musculoskeletal-articular origin.

d. Minute changes in the internal environment of cells and tissues. Examples are the state of hydration, internal temperature, blood pressure, oxygen demands, and the equilibrium of fluid and electrolytes.

2. To initiate and control the activities of tissues which require an extrinsic stimulus to function. Examples are the contraction of voluntary and smooth muscle, and the secretory processes of glands. Since positive action is produced this is called a *motor function.*

3. To coordinate the activities of units, parts, or entire organs with other structures. For example, a certain number of motor units of a muscle or muscle group may be brought into play to

smooth or regulate a primary action being performed by an antagonist. A more regional example is the contraction of neck and head muscles so that the head can follow movements of the eyes. More complex is the coordination of sequential actions calling for the correlation of the contraction of a number of muscles in a progression of movements, as in walking, violin playing, or swallowing.

4. To conduct the impulses generated in nerve cells to or away from the central nervous system. The transmission and distribution of nerve impulses originating in its cells are basic to all of the functions of the nervous system.

DEVELOPMENT OF THE NERVOUS SYSTEM

The organs of the nervous system develop from simple beginnings. A survey here of the developmental stages helps in understanding the complicated structure of the adult nervous system and the terms associated with it. Restudy of the development section in Chapter 1 will also be helpful now.

The nervous system develops from a strip of ectoderm which sinks below the surface of the embryonic body and rolls up to become the *neural tube* whose cells become the nerve cells of the brain and spinal cord. A group of ectodermal

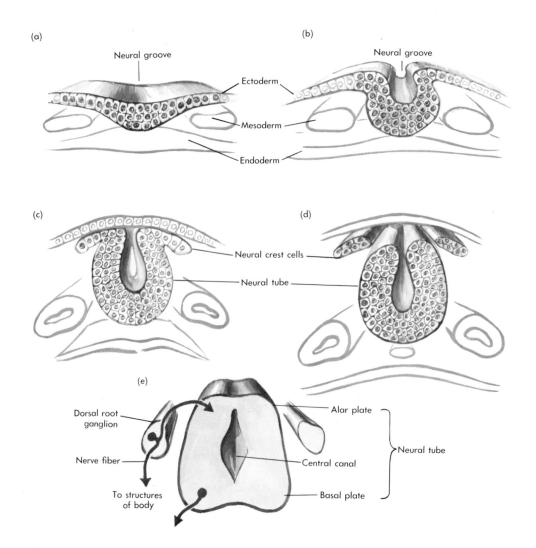

cells beside the neural tube, forming the *neural crest,* develop into collections of nerve cells outside the central nervous system known as *ganglia.* Processes of the cells of the neural tube and neural crest grow outward as *nerves* toward structures of the body to control their activities and provide the body with sensory awareness.

Expansion of the head end of the neural tube brings about the formation of the three *primary vesicles* of the brain, which are separated by constrictions where growth does not occur. The primary vesicles are the *forebrain* (prosencephalon), *midbrain* (mesencephalon), and *hindbrain* (rhombencephalon). These portions remain within the skull to become the brain. The remainder of the neural tube comes to lie within the vertebral canal as the spinal cord.

lon. The cavity of the original neural tube follows the growth of the vesicles and expands within them to form the ventricular cavities of the brain. Unexpanded portions in the areas of lesser growth form canals, foramina, and aqueducts which connect the ventricles.

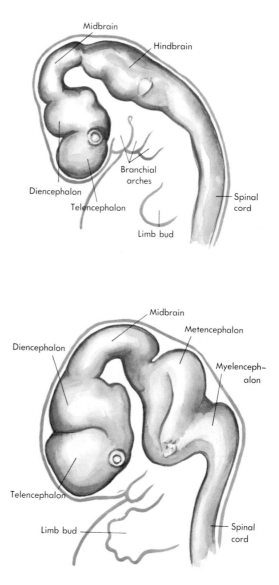

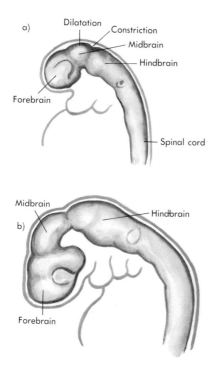

Continued growth of the primary vesicles produces secondary expansions of the walls with intervening constrictions. Five *secondary vesicles* of the brain result, from which major divisions of the brain develop. The forebrain becomes divided by the tremendous growth of its anterior end and sides into the large *telencephalon* (end brain) and the smaller *diencephalon* (interbrain). The midbrain remains undivided. The hindbrain divides into two parts, the *metencephalon* and *myelencepha-*

The caudal part of the central nervous system, which becomes the spinal cord, retains its tubular structure. A *central canal* persists from the original neural cavity and retains this form in the spinal cord. Cells of the neurectoderm surrounding the central canal specialize to form two cell types, the true *nerve cells* and the *neuroglia,* which form the special supporting tissue of the central nervous system.

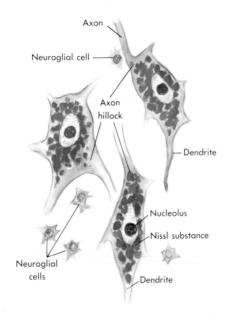

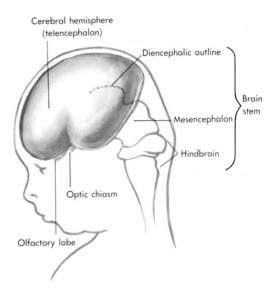

the vesicles of the hindbrain. As they grow, the cerebral hemispheres overshadow and obscure most of the rest of the brain, which is termed the *brain stem.*

The true nerve cells concentrate in the lateral walls of the growing neural tube to form the central *gray matter* of the spinal cord. Processes grow from the nerve cells to course longitudinally in the spinal cord or leave for other parts of the body. These processes form the *white matter* of the spinal cord, which lies external to the gray matter. In the cerebral hemispheres, which develop from the telencephalon, and in the cerebellum, which is a specialized outgrowth from the hindbrain, the overgrowth and folding that occur displace the nerve cells to a thin gray layer external to the columns of white matter; this cellular layer is termed the *cortex* of the adult brain.

Definitive Growth of the Brain

The following processes take place simultaneously as the five vesicled-tube develops into the parts of the adult brain.

1. Thickening of the walls of the vesicles between the constrictions which separate them.

2. Inequalities in growth in which areas of constriction remain small and unexpanded despite a relative thickening of their walls.

3. Even among the expanded vesicles some demonstrate tremendous growth in contrast to the others. The telencephalon outstrips the diencephalon. The telencephalon bulges into bilaterally symmetrical outgrowths which become the *cerebral hemispheres* of the brain. The hemispheres grow over the diencephalon, mesencephalon, and

4. The walls of the diencephalon undergo considerable thickening, a process that narrows its cavity to a cleft. Important nerve centers for sensation are located on each side of the cavity. They are collectively called the *thalamus.* The floor of the diencephalon, termed the *hypothalamus,* develops important vital centers for the regulation of many visceral functions.

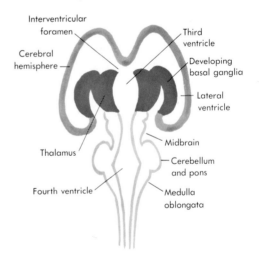

5. Great thickening of the walls of the smaller mesencephalon reduces the central cavity of this undivided vesicle to a small canal. Many sensory and motor centers are located here and through its white matter pass major nerve pathways be-

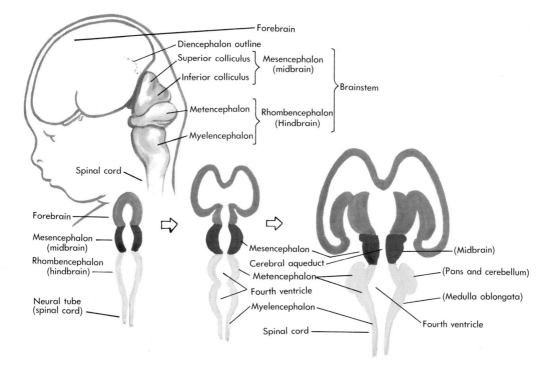

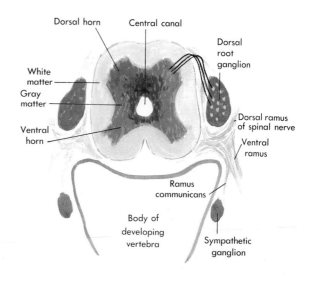

tween the portions of the central nervous system superior and inferior to it.

6. In the hindbrain similar centers develop in the metencephalon, which is characterized by the persistence of a large cavity. The part of the hindbrain which lies in the axis of the brain stem and is continuous below with the spinal cord is the myelencephalon, which, as its motor and sensory centers develop, becomes known as the *medulla oblongata.* The *cerebellum* is a subdivision of the metencephalon, which grows upward and outward from the axis of the brain stem to hide those more inferior portions which are not covered by the cerebral hemispheres. (See above illustration.)

Definitive Growth of the Spinal Cord

The increase in mass and numbers of cells in the gray matter shapes it into the form of an "H." Two *ventral horns* are seen when the spinal cord is viewed in cross section. These are really masses of the cell bodies extending upward and downward to form columns within the gray matter. The ventral horn cells form the main motor centers of the spinal cord. *Dorsal horns* are formed in a similar fashion. These, in a cross section, appear to be drawn posterolaterally on each side

of the spinal cord. While the cells of the dorsal horn columns of gray matter are sensory in nature, they are not the main sensory centers. These centers are found in *ganglia,* which are collections of nerve cells outside the spinal cord. The sensory ganglion cells have processes which extend both to the periphery of the body and into the dorsal horn area of the spinal cord where they come into a functional relationship with cells of the dorsal, sensory gray columns.

Development of Nerves

The processes of some nerve cells may course entirely within the brain and spinal cord; then they are termed *nerve tracts.* Nerve cell processes, however, may extend for great distances from their nerve cells to convey impulses to or from peripheral tissues. When the processes of a number of nerve cells, surrounded by sheaths, course beyond the brain and spinal cord they are called *nerve fibers.*

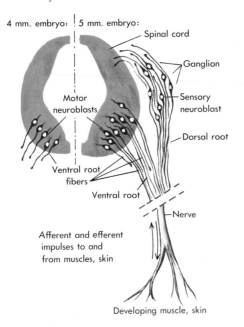

4 mm. embryo: | 5 mm. embryo:

Spinal cord

Ganglion

Motor neuroblasts

Sensory neuroblast

Dorsal root

Ventral root fibers

Ventral root

Nerve

Afferent and efferent impulses to and from muscles, skin

Developing muscle, skin

A group of nerve fibers traveling together and invested by connective tissue to form a ribbon, cable, or band is a *nerve.* When nerve fibers first begin to grow out of the ventral gray columns of the spinal cord or the associated ganglia adjacent to it, the embryo is very small. The peripherally growing processes have only short courses to reach muscles or organs. As the body grows and the elements to be innervated migrate, enlarge, lengthen, and specialize, the nerves grow and migrate along with them.

ORGANIZATION OF NERVOUS TISSUE

Nervous tissue has been defined as that fundamental tissue whose cells possess the properties of irritability and conductivity to the highest degree. Nerve cells initiate and conduct the impulses which are concerned in the major functions of the nervous system.

Unit of Construction—The Neuron

A *neuron* is a single nerve cell with all of its processes, coverings, and specialized terminations. It is both the structural and functional unit of the nervous system. (See also color plate 12.)

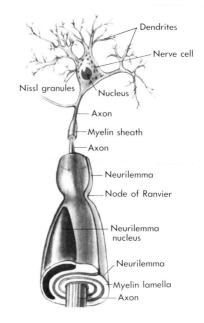

Dendrites

Nerve cell

Nissl granules

Nucleus

Axon

Myelin sheath

Axon

Neurilemma

Node of Ranvier

Neurilemma nucleus

Neurilemma

Myelin lamella

Axon

Nerve Cells. *Nerve cells* numbering in the billions in the central nervous system and in ganglia are of varied sizes and shapes. Nerve cells range from some of the smallest cells in the body (4 to 5 micra) to some of the largest (100 micra). Some are flattened by the pressure of surrounding cells and fibers whereas others are rounded, triangular or polygonal.

A typical nerve cell is that found in the ventral horn gray matter in the spinal cord. This is one of the largest of the nerve cells. Its cellular structure is typical of all types of neurons. In the illustrations note the following characteristics:

1. Nuclei are large in proportion to the mass of the cell.

2. The chromatin in the nucleus is not dispersed but instead is concentrated in a large spherical nucleolus.

3. The cytoplasm has many minute neurofibrils running through it. These are often masked by the presence of irregular prominent masses of chromatin-like material, the *chromidial* or *Nissl substance,* which is composed of nucleoprotein.

4. Both the Golgi apparatus and the mitochondria are very complex and prominent. They

have important metabolic roles in the function of the nerve cell. The clustering of mitochrondria within nerve cell processes or at their functional junctions with those of other cells indicates that they are important in the transmission of impulses.

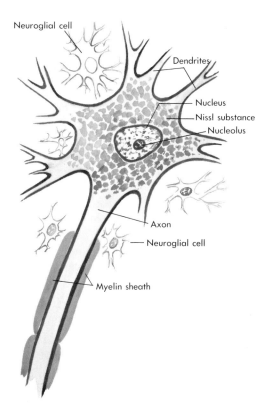

Electron microscope observations show that the interlaced *neurofibrils* of the cytoplasm are made up of groups of hollow, tubular *neurofilaments.* These tubular organelles are composed of protein. The Nissl substance which masks the neurofibrils, when magnified by the electron microscope, appears to be formed of parallel channels or cisterns of rough endoplasmic reticulum. Ribosomes stud the outer surface of the reticulum to give a granular appearance, and other ribosomes cluster together in the cytoplasm between the cisterns. The Golgi apparatus is found near the nuclear membrane, and rod-shaped mitochondria abound between the Nissl substance and the interlaced neurofibrils.

Nerve Cell Processes. The nerve cell processes are extensions of the cytoplasm which bring the nerve cell into apposition with other cells or convey its impulses to or from the periphery of the body. The processes have different names which are determined by the direction of the nerve impulse traveling upon or within them.

1. *Dendrites* are usually smaller multiple processes which fan out from one pole of a nerve cell and branch profusely. These convey impulses *into* the cell either from another cell or from peripheral parts of the body. Their branching permits one neuron to come into functional apposition with many others so that the impulses from many cells may converge upon a single neuron.

> Dendrites contain both neurofilaments and long, parallel *neurotubules,* in addition to the rough endoplasmic reticulum of Nissl material. Mitochondria are present, although compressed and elongated, and fill the dendrite even to its finest branches.

2. *Axons,* usually single, are larger and longer processes which convey a nerve impulse *away* from the nerve cell. Some axons, particularly those which run long courses to supply many structures, may branch but, as a general rule, axons do not branch except near their endings.

> The electron microscope reveals long, thin neurotubules and slender, parallel coursing channels of smooth endoplasmic reticulum in the axon. Nissl material is absent, but elongated mitochondria are present. Generally speaking, there are more neurofilaments in an axon than in a dendrite.

Coverings of the Neuron. Within the central nervous system, axons are invested by a delicate sheath formed by the neuroglial supporting cells of the central nervous system which also ensheathe the axons which are in the peripheral nerves (nerve fibers). Within the central nervous system, the axons have only the neuroglial investment. Outside of the central nervous system this sheath is more differentiated and is called the *neurolemma* or *sheath of Schwann.* Constrictions called nodes of Ranvier appear in both the sheath and the axon at regular intervals. Axons, with the exception of those confined to short courses within the gray matter and a few of the smallest caliber found in peripheral nerves, are also invested by a lipoidal material called *myelin.* This forms the fatty coat of the axon found under the neuroglial or neurolemmal sheath. The amount of myelin varies so that a nerve fiber may be large

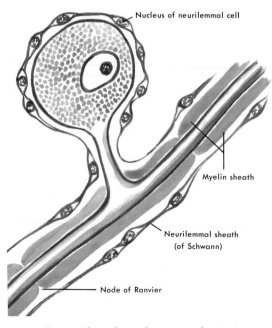

Nucleus of neurilemmal cell

Myelin sheath

Neurilemmal sheath
(of Schwann)

Node of Ranvier

or small in caliber, depending upon both the size of the axon and the amount of myelin.

Myelin has been shown by electron microscopic studies to be a part of the Schwann cell sheath rather than of the nerve process. Starting with an infolding of the unit membrane of the Schwann cell, the process of *myelinization* adds up to 50 spiral turns of the unit membrane about the axon in which layers of lipids and proteins alternate.

Polarity Types of Neurons.   The nerve cells with one axon and several dendrites are called *multipolar* neurons. Nerve cells located in ganglia have only one process. This process divides to form one process which extends to the periphery, the other passes into the central nervous system.

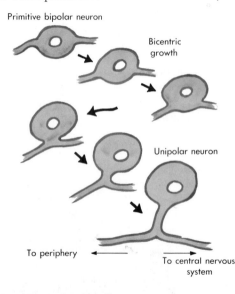

Primitive bipolar neuron

Bicentric
growth

Unipolar neuron

To periphery

To central nervous
system

This type of neuron is *unipolar*. A much rarer type has two processes, one extending from each pole of an elongated cell. These are *bipolar* neurons.

Terminations and Junctions.   The termination of a nerve cell is an integral part of the neuron. A motor neuron may end by forming a specialized terminal structure in or about a non-nervous tissue, such as muscle fibers, gland cells, or epithelia. Sensory neurons may end in special receptor structures within the skin, mucous membranes, tendons or joints, or organs of special sense. Other neurons may end by functionally joining another neuron to form a conduction chain. The point of apposition and functional junction is a *synapse.* At the synapse the transmission of a nerve impulse occurs by both structural and biochemical mechanisms. As the process of a neuron approaches a point of synapse it divides repeatedly into narrower *telodendria* which expand terminally into button-like bodies termed synaptic endings or *boutons terminaux* (terminal buttons). These indent the surface of the dendrite or the cell body of the succeeding neuron. One axon may form synaptic telodendria for another neuron's dendrite or cell body and then pass on to form similar synapses with still other neurons.

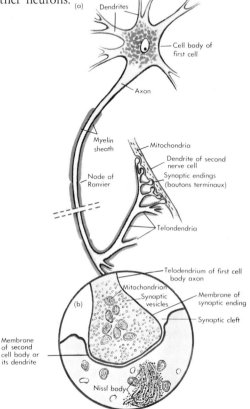

(a) Dendrites

Cell body of
first cell

Axon

Myelin
sheath

Mitochondria

Dendrite of second
nerve cell

Node of
Ranvier

Synaptic endings
(boutons terminaux)

Telondendria

Telodendrium of first cell
body axon

Mitochondrion

Synaptic
vesicles

Membrane of
synaptic ending

(b)

Synaptic cleft

Membrane
of second
cell body or
its dendrite

Nissl body

Terminal buttons contain dense clusters of mitochondria which lie amidst many minute *synaptic vesicles.* The unit membrane of the nerve ending is termed the *presynaptic membrane,* while the surface membrane of either a nerve cell body or of a dendrite of the next neuron in the conduction chain is called the *postsynaptic membrane.* There is no protoplasmic continuity between neurons. The two synaptic membranes are separated by the narrow *synaptic cleft* across which the nerve impulse is transmitted.

Nerve Impulse:
Conduction and Transmission

The incredibly complex biophysical and biochemical phenomena which together are termed the *nerve impulse* may be reduced, although oversimplified, to a change in electrical potential in the living nervous tissue. The change in potential is an expression of *irritability* as an attribute of living substance. Transmission of the change in potential is the result of the property of *conductivity* along or through protoplasm.

The neuroproteins of the nerve cell must have a significant relation to the irritability, while the hollow neurofilaments and tubules are concerned with conduction of the nerve impulse.

Transmission of the nerve impulse occurs at synapses and at the junctions of nerve fibers with effector cells, such as of muscle or of glands. Transmission also occurs within sensory nerve receptor cells or end organs where the specialized receptor cell passes the nerve impulse generated within it to the conduction chain of the nervous system. The enormous amount of research which has been, and is still being, done on transmission across adjacent neuron membranes may be reduced today to the belief that actual conveyance of the nerve impulse from one neuron to another, or from nerve to an effector tissue, is largely biochemical.

Very simply stated, it is believed that the arrival of the nerve impulse at the terminal arborization of a neuron causes the formation of a chemical *transmitter* substance. This substance may be *acetylcholine* or *norepinephrine.* Such a chemical crosses the unit membranes via synaptic clefts and troughs to enter the next link in the conduction chain or into an effector tissue. There enzymes such as *acetyl cholinesterase* inactivate the transmitter substance. In the process there is a biophysical depolarization within the receiving cell. It is this change in polarization which sets up a new electrical potential which is either conducted along the next

nerve fiber, or which itself excites the effector cell to its characteristic function such as contraction or secretion.

Support and Blood Supply. Neuroglial cells interweave among nerve cells and their processes to provide support in the place of connective tissue. The blood supply of central nervous tissue is the richest of any tissue in the body in keeping with the fact that nervous tissue has both the highest metabolic rate and the highest oxygen demand of the tissues of the body. Each nerve cell, of the billions present, is virtually surrounded by a capillary loop.

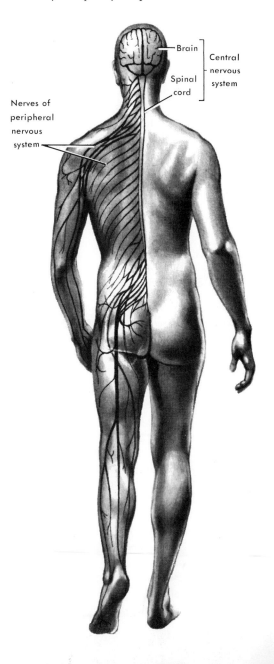

Brain

Central nervous system

Spinal cord

Nerves of peripheral nervous system

ORGANIZATION OF THE NERVOUS SYSTEM

Many principles of structure of the nervous system have been presented which can now be utilized to achieve an understanding of the nervous system as a whole. The major divisions of the nervous system are the central, peripheral, and autonomic nervous systems. The *central nervous system* consists of the brain and spinal cord. The brain lies within the cranial cavity but is continuous at the foramen magnum with the spinal cord which occupies part of the vertebral canal. The *peripheral nervous system* is composed of all of the nerves and ganglia associated with them which are external to the surface of the brain or spinal cord. Certain neurons which travel in the nerves of the peripheral nervous system are related specifically to the innervation of smooth muscle, cardiac muscle, and glands. This group of neurons is separately designated as the *autonomic nervous system* because of their particular functions and a special morphological arrangement of neurons and ganglia into a two-neuron chain. The autonomic nervous system, however, is but a special category of the peripheral nervous system.

Functional Components of the Nervous System

Neurons may be grouped functionally according to the purpose they serve. Neurons grouped in this manner are termed the functional components of the nervous system. Neurons are either motor or sensory in their general function, never both. (See color plate 12.) *Efferent fibers* carry out a *motor* function. The organs supplied are known as *effectors*. The terms motor, efferent, and effector will always refer to impulses generated in the central nervous system and traveling away from it through the peripheral nervous system. *Afferent fibers* carry out a *sensory* function. The ending of the nerve fiber in the skin, an organ of special sense, or an organ which is activated (stimulated) by an environmental or functional condition is known as a *receptor*. The terms sensory, afferent, and receptor will always refer to impulses traveling toward the central nervous system. Neurons become functional components of the nervous system as they become allied with either motor or sensory activities. Peripheral nerves are described as containing one or both of the functional components.

Motor Components. Motor components are further classified according to the nature of the effector organs or tissues.

1. *General motor fibers* (component) are the axons of cells in the ventral gray matter of the spinal cord and homologous parts of the brain stem; these fibers innervate striated voluntary muscles. (Striated muscles in the head which were associated with the gill arches of the embryo are an exception. See next component.)

2. *Special motor fibers* are found only in certain cranial nerves which travel from the brain stem and the contiguous portion of the highest part of the spinal cord. These fibers innervate only those striated muscles which migrated from the branchial (gill) arch area of the embryo to become the muscles of facial expression, mastication, swallowing, and phonation in the adult human body.

In comparative vertebrate anatomy where the idea of functional nerve components was developed, such muscles were termed "special visceral muscles" because of their role in ingestion, mastication, and swallowing of food. The term *special* as applied to this component comes from the concepts of *special* embryonic orgin and a *special* type of musculature located in a *special* area. As the adjective "visceral" becomes integrated with these concepts, it can be seen how a very specific term was created for this component in neuroanatomy: *special visceral efferent* (although many neuroanatomists would prefer the term "branchial efferent").

It is suggested for the reader's purposes that the more general term of *special motor fibers* be used.

3. *Visceral motor fibers* are the axons of certain *efferent* (motor) neurons which travel in the nerves of the peripheral nervous system in the form of a two-neuron chain. The cell body of the first neuron in the chain is located in the lateral part of the ventral gray matter of the spinal cord and homologous parts of the brain stem in the central nervous system. Its axon utilizes spinal or cranial nerves to reach the second nerve cell in the chain, which lies in a ganglion located outside of the central nervous system. The second nerve cell in this chain innervates smooth muscle, cardiac muscle, and glands.

This two-neuron chain constitutes the autonomic nervous system. Since the fibers of this component are wide spread in the body in order to innervate many essentially hollow, visceral organs, their glands, and smooth muscle, the specific term *general visceral efferent* applies. The autonomic nervous system is discussed in greater detail later in this chapter.

Sensory Components. Sensory neurons are classified according to the type and location of the receptor structures and organs.

1. *Teleceptors:* These nerve endings detect environmental changes such as light rays, sound waves, or particles in the air (odors), at a distance from the body. Receptors of the eye (vision), ear (hearing), and nose (smell) are in this category.

2. *Exteroceptive receptors:* These nerve endings detect environmental changes, such as temperature changes, touch, pressure, or painful stimuli, which affect the skin and deeper tissues of the body wall.

3. *Proprioceptive receptors:* These nerve endings detect the functional activity of voluntary muscles and their tendons, the changing position of joints, and changes in the equilibrium of the body in relation to the gravitational field.

The preceding types of receptors are related to tissues or organs of the surface of the body, the body wall, or its musculoskeletal parts. They are classified as *somatic* receptors (from *soma:* body wall). The functional components associated with somatic receptors frequently are given the modifying term of "somatic." Since they are sensory in nature, the modifying term "afferent" may also be used.

4. *Interoceptive receptors:* These receptors are located within organs which receive autonomic (general visceral motor) innervation. The degree of smooth muscle contraction, distention, peristaltic movement, and secretory activity within these visceral organs affects these receptors. They are, therefore, grouped separately as *visceral* receptors. The nerve (fiber) component conveying impulses from these organs is associated with in transit and accompanies the motor nerves of the autonomic nervous system.

Sensory components can also be classified as follows:

1. *General (somatic) sensory fibers* convey impulses from the exteroceptive receptors.

2. *Special (somatic) sensory fibers* convey impulses from the teleceptor organs and thus are concerned with the special senses. (Vision involves the retina of the eye as the receptor area. The retina is really part of the brain, as is its connection, the optic "nerve." Special sensory fibers conveying visual impulses really do not belong as a *nerve* component but, by convention, most authors classify them under this special somatic sensory component.)

Proprioceptive impulses from muscles, tendons, and joints, all of which are related to the receipt of information about the position or movement of the body, are included in the special (somatic) sensory component.

The sense of position or of movement comprises an important group of afferent components which convey impulses to the central nervous system at a level below our usual consciousness. They may be thought of as a constant flow of sensations which are acted upon almost automatically. These include:

1. The degree of stretch in muscles and tendons.

2. The degree of stretch or compression within joint structure.

3. The degree of disturbance to internal ear receptors due to movement, its start-up, acceleration, velocity, or stopping, as well as to the effect of gravity or weightlessness (in space).

4. The degree (amount) of movement in the ear mechanisms, in which a tiny membrane in inner ear receptors is finally affected by sound waves. Auditory receptors are merely modified movement receptors (item 3) which sense movement of the basilar membrane of the ear apparatus. The information sent by the nerve component becomes perceived by the person as "hearing."

3. *Special (visceral) sensory fibers* convey impulses from receptors which are related to the "chemical senses" of smell (olfaction) and taste. The receptors involved are located in the nasal cavity for smell and in the taste buds of the tongue and upper alimentary passageways.

4. *General (visceral) sensory fibers* convey sensory impulses from the visceral organs which receive an autonomic motor nerve supply. It should be noted that while the general visceral sensory fibers will travel *in association* with autonomic motor fibers, the sensory fibers do not constitute the "sensory part of the autonomic nervous system." The autonomic nervous system is purely a *motor* apparatus.

Not all nerves of the peripheral nervous system carry fibers of all of the functional components. As nerves or major pathways within the nervous system are described, their components will be identified.

The functional nerve components are reviewed in the table which appears on these pages.

CENTRAL NERVOUS SYSTEM

EXTERNAL MORPHOLOGY OF THE BRAIN AND SPINAL CORD

Grossly the brain can be seen to be composed of the *cerebrum,* the *brain stem,* and the *cerebellum.* The cerebrum consists of two large *cerebral hemispheres.* (See color plate 13.)

The Cerebrum

The folding of the cerebral cortex during development brings about the formation of *cerebral convolutions* or *gyri. Sulci* are clefts between the raised gyri. Major parts of the cerebrum are separated by deeper clefts. The deep sagittal cleft between the two cerebral hemispheres is termed the *longitudinal fissure.* Each cerebral hemisphere is divided into *lobes* which generally mediate different functions. More pronounced sulci separate these lobar divisions.

External Parts of the Cerebrum. The cerebral gyri are found on all surfaces of the cerebrum, including the surfaces bordering the longitudinal fissure. The two cerebral hemispheres are symmetrical in form and appearance even though one is considered to be dominant functionally (i.e., the left in right-handed individuals, and vice versa).

The *frontal, parietal, temporal,* and *occipital* lobes of the cerebrum are named for the bone of the cranium against which they lie. The lobes are partly demarcated by major sulci. When viewed from the side, the cerebral hemisphere looks like a clenched hand within a mitten or boxing glove. The "thumb portion" is the *temporal lobe* which lies along the side of the middle cranial fossa. The *lateral sulcus* follows the upper margin of the temporal lobe and separates it from the *parietal lobe* which is superior to it. The *central sulcus,* a most important landmark of the cerebral hemisphere, runs superiorly and backward from just above the midpoint of the lateral sulcus,

FUNCTIONAL COMPONENTS OF NERVES

Type and Impulse	Motor efferent	Motor efferent	Motor efferent	Sensory afferent
Cell Body	Main part of the ventral gray matter of spinal cord	Ventral gray matter of brain stem for cranial nerves	Lateral part of ventral gray matter of spinal cord and brain stem	Sensory root ganglia of spinal and cranial nerves
Termination/Receptor	Striated voluntary muscle excepting head	Striated voluntary muscle of head of branchial arch origin	Smooth muscle Cardiac muscle Glandular secreting units	Exteroceptive
Basic Functional Term	General motor	Special motor (or special visceral motor)—a two-neuron system	General visceral motor (autonomic motor)	General (somatic) sensory
Specific Term*	General somatic efferent	Special visceral efferent	General visceral efferent (autonomic nervous system)	General somatic afferent

*Note: The terms in this line are highly specific and are used mainly in special studies of the nervous system. They are included here for completeness. More basic terms are used in this book and, unless necessary, are abbreviated to sensory, motor, or autonomic components with modifiers used as required.

across the convex surface of the hemisphere over the medial margin of the hemisphere onto its medial surface. The central sulcus separates the parietal lobe behind from the frontal lobe.

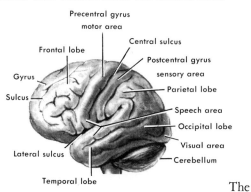

The *frontal lobe* forms the anterior portion of the cerebral hemisphere which lies upon its half of the anterior cranial fossa. The *occipital lobe* lies upon a shelf of connective tissue separating this part of the cerebral hemisphere from the cerebellum which occupies the posterior cranial fossa.

Functional Areas of the Cerebral Cortex. Several important functional areas have been identified in the cerebral hemisphere.

1. The *precentral gyrus* winds upward over the lateral surface of the hemisphere onto the medial surface in front of the central sulcus. Neurons which establish voluntary control over motor activities of the body are located in the precentral gyrus. The neurons of this *motor area* of one hemisphere control activities on the *opposite* side of the body. Neuron cell bodies which control particular motor functions (e.g., opening of mouth, movement of the arm) cluster together in centers which are aligned in an inverted order to the corresponding parts of the body. The pattern can be visualized as the figure of a small man hanging along the precentral gyrus with his knees over the longitudinal fissure and his head hanging downward. Submotor areas for the legs and feet would be in part of the precentral gyrus on the medial surface of the hemisphere. Thigh and trunk centers are high on the lateral surface of the precentral gyrus, upper limb centers lower, and head and neck centers lowest near the lateral sulcus.

2. The *postcentral gyrus* runs upward behind the central sulcus. The cortical cells here are the final station for *sensory* pathways and, therefore, this gyrus is the main *sensory area*.

3. A large cortical area of the temporal lobe

FUNCTIONAL COMPONENTS OF NERVES (Continued)

Sensory afferent	Sensory afferent	Sensory afferent	TYPE AND IMPULSE
Brain or special sense ganglia	Sensory root ganglia of cranial nerves	Sensory root ganglia of certain spinal and cranial nerves	CELL BODY
Telereceptive (special senses) Proprioceptive from muscles, tendons, joints	Chemoreceptors	Interoceptive	TERMINATION/RECEPTOR
Special (somatic) sensory	Special (visceral) sensory for smell (olfaction) and taste	General (visceral) sensory (general sensory accompanying autonomic system)	BASIC FUNCTIONAL TERM
Special somatic afferent	Special visceral afferent	General visceral afferent (autonomic nervous system)	SPECIFIC TERM*

and an area of frontal lobe cortex adjacent to the lateral sulcus make up a large area which controls *speech.*

4. The cortex of the pole of the occipital lobe is related to optical stimuli and is designated as the *visual area.*

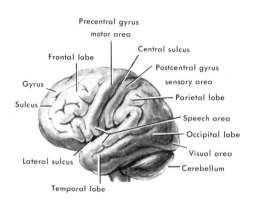

Internal Structure of the Cerebrum.   The cerebral cortex is only a thin blanket of nerve cells covering the surface of the gyri of each hemisphere. Several other cell centers, the *basal nuclei,* lie deep within the cerebrum, but the greater proportion of the cerebral hemisphere is formed by the white matter and by the internal cavities of the brain.

The *white matter* of the hemisphere consists of bands and planes of fibers which intermix and run within the mass of white matter. These fiber groups are the processes of nerve cells which are located either in the cerebral cortex or in other centers of the central nervous system. (See color plate 12.) White matter fiber groups are classified as:

1. *Short association fibers* which connect cell groups of adjacent gyri by short "U" shaped bands which bend about a sulcus.

2. *Long association fibers* which course deeper to connect one cortical area with a more distant one in the same hemisphere.

3. *Commissural fibers* which connect cortical centers of one hemisphere with similar areas of the opposite hemisphere. The most prominent of these is the *corpus callosum* which forms a broad band bridging the depths of the longitudinal fissure.

4. *Projection fibers* convey nerve impulses to and from cerebral centers and centers lower in the brain stem or spinal cord. The projection fibers are compressed into bands near the brain stem but fan through the expanded hemispheres as *radiations.*

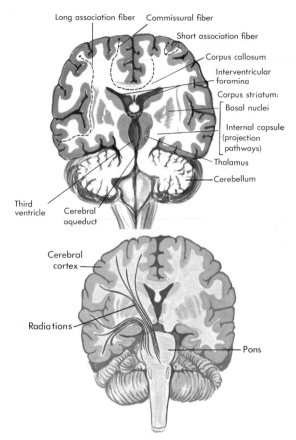

The *internal cavities* of the cerebrum are expansions of the original cavity of the prosencephalon. The midline cavity is narrowed to a cleft by the thickened walls of the diencephalon. At the forward end of this cleft, known as the *third ventricle,* an aperture leads laterally on each side into the expanded *lateral ventricle* which is the cavity within the cerebral hemisphere. The lateral ventricle extends into each of the major parts of the hemisphere.

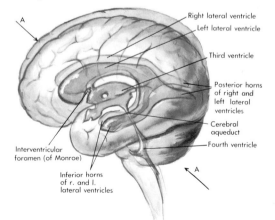

The *deep gray matter* of the cerebrum is formed by centers of nerve cell bodies, called *basal nuclei,*

which were compressed against the lateral walls of the diencephalon as the cerebral hemisphere expanded in development. These collections of nerve cells are clustered in groups which are relay stations on motor and sensory pathways between the cerebral cortex and the spinal cord. Since the basal nuclei are astride the point of entrance or exit of the projection pathways, the groups of nerve cells are permeated by white matter fibers to give a striped appearance to the base of the hemisphere. This combination of basal nuclei and white matter is described by the general term, *corpus striatum.*

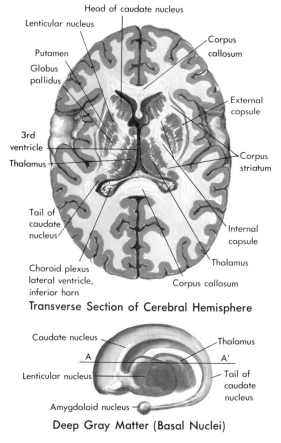

Transverse Section of Cerebral Hemisphere

Deep Gray Matter (Basal Nuclei)

The basal nuclei are divided into medial and lateral groups which are separated or surrounded by broad bands of white matter. The bands of white matter, which are formed by fibers diverging into the hemisphere or converging from its radiations, are called the *internal capsule* because the basal nuclei are encapsulated by them. The long, taillike form of the more medial nuclei gives them the group name of *caudate nucleus,* whereas the lenslike shape of the more lateral nuclei is the source of the group name, *lentiform nucleus.*

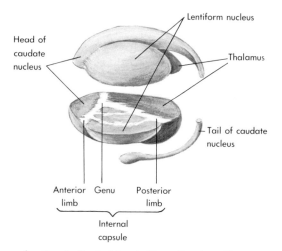

Section Indicated by A-A' on Previous Figure

Structure of the Diencephalon. The cerebral hemispheres hide the diencephalon from all but a basal view of the brain. Although overlapped by the cerebrum, the diencephalon is not fused to it. The two are separated by the *transverse cerebral fissure* which provides a route for blood vessels to reach the thin roof of the diencephalon. The walls of the diencephalon are thickened owing to a great increase in the gray matter which reduced the diencephalic portion of the neural cavity to the narrow cleft of the *third ventricle.* The diencephalon is formed mainly of gray matter which is divided as follows:

1. On each side a dense nucleus, the *thalamus,* bulges into the third ventricle. The thalamus is concerned with sensory impulses conveyed upward from the brain stem and spinal cord.

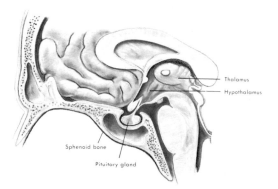

2. The floor and the lower part of the walls of the third ventricle form the *hypothalamus,* which is important in the regulation of autonomic functions. It is part of the base of the brain which can be seen between the basal surfaces of the temporal lobes. The optic nerves coming

from the orbits form a crossing, the *optic chiasm,* in front of the hypothalamus. The fibers of the internal capsule come to the basal surface of the cerebral hemispheres just behind the hypothalamic surface of the diencephalon as thick cylinders called the *cerebral peduncles.* The *hypophysis* (pituitary gland) hangs by the *hypophyseal stalk* from the floor of the diencephalon.

The Midbrain

The mesencephalon is more tubular than the derivatives of the forebrain. The midbrain lies along the midline of the base of the skull in the *tentorial notch* of the shelf of connective tissue which stretches over the posterior cranial fossa to support the posterior parts of the cerebrum. The cavity of the central nervous system is narrowed in the midbrain to the triangular *cerebral aqueduct.*

The *cerebral peduncles* converge to become flattened pillars on the ventral surface of the mesencephalon. These projection fibers have now been followed in the radiations of the cerebral hemisphere, through the internal capsule, into the cerebral peduncle at the surface of the hypothalamic region, and upon the ventral surface of the midbrain.

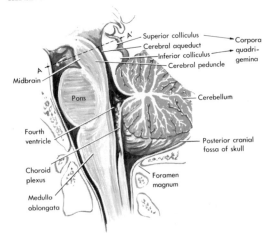

In any part of the brain stem the structures at any given level belong to either of the following:

1. Long communicating pathways between the cerebrum and centers lower in the central nervous system. The cerebral peduncles are an example.
2. Specific centers of function appearing at the particular level.

An example of the second type is found in gray matter at the floor of the cerebral aqueduct. Here is the *nucleus of the third* (oculomotor) *cranial nerve* where motor neurons for the control of most of the muscles moving the eyeball cluster together. Between this nucleus and the cerebral peduncle on the ventrolateral surface of the midbrain is the *red nucleus* within which are cells concerned with the coordination of motor impulses to voluntary muscles. The motor fibers of the oculomotor nerve stream through the red nucleus to the ventral surface of the midbrain where they assemble to form the third cranial nerve. A bit lower in the midbrain is another set of motor neurons; these, forming the *nucleus of the fourth* (trochlear) *cranial nerve,* innervate one of the eye muscles. The white matter about the cerebral aqueduct contains columns of fibers, many of which are concerned with sensory pathways rising from the spinal cord or are related to the cerebellum.

On the dorsal surface of the midbrain are four rounded eminences, the *corpora quadrigemina,* which are largely hidden from view by the occipital lobes of the cerebrum. The two superior masses, the *superior colliculi,* contain nerve cells which are related to incoming fibers of the optic tracts which have swung around the base of the brain from the optic chiasm. The two inferior quadrigeminal masses, the *inferior colliculi,* contain nerve cells which are intermediate centers for auditory sensation.

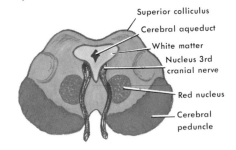

Section Indicated by A-A' on Previous Figure

The Pons and Cerebellum

The brain stem just under the tentorial shelf bulges into a portion called the *pons* (bridge). It is so named because the superficial layers of the brain stem at this level are composed of nerve fibers crossing between the two hemispheres of the cerebellum. These are applied to the ventral surface of the brain stem and cover the regional centers and the long pathways. The *cerebellum,* as a dorsal outgrowth from the metencephalon, develops much as did the telencephalon. It forms into greatly folded and fissured lobes which surround all but the ventral surface of the brain stem. The lobes fill the posterior cranial fossa of the skull.

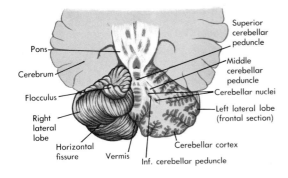

The Pons.   On the ventral surface of the brain stem the cerebral peduncles disappear at the upper border of the pons because they are covered by the transversely coursing bridge fibers of the cerebellum. The projection fibers from the peduncles are separated into scattered coarse bundles by the transverse bridge fibers and by cell nuclei in the ventral part of the pons. Some of the peduncular fibers, called *corticopontine pathway fibers,* are destined for the cerebellum. They synapse with neurons of the pons whose axons join the bridge fibers to reach the cerebellum. These fibers form on each side prominent *middle cerebellar peduncles* (brachia pontis) which spread laterally into the cerebellum. The remaining longitudinally coursing projection fibers reassemble into pillars at the inferior border of the pons to descend further in the central nervous system as the *pyramids.*

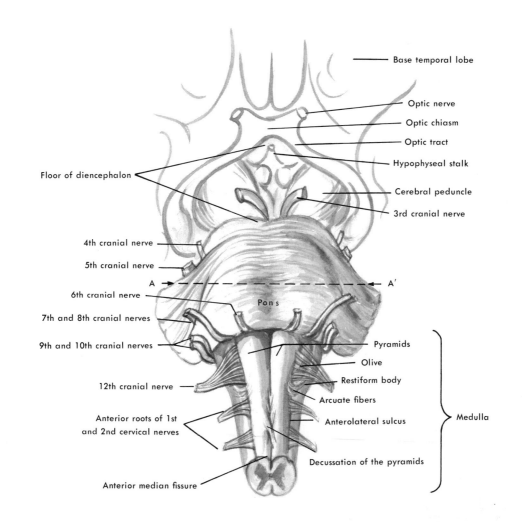

The dorsal portion of the pons forms the floor of the greatly broadened *fourth ventricle.* Gray matter is intermixed with bundles of ascending and descending fiber pathways. The gray matter is formed into successive nuclei containing the motor neurons for the fifth (trigeminal), sixth (abducens), and seventh and eighth (stato-acoustic) cranial nerves. Neurons related to the incoming sensory fibers of the fifth cranial nerve are also clustered in this area. Motor fibers of the fifth cranial nerve enroute to the muscles of mastication assemble at the lateral surface of the pons. Fibers of the sixth, seventh, and eighth cranial nerves appear at the inferior border of the pons—the sixth being closest to the midline, the eighth farthest laterally, and the seventh between.

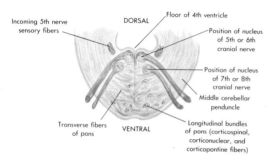

Incoming 5th nerve sensory fibers

DORSAL

Floor of 4th ventricle

Position of nucleus of 5th or 6th cranial nerve

Position of nucleus of 7th or 8th cranial nerve

Middle cerebellar penduncle

Transverse fibers of pons

VENTRAL

Longitudinal bundles of pons (corticospinal, corticonuclear, and corticopontine fibers)

Section Indicated by A-A′ on Previous Figure

The Cerebellum. The fissured cerebellum is formed into three lobes. The two lateral lobes make up the cerebellar hemispheres while the sinuous central connecting lobe is the *vermis.* The whole is molded about the dorsal part of the pons. The bulk of the gray matter is displaced, as in the telencephalon, to form a thin *cerebellar cortex* on the surface. The cerebellum is a composite of many leaves, each of which has a central core of white matter covered by the cortex. *Cerebellar nuclei* are interspersed deep amidst the white matter much like the basal nuclei of the cerebrum. Three major bands surround much of the diamond-shaped fourth ventricle. The bridge fibers connecting the cerebellar hemispheres form the *middle cerebellar peduncle.* The *superior cerebellar peduncle* (brachium conjunctivum) forming the superolateral margin of the fourth ventricle connects cerebellar nuclei with the inferior colliculus of the midbrain. The *inferior cerebellar peduncle*

connects cells of the cerebellum with centers lower in the brain stem and spinal cord. All of the nuclei, cortex, and fiber pathways of the cerebellum are concerned either with the coordination of impulses to voluntary muscle or with the equilibrium of the body.

The Medulla Oblongata

The *medulla oblongata* is the cone-shaped part of the brain stem which extends from the pons above to the spinal cord below. The demarcation from the spinal cord is not sharp, for the surface features of the spinal cord gradually expand upward onto the medulla. The medulla ends at the level of the foramen magnum. In addition to being the location of important regional sensory and motor centers including the nuclei of the remaining cranial nerves, the medulla is an area of transition between the brain stem and spinal cord. In the medulla the fibers of many major pathways change location or reassemble. Descending pathways assume patterns characteristic of the spinal cord and ascending pathways shift to patterns of the brain stem. The medulla is marked externally by fissures and by elevations which divide it externally into ventral, lateral, and posterior portions.

Ventral Surface of the Medulla. An *anterior median fissure* continues upward from the spinal cord onto the medulla. Rounded pillars on each side of the fissure mark the reappearance of the fibers of the cerebral peduncles which were buried under the bridge fibers of the pons. These pillars, now called the *pyramids,* extend downward to the lower border of the medulla where they smooth out; the fissure between them disappears temporarily because some of the projection fibers cross the midline to the opposite side in the *decussation of the pyramids.* Lateral to each pyramid is the *anterolateral sulcus* providing exit for the motor fibers of the twelfth (hypoglossal) nerve to muscles of the tongue.

Lateral Surface of the Medulla. A *posterolateral sulcus* extends upward onto the medulla from the spinal cord. An oval eminence, the *olive,* separates the fibers of the hypoglossal nerve in the anterolateral sulcus from root fibers of the ninth (glossopharyngeal), tenth (vagus), and eleventh (accessory) nerves which issue from the posterolateral sulcus in a linear series.

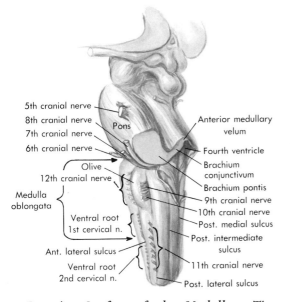

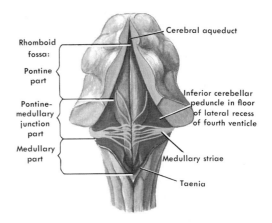

The open part of the fourth ventricle is thinly roofed by non-nervous tissue. If this membrane is removed, the floor of the fourth ventricle, shared by the pons and medulla, is exposed. It is called the *rhomboid fossa* because of its elongated diamond shape.

Posterior Surface of the Medulla.

The diamond-shaped fourth ventricle tapers inferiorly so that in the lower part of the medulla the posterior aspect of the ventricle is gradually closed. The ventricle itself funnels into a *central canal* which is continuous with that of the spinal cord. From the point of closure of the fourth ventricle the lower medulla presents a *posterior median fissure* which is continuous inferiorly with the same fissure on the posterior surface of the spinal cord. Adjacent to the fissure on each side of the closed part of the medulla is a column of ascending fibers, the *fasciculus gracilis.* This vertical column is separated from a companion group at its lateral side, the *fasciculus cuneatus,* by a *posterior intermediate sulcus.* Gray matter centers found at the heads of the columns—the *nuclei gracilis* and *cuneatus*—are relay points on ascending pathways.

The inferior, medullary part of the floor is roughly triangular and contains three raised masses of gray matter. These cellular areas, in order from the midline laterally in the floor of the fourth ventricle, are the *nucleus of the hypoglossal nerve,* the *nucleus of the vagus nerve,* and the *area acustica* which also extends upward onto the pontine surface of the floor of the fourth ventricle. *Vestibular nuclei* compose the area acustica which is sensory in nature in contrast to the hypoglossal and vagal nuclei which are motor in function.

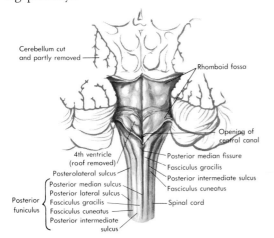

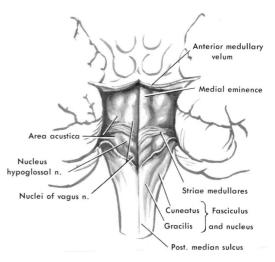

Motor nuclei for the ninth (glossopharyngeal) and eleventh (accessory) nerves are located in a heterogeneous area of cell bodies and nerve fibers deep in the medulla. This area, known as the *reticular formation,* contains scattered neuronal cell bodies which are not well organized into nuclei but do serve vital processes. Respiration, heart beat, swallowing, and salivation are among the functions controlled by neurons in the reticular formation whose fibers travel along tracts of the brain stem to emerge in pertinent cranial or spinal nerve roots. In addition to such neurons there is the *nucleus ambiguus* which is a vaguely defined column of motor neurons running longitudinally through the area. Axons from this column enter the ninth, tenth, and eleventh cranial nerves to innervate muscles of the pharynx and larynx.

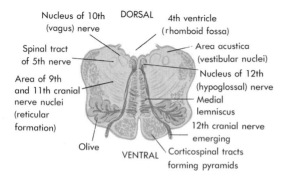

One more feature of the medulla deserves mention. The fifth cranial nerve is identified with the pons. Here is the point of emergence from the brain stem and the location of its motor nucleus to the muscles of mastication. The greater part of this nerve is sensory, however, and the incoming fibers carry sensory impulses from the skin of the greater part of the head. Sensory centers for these fibers are too large to be accommodated entirely in the pons. Many of the fifth nerve sensory fibers and the gray matter of the neurons with which they make connections are displaced inferiorly into the medulla. They form the *spinal tract* and *nucleus of the spinal tract of the fifth nerve,* often described simply as the descending tract of the fifth nerve.

The Spinal Cord

Many of the features of the spinal cord have already been mentioned in describing the organization of the nervous system.

External Appearance of the Spinal Cord. The spinal cord is the central occupant of the

vertebral canal. It continues from the brain stem caudally (inferiorly) at the foramen magnum. (See color plate 13.) Contained within its covering sheaths, the spinal cord descends to the level of the intervertebral disk between the first and second lumbar vertebrae where nervous tissue ends.

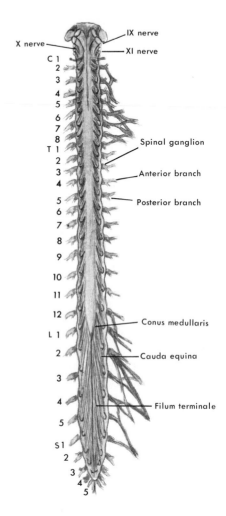

The spinal cord ceases its linear growth prior to birth and long before the trunk has attained adult dimensions. Therefore, only the roots to nerves which extend to the lower parts of the body, cord coverings, and a non-nervous strand (*filum terminale*) continue below the end of the cord.

On the posterior surface of the spinal cord is the shallow *posterior median sulcus.* A fusion of the superficial coverings of the cord continues the line of the sulcus inward toward the central canal as the *posterior median septum*. The *posterolateral sulcus* is well defined by the entrance of the posterior roots. The *posterior funiculi* are rounded vertical columns of white matter on each side of the posterior median sulcus, limited laterally by the posterolateral sulcus. A shallow posterior intermediate sulcus divides the posterior funiculus into the more medial *fasciculus gracilis* and the more lateral *fasciculus cuneatus* whose upper ends were found on the medulla. The lateral surface of the spinal cord is formed into the *lateral funiculus* of fiber tracts. It is located between the posterior roots and posterolateral sulcus behind and the line of emergence of the anterior roots on the anterolateral surface. The deep *anterior median fissure* indents the anterior surface and provides a channel of entry for the spinal arteries. The white matter on each side of the fissure is rounded laterally to the line of anterior roots to form the *anterior funiculus.*

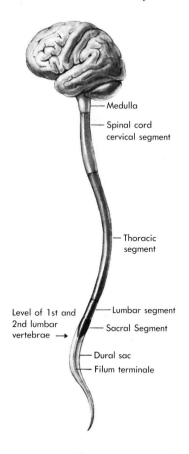

Medulla

Spinal cord cervical segment

Thoracic segment

Level of 1st and 2nd lumbar vertebrae →

Lumbar segment

Sacral Segment

Dural sac

Filum terminale

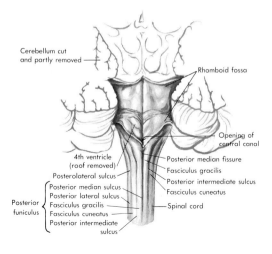

Cerebellum cut and partly removed

Rhomboid fossa

Opening of central canal

4th ventricle (roof removed)

Posterolateral sulcus

Posterior median sulcus

Posterior lateral sulcus

Posterior funiculus { Fasciculus gracilis

Fasciculus cuneatus

Posterior intermediate sulcus

Posterior median fissure

Fasciculus gracilis

Posterior intermediate sulcus

Fasciculus cuneatus

Spinal cord

Each region of the spinal cord displays characteristic features depending upon the size and nature of the gray columns, the number and disposition of fiber tracts, and the number or nature of motor units to be innervated from the region or its component segments. For instance, the cell bodies of motor neurons and the connections of sensory neurons for the upper and lower extremities extend across several segments of the cervical and lumbar regions respectively. The consequent increase in the mass of nervous tissue required to serve the extremities results in *cervical* and *lumbar enlargements* of the spinal cord. (See also color plate 13.) The eleventh (accessory) cranial nerve, whose motor nucleus has been identified in the brain stem, supplies, in addition to the muscles of the pharynx, the trapezius and the sternocleidomastoid muscles. The motor nerve cells for these two muscles are located within the upper segments of the spinal cord. The motor fibers to these muscles emerge from the lateral surface of the spinal cord just anterior to the posterior roots. The filaments join as the *spinal root of the accessory nerve* which ascends through the foramen magnum to join the cranial root.

EXTERNAL ENVIRONMENT OF THE CENTRAL NERVOUS SYSTEM

The central nervous system is covered and protected by a series of membranes which are called collectively the *meninges*. Extensions of the membranes hold the brain and spinal cord in position. The coverings also provide for a fluid space within which the entire central nervous system is suspended.

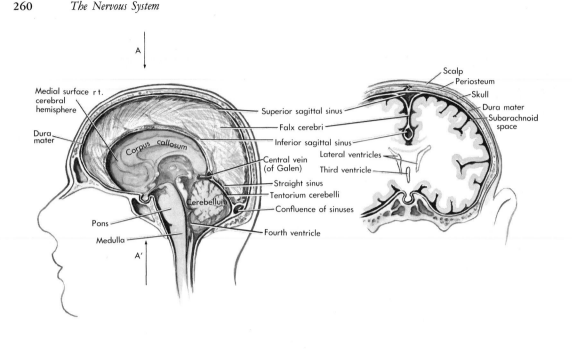

Coverings of the Brain

A delicate connective tissue, the *pia mater,* invests the brain and spinal cord and is adherent to their surfaces. The pia mater follows every surface feature and dips into all fissures and sulci of the cerebral and cerebellar hemispheres, brain stem, and spinal cord. The blood vessels of the brain and spinal cord branch within the pia mater. As the blood vessels enter the brain or spinal cord, they are surrounded by *perivascular spaces* which are lined by inward extensions of the pia mater.

The *arachnoid* membrane is a gauzelike layer of connective tissue outside the pia mater which fits about the entire central nervous system like a loose sleeve. It bridges fissures, sulci, and irregular surface features. The arachnoid is loosely attached to the pia mater by webs (trabeculae) of arachnoidal tissue. The two membranes are not in contact, for the *cerebrospinal fluid* seeps between them to form a fluid cushion about the central nervous system. (See above illustration.)

The *dura mater* is a strong collagenous membrane which is external to the arachnoid and is separated from it by the potential *subdural space.* Normally the dura and arachnoid are in intimate contact. Of the three coverings only the dura is a membrane of considerable substance. It forms a dense fibrous sac which protects and encloses the entire central nervous system. Within the skull the dura also acts as the internal periosteum of the cranial bones. It consists of two layers. The external *endosteal layer* lines the cranial cavity, its vault, and the cranial fossae. (See illustration at top of following page.) The internal *meningeal layer* is smooth and faces the arachnoid. At the foramen magnum the cranial dura mater becomes continuous with the spinal dura mater.

Processes of the Cranial Dura Mater

Extensions of the cranial dura, called *dural processes,* form partitions and shelves between parts of the brain. These processes hold the brain in position, support some of its parts, and provide passageways for venous blood. The venous passageways, formed by the splitting away of the meningeal layer from the endosteal layer of the dura, are not actually veins although they are lined by endothelium. Continuous with the cerebral veins whose blood they receive, these passageways, the *dural venous sinuses,* form a confluent network which drains into the internal jugular vein at the base of the skull.

Along the midline of the calvarium, the meningeal layer of the dura on each side splits off from the endosteal layer to form a vertical partition, the *falx cerebri,* which hangs down between the two cerebral hemispheres in the lon-

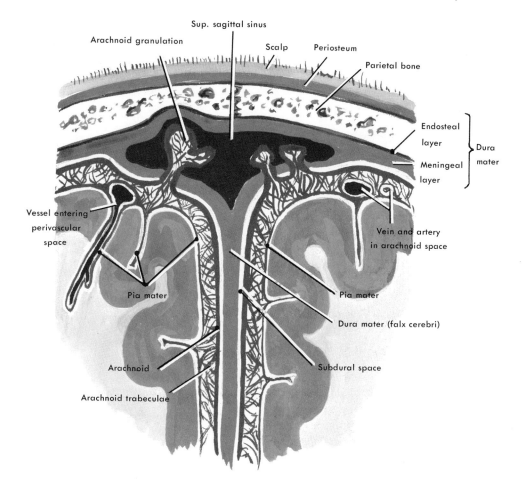

Sup. sagittal sinus

Arachnoid granulation

Scalp Periosteum

Parietal bone

Endosteal layer

Meningeal layer

Dura mater

Vessel entering perivascular space

Vein and artery in arachnoid space

Pia mater

Pia mater

Dura mater (falx cerebri)

Arachnoid

Subdural space

Arachnoid trabeculae

gitudinal fissure. The falx cerebri ends anteriorly in a free border in the depths of the longitudinal fissure above the corpus callosum.

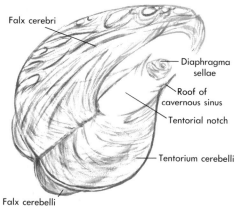

Falx cerebri

Diaphragma sellae

Roof of cavernous sinus

Tentorial notch

Tentorium cerebelli

Falx cerebelli

Posteriorly the free border of the falx cerebri ends as the partition joins a horizontal dural process, the *tentorium cerebelli,* at right angles. The tentorium cerebelli is split off from the cranial dura along a convex border on the occipital and parietal bones. This process spreads transversely across the superior surface of the cerebellum and forms a shelf which supports the occipital lobes of the cerebrum above. The tentorium is indented anteriorly by the *tentorial notch* through which the brain stem passes. Lateral to the notch, the tentorium cerebelli attaches to the posterior margins of the middle cranial fossae. The sickle-shaped *falx cerebelli* is another vertical partition of dura which hangs downward in the midline from the tentorium cerebelli between the cerebellar hemispheres. It is attached to the occipital bone posteriorly and ends in a concave free border anteriorly. A small circular sheet of dura mater covers the sella turcica. This *diaphragma sellae* forms a roof over the pituitary gland and is perforated by the stalk of the gland.

Coverings of the Spinal Cord

The same three layers of meninges cover the spinal cord. The spinal dura mater does not form

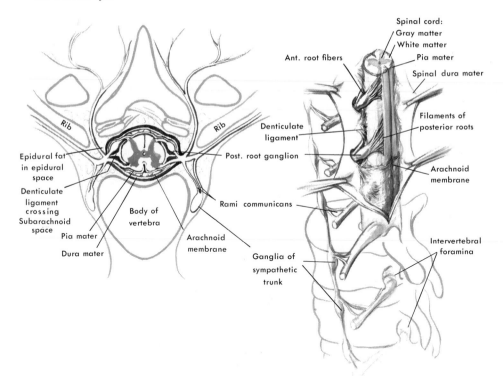

the internal periosteum of the vertebral canal but is a tough envelope surrounding the spinal cord and suspended with it. A space exists between the walls of the vertebral canal and the dural envelope enclosing the spinal cord. This *epidural space* is filled with fat and blood vessels to the spinal cord. The epidural fat is a semifluid cushion which supplements the dura and the spinal cerebrospinal fluid in protecting the spinal cord.

Lateral extensions of the spinal pia mater bridge the subarachnoid space as a series of toothed shelves located midway between the dorsal and ventral spinal nerve roots. The *denticulate ligament* so formed assists in positioning and suspending the spinal cord within its coverings. At the lower end of the spinal cord, the pia mater is extended into a collagenous strand, the *filum terminale*. This strand is surrounded by the lower spinal nerve roots. These roots, descending to their respective lower lumbar and sacral intervertebral foramina, hang within the dural sac like a horse's tail or *cauda equina*. The dural sac, lined internally by arachnoid, ends blindly at the level of the second sacral vertebra. (See color plate 13.) A combined strand of dural connective tissue and the filum terminale continues to the back of the coccyx as the *filum of the dura mater* where it anchors the spinal cord and its coverings to the end of the vertebral canal.

Cerebrospinal Fluid System

The *cerebrospinal fluid* fills the internal cavities of the central nervous system and passes through communicating openings in the fourth ventricle into the subarachnoid space about the brain and spinal cord to form a protective fluid cushion. (See above illustration.) The fluid within the brain cavities helps maintain the shape of the brain. The cerebrospinal fluid system takes the place of lymphatic vessels within the brain, for tissue fluids may seep along perivascular spaces to reach the subarachnoid space for ultimate return to the venous system.

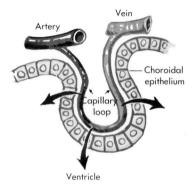

Formation of Cerebrospinal Fluid. The cerebrospinal fluid is a clear, colorless fluid which contains electrolytes similar to those in the blood but is almost devoid of proteins. It is formed at sites in the two lateral ventricles, and the third and fourth ventricles where capillaries invaginate thin epithelial areas in the walls to form tufted projections, called the *choroid plexuses,* through which the fluid passes into the ventricles from the blood.

Internal Cavitary System.   A circulatory flow

of the cerebrospinal fluid occurs from the lateral ventricles through the interventricular foramina (of Monro) on each side into the third ventricle and then through the cerebral aqueduct (of Sylvius) into the fourth ventricle. Although there is cerebrospinal fluid within the central canal of the spinal cord, it is not believed that there is a significant circulation within its narrow confines.

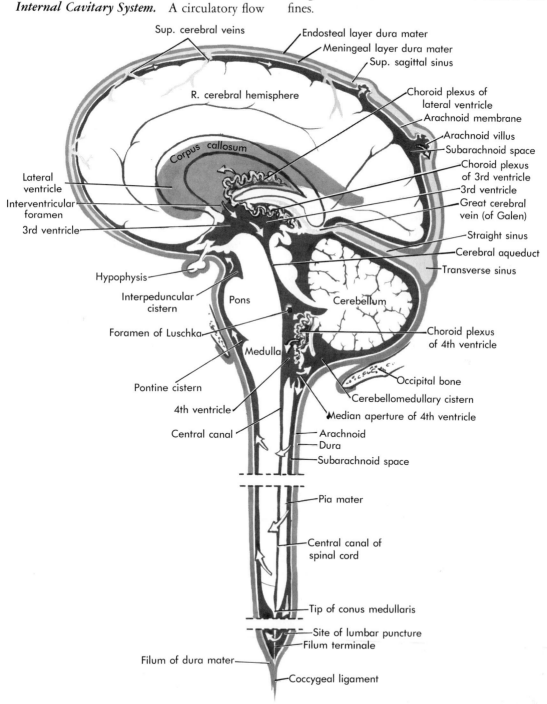

Subarachnoid Fluid Spaces. Cerebrospinal fluid passes into the subarachnoid space through two lateral apertures and one median aperture in the roof of the fourth ventricle. These sites of entry into the subarachnoid space are overhung by the bulge of the cerebellum. The arachnoid is draped loosely from the medulla below to the cerebellum above, forming a widening of the subarachnoid space in which a considerable pool of fluid collects. This is known as the *cerebellomedullary cistern* (cisterna magna) of the subarachnoid space. Other such cisterns are formed about the brain wherever the arachnoid stretches over large irregularities of the surface. Major ones are the *pontine cistern* inferior to the bulge of the pons, the *interpeduncular cistern* between the cerebral peduncles at the base of the brain, and the *lateral cerebral cistern* about the large lateral fissure of the cerebrum. The fluid circulates upward about the base of the brain and then over the surfaces of the cerebral hemispheres. The fluid also circulates downward in the subarachnoid space about the spinal cord. Beyond the lower limit of the cord the fluid seeps between the lower spinal nerve roots within the dural-arachnoid sac. The subarachnoid space can be entered with a needle (lumbar puncture) below the level of the second lumbar vertebra without the danger of injury of the spinal cord in order to remove fluid for diagnostic purposes or to reduce an increase in cerebrospinal fluid pressure. (See illustration on preceding page.)

BLOOD SUPPLY OF THE CENTRAL NERVOUS SYSTEM

Arterial blood reaches the brain and spinal cord by way of the internal carotid and vertebral arteries of the neck. See pages 333 and 334.

Arteries to the Brain

The *internal carotid artery* rises through the foramen lacerum of the cranial floor to enter one of the venous sinuses between the layers of the dura. The artery then turns upward through the dura into the subarachnoid space and finally bends backward under the optic chiasm. Branches of the internal carotid artery supply the orbit via the *ophthalmic artery,* the pituitary gland via *hypophysial arteries,* and much of the brain via *anterior* and *middle cerebral arteries.*

The *vertebral arteries* arch over the first cervical vertebra to pass through the foramen magnum.

They ascend on each side in front of the medulla within the subarachnoid space and usually join to form the single *basilar* artery on the anterior surface of the pons. The vertebral and basilar arteries supply the spinal cord, lower brain stem, cerebellum, and posterior portion of the cerebrum.

Distribution of Intracranial Arteries. The table on p. 266 summarizes the main patterns of blood supply to the central nervous system.

Arterial Anastomoses. The internal carotid and vertebral arterial systems to the brain are connected through communicating arteries to form the *arterial circle* of the base of the brain which is more familiarly known as the *circle of Willis.* The connection is made anteriorly by an *anterior communicating artery* between the two anterior cerebral arteries, and posteriorly by a *posterior communicating artery* on each side between the internal carotid and the posterior cerebral arteries.

Blood Supply to the Meninges. The pia mater and arachnoid are supplied with blood by small branches from the arteries of the various areas as they traverse the subarachnoid space. The cranial dura mater requires a considerable blood supply which is derived from the following sources:

1. *Middle meningeal arteries* which on each side ascend from the maxillary artery of the external carotid system to comprise the major supply to the dura mater.

2. *Meningeal branches* from the *occipital artery* passing through minute, unnamed foramina of the posterior part of calvarium to supply the posterior portion of the dura.

3. *Accessory meningeal arteries* passing through other minute foramina from the network of vessels supplying the scalp.

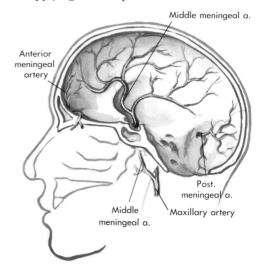

Blood Supply to the Vertebral Canal. The spinal cord receives its major arterial supply from the anterior and posterior spinal arteries. Other arteries enter the vertebral canal at each segmental level through the intervertebral foramina. These vessels arise from the vertebral artery, the posterior intercostal arteries, and the lumbar and sacral parietal arteries. The arteries pass along the spinal nerve roots to supply the nerve roots and dorsal root ganglia by way of *anterior* and *posterior radicular arteries,* but they also give off *anterior* and *posterior medullary branches* which anastomose upon the surface of the spinal cord with the spinal arteries. A large plexus of veins is located in the fat of the epidural space. This *internal vertebral venous plexus* drains blood from the epidural fat, the meninges, and the spinal nerve roots. *Spinal veins* draining the substance of the spinal cord join the vertebral plexus to form *intervertebral veins* which leave the vertebral canal through the intervertebral foramina.

Venous Drainage of the Brain

The venous sinuses serve as the main collecting channels within the cranium. They receive not only the veins of the brain but also veins from the cranial bones, meninges, and *emissary veins* which traverse minute unnamed foramina of the skull conveying blood from superficial tissues.

The large *superior sagittal sinus* lies within the junction of the cranial dura with the falx cerebri. This long sinus runs backward to a point near the internal occipital protuberance. Irregular extensions to each side form the *lateral lacunae.* The walls of both the sinus and the lacunae are indented by invaginating tufts of the arachnoid which project into the blood stream within. These *arachnoid granulations* are the means whereby cerebrospinal fluid seeps into the venous system. The *inferior sagittal sinus* lies in the free margin of the falx cerebri. It is joined by the *great cerebral vein* (of Galen) to form the *straight sinus* which extends posteriorly along the line of juncture of the falx cerebri with the tentorium cerebelli.

The superior sinus arching down from above, the straight sinus approaching from in front, and the *occipital sinus* curving upward along the line of attachment of the falx cerebelli terminate in a common pool of venous blood in a broadened pocket of dura called the *confluence of sinuses.* The

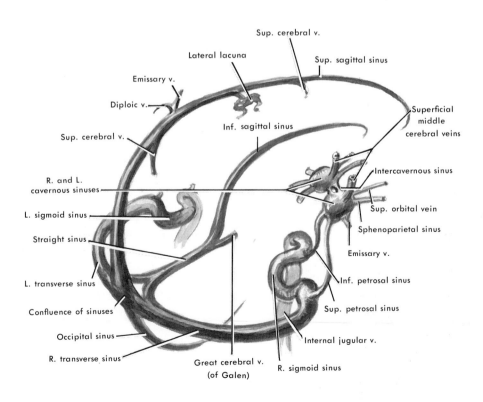

BLOOD SUPPLY TO THE CENTRAL NERVOUS SYSTEM

MAIN VESSEL	MAJOR BRANCH	GENERAL LOCATION	AREA OF SUPPLY
	Anterior cerebral	Longitudinal fissure	*Central branches* to basal nuclei, hypothalamus and internal capsule *Cortical branches* to superior and medial surfaces of frontal and parietal lobes
Internal carotid arteries	Middle cerebral	Lateral fissure	*Central branches* to thalamus, corpus striatum, and internal capsule *Cortical branches* to lateral surface of frontal and parietal lobes; superior and lateral surface of temporal lobe; and lateral surface of occipital lobe
	Anterior choroid	Parallels optic tract	Internal capsule, cerebral peduncle, choroid plexus of lateral ventricle
Vertebral arteries before junction	Anterior spinal	From each side join in front of medulla, forming a median vessel	Anterior part of medulla and spinal cord
	Posterior inferior cerebellar	On each side winds posteriorly around medulla	Posterior part of medulla, fourth ventricle, cerebellum
	Posterior spinal	From either vertebral or above vessel; on each side descends along side of medulla	Posterior part of spinal cord
	Pontine	On pons	Pons
Vertebral arteries through basilar	Anterior inferior cerebellar	Spread laterally toward cerebellum higher than posterior inferior arteries	Anastomose with posterior inferior arteries to supply pons and cerebellum
	Labyrinthine (internal auditory)	Into internal acoustic meatus	Internal ear
	Superior cerebellar	Laterally, highest of cerebellar arteries	Superior surface of cerebellum
	Posterior cerebral	Between temporal and occipital lobes	Medial and inferior portions of occipital and temporal lobes

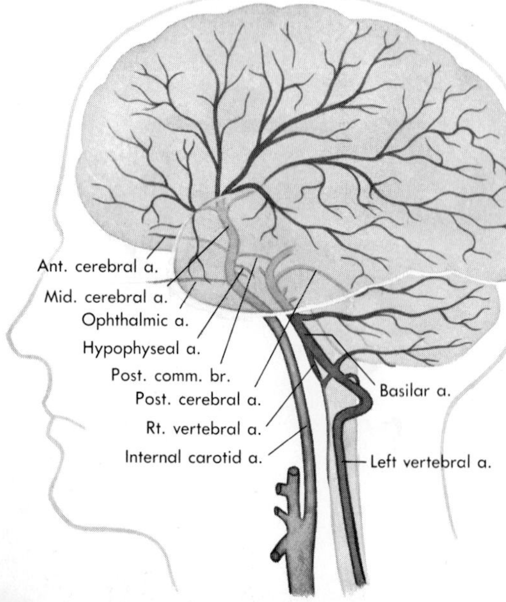

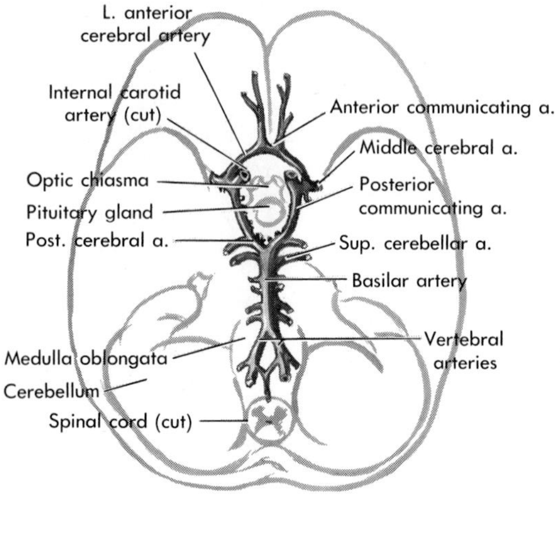

transverse sinus swings laterally on each side from the confluence in the line of attachment of the tentorium cerebelli. The sinus turns downward at the margin of the petrous portion of the temporal bone. The same sinus is now called the *sigmoid sinus* as it leaves the tentorium cerebelli on an "S" shaped course to the jugular foramen. Below the jugular foramen the sigmoid sinus is continuous with the internal jugular vein whereby the venous blood from the system of sinuses is returned toward the heart.

The *cavernous sinus* is a venous pocket between the endosteal and meningeal layers of dura mater which extends along the side of the body of the sphenoid bone from the superior orbital fissure to the apex of the petrous portion of the temporal bone. This passageway receives blood from the superior orbital veins, from a small *sphenoparietal sinus,* and from several cerebral veins. The sinuses of the two sides are connected by one or two *intercavernous sinuses.* The cavernous sinuses are drained by *superior* and *inferior petrosal sinuses* which cross the petrous bone. The superior sinus swings laterally to join the transverse sinus just before it becomes the sigmoid sinus.

The inferior sinus angles sharply downward

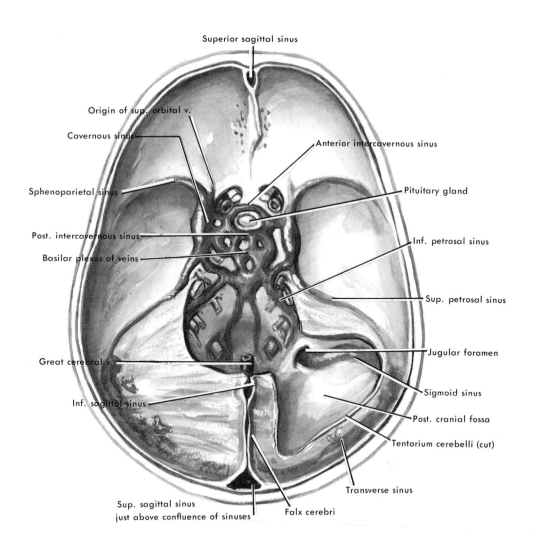

nearer the midline to join either the sigmoid sinus in the jugular foramen or the internal jugular vein.

PERIPHERAL NERVOUS SYSTEM

The peripheral nervous system is composed of the nerves and the ganglia associated with them. Although an arbitrary distinction may be made at the surfaces of the brain and spinal cord between the central and peripheral divisions of the nervous system, they are continuous and are unified functionally. The peripheral nervous system consists of a series of nerves which, to the eye, appear to issue from the surface of the brain stem and spinal cord at regular intervals. (See color plate 13.) Many fibers within those nerves, however, are *entering* the central nervous system in the dorsal roots just as others are *leaving* it in the ventral roots. The peripheral nervous system is divided into cranial and spinal parts, containing respectively, *cranial* and *spinal nerves*. A further part of the peripheral nervous system is the autonomic nervous system which is not a separate morphological entity but a grouping of special nerve components which travel within certain spinal and cranial nerves to innervate smooth muscle, cardiac muscle, and glands.

SPINAL NERVES

There are 31 pairs of spinal nerves which are grouped in accordance with the vertebrae and the intervertebral foramina between them. There are seven cervical vertebrae but there are eight pairs of cervical spinal nerves. This discrepancy results because the first cervical nerve leaves the upper end of the vertebral canal between the base of the skull and the first cervical vertebra. The second to seventh cervical nerves also leave the vertebral canal *above* their corresponding vertebra which causes the eighth cervical nerve to leave between the seventh cervical and first thoracic vertebrae. A more standard pattern begins with the thoracic nerves. Twelve pairs of thoracic and five pairs of lumbar spinal nerves arise from corresponding segments of the spinal cord in order, and leave the vertebral canal *below* their correspondingly numbered vertebrae. There are five pairs of sacral nerves corresponding to the five fused segments of the sacrum. Only one coccygeal nerve is usually found.

Spinal Nerve Roots

The disparity between the length of the spinal cord and the length of the vertebral column influences the disposition and length of the spinal nerve roots. Only the uppermost cervical roots have short, transverse courses to their respective intervertebral foramina. The roots below this level become increasingly longer and descend at sharper angles to attain the level of their foramina of exit.

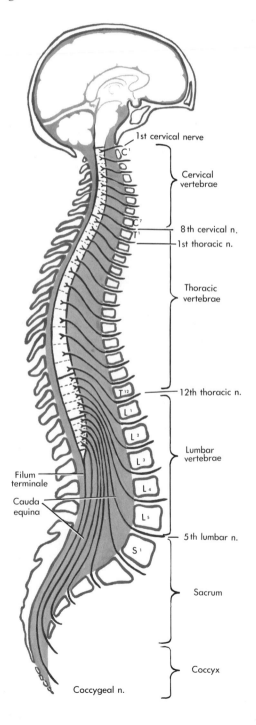

The lower end of the spinal cord issues lower lumbar and sacral roots at segmental intervals but is also shrouded by roots descending within the vertebral canal from levels above to reach their exit levels. Part of the cauda equina is formed by lumbar roots, as well as sacral roots, traveling far below the end of the spinal cord.

Plan of a Typical Spinal Nerve

A thoracic spinal nerve contains all of the elements and exhibits the structural pattern of spinal nerves in general.

Formation of the Spinal Nerve.   The axons of ventral gray column neurons issue as a series of fine filaments which join to form larger rootlets. The anterior rootlets join to form the anterior motor root which passes outward and descends within the dural sac. On the opposite side of the denticulate ligament the fibers in the posterior sensory root proceed upward and centrally to enter the same segmental level of the spinal cord. The dorsal root ganglion is located in or near the intervertebral foramen. At this point the two roots are outside the dural sac but are covered by a prolongation of dura mater. The anterior root is applied to the surface of the dorsal root

ganglion in the foramen, as are blood vessels entering and leaving this level of the vertebral canal. The anterior and posterior roots blend peripheral to the ganglion to form the single trunk of the *spinal nerve.*

Peripheral Course and Branches.   Spinal nerves are also called *mixed nerves* because of the mixture of sensory and motor fibers within them. Just after the spinal nerve is formed it divides into two major branches. The *posterior primary division* of the spinal nerve (posterior ramus) turns sharply backward into the long muscles of the back to supply these muscles and the superficial tissues overlying them. The major portion of the spinal nerve continues peripherally as the *anterior primary division* (anterior ramus).

Just after the common spinal nerve trunk is formed, thoracic and lumbar spinal nerves give off a slender strand of myelinated autonomic nerve fibers called the *white ramus communicans.* These fibers pass to a long strand of special ganglia and nerve fibers, which courses down along the anterolateral aspect of the bodies of the thoracic and upper lumbar vertebrae. This strand is the *sympathetic trunk,* a part of the autonomic system. In some cases all of the nerve fibers which leave the spinal nerve to form the white ramus remain together. They pass into the anterior pri-

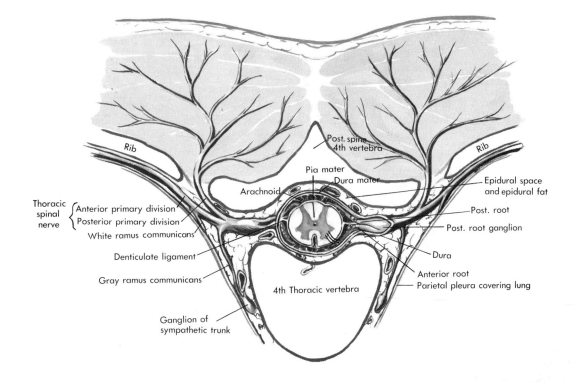

mary division so that the white ramus appears to branch from the anterior division instead of from the common spinal nerve trunk. The ultimate effect is the same. Usually a slender *gray ramus communicans* of unmyelinated autonomic fibers returns from the sympathetic trunk, at the same segmental level, to bring fibers for the innervation of smooth muscle and glands of the body part supplied by that particular spinal nerve. The gray ramus communicans usually rejoins the common spinal nerve trunk closer to the union of dorsal and ventral roots than the more distal departure of the white ramus. In this way the returning gray ramus fibers can pass into both anterior and posterior primary divisions, both of which distribute autonomic fibers to the parts of the body which they serve. It might be helpful to refer to the description of autonomic neurons on pages 282 to 285 at this time.

The anterior division then passes outward. The thoracic spinal nerve is a *peripheral nerve,* a term applied only when a nerve has acquired all the component fibers required for innervation of its area of distribution. The thoracic spinal nerves curve about the trunk supplying structures of the body wall. The anterior divisions of the thoracic nerves are called *intercostal nerves* because of their close relationship to the ribs in part or all of their courses.

Plexuses

Spinal nerve roots contain all the fibers leaving or entering a particular segment of the spinal cord. Gray matter, however, forms long columns within the cord. Sensory connections for posterior root fibers may extend across several segments. Similarly the anterior horn motor cells for a muscle, or a related group of muscles, may extend over several segments of the ventral gray column and the axons may issue from the spinal cord in several spinal nerves. It is necessary for all the component fibers concerned with the innervation of structures in a local part of the body to be gathered into a nerve destined for that area. This is accomplished in several great interweavings of nerves which are found in the cervical and lumbosacral regions. Each intermingling of fibers where nerve components are sorted and channeled into a peripheral nerve is called a *nerve plexus.*

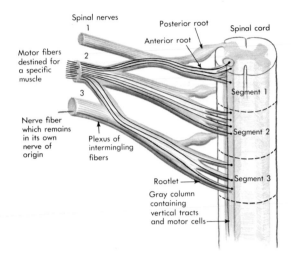

Nerve fibers traveling in the lower cervical and upper thoracic spinal nerves are usurped from destinations in the neck or thorax during development for innervation of the upper extremity. These fibers are sorted and channeled into the major nerves of the limb in the *brachial plexus.* A similar *lumbosacral plexus* groups lumbar and sacral nerves which innervate the lower limb. A minor plexus, the *cervical plexus,* is formed in the neck from upper cervical spinal nerves.

Segmentation in Peripheral Nerves

A remarkable persistence of the segmental distribution of nerve fibers is seen in the sensory supply of the skin. The skin is innervated in a pattern of strips and bands winding obliquely about the body. Such a skin area supplied mainly from one spinal cord segment through a spinal nerve is called a *dermatome.* Although one spinal nerve is the main nerve conveying sensory impulses from a dermatome, each dermatome receives sensory innervation from the nerve of the dermatomes on each side of it. The overlap of innervation means that if the function of one spinal nerve is impaired, the sensation of its corresponding skin area is rarely lost completely. Segmentation of motor supply also exists but it is not as sharply demarcated. Nevertheless, muscles of similar location and related function are frequently innervated from the same or several adjacent segments of the spinal cord. The motor fibers from these segments reassemble through plexuses into the same peripheral nerve.

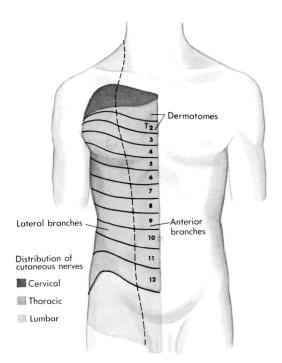

Dermatomes

T2
3
4
5
6
7
8
9
10
11
12

Lateral branches

Anterior branches

Distribution of cutaneous nerves

■ Cervical

▨ Thoracic

▧ Lumbar

the long posterior cervical muscles themselves are supplied by a series of posterior rami or *posterior cervical nerves* derived from the second to the eighth cervical nerves.

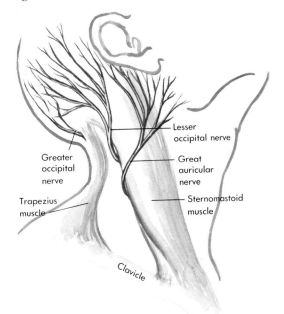

Lesser occipital nerve

Greater occipital nerve

Great auricular nerve

Trapezius muscle

Sternomastoid muscle

Clavicle

DISTRIBUTION OF THE SPINAL NERVES

Cervical Nerve Distribution

During development the upper limb bud draws muscle masses and their accompanying nerves from the lower cervical and first thoracic segments of the body. This results in the drawing of most of the fibers of the anterior primary divisions of cervical nerves 5 to 8 and the first thoracic nerve into the upper limb. Only the posterior primary divisions of the cervical nerves and the anterior primary divisions of the first four cervical nerves remain to supply neck structures.

Posterior Cervical Nerves. The posterior divisions of the first and second cervical nerves curve about the base of the skull, innervating the suboccipital muscles. The posterior division of C1 is known as the *suboccipital nerve.* The posterior division of C2, a branch known as the *great occipital nerve,* supplies the skin over the back of the head as high as the vertex of the skull. The posterior divisions of C2 and C3 join to supply the skin of the back of the ear, the mastoid area, and the upper part of the back of the neck. The skin over the lower posterior cervical region and

Cervical Plexus. The anterior primary divisions of the first four cervical nerves join to form the *cervical plexus,* from which cervical nerves emerge to supply the anterior cervical muscles as well as an extensive skin area. A complicated series of branches and loops results, because the nerves of the cervical plexus supply not only their own upper cervical segmental muscles and overlying skin, but also the anterior part of the lower cervical segments and the first thoracic segment, whose nerve supply was usurped by the upper limb in development. The following peripheral nerves are formed from the cervical plexus:

1. *Lesser occipital nerve:* This nerve winds upward around the posterior margin of the sternocleidomastoid muscle to supply the skin of the posterior (cranial) surface of the external ear and adjacent skin of the back of the head.

2. *Great auricular nerve:* This is a similar nerve, which winds upward across the sternocleidomastoid muscle to supply the skin just anterior to the external ear over the parotid gland and the skin of both surfaces of the lower part of the external ear.

3. *Supraclavicular nerve:* The posterior, middle, and anterior supraclavicular nerves spread out over the lateral and anterior aspects of the neck.

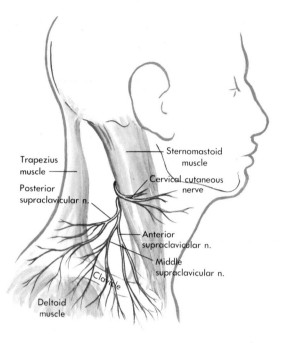

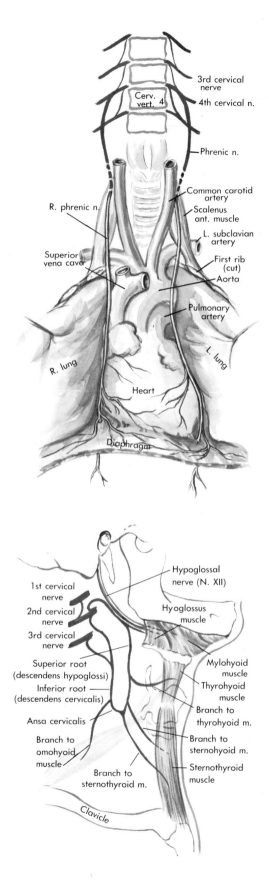

These nerves supply the skin of the side of the neck as well as that of the front of the body from the lower margin of the jaw to the upper thoracic region. The lower fibers of these nerves descend below the level of the clavicle to innervate the skin of the deltoid region, the clavicular area, and the pectoral region, as low as the level of the second rib.

4. *Phrenic nerve:* Branches from the third, fourth, and fifth cervical nerves are joined into a long strand carrying fibers which descend to innervate the musculature of the diaphragm and convey sensory impulses from part of its pleura. (See illustration at top of the right column.)

5. *Ansa cervicalis:* A long loop of motor fibers is formed from the first, second, and third cervical nerves. This loop, from which motor branches pass to the strap muscles of the neck, the scalene muscles, and the prevertebral muscles, is called the *ansa cervicalis.* Two roots comprise the loop. The *superior root,* sometimes called the descendens hypoglossi, seems to be a branch of the twelfth (hypoglossal) cranial nerve. The fibers in the superior root, however, are really from the first cervical nerve which traveled for a short distance with the cranial nerve and then descended from it. The *inferior root,* sometimes called the *descendens cervicalis,* receives fibers from the second and third cervical nerves. This strand descends to join the superior root in a U-shaped loop, often separately called the *ansa hypoglossi.*

The Brachial Plexus

The anterior primary divisions of the fifth to eighth cervical nerves and of the first thoracic nerve enter into the formation of the *brachial plexus* which forms peripheral nerves that innervate specific structures of the upper limb. Nerves to the shoulder and pectoral muscles also arise from the plexus.

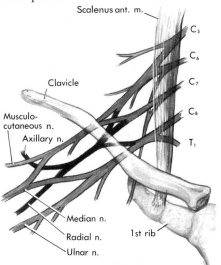

The brachial plexus is best understood if it is analyzed into portions as it forms roots, trunks, divisions, cords, and terminal branches. Branches that are given off before the terminal branches are formed are called collateral branches.

Roots

The *roots* of the brachial plexus are the anterior primary divisions of cervical nerves 5 to 8 and the first thoracic. Sometimes a C4 contribution is found. These come out into the neck between the anterior and middle scalene muscles. Collateral muscular branches pass from each of the cervical roots to the scalene and longus colli muscles. Similarly the short first thoracic intercostal nerve proceeds from the first thoracic nerve to follow the first rib. The *dorsal scapular nerve* passes from the C5 root to the rhomboid muscles. Fibers leave each of the cervical roots to form a long strand, the *nerve to the serratus anterior* (also known as the long thoracic nerve).

Trunks

The fifth and sixth cervical roots join to form the *upper trunk* of the brachial plexus. A collateral branch, the *nerve to the subclavius muscle,* leaves this trunk. The *suprascapular nerve* also leaves the upper trunk to the supraspinatus and infraspinatus muscles.

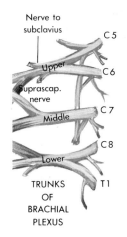

The seventh cervical root continues outward alone to become the *middle trunk*. The eighth cervical and first thoracic roots unite to form the *lower trunk*. The brachial plexus now consists of the three large trunk cables.

Divisions

Each of the trunks splits into an *anterior* and a *posterior division*. This is one of the most important features of the brachial plexus, for in the anteroposterior splitting, fibers are sorted into groups from each trunk which will innervate structures of either the preaxial (flexor, anterior) or the postaxial (extensor, posterior) compartments or surfaces of the upper extremity. The three posterior divisions join to form the posterior cord through which fibers from all roots of

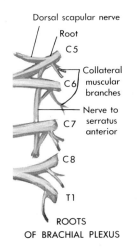

ROOTS
OF BRACHIAL PLEXUS

the plexus pass to innervate the posterior muscles of the brachium and the extensor muscles of the upper extremity. The anterior divisions of the upper and middle trunks, carrying fibers from C5, C6, and C7, join to form the *lateral cord* while the anterior division of the lower trunk, carrying C8 and T1 fibers, continues as the *medial cord.* The anterior divisions carry fibers which innervate the anterior muscles of the brachium and the flexor mechanisms of the upper limb.

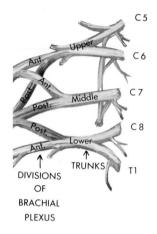

Cords

The *cords* of the brachial plexus are located in the axilla surrounding the axillary artery in positions indicated by their names. The cords give rise to terminal branches which are the main peripheral nerves of the upper limb, but before doing so they send out collateral branches.

The *lateral cord* gives rise to the *lateral pectoral nerve* containing C5, C6, and C7 fibers. The *medial cord* contains fibers of C8 and T1 origin which leave the cord as the *medial pectoral nerve*. These two nerves usually form a loop from which branches pass to the pectoralis major and minor muscles.

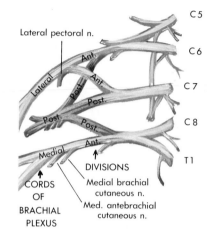

The medial cord also forms two collateral nerves, the *medial brachial cutaneous* and *medial antebrachial cutaneous nerves,* which supply the skin of the medial aspect of the arm and forearm. The fibers of the posterior cord divide as some pass into the *axillary nerve* and others continue distally as the *radial nerve*. The axillary nerve, sometimes known as the circumflex nerve, is made up of C5 and C6 fibers. It curves posteriorly to supply the deltoid and teres minor muscles.

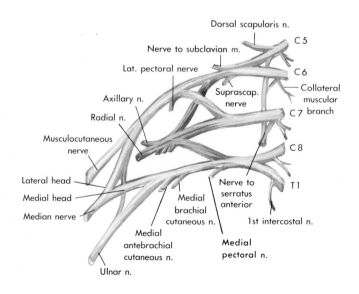

Nerves of the Upper Limb

The musculocutaneous, median, ulnar, and radial nerves leave the brachial plexus in the axilla to innervate all of the structures of the upper limb.

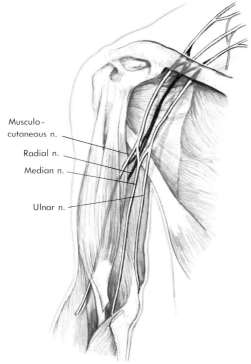

Musculo-cutaneous n.
Radial n.
Median n.
Ulnar n.

Musculocutaneous Nerve. The *musculocutaneous nerve* (C5 and C6) pierces the coracobrachialis muscle, crosses the brachium obliquely laterally between the biceps brachii and coracobrachialis muscles, and becomes a cutaneous nerve of the lateral surface of the forearm. It supplies motor fibers to the coracobrachialis, biceps brachii, and brachialis muscles.

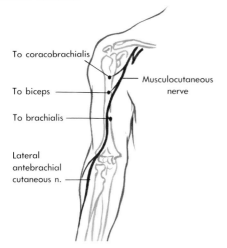

To coracobrachialis
To biceps
To brachialis
Musculocutaneous nerve
Lateral antebrachial cutaneous n.

Median Nerve. The *median nerve* has no branches in the brachium. It passes deep to the bicipital tendon in the antecubital fossa and then courses in the midline of the forearm on the deep surface of the flexor digitorum superficialis muscle to the hand. The median nerve carries fibers from all segments entering into the brachial plexus. This nerve is the great flexor-pronator nerve of the forearm and hand and the nerve of motor supply to the thumb muscles.

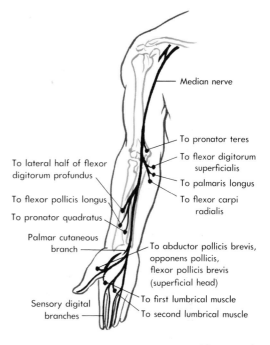

Median nerve
To pronator teres
To flexor digitorum superficialis
To palmaris longus
To flexor carpi radialis
To lateral half of flexor digitorum profundus
To flexor pollicis longus
To pronator quadratus
Palmar cutaneous branch
To abductor pollicis brevis, opponens pollicis, flexor pollicis brevis (superficial head)
To first lumbrical muscle
To second lumbrical muscle
Sensory digital branches

The muscles supplied by the median nerve in the forearm include both pronators and all of the flexor muscles of the front of the forearm *except* the flexor carpi ulnaris and the medial half of the flexor digitorum profundus. In the hand it innervates the majority of the thenar muscles to the thumb: the abductor pollicis brevis, opponens pollicis, superficial head of the flexor pollicis brevis as well as the lateral two lumbrical muscles. The digital branches convey sensory fibers from the palmar aspect of the thumb: the second and third fingers and the lateral half of the fourth finger; *and* the dorsal surface of the tips of the same fingers. Sensation of the lateral half of the palm up to the wrist crease is conveyed by the median nerve.

Ulnar Nerve. The *ulnar nerve* (C8 and T1) leaves the axilla in the neurovascular compartment of the brachium but, midway down the upper arm, pierces the medial intermuscular septum to enter the posterior compartment. It does not carry sensory fibers from the arm or forearm. The medial brachial cutaneous and medial antebrachial cutaneous nerves serve this function along the medial aspects of the upper limb. The nerve winds behind the medial epicondyle of the humerus and then courses distally on the medial side of the forearm in close association with the ulnar artery and flexor carpi ulnaris tendon. The ulnar nerve provides motor fibers for the flexor carpi ulnaris muscle and the medial half of the flexor digitorum profundus muscle. It innervates the interosseous muscles and the medial two lumbrical muscles.

along the radial groove into the posterior compartment of the brachium. The radial nerve innervates the triceps and anconeus muscles and a part of the brachialis. A sensory posterior cutaneous branch carries sensory fibers from the posterior aspect of the brachium. The nerve innervates the muscles arising from the lateral humeral epicondyle which includes the extensor muscles of the wrist and the supinator. As the nerve passes down the forearm it completes the extensor nerve supply through branches to the extensor muscles of the fingers, the thumb extensors, and the long abductor muscle of the thumb.

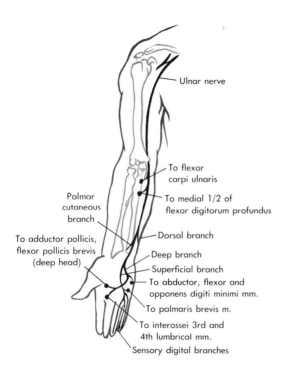

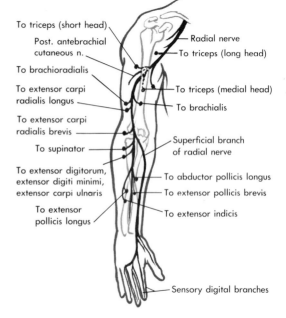

All of the hypothenar muscles, the palmaris brevis, the deep head of the flexor pollicis brevis, and the adductor pollicis muscle are supplied by this nerve. The role of the ulnar nerve in innervating so many of the intrinsic muscles of the hand makes it extremely important in the delicate movements of the hand. Sensory fibers from the ulnar side of the palm, both surfaces of the fifth finger, and the adjacent side of the fourth finger are carried by the ulnar nerve.

Radial Nerve. The *radial nerve* is the great extensor-supinator nerve of the upper limb. Carrying fibers from all of the brachial plexus segments, the nerve winds around the humerus

The radial nerve carries sensory fibers from the posterior aspect of the forearm; half of the dorsum of the hand; and the dorsal aspect of the thumb, second finger, and the lateral half of the fourth finger *except* for the skin over the distal phalanges (supplied by the median nerve).

DISTRIBUTION OF THE THORACIC NERVES

The posterior primary divisions of each thoracic spinal nerve pass posteriorly with medial and lateral branches forming a linear segmental series supplying the long muscles of the back and the skin of the back as far laterally as the midscapular line. The anterior primary division courses laterally from the vertebral column as an *intercostal nerve*. The intercostal nerve follows the upper thoracic interspaces around the trunk innervating the intercostal and other thoracic muscles and the parietal pleura. The lower intercostal nerves leave to supply the parietal structures of the abdomen. Each nerve has lateral and anterior cutaneous branches which convey sensory fibers from the skin of its dermatome.

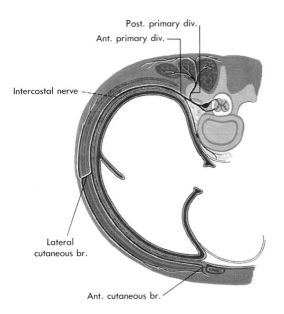

The intercostal nerve distribution varies at certain specific levels as follows:

1. The *first thoracic* nerve sends most of its fibers to become the lowest root of the brachial plexus.

2. The *second thoracic nerve* forms a large lateral cutaneous branch which bridges the axillary fossa. It carries sensory fibers from the skin of the axillary fossa and the uppermost medial aspect of the brachium. The lateral cutaneous branch of the *third thoracic nerve* similarly supplies the skin of the medial wall of the axilla.

3. The *third to sixth thoracic nerves* are the most typical as they pass around their interspaces close to the under surface of their rib, until they divide to form the lateral cutaneous branches and nerves to the intercostal and thoracic muscles or parietal pleura. These nerves proceed almost to the lateral border of the sternum, finally sending anterior branches through the attachments of the pectoralis major muscle to form a linear anterior cutaneous series near the midline.

4. The *seventh to eleventh thoracic nerves* are called the *thoracoabdominal nerves* because they supply trunk muscles which extend across or around the two regions. These nerves pass from the intercostal spaces at the medial ends of the ribs either where the fused costal cartilages begin or where the caps of the lowest unattached ribs appear. The nerves course between the transversus abdominis and internal oblique muscles, innervating them and sending branches to the external oblique muscle. These nerves end near the midline by supplying the rectus abdominis muscle. Each nerve sends lateral and anterior cutaneous branches to the skin.

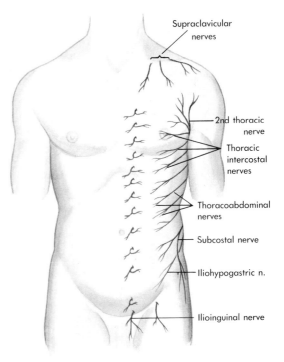

5. The *twelfth thoracic nerve,* curving laterally under the short twelfth rib, is termed the subcostal nerve because it has no intercostal course. This nerve sweeps low on the anterolateral aspect of the abdomen.

LUMBAR NERVES AND THE LUMBOSACRAL PLEXUS

The lumbar nerves, with the exception of the first, enter into association with the upper sacral nerves to innervate the lower limb.

First Lumbar Nerve

The posterior primary division joins the segmental series innervating the back muscles and the skin overlying them. The anterior primary division divides to form the *iliohypogastric* and *ilioinguinal nerves*. These nerves sweep around the abdomen as sensory nerves of the anterolateral abdominal wall and the genital organs below the level of distribution of the subcostal nerve. The *iliohypogastric nerve area* is the region just above the pubic bone.

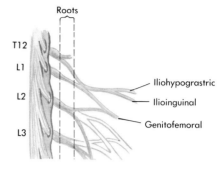

These are the *roots* of the lumbar plexus. A branch of the first lumbar nerve joins with a branch from the root of the second to form the *genitofemoral nerve* (L1 and L2) which supplies the spermatic cord in the male, the roull ligament and labium majus of the female, and skin of the front of the thigh over the femoral triangle.

The roots from the second, third, and fourth lumbar nerves divide into *anterior* and *posterior*

The *ilioinguinal nerve* curves around into the inguinal canal and, emerging from the superficial inguinal ring, supplies the skin of the root of the penis and upper scrotum of the male or the labium majus of the female.

Lumbosacral Plexus

The anterior primary divisions of the five lumbar and the first four sacral nerves form the *lumbosacral plexus* from which terminal nerves spring to innervate the lower limb and pelvic structures. The intermingling of fibers is not as regular as in the brachial plexus. The simplified scheme illustrated in the accompanying figures should be followed as the main points are presented.

The lumbar portion of the plexus, sometimes called the *lumbar plexus*, is formed as the first four lumbar nerves pass into the psoas major muscle.

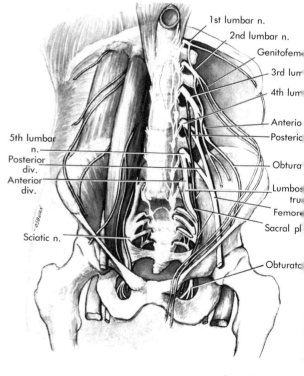

from Healey

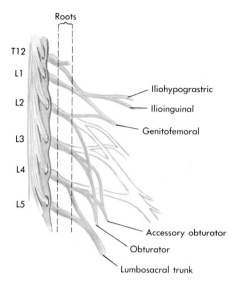

Roots

T12
L1
L2
L3
L4
L5

Iliohypograstric
Ilioinguinal
Genitofemoral

Accessory obturator
Obturator
Lumbosacral trunk

nerve also carries sensory fibers from the skin of the front of the thigh below the femoral triangle, the medial side of the thigh and leg, and the hip and knee joints. Branches from the posterior divisions of the second and third lumbar nerves form the *lateral cutaneous nerve of the thigh*.

The *anterior divisions* of the second, third, and fourth lumbar nerves unite to form the *obturator nerve* (L2, L3, L4) which follows the lateral wall of the pelvis to the obturator foramen. This nerve enters the thigh to supply the adductor group of muscles and the gracilis. Sensory fibers come from the skin of the medial aspect of the thigh and from the hip and knee joints.

divisions. The *posterior divisions* of these nerves unite to form the *femoral nerve* (L2, L3, L4) which passes under the inguinal ligament and breaks up into a fan of individual nerves which innervate the muscles of the front of the thigh: the quadriceps femoris group, sartorius, and pectineus. The

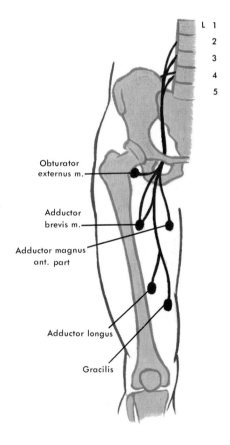

L 1
2
3
4
5

Obturator externus m.
Adductor brevis m.
Adductor magnus ant. part
Adductor longus
Gracilis

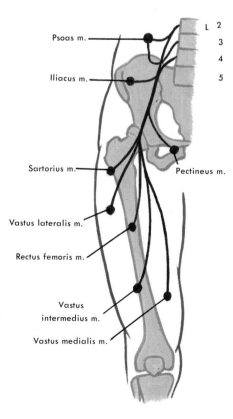

Psoas m.
Iliacus m.

L 2
3
4
5

Sartorius m.
Pectineus m.
Vastus lateralis m.
Rectus femoris m.
Vastus intermedius m.
Vastus medialis m.

The fourth lumbar root enters into a second plexus as it sends fibers to join with the fifth lumbar nerve in the formation of the *lumbosacral trunk*. The lumbosacral trunk then joins with the anterior primary divisions of the upper four sacral nerves to form the *sacral plexus*. The lumbosacral trunk and the upper *three* sacral nerves divide into anterior and posterior divisions which are somewhat obscured, since all of the fibers are bundled together to form the broad flattened band of the *sciatic nerve* (L4, L5, S1, S2, S3).

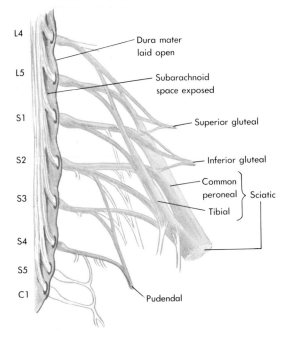

The divisions remain separate in the sciatic nerve, however, and when the sciatic nerve divides, the fibers of the posterior divisions pass into the lateral popliteal nerve and the anterior division fibers pass into the medial popliteal nerve.

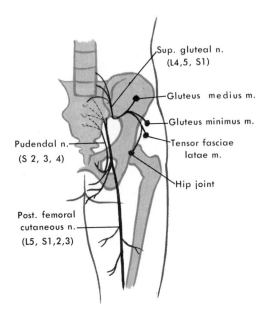

Other major nerves formed from the sacral plexus are:

1. *Pudendal nerve* (S2, S3, S4) which sends an *inferior rectal nerve* to the external sphincter muscle of the anus and surrounding skin, and the *perineal nerve* which innervates the muscles closing the pelvic outlet.

2. *Gluteal nerves, superior* (L4, L5, S1) and *inferior* (L5, S1, S2), wind backward through the greater sciatic foramen to the gluteal muscles.

3. Other branches course from the sacral plexus to the small muscles of the hip joint.

4. Pelvic autonomic fibers contribute to autonomic plexuses which innervate the viscera of the pelvis.

5. The posterior divisions of the upper three sacral nerves contribute fibers which join to form the *posterior femoral cutaneous nerve.* This nerve carries sensory impulses via many branches from the skin of the buttock, perineum, and back of the thigh and calf of the leg.

Nerves to the Posterior Abdominal Muscles

The psoas major muscle is innervated by nerves from the L2 to L4 roots to the lumbar plexus. The iliacus portion of the iliopsoas muscle is supplied from the femoral nerve. Nerves from the L1 to L4 roots to the lumbar plexus pass to innervate the quadratus lumborum muscle which is also supplied by the subcostal nerve.

Nerves to the Lower Limb

Three major nerves provide for the innervation of the muscles of the lower limb—the femoral, obturator, and sciatic nerves. Sensory fibers are also carried in these nerves.

Femoral and Obturator Nerves. The muscles supplied by the femoral and obturator nerves have been listed in the preceding section. Loss of function of the femoral nerve will principally impair extension of the knee joint, whereas paralysis of muscles innervated by the obturator nerve will affect adduction of the hip joint.

Sciatic Nerve. The *sciatic nerve* is the great nerve of the lower extremity. After passing out of the greater sciatic foramen, the nerve courses between the greater trochanter of the femur and the ischial tuberosity. When the nerve divides in the thigh, the lateral popliteal division continues as the *common peroneal nerve,* while the medial division becomes the *tibial nerve.* Within the thigh the sciatic nerve sends branches to the posterior thigh muscles which are collectively called the *nerve to the hamstring muscles.*

The *tibial nerve* retains the posterior position of the parent sciatic nerve. It descends through the popliteal fossa to take a position deep to the calf muscles. Among the major distributions of the tibial nerve are:

1. Motor fibers to the posterior leg muscles, including the triceps surae group, plantaris, popliteus, tibialis posterior, and the long flexor muscles to the toes.

2. Sensory fibers coming from the skin over the calf, heel, and sole of the foot.

3. The terminal branches forming the *lateral* and *medial plantar nerves* whose motor fibers supply all of the plantar muscle groups of the sole of the foot.

Disturbances in function, which depend on the level of injury to the tibial nerve, may impair plantar flexion of the foot, flexion of the toes, and sensation of the back of the leg and sole of the foot.

The *common peroneal nerve* descends to the popliteal fossa where it divides into the *deep* and *superficial peroneal nerves,* which supply structures of the lateral and anterior aspects of the leg and the dorsum of the foot. The *deep peroneal nerve* follows the interosseous membrane, supplying the tibialis anterior muscle and the long extensor muscles of the toes. The extension of the deep peroneal nerve into the dorsum of the foot is sensory except for innervation of the extensor digitorum brevis muscle.

The superficial peroneal nerve descends laterally in front of the fibula to supply the peroneus longus and brevis muscles, whereupon it crosses the front of the foot, thus providing cutaneous nerves to the dorsum of the foot and to the toes with the assistance, medially, of the terminations of the deep peroneal nerve. Functional disturbances in the case of peroneal nerve injuries include impaired dorsiflexion of the foot, reduced or lost eversion of the foot, and sensory loss over the lateral aspect of the leg and dorsum of the foot.

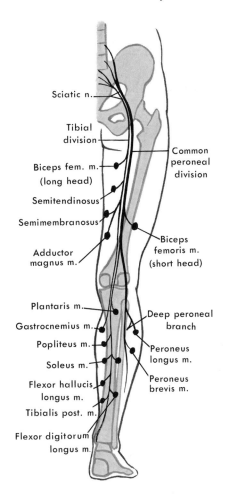

AUTONOMIC NERVOUS SYSTEM

Despite its name, the autonomic nervous system is neither a separate nervous system nor a sharply demarcated part of the central and peripheral nervous systems. The autonomic nervous system is a two-neuron chain, with the nerve cell centers for the first neuron in the brain and spinal cord, and the cell bodies of the second neuron located in ganglia outside of the central nervous system. The fibers of the second neuron innervate smooth muscle, cardiac muscle, and glands. (See color plate 14.) Autonomic nerve fibers reach these tissues by entering the nerves of the peripheral nervous system.

The occurrence of smooth muscle and glandular tissue in the circulatory, respiratory, digestive, and genitourinary systems indicates that these visceral systems are innervated by autonomic nerves. The functions of the autonomic nervous system, therefore, include the control of circulation, heart rate, respiration, secretion, and motility of organs which contain smooth muscle.

Autonomic Outflows

The hypothalamus, with its connections to the pituitary gland, controls major centers regulating respiration, circulation, heart rate, smooth muscle contraction, and secretion which are located in the brain stem and in the lateral gray matter of the spinal cord. Two great groups of fibers, called *outflows,* enter the peripheral nervous system from the centers in the brain stem and spinal cord. (See color plate 14.) The *thoracolumbar outflow* (formerly called the *sympathetic nervous system*) enters the sympathetic trunk of the thorax and abdomen from the thoracic and upper lumbar segments of the spinal cord. The *craniosacral outflow* (formerly called the *parasympathetic nervous system*) is not associated with the sympathetic trunk. Fibers are carried in certain cranial nerves and in the second and third sacral nerves. Its fibers, upon reaching the region of the organs to be supplied, enter into complex ganglia and plexuses.

Autonomic Neurons

The autonomic nervous system differs from the rest of the peripheral nervous system in that its fibers comprise a two-neuron chain with an intermediate synapse between the central nervous system and the effector organ. Basic differences exist in the neuronal paths and synapses of the two outflows.

Thoracolumbar Neurons. The first motor cell body is located within the lateral gray column of the thoracic or lumbar spinal cord. Its axon leaves the cord in the anterior root. This *preganglionic fiber* utilizes the white ramus communicans to reach the sympathetic trunk.

The *sympathetic trunk* is a long chain of ganglia, connected by strands of fibers, which is located beside the bodies of thoracic and lumbar vertebrae. (See color plate 14.) The sympathetic trunk is continued into the neck by connecting strands between the first thoracic sympathetic ganglion and the inferior cervical ganglion. These two ganglia may be fused into one cervicothoracic or *stellate* ganglion. A middle and a

superior cervical sympathetic ganglion, located higher in the neck, are connected to the lower cervical ganglion by strands of fibers. (See illustration on opposite page.)

Upon attaining the sympathetic trunk, the preganglionic fiber may follow one of several patterns. (1) It may synapse with the second neuron of the chain in the sympathetic ganglion of the same level of its entrance to the sympathetic trunk. (2) It may pass up or down the sympathetic trunk for several segments before synapsing with the second neuron at the level of exit from the trunk. (3) Preganglionic fibers which will innervate abdominal organs leave the sympathetic trunk without synapse. These preganglionic fibers form the *splanchnic nerves* which pass out of the thorax through the diaphragm. Eventually these fibers reach one of the collateral sympathetic ganglia in the abdomen where they synapse with the second neuron. These ganglia, also called *preaortic ganglia,* from their location near main branches of the aorta, are typified by the great *celiac ganglion* at the root of the celiac artery (often called the *solar plexus* from its radiating postganglionic fibers). (4) The thoracolumbar fibers to the upper limb must pass upward to the inferior cervical ganglion in the neck because there is no outflow from the cervical spinal cord. (5) Head structures receiving a thoracolumbar innervation are supplied by preganglionic fibers ascending to the superior cervical ganglion for synapse.

Regardless of the location of the sympathetic ganglion, the fiber leaving the second cell body in the ganglion is termed the *postganglionic fiber.* Those which enter spinal nerves utilize the gray ramus communicans. The postganglionic fibers from collateral ganglia, such as the preaortic ganglia of the abdomen, travel in autonomic plexuses along blood vessels to reach the viscera. Postganglionic thoracolumbar fibers to head structures form networks on the walls of the internal carotid artery called the *internal carotid plexus.* The fibers use the arteries as highways to reach the structures they supply.

Craniosacral Neurons. The cell body of the first neuron is located within the brain stem in association with the nucleus of a cranial nerve which carries the autonomic component (cranial nerves 3, 7, 9, 10, and 11) in the case of the cranial portion. (See color plate 14.) The first cell body of the sacral portion is located in the

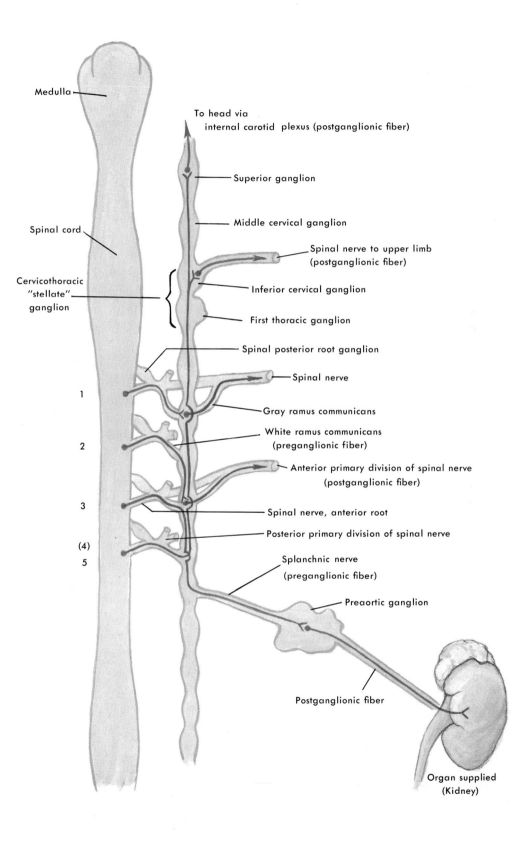

Medulla

To head via
internal carotid plexus (postganglionic fiber)

Superior ganglion

Middle cervical ganglion

Spinal cord

Spinal nerve to upper limb
(postganglionic fiber)

Cervicothoracic
"stellate"
ganglion

Inferior cervical ganglion

First thoracic ganglion

Spinal posterior root ganglion

Spinal nerve

1

Gray ramus communicans

White ramus communicans
(preganglionic fiber)

2

Anterior primary division of spinal nerve
(postganglionic fiber)

3

Spinal nerve, anterior root

Posterior primary division of spinal nerve

(4)

Splanchnic nerve
(preganglionic fiber)

5

Preaortic ganglion

Postganglionic fiber

Organ supplied
(Kidney)

lateral gray column of the second and third sacral segments. The craniosacral *preganglionic* neurons are long compared to those of the thoracolumbar outflow. The ganglia associated with the third, seventh, ninth, and eleventh cranial nerves are located along the course of the cranial nerve with a relatively short *postganglionic* neuron. The preganglionic fibers of the tenth cranial nerve (vagus) run a long course through the neck, thorax, and abdomen to synapse finally with the second neuron only in the wall of the organ supplied. The same is true of the sacral preganglionic fibers which synapse with the second neuron only in the wall of the pelvic organ supplied.

Function of the Autonomic Nervous System

Both outflows of the autonomic nervous system provide for the regulation of the visceral functions of the body. From hypothalamic and brain stem control centers flow impulses which coordinate the secretion of endocrine and exocrine glands; control temperature, circulation, and respiration; and regulate digestion and fluid intake and output. The result of all of these functions is the maintenance of the internal chemical environment of the body.

Thoracolumbar Functions. The thoracolumbar (sympathetic) outflow, in addition to inner-

DUAL AUTONOMIC INNERVATIONS

ORGAN	TISSUE	SYMPATHETIC (THORACOLUMBAR OUTFLOW)	PARASYMPATHETIC (CRANIOSACRAL OUTFLOW)
Eye	Smooth muscle of pupil	Pupil dilated	Pupil constricted
	Smooth muscle of lens	Lens made thinner for far vision	Lens made thicker for near vision
Lacrimal gland	Glands	Stimulates secretion and dilates blood vessels	Inhibits secretion and constricts blood vessels
Salivary glands	Glands	Inhibits secretion and constricts blood vessels	Stimulates secretion and dilates blood vessels
Skin	Sweat glands	Stimulates secretion	Inhibits secretion
	Muscles of hairs	Contraction to make hairs stand up	Inhibits contraction
Gastrointestinal tract	Mucous and digestive glands	Inhibits secretion	Stimulates secretion
	Smooth muscle	Inhibits peristaltic contraction	Stimulates peristaltic contraction
Heart	Cardiac muscle	Increases heart rate and blood pressure and dilates coronary arteries	Slows heart rate and constricts coronary arteries
Lung	Bronchial glands	Inhibits secretion	Stimulates secretion
	Bronchial smooth muscle	Inhibits contraction (bronchodilatation)	Stimulates contraction (bronchoconstriction)
Pelvic organs	Muscle of organ walls and alimentary, bladder, and reproductive sphincters	Inhibits contraction of organ walls; stimulates contraction of sphincters; inhibits secretion	Stimulates contraction of organ walls; relaxes sphincters; stimulate secretion
Blood vessels in general	Smooth muscle of walls	Constricts peripheral blood vessels	Dilates peripheral blood vessels

vating the smooth muscle and glands in its area of distribution, contributes major nerves to the heart, lungs, and respiratory muscle. Impulses from this outflow increase the heart and respiratory rates. Sympathetic connections to the adrenal gland increase the secretion of the adrenal hormone, epinephrine. The combination of physiological responses results in the readiness of the body to react to stress or danger.

Craniosacral Functions. The parasympathetic outflow is more concerned with specific visceral functions such as digestion, secretion of exocrine glands, and contraction of smooth muscle. Many of these functions are mediated by the long vagus nerve which has a wide distribution to alimentary organs. The sacral portion sends fibers to pelvic organs whereby excretory and reproductive mechanisms are regulated.

Dual Autonomic Supply. Certain organs of the body which contain smooth muscle and glands, as well as the cardiac muscle of the heart, are innervated by both outflows of the autonomic nervous system. The thoracolumbar (sympathetic) and craniosacral (parasympathetic) outflows may have opposite effects on smooth muscle or glands within the same organ. Furthermore, the effect of the outflows upon one organ may be opposite to that of the same outflows upon another. The responses of some typical organs are summarized in the table on opposite page.

Visceral Sensory Pathways

Autonomic motor activities are dependent upon awareness of the activity of the organs innervated. Visceral sensory impulses are carried by visceral afferent fibers in the same nerves carrying autonomic outflow fibers. The thoracolumbar sensory fibers enter the sympathetic trunk at the same level that fibers leave to innervate an organ. The cell bodies of the sensory fibers are in the dorsal root ganglia, indicating that visceral sensory fibers constitute a one-neuron chain. Craniosacral sensory fibers travel in either sacral or certain cranial nerves. The cell bodies are in the dorsal root ganglia or in the sensory ganglia of the cranial nerves. Visceral sensations relate to the degree of fullness of viscera, the amount of smooth muscle contraction, and the state of traction upon organs. Most impulses remain at the subconscious level and are related to reflex activities. Violent smooth muscle contraction or over-

distention of organs results in painful or distressing stimuli which do penetrate the conscious level to prompt voluntary responses. Remember that most anatomists ascribe only *motor* activities to the autonomic nervous system, so that visceral sensory impulses utilize afferent fibers which only *accompany* the autonomic component.

CRANIAL NERVES

The *cranial nerves* are peripheral nerves whose fibers arise from or course to the brain stem. Each cranial nerve varies in the nerve components which it carries, depending upon the structures innervated. Some of the cranial nerves are related to the organs of special sense or to muscles of special origin. These nerves carry component fibers which differ from those in spinal nerves or in other cranial nerves. Each of the cranial nerves will be described in their numbered serial order.

First (Olfactory) Cranial Nerve

Olfactory cells form the olfactory area of the nasal mucosa and act as receptor cells which initiate impulses associated with the sense of smell. Central processes of these cells extend through the roof of the nasal cavity as a series of olfactory filaments which traverse the cribriform plate. The *olfactory bulb* lies upon the cribri-

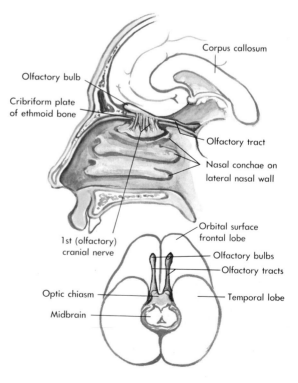

form plate. The olfactory processes synapse with the second neuron in this bulb which is the dilated peripheral end of the *olfactory tract* of the brain. (See also color plate 16.) Fibers of the olfactory tract carry the special sensory impulses backward to the tip of the temporal lobe where cortical areas are located in which the sensations of smell are perceived.

Second (Optic) Cranial Nerve

The *optic nerve* is another tract of the brain which runs peripherally as a nerve. The retina of the eye contains cell bodies of a first neuron whose peripheral processes act as light receptors. (See also color plate 15.) The central processes form the optic nerve which passes through the optic foramen. The two optic nerves enter into a partial crossing in the *optic chiasm*. The illustration shows that the special sensory fibers of an optic nerve represent the temporal and nasal halves of an individual's *field of vision*. In the optic chiasm the fibers of the nasal side of each retina cross to the opposite side, whereas the fibers from the lateral side continue posteriorly without crossing. Therefore, the fibers from the same lateral side and those from the opposite nasal side continue posteriorly as the *optic tract*.

The optic tract passes backward carrying three types of fibers which enter different parts of the brain.

1. *Visual fibers* pass into the lateral geniculate body, a part of the thalamic nuclei. Following synapse, fibers of the next neuron pass posteriorly to the visual area of the cortex in the occipital lobe.

2. *Light reflex fibers* pass into the midbrain where they synapse. Fibers of the next neuron connect with nuclei of the third cranial nerve whereby reflex impulses are conveyed to smooth muscle controlling the size of the pupil.

3. Skeletal or *somatic reflex fibers* enter the superior colliculus of the midbrain to synapse. Second neuron fibers descend to the cervical spinal cord to synapse about the anterior horn cells of nerves to the suboccipital and cervical muscles. Reflex movements can occur to move the head from danger. Other fibers pass upward to the nuclei of the third, fourth, and sixth cranial nerves to institute reflex movements of the eyes so that one can follow a moving object. Many fine coordinations of eye and neck movements go on constantly.

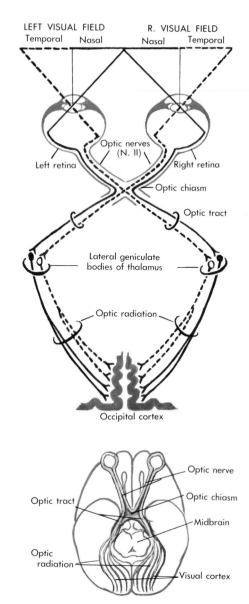

Third (Oculomotor) Cranial Nerve

This nerve is so named because it is the chief motor nerve for movement of the eye. Its fibers pass forward to leave the skull through the superior orbital fissure. It has the following components:

1. *General motor fibers* to the superior rectus, medial rectus, inferior rectus, and inferior oblique muscles of the eyeball (see section on the eye). The levator palpebrae superioris muscle is also innervated by this nerve.

2. *Parasympathetic fibers* to the muscles of the pupil and lens. These fibers are preganglionic and leave the nerve to synapse in the *ciliary ganglion* which is the parasympathetic ganglion for the

eye. Postganglionic fibers from the ganglion pass forward through the orbit in ciliary nerves to the eye.

3. *Sympathetic fibers* from the internal carotid nerve plexus pass through the ciliary ganglion as postganglionic fibers enroute to the eye.

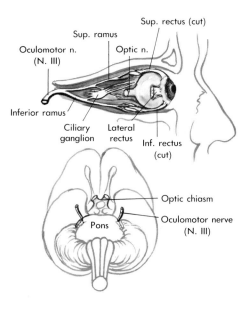

Fourth (Trochlear) Cranial Nerve

The trochlear nerve only carries general motor fibers to the superior oblique muscle of the eyeball. The slender nerve enters the orbit through the superior orbital fissure.

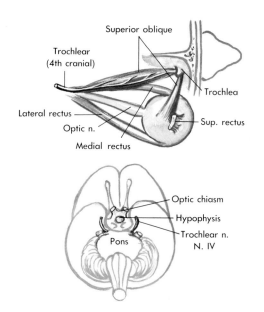

Fifth (Trigeminal) Cranial Nerve

The fifth cranial nerve is a mixed nerve which carries the general sensory fibers from the skin, teeth, and mucous membranes of the head. It also conveys special motor fibers to the muscles of mastication. The nerve emerges as motor and sensory roots from the side of the pons. The two roots travel forward together to the middle cranial fossa. The sensory root dilates and indents the side of the cavernous sinus. The broadening is the *trigeminal* or *semilunar ganglion* of the nerve which contains the cell bodies of the fibers of the sensory root.

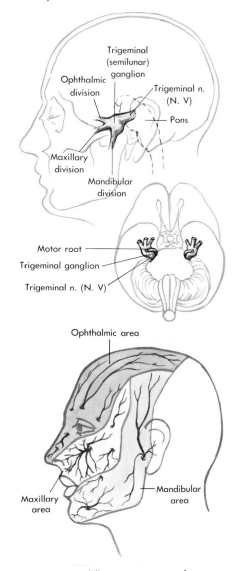

The motor root does not pass through the ganglion but is plastered to it. The sensory root is formed from three large converging branches; these are the ophthalmic, maxillary, and mandibular divisions of the nerve. The motor root joins the mandibular division to distribute motor branches to the masticatory muscles. The distribution of the sensory divisions is as follows:

1. *Ophthalmic nerve:* This branch enters the orbit through the superior orbital fissure. Its branches convey sensory impulses from the eyeball; the conjunctival sac lining the eyelids; the skin of the anterior half of the scalp, forehead, and upper eyelid; the nasal mucosa; and the frontal sinus. It innervates the lacrimal gland.

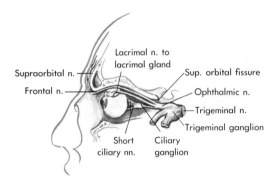

2. *Maxillary nerve:* This branch passes through the foramen rotundum into the pterygopalatine fossa. From there the nerve passes into the orbit through the infraorbital fissure to the infraorbital foramen and the face.

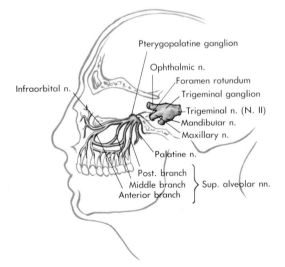

From the many branches of this nerve come general sensory impulses from the upper teeth and gums, the maxillary sinus, the mucous membrane of the palate, and the skin

of the face and upper lip. *Parasympathetic* secretory fibers to the nasal, palatine, and lacrimal glands synapse in the *pterygopalatine ganglion* which is related to the maxillary nerve in the pterygopalatine fossa. The postganglionic fibers traverse any fifth cranial nerve branches distributing to their destination. Postganglionic sympathetic fibers from the internal carotid plexus use the same routes.

3. *Mandibular nerve:* The third sensory division, together with the motor root of the nerve, passes through the foramen ovale into the infratemporal fossa. Sensory branches convey general sensory impulses from the scalp behind the ear; the lower teeth and gums; the skin of the chin, jaw, and lower lip; the mucous membrane of the oral cavity; and the anterior two-thirds of the tongue. The motor root innervates the masticatory muscles and the anterior root of the digastric and mylohyoid muscles. Special features of the mandibular nerve include parasympathetic relations and a morphological linkup with the seventh cranial nerve. Special sensory fibers for taste bud receptors of the anterior two-thirds of the tongue leave the tongue with the lingual branch of the mandibular nerve.

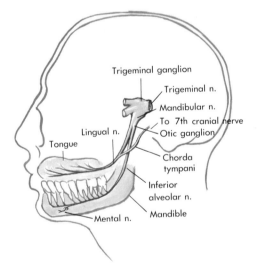

The taste fibers depart into the *chorda tympani* branch of the seventh cranial nerve. The *otic ganglion,* associated with the nerve in the infratemporal fossa, contains the cell bodies of the second neuron for parasympathetic innervation of the parotid gland. Postganglionic sympathetic fibers pass through the otic ganglion enroute to the parotid gland.

Sixth (Abducens) Cranial Nerve

This third nerve supplying the musculature of the eye enters the orbit through the superior orbital fissure to innervate the lateral rectus muscle. In addition to the general motor fibers, the sixth cranial nerve carries postganglionic sympathetic fibers.

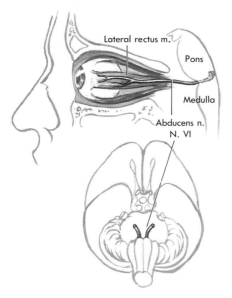

Seventh (Facial) Cranial Nerve

The *facial nerve* has been identified as the nerve of facial expression. It is a mixed nerve which emerges from the brain stem in close association with the eighth cranial nerve. Both nerves enter the petrous portion of the temporal bone through the internal acoustic meatus. The facial nerve curves above the internal ear through the *facial canal*. As it turns sharply downward and backward behind the middle ear, the *geniculate* or *facial ganglion* is located on the nerve. The nerve emerges onto the side of the face through the stylomastoid foramen. It gives rise to a number of peripheral branches between the lobes of the parotid gland. The components carried in this nerve are:

1. Special motor fibers which innervate the muscles of facial expression and scalp. Motor fibers also innervate the buccinator, platysma, posterior digastric, and stylohyoid muscles.

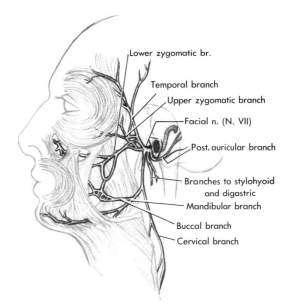

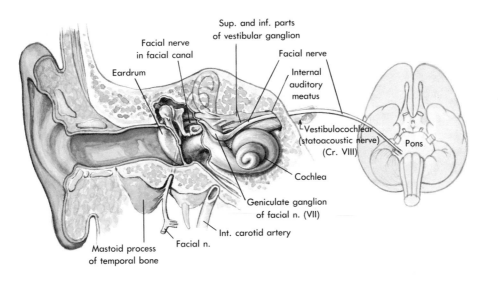

2. Special visceral sensory fibers from the taste buds of the anterior two-thirds of the tongue have their cell bodies in the geniculate ganglion. The chorda tympani nerve, running from the ganglion, joins the lingual branch of the mandibular nerve to reach the tongue. The taste fibers are carried along this route.

3. Parasympathetic fibers to the lacrimal gland and nasal mucous membrane have second cell bodies in the geniculate ganglion. The postganglionic fibers course in the *greater petrosal nerve* toward the foramen lacerum in a groove of the temporal bone. There the *deep petrosal nerve,* carrying sympathetic fibers from the internal carotid nerve plexus, joins to form the *nerve of the pterygoid canal.* The fibers reach and pass through the pterygopalatine ganglion to be distributed along fifth cranial nerve branches.

Eighth (Vestibulocochlear) Cranial Nerve

This nerve (also known as the auditory, acoustic, or statoacoustic nerve) consists of two parts, the *vestibular division* and the *cochlear division.* (See

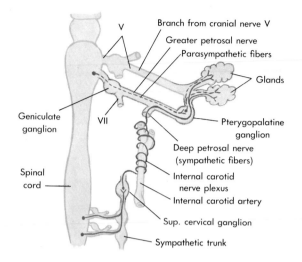

also color plate 15.) These fibers convey equilibratory and auditory sensations, respectively, special sensory in type, from the internal ear with which they are described.

Ninth (Glossopharyngeal) Cranial Nerve

The *glossopharyngeal nerve* is closely associated with parts of the tenth and eleventh nerves. The

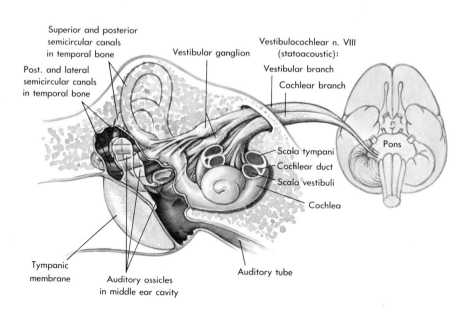

Ninth Cranial Nerve

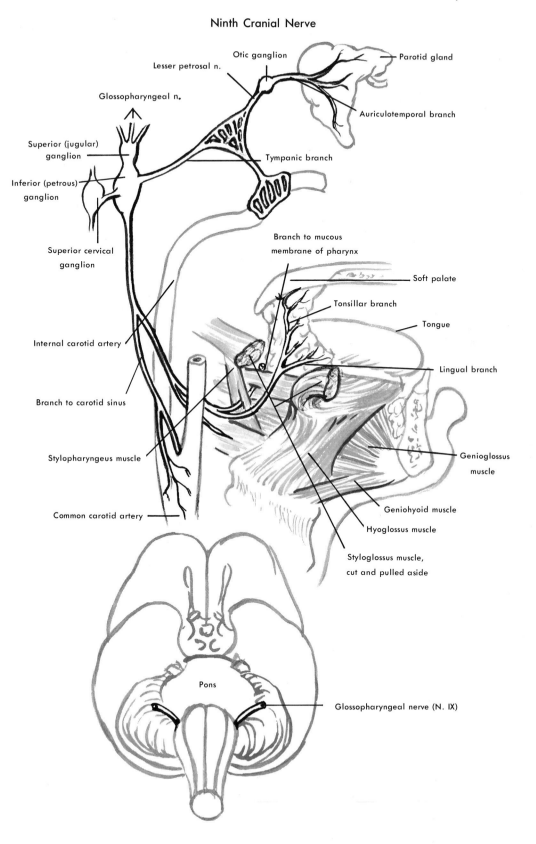

Otic ganglion

Lesser petrosal n.

Parotid gland

Glossopharyngeal n.

Auriculotemporal branch

Superior (jugular) ganglion

Tympanic branch

Inferior (petrous) ganglion

Superior cervical ganglion

Branch to mucous membrane of pharynx

Soft palate

Tonsillar branch

Tongue

Internal carotid artery

Lingual branch

Branch to carotid sinus

Stylopharyngeus muscle

Genioglossus muscle

Common carotid artery

Geniohyoid muscle

Hyoglossus muscle

Styloglossus muscle, cut and pulled aside

Pons

Glossopharyngeal nerve (N. IX)

ninth cranial nerve is a predominantly sensory nerve which carries afferent fibers from the tongue and pharynx, deriving its name from these areas. The nerve descends into the jugular foramen with the tenth and eleventh nerves. Two ganglia, the *superior* (or jugular) and the *inferior* (or petrous), swell its course and contain the cell bodies of the sensory fibers. The nerve descends into the neck carrying the following components:

1. Special visceral sensory fibers from taste buds of the posterior third of the tongue. These fibers travel a lingual branch.

2. General sensory fibers from the mucous membrane of the tonsils and pharynx.

3. Special motor fibers to the stylopharyngeus muscle.

4. Parasympathetic fibers which travel, via the tympanic branch and the lesser petrosal nerve (with synapse in the otic ganglion), into the middle ear. These fibers convey secretory fibers for the mucosal glands of the middle ear, mastoid air cells, and the parotid gland.

5. A group of fibers, called the *branch to the carotid sinus,* are carried to the special receptors in the carotid sinus. Impulses from these receptors at the bifurcation of the common carotid artery pass to the circulatory centers in the medulla.

Tenth (Vagus) Cranial Nerve

The *vagus nerve* was once called the pneumogastric nerve because of its innervation of the thoracic organs and gastrointestinal tract. Its present name is taken from its vagabond course through the body. This nerve is the important cranial parasympathetic motor nerve to structures outside the head. The nerve descends through the jugular foramen, presenting two ganglia, the *superior* (or jugular) and the *inferior* (or nodose) *ganglia*. These contain the cell bodies of the sensory fibers of the vagus nerve. (See illustration on opposite page.)

The course of the vagus nerve is a long one. It enters the carotid sheath with the internal jugular vein and the internal carotid artery. After giving off branches in the neck, the vagus descends into the thorax. The *right* vagus nerve gives off the *right recurrent laryngeal nerve* which ascends back into the neck around the right subclavian artery. The *left* vagus nerve gives off the *left recurrent laryngeal nerve* which curves under the arch of the aorta to ascend back into the neck.

In the mediastinum each vagus nerve breaks up into a *pulmonary plexus* from which branches pass into the lungs. Below this level each vagus nerve forms an *esophageal plexus* but reassembles only as a series of strands. *Anterior vagal trunks* pass anterior to the esophagus through the diaphragm. They are composed of fibers from both vagus nerves with the left predominating. *Posterior vagal trunks* pass posterior to the esophagus with the right nerve predominating. The vagal fibers pass through the preaortic ganglia without synapsing. Vagal branches reach the walls of the stomach, intestines, and associated glands and organs, where they synapse with the second neurons in *submucosal* (glands) and *myenteric* (smooth muscle) *plexuses.*

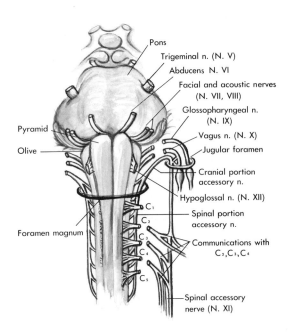

Component fibers conveyed in the vagus nerve are the:

1. Special motor fibers to the muscles of the larynx. These fibers are carried in the superior laryngeal (cricothyroid muscle only) and recurrent laryngeal nerves.

2. Special motor fibers to the muscles of the soft palate and the pharynx.

3. General sensory fibers from the mucous membrane of the pharynx, larynx, esophagus, bronchi, lungs, and abdominal viscera.

Tenth Cranial Nerve

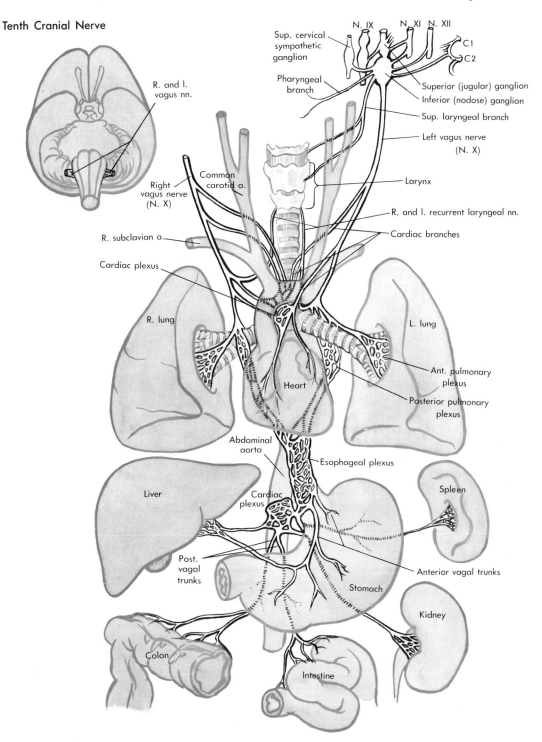

R. and l. vagus nn.

N. IX N. XI N. XII

Sup. cervical sympathetic ganglion

Pharyngeal branch

C1
C2

Superior (jugular) ganglion

Inferior (nodose) ganglion

Sup. laryngeal branch

Left vagus nerve (N. X)

Right vagus nerve (N. X)

Common carotid a.

Larynx

R. and l. recurrent laryngeal nn.

R. subclavian a.

Cardiac branches

Cardiac plexus

R. lung

L. lung

Ant. pulmonary plexus

Heart

Posterior pulmonary plexus

Abdominal aorta

Esophageal plexus

Liver

Spleen

Cardiac plexus

Post. vagal trunks

Anterior vagal trunks

Stomach

Kidney

Colon

Intestine

4. Special sensory fibers from a few taste buds of the pharynx and larynx.

5. Parasympathetic fibers which pass to the smooth muscle and glands of the thoracic viscera (via esophageal and pulmonary plexuses) and the abdominal organs (via alimentary plexuses). (See color plate 14.) Fibers to the cardiac muscle of the heart leave the vagus nerve by way of *superior cardiac nerves* from the neck and by *thoracic cardiac nerves* which pass through sympathetic plexuses on the walls of the great vessels. Ganglion cells of the second neurons to the heart may be located

among these plexuses or in the wall of the heart itself.

Eleventh (Accessory) Cranial Nerve

The accessory nerve is partly a cervical nerve and partly a displaced portion of the vagus nerve. It consists of spinal and cranial parts. The *spinal part* ascends from the cervical spinal cord to join the *cranial part* whose fibers emerge from the medulla just below the vagus nerve. The two parts run together to the jugular foramen. At this level the cranial part joins the vagus nerve, adding to it the special motor fibers to the muscles of the soft palate and the pharynx. The spinal part descends into the neck carrying general motor fibers to the trapezius and sternocleidomastoid muscles.

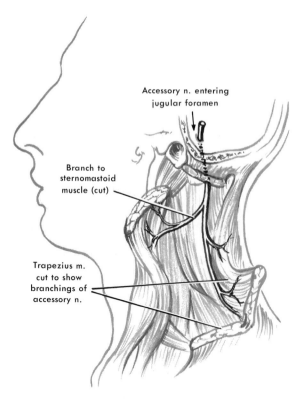

Accessory n. entering jugular foramen

Branch to sternomastoid muscle (cut)

Trapezius m. cut to show branchings of accessory n.

Twelfth (Hypoglossal) Nerve

This motor nerve to the muscles of the tongue leaves the medulla to descend through the hypoglossal canal. The hypoglossal nerve swings laterally and inferiorly, being joined for a distance by cervical nerve fibers destined for the ansa cervicalis before passing into the tongue. The hypoglossal nerve carries general motor fibers to the tongue muscles. (See illustration on opposite page.)

MAJOR NERVOUS PATHWAYS

Nervous pathways are routes along chains of neurons of the nervous system whereby sensory awareness reaches the cerebral cortex and a motor response is initiated. Myriads of pathways are possible. Those described are the most commonly involved in the functions of the body.

SENSORY PATHWAYS

Certain exteroceptive pathways mediating general sensation are of importance. General sensibility includes the perception of touch, pressure, pain, and temperature. Another important sensory pathway is that for proprioceptive stimuli.

Sensory Receptors

Various receptors are sensitive to particular stimuli. The simplest is a free nerve ending in which an axon divides into terminal branches which wind between epithelial or connective tissue cells or about a hair follicle. Free nerve endings are usually associated with painful stimuli. Arborizations about hair follicles are touch receptors in hairy areas. The combination of swellings on terminal branches with a swirled connective tissue capsule forms the *Meissner corpuscle* of the dermis of the skin which is related to touch.

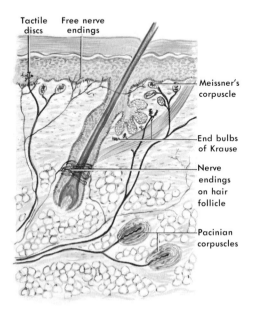

Tactile discs

Free nerve endings

Meissner's corpuscle

End bulbs of Krause

Nerve endings on hair follicle

Pacinian corpuscles

Simple connective tissue capsules about nerve endings in the dermis, the *end bulbs of Krause,* are sensitive to temperature changes. Pacinian corpuscles are more complicated, laminated cor-

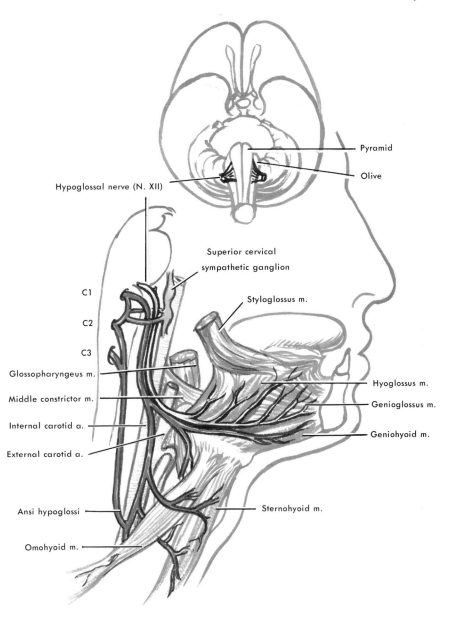

Pyramid

Olive

Hypoglossal nerve (N. XII)

Superior cervical
sympathetic ganglion

Styloglossus m.

C1

C2

C3

Glossopharyngeus m.

Middle constrictor m.

Internal carotid a.

External carotid a.

Hyoglossus m.

Genioglossus m.

Geniohyoid m.

Ansi hypoglossi

Sternohyoid m.

Omohyoid m.

puscles of the deep layers of skin and fasciae which are sensitive to deep touch and pressure. The type of receptor is less important than the area of cortex in which the pathway ends. Regardless of the receptor, nerve impulses are alike. It is the cortical perception which makes it possible to distinguish between the types of stimuli.

Pain and Temperature Pathways

The sensory or afferent pathways for pain and temperature are similar. From a skin receptor the impulse travels centrally into the posterior root

to the posterior horn gray matter of the spinal cord. (See color plate 12.) Two pathways, *reflex* and *ascending,* are open to the impulse through synapses with a second neuron in the posterior horn gray matter.

Spinal Reflex Pathways. A *spinal reflex arc* is the simplest, quickest pathway by which action can be taken in response to a stimulus. *Reflexes* are quick automatic motor responses which usually take place at the level of entry of the fibers. They move the body or one of its parts away from threatening stimuli, such as heat from a stove or the prick of a sharp instrument. The *reflex pathway* (arc) involves the synapse of the entering fiber with the cell body of a very short *association neuron* whose axon curves through the gray matter to the anterior horn.

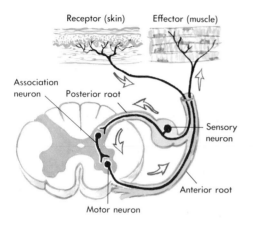

The association neuron synapses with a third or motor neuron in the anterior gray column. A motor impulse speeds through the anterior root of the spinal nerve to activate a series of voluntary motor units in a skeletal muscle.

Ascending Pain and Temperature Pathways. The ascending pathway carries the impulse to higher centers in the thalamus or the cortex. If the impulse follows this alternative the entering fiber synapses in the posterior column gray matter with a second neuron. The axon of this neuron *crosses* the gray matter of this segment to the *opposite* side of the spinal cord. This axon, with many others like it, forms a tract in the lateral funiculus of white matter which is termed the *lateral spinothalamic tract.* As the axons ascend, this tract grows larger as more pain or temperature fibers cross the gray matter of upper levels to ascend. The spinal tract of the fifth cranial nerve carries the first neurons of sensory pathways from the head. The second neurons of these pathways cross the midline to join the lateral spinothalamic tract which, as its name implies, continues upward through the brain stem to the thalamus. Synapse with a third neuron occurs in the thalamus, the axon of which passes to the cortex of the postcentral gyrus of the cerebral hemisphere. Only at the cortical level is the impulse actually perceived as pain or a difference in temperature.

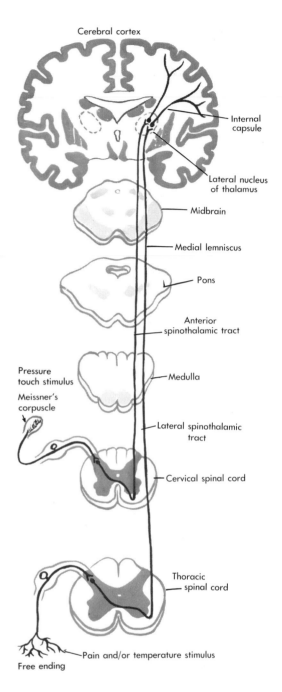

Touch and Pressure Pathways

Touch sets up impulses which enter the spinal cord where even more pathways are open to impulses upon arrival in the posterior horn.

Spinal Reflexes for Touch. These follow the same spinal reflex arc patterns as described for pain reflexes.

Uncrossed Ascending Pathways. The first neuron synapses in the posterior column gray matter with a second neuron, or ascends a level or so in the gray column for this purpose. In either case the axon of the second neuron enters the posterior funiculus of spinal cord white matter and ascends to the medulla in either the fasciculus gracilis (lower spinal nerves) or fasciculus cuneatus (upper spinal nerves). The second neuron axon ends by synapsing in either the nucleus gracilis or nucleus cuneatus of the medulla.

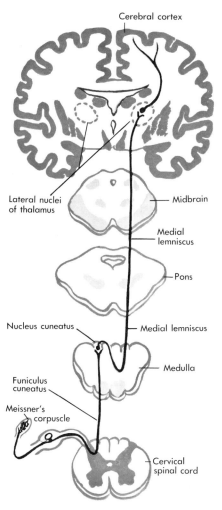

Only at the level of the medulla do the axons of the third neuron cross the midline to enter into a prominent band of white matter, the *medial lemniscus*. The pathway fibers ascend in this tract to the thalamus from which a fourth neuron ascends to the cerebral cortex of the postcentral gyrus.

Crossed Ascending Pathways. Another tactile pathway leads from the synapse of the first neuron in the posterior gray column. The axons of the second neuron immediately *cross* to the opposite side of the cord to enter the *anterior spinothalamic tract* of the white matter of the anterior funiculus. The axons ascend to the medulla in this tract, being joined by others at higher levels of the cord. In the medulla these fibers pass into the medial lemniscus of that side without further synapse to rise to the thalamus for synapse with a third relay neuron which ascends to the cortex following the path of the originally uncrossed route. Touch pathways of the head enter the main sensory nucleus of the fifth cranial nerve in the pons, synapse, and cross to the opposite medial lemniscus.

It may be seen, therefore, that tactile pathways are both uncrossed and crossed in the spinal cord so that injuries to the spinal cord will result in less disturbance of touch than of pain pathways which are entirely crossed. Above the level of the medulla the touch fibers, like the pain fibers, are entirely crossed.

Pressure Pathways. Impulses traveling along the spinal nerves follow pressure pathways in the spinal cord, brain stem, and thalamus to the cortex which are identical to the uncrossed tactile pathway. Reflex arcs are also similar.

Proprioceptive Pathways

Proprioceptive receptors send impulses from joints, tendons, and muscles which refer to the position of body parts, the state of muscular contraction, or the degree of stretch placed upon tendons. These impulses follow the spinal nerves into the gray column of the spinal cord. A simple reflex pathway of automatic proprioception may be followed at the segmental level, or the impulse may ascend to the cortex where conscious proprioception occurs. In the former case the pathway is as for all reflex arcs. In the latter case the conscious proprioceptive pathway is the same as for *uncrossed* tactile pathways of the spinal cord.

Timing and coordination are important in muscular activity, as is seen in the inhibition of an antagonistic extensor group of muscles to permit a flexor group to act, or the gradual relaxation of the antagonistic group to smooth the otherwise jerking action of the prime movers. The cerebellum is an important coordination center which depends greatly upon proprioceptive sensation. Its incoming proprioceptive fibers syn-

Labels for figure: Cerebral cortex; Lateral nuclei of thalamus; Midbrain; Medial lemniscus; Pons; Nucleus cuneatus; Medial lemniscus; Medulla; Funiculus cuneatus; Meissner's corpuscle; Cervical spinal cord

apse in the posterior gray column of the spinal cord. The axons of the second neurons ascend in either the *anterior* or *posterior spinocerebellar tracts* which enter the cerebellum by way of the inferior cerebellar peduncles. Connections are made with neurons of the cerebellar cortex whereby motor impulses are returned to segmental levels of the spinal cord which influence anterior horn motor cells. In this way the impulses going to muscle groups are modified.

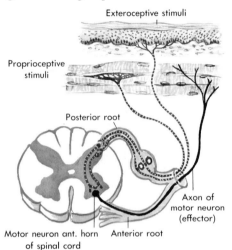

Exteroceptive stimuli

Proprioceptive stimuli

Posterior root

Motor neuron ant. horn of spinal cord

Anterior root

Axon of motor neuron (effector)

Proprioceptive Feedback in Muscular Function.

The gamma efferent system was introduced in Chapter 4 with the mention of alpha and gamma motor fibers and the muscle spindle–tendon organ complex. The proprioceptive pathway which provides feedback to modify muscular contraction via a special reflex arc is described more fully here.

Structure of the Muscle Spindle: The underdeveloped contractile elements of the muscle spindle are sensory structures functionally. The contractile components, the *intrafusal fibers,* respond to gamma motor fiber nerve impulses. As the intrafusal fibers contract, other elements within the spindle are stimulated to set up sensory (proprioceptive) impulses.

The muscle spindle has an outer connective sheath which stretches over the other components at the center of the spindle and tapers to slender poles at each end.

The sheath contains fluid in which the other elements are situated.

Two types of intrafusal fibers lie within the fluid-filled space. The larger and more numerous are the *nuclear chain fibers;* the less dominant are the *nuclear bag fibers.* The nuclei of the chain fibers are arranged in a linear row at the bulge of the spindle. The nuclei of the bag fibers comprise an irregular bunch at the equator of the spindle.

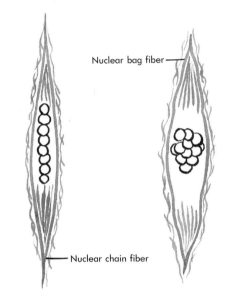

Nuclear bag fiber

Nuclear chain fiber

Underdeveloped contractile myofibrils lie at the poles of both types of intrafusal fibers. The myofibrils are innervated by the gamma motor nerve fibers.

Motor Innervation of the Muscle Spindle: The extrafusal, striated voluntary muscle fiber receives its nerve impulses to contract by way of the large, alpha motor neurons which end upon typical motor end plates. The gamma neurons with smaller fibers bring motor innervation to the elements of the muscle spindle.

Within the gamma group of fibers are two subtypes. *Gamma 1 fibers* are slightly larger and terminate upon intrafusal fibers by way of miniature nerve end plates. *Gamma 2 fibers* end diffusely in a long path along the surface of an intrafusal fiber, thus being given the name of *a trail ending.* (See illustration on opposite page.)

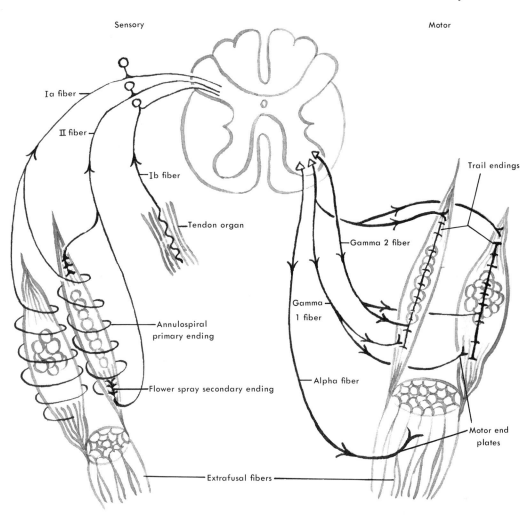

Sensory

Motor

Ia fiber

II fiber

Ib fiber

Tendon organ

Annulospiral
primary ending

Flower spray secondary ending

Extrafusal fibers

Trail endings

Gamma 2 fiber

Gamma
1 fiber

Alpha fiber

Motor end
plates

Sensory Innervation of the Muscle Spindle: Each intrafusal fiber, whether bag or chain type, has a sensory ending wound spirally around its equator, which is sometimes called an *annulospiral ending* but more usually is termed the *primary sensory ending*. These endings begin the proprioceptive sensory axons, which enter the spinal nerve as *Group Ia sensory fibers*. The nuclear chain fibers also possess *secondary endings* which spread out along each side of the primary ending. These secondary endings, aligned like a spray of flowers (and thereby called *flower spray endings*), also send their fibers to enter the spinal nerve as *Group II sensory fibers*. Both groups of sensory axons journey to the spinal cord by way of the dorsal root of the spinal nerve. The Golgi tendon organ sends afferent fibers (Group Ib) also to the dorsal root via the spinal nerve.

The Gamma Loop: When all the sensory and motor elements of the muscle spindle are put together, the result is a motor-sensory reflex arc which is called the *gamma loop*. The basic functional patterns of the gamma loop are as follows.

1. *In response to postural changes:* A muscle and its fascicular groups may be put on stretch by gravity or by shifts in the position of a joint. The pull on the muscle spindle causes stimulation of the sensory nerve endings of the muscle spindle. Afferent impulses are set up, especially by the primary sensory nerve endings, to pass centrally along Group Ia sensory fibers.

The cell bodies of the Ia sensory fibers are located within the dorsal root ganglion, from which a central process conveys the impulses into the dorsal horn gray matter of the spinal cord. The Ia fibers pass directly to the ventral horn gray matter without an association neuron being interposed.

The Ia sensory fibers end around alpha motor neurons. The reflex arc is completed by the passage of the impulse across the synaptic space to stimulate the alpha motor neurons. Impulses of a voluntary motor type spread peripherally through the spinal nerve to reach the extrafusal fibers via the motor end plate. Voluntary muscle contraction ensues, in which purposeful action or the adjustment of the limb to postural situations occurs. The foregoing are the basis of the *stretch reflex* and are responsible for the maintenance of reflex muscle tone, as well as for postural responses.

2. *In voluntary muscle contraction:* The gamma motor system is under the control of higher nerve centers by way of the upper motor neuron pathway. The impulses from such centers above the spinal cord stimulate the gamma motor neurons of the spinal cord ventral horn. Nerve impulses pass peripherally via gamma 1 or 2 nerve fibers to end by the miniature end plates or trail endings upon the nuclear bag or chain intrafusal fibers of the muscle spindle. The gamma motor stimuli can be considered as a pilot mechanism to induce the intrafusal fibers to contract. The intrafusal fiber contraction stimulates the sensory structure of the spindle. A swift volley of impulses flashes over the afferent fibers to reach the spinal cord ventral horn cells. Alpha motor neurons respond with another volley of nerve impulses, which pass over the gamma 1 fibers to cause powerful contraction of the motor units of the extrafusal fiber mass of the whole muscle.

Students in the health professions, particularly physical restoration therapists, will find that this structural and functional summary will be built upon in courses on physiology and kinesiology.

Visceral Sensory Pathways

Sensations from the visceral organs of the body were described in the discussion of the autonomic nervous system.

MOTOR PATHWAYS

Pathways for the control of visceral activities have been described with the autonomic nervous system. Only three major motor pathways will be described: those initiating conscious control of purposeful voluntary movements, those concerned with the regulation and coordination of muscular activity, and those which convey the actual impulses to the effector muscles. It is the last-mentioned pathway upon which the others impinge and it is often called the *final common pathway*.

The Final Common Pathway

All motor responses depend upon impulses being initiated in the motor nerve cells whose axons reach out to the voluntary muscles. These motor neurons are located in the anterior gray column of the spinal cord or the motor nuclei of cranial nerves in the brain stem. The nerve cells are highly specific in that they innervate only a number of striated muscle fibers (a motor unit) in an individual muscle. A related group of these cells supply a named muscle. Such neurons have no relation to any other muscle and the response to their impulses is a specific muscular action.

The Pyramidal System

The pathway for initiating voluntary movements under conscious rather than reflex control begins with the cortical cells of the precentral gyrus. The axons descend through the central nervous system in the corona radiata, internal capsule, cerebral peduncle, and pyramid of the medulla. The axons as a group form the *corticospinal tract* of the brain stem, but because they also form the pyramids the pathway is known as the *pyramidal system.* At the lower level of the medulla the majority of the axons cross the midline in the *decussation of the pyramids* to the opposite side of the spinal cord where they descend in the lateral funiculus as the *lateral corticospinal tract.* This crossing is an example of the fact that one side of the body is controlled by the opposite cerebral hemisphere. A minority of the descending axons do not cross but continue downward on the same side in *anterior corticospinal tract* located between the anterior horn and the surface of the spinal cord. At segmental levels axons peel off the corticospinal tracts to enter the gray matter of the anterior horn. The cortical (upper motor) neuron is thus brought into a position to initiate activity of the spinal (lower motor) neuron. From this point the pattern of the final common pathway begins.

The musculature of the head and neck is innervated from a central nervous system level above the pyramids. The cortical axons which descend to the nuclei of cranial nerves which innervate muscles of the head are termed *corticobulbar fibers* because they end in the brain stem. The final common pathway is from the motor cells of the cranial nerve nucleus.

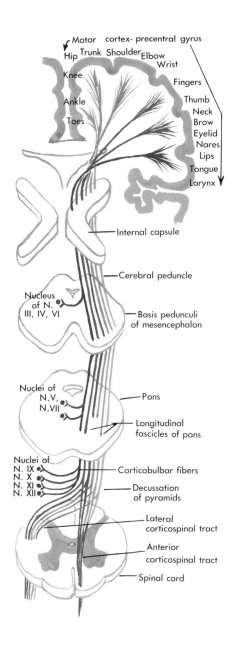

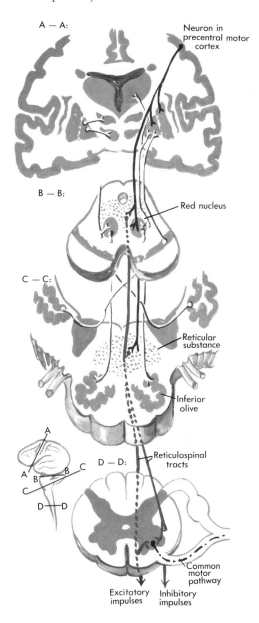

or *inhibitory mechanisms.* Nerve fibers descend from each into the spinal cord to end in relation to final pathway motor neurons.

The Extrapyramidal System

The pathways which coordinate muscular activity do not travel in the pyramidal pathways. They, therefore, are described as making up the *extrapyramidal system.* The cell bodies are located in the cerebral hemisphere in front of the motor areas of the precentral gyrus. The axons descend in the internal capsule to the basal nuclei. A tremendous number of pathways involving centers in the brain stem and the cerebellum are involved which are too extensive to describe except in a textbook of neuroanatomy. The pathways are divided into those related to *excitatory*

The balance between the two mechanisms coordinates muscle activity.

THE EYE

Vision is stimulated by light waves which originate from sources of energy located at a distance or are reflected from other objects. The visual apparatus receives the light rays and transforms them to nerve impulses which in the visual areas of the cerebral cortex are perceived as vision. The eye is the receptor organ for vision. Only one portion of the eye, the *retina,* contains the actual receptor cells and is connected with the optic nerve. All other portions of the eye are concerned with gathering the light rays and focusing them upon the retina. Structures around the eye serve to protect the eyeball or to move it to follow objects.

Retina and Generation of Impulse

The eyeball consists of three layers. (See color plate 15.) The outer *sclera* is a heavy collagenous bulb which gives the globe shape to the eyeball, protects its contents, and receives the attachment of ocular muscles. The middle layer, the *choroid,* is a vascular layer which nourishes the eyeball and is heavily pigmented to reduce internal reflections. The innermost layer, the *retina,* is actually a part of the brain which grew out of the forebrain to become incorporated into the other components of the eye. The *optic nerve* connecting this layer with the brain is a nerve tract of the brain which is displaced to function as a cranial peripheral nerve.

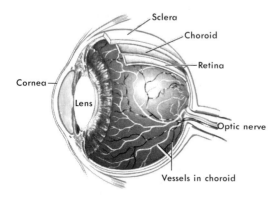

The retina consists of many cellular layers. The layers of cells which are related to the reception of visual stimuli are:

1. *Rods and Cones:* These cells, named for their shapes, are located deepest in the retina next to the choroid layer. These are the cells which are receptive to light rays, which must penetrate all of the other layers of the retina to reach them. There are about four times as many rod cells as cone cells, but in the *macula lutea,* the location of sharpest vision, cones are in the majority. A depression in the exact center of the macula lutea,

called the *fovea,* is formed exclusively of cone cells. Both rod and cone cells contain a visual pigment which reacts photochemically with light rays. The pigment in rod cells is known as *rhodopsin* or visual purple; that in cone cells has not been exactly identified. When light rays strike through the retinal layers upon the rods and cones, a chemical reaction occurs during which an impulse is generated and the visual pigment is transiently bleached. The process is reversible for the pigment re-forms constantly.

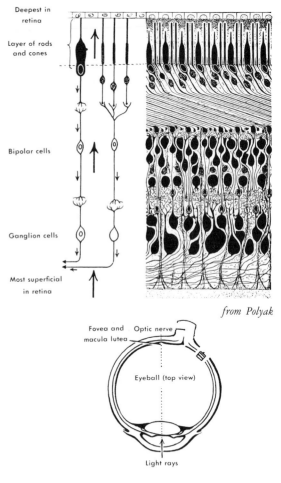

from Polyak

The pigment is "used up" in intense sunlight but it re-forms during the period of *dark adaptation.* It is present in greatest amounts in darkness when the number of available light rays to be acted upon is greatly reduced.

It is believed that only cone cells are sensitive to colors, different cones being sensitive to each of the primary colors of the spectrum. Blending of receptor responses is said to be responsible for the color perception of varying impulses in the visual cortex. *Color vision,* therefore, is a matter of color perception, and defects in "color blindness," usually a sex-linked defect in 4 percent of males, indicates a defect in the visual perception area of the brain rather than a retinal defect.

2. *Bipolar cells:* The bipolar cells are neurons which act as an intermediate link between the receptor cells and the optic nerve.

3. *Ganglion cells:* These cells, located at the surface of the retina, synapse with the bipolar cells and are the cell bodies whose axons pass across the retina to form the optic nerve. Just medial to the macula lutea all of these fibers converge at the *optic disk* to turn posteriorly as the optic nerve.

Focusing Apparatus

Light rays from distant sources diverge through the atmosphere. They must be collected and made parallel, and then focused upon the retina at the macula. Focusing is accomplished by two portions of the eye which have curved surfaces. The front fifth of the scleral coat of the eye bulges into a transparent curved surface, the *cornea,* through which light is admitted to the eye and by which initial focusing is accomplished. (See color plate 15.) If the cornea is smoothly curved, light rays are refracted toward the lens. If irregularities exist in the curvature of the cor-

nea, the light rays are not properly refracted and *astigmatism* occurs—one cause of blurred vision.

After passing through the cornea the light rays traverse the *lens* of the eye which is an elastic, doubly curved transparent body suspended in the light path. The shape of the lens is changed by the contraction of smooth muscle fibers of the *ciliary muscle* which pull upon ligaments attached to the capsule of the lens. The light rays are directed through the eye to fall into focus upon the retina. In the viewing of distant objects the light rays are more parallel and less refraction is necessary; thus ciliary muscles maintain greater tension to keep the lens thinner. In near vision, the light rays from objects close to the eye are diverging and considerable refraction is necessary. In this process of *accommodation* the tension upon the ligaments of the lens is relaxed to permit the lens to become thicker. (See illustration at bottom of page.)

In some cases the focusing apparatus is insufficient to bring the light rays to a sharply focused image upon the retina. In *myopia* (nearsightedness) elongation of the eyeball allows light rays to be brought into focus in front of the retina. In *hypermetropia* (far-sightedness) shortening of the eyeball results in focusing at a point behind the retina. In both cases, as well as in astigmatism, corrective artificial lenses in eyeglasses or contact lenses supplement the focusing apparatus of the eye.

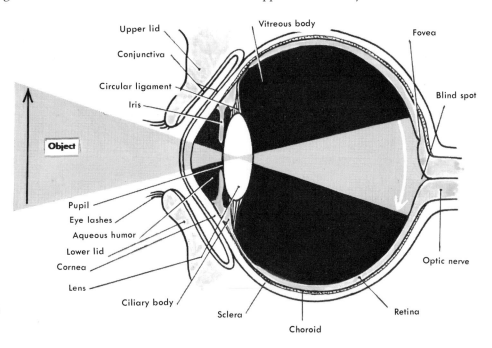

Light Regulatory Apparatus

The *iris* is a circular curtain hanging behind the cornea and in front of the lens. The iris is the colored portion of the front of the eye with an opening, the *pupil,* in its center. (See color plate 15.) Radiating smooth muscle fibers, forming the *dilator pupillae muscle,* and circular fibers, forming the *sphincter pupillae muscle,* work within the iris to change the diameter of the aperture of the pupil. This mechanism, under autonomic nervous control, regulates the amount of light admitted to the eye. The pupil appears dark because of the light-absorbing quality of the choroid coat within the eye. Pigment cells of the iris produce the color of the eye. Less pigment results in a blue iris, more in a brown iris.

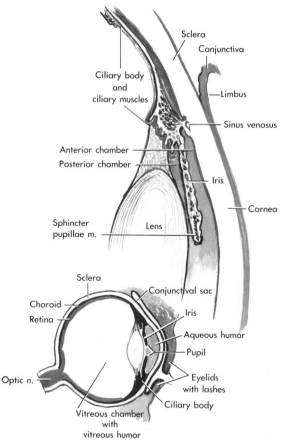

Chambers of the Eye

The anterior surface of the lens is loosely draped by the pupillary margin of the iris. In front of these structures and behind the cornea is the *anterior chamber* of the eye, which is filled with the liquid *aqueous humor.* The *posterior chamber* behind the iris is also filled with aqueous humor. The large posterior five-sixths of the globe is the *vitreous chamber,* which is filled with the gelatinous *vitreous humor.* These two colloidal humors are completely transparent to light rays and serve to nourish the interior of the eye and to preserve its globular shape. The aqueous fluid, similar to cerebrospinal fluid, is continuously formed by filtration from the blood and is reabsorbed into the venous circulation whereas the vitreous humor is a permanent interior body of the eye.

Protective Structures

The eyeball fits into a smooth fibrous socket, the *fascia bulbi,* which is continuous with the fasciae of the ocular muscles. The eyeball and its fascia are positioned by, and rotate in, the *retrobulbar fat* which fills all of the bony orbit not occupied by the eyeball or its accessory structures. The orbit protects all but the anterior fifth of the eyeball. It is this portion, termed the presenting part of the eye, which must face the environment without interference to the passage of light rays. Flexible protective structures guard the presenting parts of the eye.

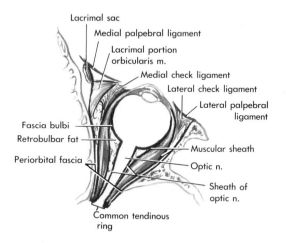

Eyelids. The *eyelids* are flaps of skin strengthened by *tarsal plates* of dense connective tissue. The free margins of the eyelids bound a slit between them called the *palpebral fissure.* The levator muscle of the upper eyelid, acting with gravity upon the lower eyelid, widens the palpebral fissure to "open the eye" and admit more light. Contraction of the orbicularis oculi muscle "closes the eye" by pressing the free margins of the eyelids together to shut out light or to protect the front of the eye. The free margin presents two or three rows of hairs, the *cilia* or *eyelashes*. Both sebaceous and sweat glands lubricate the margins.

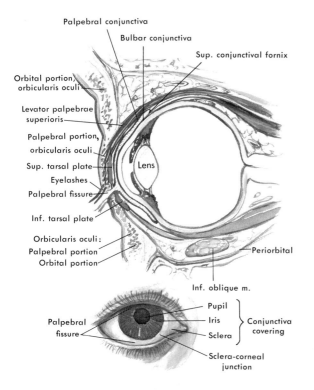

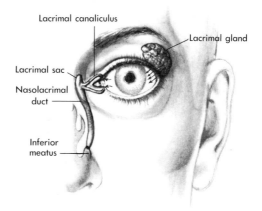

Conjunctiva. A potential space, the *conjunctival sac,* intervenes between the posterior surface of the eyelids and the front of the eyeball. This space is lined by a mucous membrane adaptation of the skin. The *palpebral conjunctiva* covers the posterior surface of each lid from the free margins to crevices termed the *superior* and *inferior fornices* of the conjunctival sac. At these points the mucous membrane reflects onto the front of the eyeball as the *bulbar conjunctiva* to follow the sclera over the "white of the eye" to the junction of the sclera and cornea. At this point the bulbar conjunctiva is attached to the sclerocorneal junction to become the external layer of the cornea. The conjunctivae form a moist sac, lubricated by tears, which helps protect the front of the eye, keeps the transparent cornea moist, and provides for free movement of the eyelids.

Lacrimal Apparatus. Tears, usually thought of only in association with the emotions, are a watery fluid secreted by the lacrimal gland which moistens the front of the eye and washes away foreign particles which might injure the delicate corneal surface. The *lacrimal gland* is located at the superolateral angle of the orbit. (See color plate 15.) Its multiple *lacrimal ducts* open into the superior fornix of the conjunctival sac. The tears wash across the front of the eye, prevented usually from overflowing by the oily secretions on the free margins of the eyelids.

About half the volume of the tears evaporates. The balance enters a *lacrimal canaliculus* at the medial extremity of each eyelid. The canaliculi drain into the *lacrimal sac* which is the dilated upper end of the *nasolacrimal duct.* The duct courses vertically downward in a canal through the lacrimal, maxillary, and inferior nasal conchal bones to enter the inferior meatus of the nose.

Eye Muscles and Eye Movement

The eyeball, in its sling of fascia bulbi, is balanced in the orbital fat by the pull of six extraocular muscles. These muscles work together under the control of the central nervous system. The degree and direction of movement depend upon the increase in contraction of some of the muscles accompanied by the relaxation of antagonistic muscles. Most of the ribbon-like muscles arise from the back of the orbit and from a central tendinous ring which surrounds the optic canal. The muscles pass forward to attach at certain points on the sclera in relation to the equator of the eyeball. The muscles are summarized in the table on the next page.

MUSCLES MOVING THE EYES

MUSCLE	Superior rectus	Inferior rectus	Medial rectus	Lateral rectus	Superior oblique	Inferior oblique
ORIGIN	In common from the central tendinous ring				Sphenoid bone medial to optic canal	Front of orbit lateral to naso-lacrimal duct canal
COURSE	Along orbital roof	Along orbital floor	Along medial orbital wall	Along lateral orbital wall	Forward above superior rectus; turns laterally downward through a pulley	Laterally and backward under inferior rectus to wind upward
ATTACHMENT	Into sclera in front of equator of eyeball at location indicated by the muscle name				Posterolaterally into sclera behind the equator of the eyeball under superior rectus	Posterior and lateral aspect of sclera behind equator under lateral rectus
NERVE	Third	Third	Third	Sixth	Fourth	Third
PRINCIPAL ACTION	Elevates eye	Depresses eye	Adducts eye medially	Abducts eye laterally	Depresses adducted eye	Elevates adducted eye

Sup. oblique
Medial rectus
Lat. rectus
Inf. rectus
Inf. oblique
Superior rectus m. elevates eye in primary and abducted position

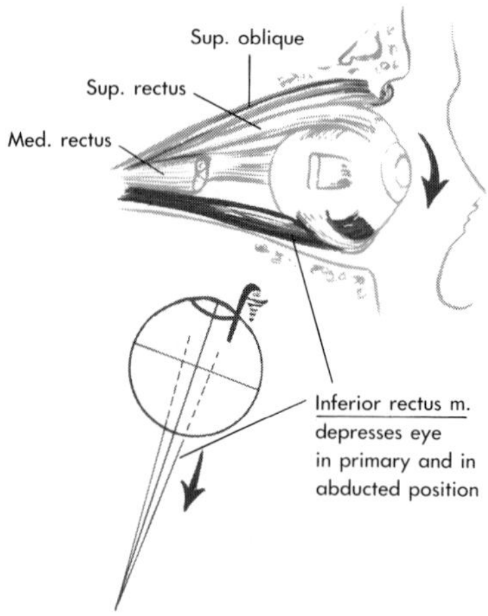

Sup. oblique
Sup. rectus
Med. rectus
Inferior rectus m. depresses eye in primary and in abducted position

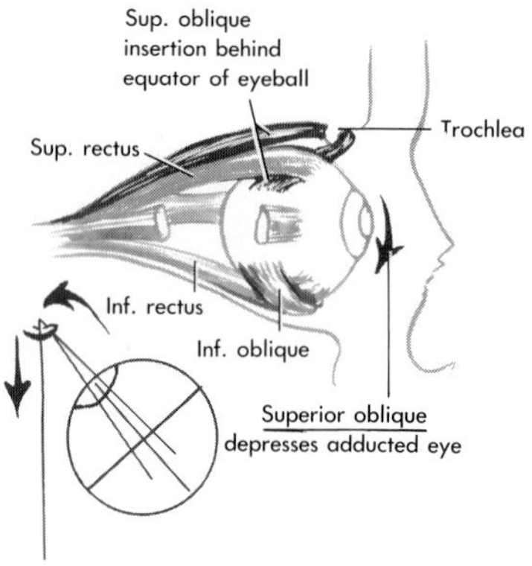

Sup. oblique insertion behind equator of eyeball
Trochlea
Sup. rectus
Inf. rectus
Inf. oblique
Superior oblique depresses adducted eye

Sup. oblique
Sup. rectus
Lateral rectus
Medial rectus abducts eye medially

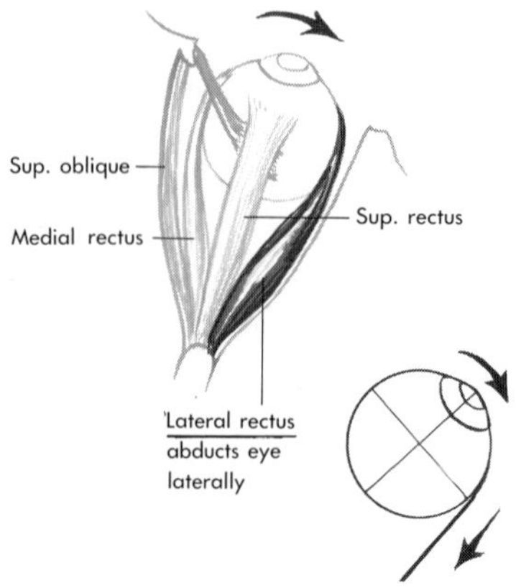

Sup. oblique
Sup. rectus
Medial rectus
Lateral rectus abducts eye laterally

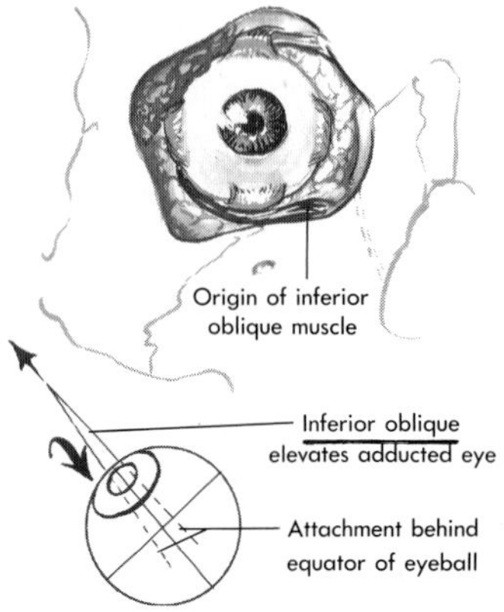

Origin of inferior oblique muscle
Inferior oblique elevates adducted eye
Attachment behind equator of eyeball

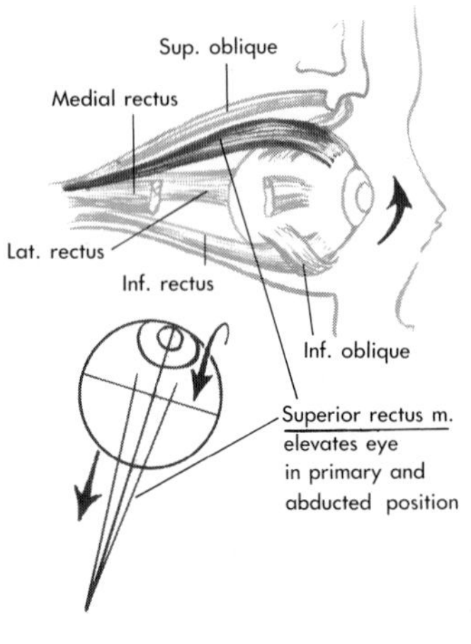

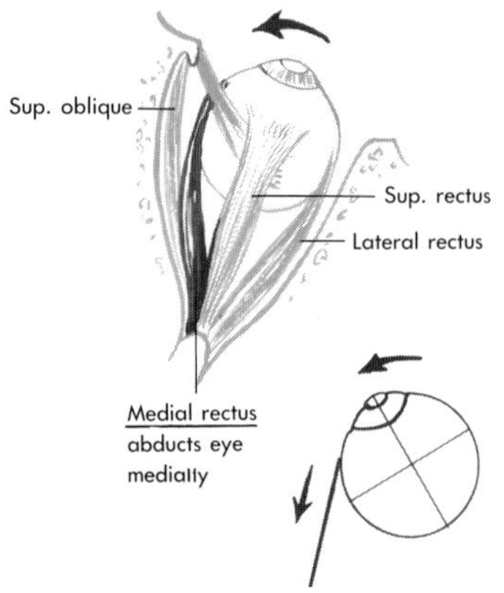

Coordinated Action of Eye Muscles. The eyeball can be turned in every direction of forward gaze. Except in adduction and abduction the extraocular muscles do not work alone. While the superior rectus pulls to elevate the eye when the gaze is straight forward, or the inferior rectus depresses the eye in this primary position, neither muscle's line of pull permits these movements when the eye has been adducted. The illustrations show that when the eyeball has already been adducted by the medial rectus, the oblique muscles are in position to perform these movements. The lateral course of the superior oblique after it leaves its pulley and its insertion *behind* the equator of the eyeball permit it to depress the adducted eye. Similarly, the completely lateral course of the inferior oblique pulling the posterior part of the eyeball from below will elevate the adducted eye. Note, however, that the eyeball, already abducted by the lateral rectus, is in an even better position for the superior rectus to elevate and the inferior rectus to depress it.

Conjugate movements of the eyeballs involve the simultaneous movements of both eyes. Some muscles contract while others are inhibited from acting in each orbit. Both eyes, for example, turn together to the right by the simultaneous action of the right lateral rectus and the left medial rectus muscles. Binocular vision requires the fusing of images from the two retinas. This is important in focusing for near vision, accomplished by convergence whereby both eyes turn inward by simultaneous action of both medial rectus muscles.

THE EAR

Hearing is a special sense whose stimulus arises at a distance from the body in objects which produce physical vibrations in the atmosphere. These vibrations impinge upon the surface of the body as sound waves. Nerve impulses, set in motion by the vibrations, traverse the auditory pathways of the eighth cranial nerve to reach the auditory area of the cerebral cortex where sound perception occurs. The ear is the portion of the auditory mechanism which acts as the receptor organ for hearing. The process may be divided into the following parts:

1. Collection of the sound waves by the *external ear.*

2. Conversion of physical vibrations in the air to mechanical vibrations by the structures of the *middle ear.*

3. Conversion of mechanical vibrations to vibrations in fluid-filled cavities at the boundary of the middle ear with the internal ear.

4. Conversion of fluid vibrations to auditory nerve impulses by the *internal ear.*

Collection of Sound Waves

The *external ear* consists of the *auricle* (*pinna*) and a passageway into the head, the *external acoustic meatus.* (See color plate 15.)

Auricle. The auricle is a flap of skin spreading out on the lateral surface of the head and stiffened and formed into margins, elevations, and depressions by a plate of elastic cartilage. The *helix* is the external rim, and the *anthelix* is a broad surface ridge paralleling the rim and separated from it by the *scaphoid fossa.* The *lobule* is a soft vascular flap of skin hanging downward from the auricle. The central depressed area is the *concha,* guarded anteriorly by a surface projection, the *tragus.*

External Acoustic Meatus. Sound waves, collected by the auricle, enter the opening of the *external acoustic* (auditory) *meatus* at the depths of the concha. This tubular passageway leads into the skull. The outer third, lined by skin containing coarse hairs and ceruminous (wax-producing) glands, is supported by tongues of elastic cartilage extending inward from the auricle. The inner two thirds is in the temporal bone with the skin adhering tightly to the periosteum of the bony canal.

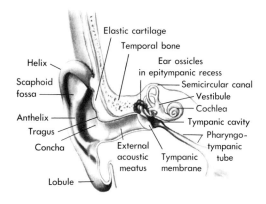

Conversion to Mechanical Vibrations

A fibrous membrane, the *eardrum* or *tympanic membrane,* is set obliquely between the medial end of the external acoustic meatus and the *middle ear* (*tympanic cavity*). The lateral side of the eardrum is covered by skin while the medial side is covered by the mucous membrane of the tympanic cavity. Sound waves traveling in the air of the external acoustic meatus cause the eardrum to vibrate.

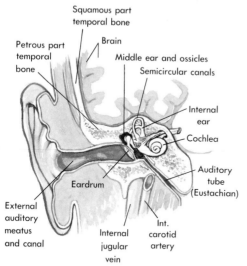

cartilage and bone, establishes a connection to the atmosphere through the nasal cavity so that air pressure is equalized on each side of the eardrum. The tympanic cavity extends above the level of the eardrum as the *epitympanic recess* with only a thin shelf of temporal bone between it and the temporal lobe of the brain in the middle cranial fossa. The *hypotympanic recess* extends downward below the level of the eardrum. The middle ear communicates posteriorly with mucous membrane-lined *mastoid air cells.* (See also color plate 15.)

Ossicular Chain.   Three tiny ear bones arch upward and medially across the tympanic cavity. These auditory ossicles, connected by tiny ligaments and miniature synovial joints, are set into mechanical vibration by the eardrum and transmit these vibrations across the middle ear. The *malleus* (hammer bone) has a handle-like process attached to the eardrum to pick up the vibrations. Its head transfers the vibrations to the *incus* (anvil bone) which articulates with the head of the *stapes* (stirrup bone). The foot plate of the stapes is set into the *oval window* on the medial wall of the middle ear. Excessive vibration of the auditory ossicles, as with loud sounds, is prevented by tension upon the malleus and stapes by the pull of two miniature muscles, respectively the tensor tympani and stapedius muscles. (See also color plate 15.)

Tympanic Cavity.   The middle ear is a box-like cleft within the temporal bone. This cleft is lined by a mucous membrane similar to that of the nasal cavity and contains air because it is linked to the nasopharynx by the *pharyngotympanic tube* (auditory tube, eustachian tube). This tube, running forward and downward from the anterior wall of the tympanic cavity through

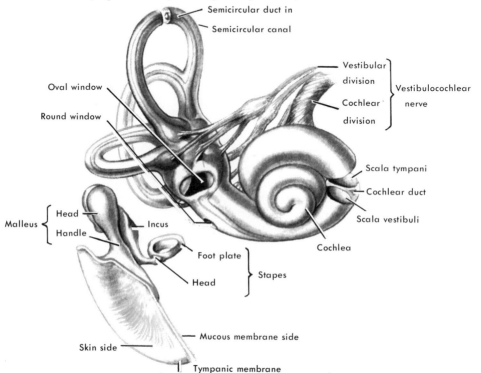

Conversion to Fluid Vibrations. The medial wall of the middle ear presents two openings in the bony wall. The oval window holds the footplate of the stapes. The round window is closed by the flexible secondary tympanic membrane. Medial to the bony wall, in the interior of the petrous portion of the temporal bone, is the *internal ear* or labyrinth. (See color plate 15.) The internal ear is composed of a fluid-filled duct system which in turn is suspended within fluid-filled bony passageways. At the oval window, vibration of the foot plate of the stapes converts the mechanical vibrations into fluid waves.

Conversion to Nerve Impulses

The *internal ear* is a complicated structure of miniature tubes and sacs suspended within rather similar bony passageways. It is understood easily if a simple basic plan, as illustrated, is kept in mind. The *otic capsule* is the bony structure of the petrous portion of the temporal bone. It is permeated by irregular channels and cavities lined by an endosteal membrane which comprise portions of the *periotic labyrinth*. The periotic labyrinth is filled by the *periotic fluid* (perilymph) which seeps between delicate strands which are much like arachnoid trabeculae. The trabeculae suspend an epithelial tubular system within the periotic labyrinth. The parts of the tubular system are bathed by the periotic fluid. The epithelial tube system makes up the *otic* (membranous) *labyrinth*. This is filled with *otic fluid* (endolymph). No matter how complex the portions of each labyrinth, the basic plan is present.

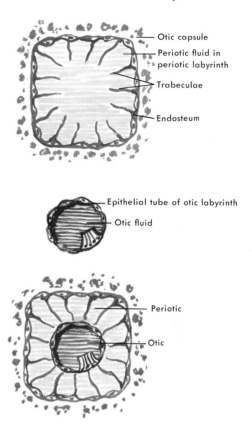

Periotic Labyrinth. A large irregular chamber, the *vestibule,* is the middle part of the periotic labyrinth. Vibrations of the stapes are transformed into fluid waves of the vestibular periotic fluid. Three *semicircular canals* loop at right angles to each other from the vestibule. These are the *anterior* (superior), *lateral,* and *posterior semicircular canals.* The anteromedial extremity of the vestibule projects into a coiled bony canal which, since it resembles a snail's shell, is termed the *cochlea.*

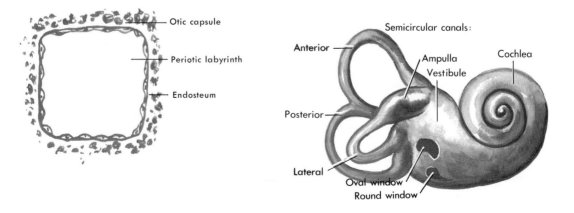

The cochlear passageway winds two and one-half times around a central bony core, the *modiolus*, which contains the spiral ganglion of the cochlear part of the eighth cranial nerve.

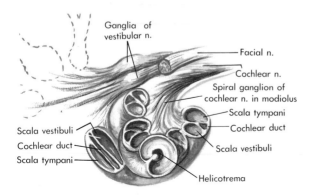

Ganglia of vestibular n.

Facial n.

Cochlear n.

Spiral ganglion of cochlear n. in modiolus

Scala tympani

Cochlear duct

Scala vestibuli

Scala vestibuli

Cochlear duct

Scala tympani

Helicotrema

(See also color plate 15.) One wall of the epithelial tube within the cochlea joins the bony ledges (spiral lamina) of the modiolus to divide the cochlea into two channels, the scala vestibuli and the scala tym-

pani, which are interconnected only at the tip of the cochlea, the *helicotrema*. All the parts of the periotic labyrinth named are bony chambers filled with periotic fluid through which fluid waves pass.

Otic Labyrinth. The epithelial tube system of the otic labyrinth presents local dilatations as it hangs suspended in the periotic labyrinth. Within the vestibule is a saclike dilatation of the epithelial tube termed the *saccule.* On one wall of the saccule is a plaque of tall columnar cells bearing hairlike cilia which project into the otic fluid capped by a gelatinous substance. This is the *macula sacculi,* one of the receptors for the vestibular portion of the eighth cranial nerve. A short remnant of the original epithelial tube, the *ductus reuniens,* permits otic fluid to flow into a long coiled dilatation of the epithelial tube within the cochlea. This is the *cochlear duct,* more triangular than circular in shape. One wall, the *basilar membrane,* is an extension of the spiral lamina of the modiolus that divides the periotic cochlea into the two scala mentioned before.

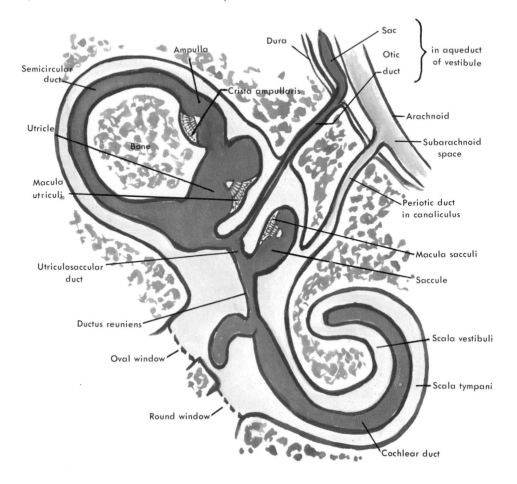

Semicircular duct

Ampulla

Dura

Sac

Otic

duct

in aqueduct of vestibule

Crista ampullaris

Arachnoid

Utricle

Bone

Subarachnoid space

Macula utriculi

Periotic duct in canaliculus

Macula sacculi

Saccule

Utriculosaccular duct

Ductus reuniens

Oval window

Scala vestibuli

Round window

Scala tympani

Cochlear duct

Another wall of the cochlear duct is modified into a long coiled neuroepithelial strip whose specialized cells form the *organ of Corti,* the receptor end organ for hearing.

If we return to the saccule within the vestibule as a central point of reference, we see a remnant of the original epithelial tube that forms the *utriculosaccular duct* which joins the saccule to another dilated portion of the tube, the *utricle.* The utricle also bears a neurosensory epithelial plaque, the *macula utriculi.* Three *semicircular ducts* branch from the utricle to traverse the periotic semicircular canals and return to the utricle. The anterior and posterior ducts return by a common duct. Each semicircular duct presents a local dilatation, the ampulla. An epithelial area, the *crista ampullaris,* is formed by cells much like those in a macula. These are also receptors for the vestibular division of the eighth cranial nerve. All of the named portions of the otic duct system are interconnected and filled with otic fluid. Fluid waves in the periotic fluid, pressing against the epithelial membrane, produce similar waves in the otic fluid which stimulate the receptor cells.

Pressure Release Mechanisms.   Both the otic and periotic fluids are constantly forming from the epithelial linings of each labyrinth. The fluids are under pressure to facilitate the passage of waves through a compressible medium. Excessive pressure in the periotic labyrinth can be relieved by bulging of the secondary tympanic membrane at the round window. This labyrinth, however, is also connected to the subarachnoid space by the *periotic* (perilymphatic) *duct,* which runs from the cochlea through the *cochlear canaliculus* of the petrous bone to a point near the internal auditory meatus. Excess periotic fluid can enter the subarachnoid space through an opening in the dura at this point. There is much less otic than periotic fluid. An *otic* (endolymphatic) *duct* branches from the utriculosaccular duct to run through the bone in the *aqueduct of the vestibule.* This duct ends blindly in the *endolymphatic sac* which lies under the dura on the posterior surface of the petrous bone. It functions as a pressure expansion device.

Nerve Pathways for Hearing.   Waves of vibration in the otic fluid of the cochlea stimulate the hairs of specialized receptor cells of the organ of Corti. These cells set up nerve impulses in the peripheral processes of bipolar nerve cells which are located in the *spiral ganglion* of the cochlear division of the eighth cranial nerve. Central processes become the nerve axons of the cochlear division of the vestibulocochlear (eighth cranial) nerve. These fibers pass to the brain stem to reach *cochlear nuclei* where they synapse. Second neuron fibers ascend to auditory centers of the temporal lobe via a further synapse in the thalamus. Other reflex pathways connect with motor nuclei for quick turning of the head to follow or to avoid sources of sound. (See illustration on following page and color plate 15.)

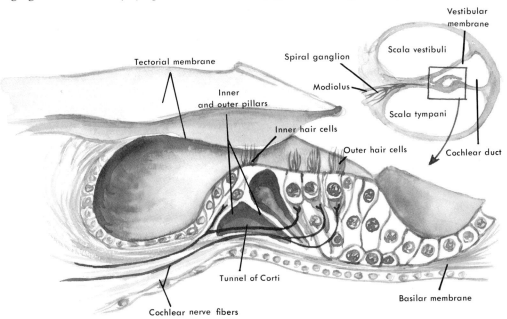

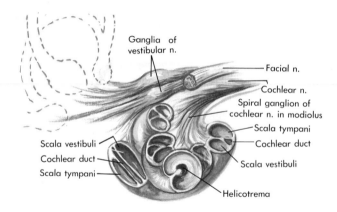

Ganglia of
vestibular n.

Facial n.

Cochlear n.

Spiral ganglion of
cochlear n. in modiolus

Scala tympani

Cochlear duct

Scala vestibuli

Helicotrema

Scala vestibuli

Cochlear duct

Scala tympani

Nerve Pathways for Equilibrium. Movements of the head or body as a whole produce fluid movements within the semicircular ducts, utricle, and saccule. The gelatinous caps at the neurosensory areas are moved to stimulate hairs of the receptor cells. These, in turn, set up nerve impulses in the peripheral processes of the nerve cells within the *vestibular ganglion* of the vestibular division of the eighth cranial nerve. The impulses travel to the brain stem as fibers of the vestibular division of the nerve to synapse with other neurons in the *vestibular nuclei*. Many pathways, reflex in type, convey impulses to the motor centers of the spinal cord where reflex adjustments of posture and locomotion are made. Other pathways provide reflex connections to the eyes and to the salivatory and vomiting centers. Excessive stimulation of the vestibular receptors frequently results in motion sickness in which dizziness, nausea, salivation, vomiting, and eye disturbances result from reflex vestibular connections.

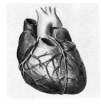

6

The Circulatory System

The circulatory system consists of a network of interconnecting tubular structures through which a muscular organ, the heart, pumps the blood to all parts of the body. The circulatory system is associated with the spleen, liver, and bone marrow in the formation and replacement of blood cells.

FUNCTIONS OF THE CIRCULATORY SYSTEM

The following are functions of the circulatory system:

1. It contains the blood.

2. It causes the blood to move by the pumping action of the heart and distributes the blood to the regions of the body by a branching set of vessels known as the *arteries*.

3. It makes possible the diffusion of nutritive substances and oxygen from the blood into the intercellular (tissue) spaces of the body by way of a continuing network of thin-walled vessels, the *capillaries*. The same network receives wastes, carbon dioxide, and part of the excess tissue fluid from the tissues.

4. It returns the blood to the heart through the *veins*.

5. It routes the blood to such organs as the kidneys, liver, and lung where the blood can be purified of the wastes of metabolic activity of the body tissues.

As the result of the existence of structures to provide for these functions, the circulatory system contributes to:

1. The maintenance of life, since metabolism of the vital organs of the body is impossible without nourishment, oxygen, waste removal, and the receipt of nutritive substances which have been absorbed by organs of the digestive system. The circulatory system transports and distributes these substances to organs which modify, store, or use them in their metabolism.

2. The regulation of body temperature by the increase or decrease in the amount of blood circulated to the subcutaneous tissues, skin, and sweat glands.

3. The distribution of important hormones from the endocrine glands.

4. The transfer of body reserves of mineral salts.

5. The dissemination of substances involved in the immune and allergic responses of the body.

6. The provision of a mobile reserve defense force of the body against infection, allergens, and particulate matter. White blood cells circulate constantly in the blood vessels. They are instantly

available to support the phagocytic cells of tissues by passing through the thin endothelial walls of the capillaries.

BLOOD

Blood is the fluid tissue which is contained within the vessels of the circulatory system. Blood consists of a fluid, the *blood plasma,* in which the *blood cells* are suspended. The body of a person of average size contains 5 to 6 liters of blood. Blood is a slightly viscous fluid whose color varies between bright red and a darker bluish red, depending upon the amount of oxygen it is carrying.

Blood Plasma

Ninety per cent of blood plasma is water, which provides a vehicle for the transportation of the cells suspended in this intercellular medium and of substances dissolved in it. The water also provides for the fluid needs of the cells and tissues. The balance of the blood plasma is made up of the *plasma proteins* and dissolved electrolytes, nutritive elements, and waste products. The plasma proteins are albumin, globulin, gamma globulins, and fibrinogen; these are large molecules which cannot escape through the walls of the capillaries. The proteins are responsible for the viscosity of the blood, carry immune substances, and produce the osmotic pressure needed to attract fluids from the tissues into the capillaries at the beginning of the venous return system. Many substances, such as the fatty materials of fatty acids, glycerides, and lipoproteins which will not dissolve in the watery blood plasma, are transported by their loose binding to the blood protein molecules. This is also true for most of the hormones that are distributed by the blood stream. Fibrinogen is the blood protein which, in bleeding, is transformed into an insoluble form, *fibrin.* Fibrin forms a meshwork of strands in a wound into which blood cells are trapped to form a *blood clot.* If all the fibrin is removed from a blood sample, the remaining elements of the blood plasma form *blood serum.*

Blood Cells (Formed Elements)

Blood cells are divided into two groups, the *red blood cells* (erythrocytes) and the *white blood cells* (leukocytes).

Erythrocytes.   Red blood cells are minute disks of cytoplasm, measuring about 7.5 micra in diameter, whose surfaces are concave so that they are dumbbell shaped when viewed on edge. The nucleus of the erythrocyte disappears as the cell develops. The cytoplasm which remains is converted into a framework (stroma) for the carrying of oxygen as the erythrocyte is swept passively through the circulatory system. A pigment, *hemoglobin,* is held in the stroma. Hemoglobin is a complex molecule of iron combined with protein. It has a remarkable affinity for oxygen, which it picks up readily during circulation of the blood through the lungs and binds to itself for transport, but which it yields as readily to the tissues of the body. Some of the carbon dioxide is carried similarly, with the rest dissolved in the blood plasma. An astronomical number of red blood cells are constantly circulating. In women each milliliter of blood contains 4.5 to 5 million erythrocytes, whereas in men there are 5 to 6 million cells per milliliter of circulating blood. Red blood cells live only three months. Aged cells are removed from circulation by the spleen to be replaced in the adult by new cells formed in the marrow of the long bones, vertebrae, sternum, and iliac bone.

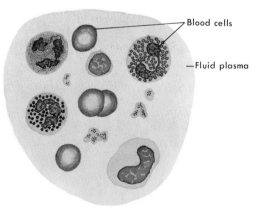

Blood cells

Fluid plasma

Leukocytes.   The leukocytes differ from the erythrocytes in being active ameboid forms which possess nuclei. There are only 5000 to 10,000 white blood cells per milliliter of circulating blood, but greater numbers can be released from storage when the body is threatened by infection, allergy, or stress.

The leukocytes, averaging several micra larger in size than erythrocytes, are divided into two main groups: *granulocytes,* which have granules in their cytoplasm, and *agranulocytes,* which do not. Three types of *granulocytes* are distinguished by the nature of the granules when stained by dyes during microscopic examination of the blood. *Neutrophils,* comprising between 65 to 75 per cent of leukocytes, contain fine, pinkish purple granules. These cells are phagocytic and respond in great numbers to infections anywhere in the body where they actively engulf bacteria and particulate matter. Neutrophils, in common with other leukocytes, reach the site of infection after transportation through the blood by passing through the endothelial capillary walls in ameboid fashion (diapedesis). *Eosinophils* (2 to 5 per cent) are characterized by large, bright red cytoplasmic granules. These granulocytes appear in large numbers at the sites of parasitic infections and allergic reactions. *Basophils* (0.5 per cent) possessing large purple granules, are believed to be related to the production of heparin, which prevents clotting of the blood within the body. All the granulocytes possess one common characteristic—nuclei, which show great variety of form. Therefore the term *polymorphonuclear leukocytes* is a common synonym for these cells. The nucleus of the neutrophil is formed into three to five lobes connected by chromatin filaments. The nucleus of the eosinophil is characteristically bilobar and that of the basophil is irregularly sinuous.

The *agranulocytes* are composed of the *lymphocytes* and the *monocytes. Lymphocytes* (20 to 25 per cent) are small round cells only slightly larger than the red blood cell. A large nucleus almost fills the cell, leaving only a rim of cytoplasm. Lymphocytes are believed to be related to immunological responses of the body and to be active in the detoxification of noxious substances. The *monocytes* (3 to 8 per cent) are the largest of the leukocytes. More cytoplasm is present than in a lymphocyte and it surrounds a kidney or horseshoe-shaped nucleus. Monocytes move in large numbers from the blood to the site of infection where they support neutrophils in a phagocytic action, including the scavenging of debris and dead neutrophils.

Blood platelets are not true cells but are tags of cytoplasm containing some chromatin material. Platelets tend to clump together and, by their adhesive qualities, plug wounds of blood vessels. It is believed that they also form thromboplastin, an integral component of the clotting process.

OVERVIEW OF THE CARDIOVASCULAR SYSTEM

Main Parts

The closed circulatory system consists of a ramifying system of tubes called the *arteries.* These diverge from the muscular pumping organ, the *heart,* which is part of the central partition (mediastinum) of the thorax. The arteries become thinner, smaller, and of simpler tissue composition as they diverge and repeatedly branch within all parts of the body. (See color plate 8.)

The arteries continue as microscopic vessels, the *arterioles,* which finally branch into a network of endothelial tubes, the *capillaries,* within the tissues and in contiguity with the intercellular substance. It is through capillary–tissue space relationship that the major functions of the circulatory system are achieved.

Capillaries gather into the tubular beginnings of a closed return system with the formation of the *venules.* Venules converge to form larger vessels, the *veins,* which run parallel to the arteries. Two or more veins of equivalent size approach each other, usually at an acute angle, to form a larger one. The same principle applies to the formation of the larger and larger veins of the extremities, head, and trunk until two major venous vessels are formed. These are the *venae cavae,* which conduct the blood from superior and inferior body regions into the heart.

Cardiovascular Circuits

The blood vessels are grouped into several functional circuits. A great, ramifying *systemic arterial circuit* leads out from the heart. The systemic arteries of the body distribute oxygenated arterial blood, rich in nutritive substances, to the organs and tissues of all the regions of the body. Key arterial vessels lead through the neck to the head, through the cervicoaxillary canal and axilla to the upper extremity, and through the thorax and abdomen to the major organs and wall of the trunk. At the brim of the pelvis, arteries proceed into the pelvis, perineum, and lower extremity.

The companion system of systemic veins makes up the *systemic venous circuit,* which returns blood, depleted of oxygen and nutriments, to the heart

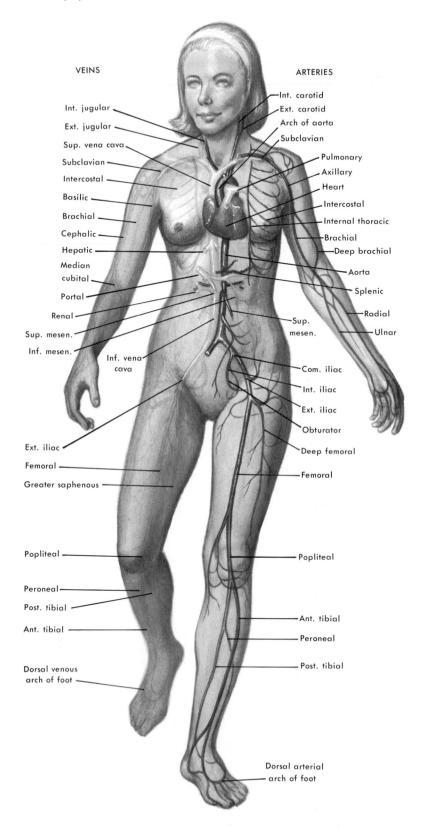

VEINS

ARTERIES

Int. jugular

Ext. jugular

Sup. vena cava

Subclavian

Intercostal

Basilic

Brachial

Cephalic

Hepatic

Median
cubital

Portal

Renal

Sup. mesen.

Inf. mesen.

Inf. vena
cava

Ext. iliac

Femoral

Greater saphenous

Popliteal

Peroneal

Post. tibial

Ant. tibial

Dorsal venous
arch of foot

Int. carotid

Ext. carotid

Arch of aorta

Subclavian

Pulmonary

Axillary

Heart

Intercostal

Internal thoracic

Brachial

Deep brachial

Aorta

Splenic

Radial

Sup.
mesen.

Ulnar

Com. iliac

Int. iliac

Ext. iliac

Obturator

Deep femoral

Femoral

Popliteal

Ant. tibial

Peroneal

Post. tibial

Dorsal arterial
arch of foot

from these regions. The blood is loaded with carbon dioxide and other body wastes but carries the secretory products of the endocrine system.

A separate but integral *pulmonary arterial circuit* takes such venous blood from the heart to the lungs for removal of volatile wastes and for reoxygenation. The *pulmonary venous circuit* returns this blood to the heart.

The terminology related to the two main paths of circulation within the human body is clarified by the table on this page.

VESSELS OF THE CARDIOVASCULAR SYSTEM

The simplest blood vessel, the capillary, is an endothelial tube whose embryologic origin and development were described in Chapter 1. Arteries and veins are larger and more complex in structure because of the addition of other tissues around the simple tube.

Capillaries

The number and density of capillaries vary from tissue to tissue in direct proportion to the metabolic activity of the tissue. Tissues with high metabolic rates and a great need for oxygen may have one capillary for each functioning unit. For example, each nerve cell in the brain is said to be located within the loop of one capillary so that it may be considered to be bathed in or separated from a pool of blood only by the thin endothelial membrane of the capillary. A cardiac muscle fiber may have a capillary paralleling it on each side. All types of muscle, nervous tissue, and the glands of the body are richly supplied with capillaries. In contrast hyaline cartilage is almost devoid of capillaries.

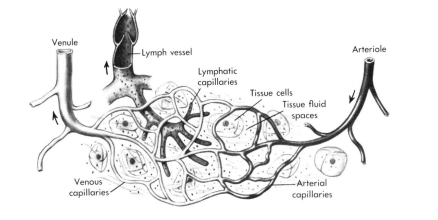

Arteries

From the lumen outward, three layers will be found making up the wall of an artery. These are called the *tunics* (coats) of the vessel.

1. *Intima:* a composite layer which consists of the endothelium, a delicate layer of subendothelial (intimal) connective tissue, and an internal elastic membrane.

2. *Media:* a layer of intermixed and interwoven elastic connective tissue fibers and smooth muscle cells. An external elastic membrane separates the media from the outer tunic.

3. *Adventitia:* a loose layer of areolar connective tissue which invests the vessel and gradually blends into the connective tissue of the area in which the vessel courses.

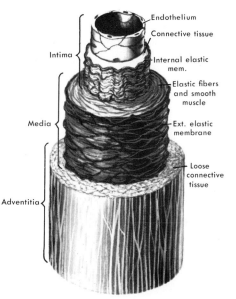

CIRCUIT	DISTRIBUTING VESSELS	CONVEYS	FOR
Systemic arterial	Systemic arteries	"Arterial" blood	Tissue metabolism
Systemic venous	Systemic veins	"Venous" blood	Return to the heart
Pulmonary arterial	Pulmonary arteries	"Venous" blood	Volatile waste removal and reoxygenation
Pulmonary venous	Pulmonary veins	"Arterial" blood	Delivery to the systemic arteries

Arteries are classified according to size as large, medium, and small (to which are added the smallest of all, the arteriole). A more functional classification of arterial vessels is the following:

1. *Conducting arteries:* These are the largest arteries, including those which spring from the heart and branch into large regional vessels to the major body regions. They are characterized by a large lumen and thick fibromuscular walls in which the elastic tissue component is prominent. Examples are the aorta and the pulmonary, brachiocephalic, and carotid arteries. (See also color plate 8.)

2. *Distributing arteries:* Names are given to these vessels according to the region the arteries traverse or the part of the body which they supply with blood. Examples are the femoral, brachial and hepatic arteries.

3. *Small arteries:* These are either unnamed or are given the name of a muscle or gland which they "nourish" or "supply." The vessels branch repeatedly within the organ and ultimately are lost to the eye as they pass into the microscopic zone.

4. *Arterioles:* These microscopic vessels ramify amid the tissue components of an organ. Arterioles are the smallest arterial vessels to possess smooth muscle. Since the walls are thin, the contraction of the smooth muscle can shut off arterial circulation into the capillary network beyond. In this way the amount of blood supplied, and thus the degree of function of the structures dependent upon arteriole or group of arterioles, can be controlled. (See also color plate 8.)

Veins

Blood pressure is not high in the venous channels which return the blood from capillary networks to the heart. Veins tend to be larger in diameter than arteries, but are irregular and have thinner walls. The elastic tissue component of the middle and outer tunics is less and the muscular content of the middle tunic is smaller in amount as well as more loosely organized.

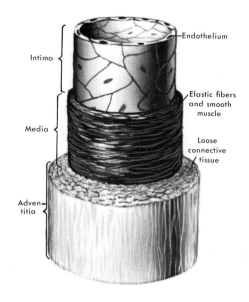

Veins are classified as large, medium, and small but do not in general show marked differences in structure. *Venules* are merely the smallest microscopic vessels that extend centrally from the capillary network. Veins, however, possess a feature not found in arteries. Their *valves* are flaps of endothelium, reinforced by connective tissue, which form pockets within the lumen. If venous blood tends to flow peripherally, the current causes the valves to spread out and restrain the retrograde flow. (See also color plate 7.)

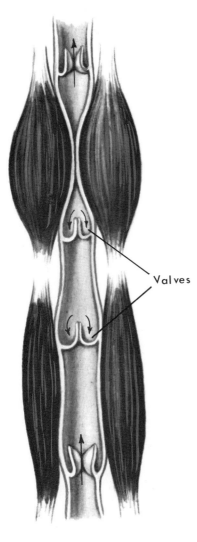

Valves

THE HEART AND MEDIASTINUM

The *heart* is located centrally within the thorax, within a central connective tissue partition, the *mediastinum,* which divides the thorax into right and left compartments which contain the lungs. Closely associated with the heart in the mediastinum are the large conducting arteries and the great veins which return venous blood from the head, neck, trunk, and limbs. Organs of the respiratory system (trachea, bronchi), digestive system (esophagus), lymphatic system, and nervous system also traverse the thorax en route to thoracic organs or to the abdomen or neck. These, with the heart, become the dominant structures of the mediastinal partition which overshadow the connective tissue framework. The individual shapes of these organs contained within the mediastinum produce characteristic conformations of the parts of this partition in which they lie or course. The organs are so closely packed within the mediastinum that a change in size or function or disease of one may have adverse effects upon the others.

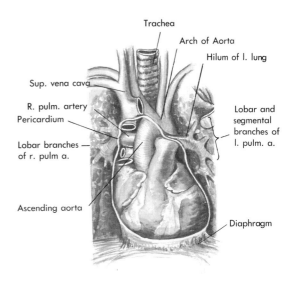

Middle Mediastinum

The mediastinum can be divided for descriptive purposes into superior and inferior portions. The *superior mediastinum* lies above the level of the heart. It contains the great arterial vessels; the superior vena cava which returns venous blood from the head, neck, and upper limbs; major nerves; the esophagus and trachea; and the main tube of the lymphatic system (thoracic duct). The *inferior mediastinum* is immediately subdivided into the anterior, middle, and posterior mediastina. The *anterior mediastinum* lies in front of the heart; the *posterior mediastinum* lies posterior to the heart. The anterior portion is relatively insignificant, but the posterior mediastinum contains a continuation of many of the structures found in the superior subdivision.

The *middle mediastinum* is occupied largely by the heart and its enclosing envelope, the *peri-*

cardium. This subdivision of the central thoracic partition bulges, because of the presence of the globular heart, to become the widest, most dominant portion.

Pericardium

The heart beats or contracts with a pulsatile wringing action that requires freedom of movement within the close relationships of the crowded mediastinal partition. It needs a structure to separate it from its neighboring structures and to reduce friction. The pericardium is a fibroserous envelope which meets these needs. The pericardium is separated from the heart by the *pericardial cavity.* A small amount of straw-colored *pericardial fluid* occupies this cleft between the outer surface of the heart and the inner surface of the pericardium.

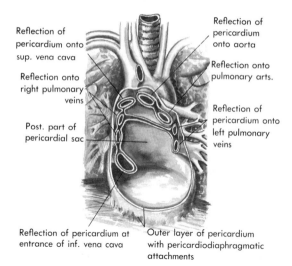

Reflection of pericardium onto sup. vena cava

Reflection onto right pulmonary veins

Post. part of pericardial sac

Reflection of pericardium onto aorta

Reflection onto pulmonary arts.

Reflection of pericardium onto left pulmonary veins

Reflection of pericardium at entrance of inf. vena cava

Outer layer of pericardium with pericardiodiaphragmatic attachments

The pericardium consists of two layers. An inner *serous layer* is comprised of a thin sheet of flat mesothelial cells, which at the base of the pulmonary vessels and aorta is reflected down to become the outer covering of the heart. The outer *fibrous layer* of the pericardium is a strong areolar connective tissue envelope about the heart which extends superiorly to blend into the outer tunic of the great vessels.

THE HEART

The heart is a muscular organ derived from the sinuous bending and partitioning of the cardiac tube of the embryo. Popular science descrip-

tions lend the impression that the heart resembles a house of four rooms with walls wide apart and space between them. This viewpoint is inaccurate. It is more realistic to consider the heart as a hollow contracting muscle which propels the blood passing through its lumen. The manner of its contraction is to be described shortly. The heart generates its own impulse to contraction. Nerves to the heart modify but do not institute cardiac contraction.

Structure of the Heart

Like all large blood vessels the heart has three layers, which are termed epicardium, myocardium, and endocardium. The *myocardium* is the thick muscular middle layer. The *epicardium* is the fibroserous outer layer which faces the pericardial cavity. The *endocardium* is the thin, smooth internal layer which lines the interior surfaces of the heart and is bathed by the blood which passes over them. Both the latter layers are thin in comparison to the myocardium, which dominates the organ.

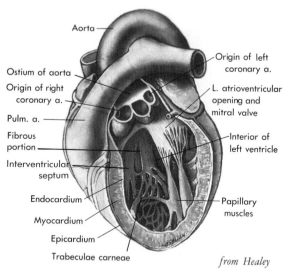

Aorta

Ostium of aorta

Origin of right coronary a.

Pulm. a.

Fibrous portion

Interventricular septum

Endocardium

Myocardium

Epicardium

Trabeculae carneae

Origin of left coronary a.

L. atrioventricular opening and mitral valve

Interior of left ventricle

Papillary muscles

from Healey

Myocardium. The myocardium is made up of spirally wound sheets of cardiac muscle which may be described as starting at a central point, winding around the heart, and returning to the point of origin. In the process these sheets surround the "chambers" of the heart and, upon contraction, propel the blood through them. The sheetlike nature of the myocardial musculature is evident only when special techniques are used to loosen the connective tissue which binds the sheets together. Then the myocardium can be unrolled to display its true form and disposition about the heart.

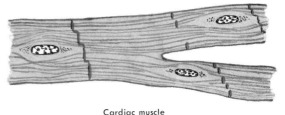

Cardiac muscle

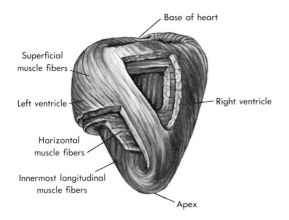

Base of heart

Superficial muscle fibers

Left ventricle

Right ventricle

Horizontal muscle fibers

Innermost longitudinal muscle fibers

Apex

Cardiac muscle is a striated form of muscle. It differs from striated voluntary (skeletal) muscle in the following ways:

1. It is not under the direct control of the nervous system.

2. It lacks definite motor unit organization and motor end plates.

Cardiac muscle fibers are branched. The myofibrils, however, do not branch but pass into one fiber branch or another. Cardiac muscle fibers are broader than those of voluntary muscle and are distinguished by brick-shaped nuclei. There is a narrow clear zone about each nucleus around which the striated myofibrils bend. A characteristic appearance is made by short bands, the *intercalated disks,* which cross the fiber in a step-like formation and represent the ends of cellular units. A more primitive type of cardiac muscle persists in the cardiac conduction system which initiates and propagates the impulse for contraction of the muscle fibers.

Cardiac muscle has an ultrastructural configuration similar to that of voluntary striated muscle. The unit membrane, thick and thin myofilaments, and banding are much the same. The clear area about the nucleus is loaded with mitochondria, and many mitochondria are embedded amidst the tubular endoplasmic reticulum, known in muscle as the *sarcoplasmic reticulum.* The abundant mitochondria, droplets of lipids, and many granules of glycogen indicate that energy reactions incident to muscle contraction are important to this contracting tissue.

Epicardium. The mesothelium of the serous membrane on the surface of the heart is firmly attached to the myocardium by an intermediate zone of delicate connective tissue. In certain grooves and about blood vessels, fat collections appear. The amount of fat increases in obese persons and in certain diseases until there is a great admixture of fat into the epicardial connective tissue.

Endocardium. The internal lining of the heart is made of a smooth thin endothelium which covers all the projections and irregularities. It is modified to form the valves of the heart. The endocardium is bound to the myocardium by a network of subendocardial connective tissue. This smooth lining is perforated by the openings of the *thebesian vessels,* which are described with the blood supply of the heart.

External Morphology of the Heart

The heart is an elongated globular organ which is suspended within the mediastinum so that it may pulsate freely.

External Form of the Heart. The heart constantly changes its form during life as the chambers within it fill or empty in response to the contraction of its muscular wall and the opening and closing of its valves. The heart of a young child is globular or pear shaped. It lengthens with growth of the trunk to assume an elongated cone shape and is directed obliquely to the left. In persons with broad chests the heart acquires a more horizontal disposition. This also occurs in certain heart diseases in which the heart increases in size. Persons with long narrow chests have long narrow hearts which do not bulge as much to the left. The statement often encountered that a person's heart is the size of his closed fist is only relatively true; it does not take into account the pericardium or the great vessels which spring from the heart or return blood to it.

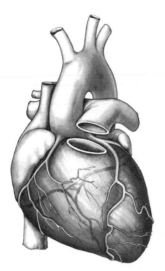

Parts of the Heart.   In form the heart is a blunt rounded cone. The base of this cone (and the *base* of the heart) is formed by the two *atria,* which are the receiving chambers for blood returning from all areas of the body or from the lungs. Between these chambers are the two great vessels, the pulmonary artery and the aorta, which conduct blood away from the heart. The base is directed superiorly, posteriorly, and to the right. The main body of the heart is formed by both the *ventricles,* the *apex* of the heart being formed by the left ventricle. The apex is directed inferiorly, anteriorly, and to the left.

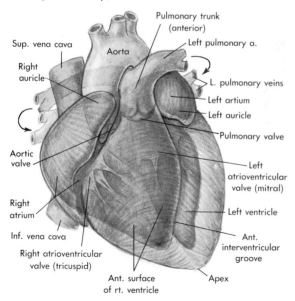

Sup. vena cava
Aorta
Pulmonary trunk (anterior)
Left pulmonary a.
Right auricle
L. pulmonary veins
Left artium
Left auricle
Pulmonary valve
Aortic valve
Left atrioventricular valve (mitral)
Right atrium
Left ventricle
Inf. vena cava
Ant. interventricular groove
Right atrioventricular valve (tricuspid)
Ant. surface of rt. ventricle
Apex

The heart contains four cleftlike spaces. When these spaces are filled transiently with blood between contractions of the heart, they are called the *"chambers"* of the heart. At the end of con-

traction of the myocardium which surrounds a chamber, the endocardial surfaces approach or touch. The chamber then temporarily disappears.

The cardiac chambers are only imperfectly delineated on the external surfaces of the heart. Examination of the epicardial surfaces reveals an *atrioventricular groove* crossing the *anterior* (sternocostal) *surface* of the heart between the upper third (base) and the lower two thirds. This groove separates the atria from the ventricles. The oblique position of the heart influences the position and relationships of the chambers. The *right atrium* is directed posteriorly toward the esophagus and descending aorta of the posterior mediastinum. The *left atrium* is turned forward to form part of the outline of the left side of the heart. The bulk of the heart projecting to the left and anteriorly is formed by the *left ventricle.* This portion, during contraction, raises so that the apex strikes against the internal surface of the left anterior thoracic wall. The impulse is frequently seen and felt here as the *apex* (apical) *beat* of the heart. Only the right border of the heart below the base (but much of the posterior surface) is formed by the right ventricle.

Internal Morphology of the Heart

Each chamber of the heart and its myocardial walls have a characteristic appearance, structure, and associated vessels.

Right Atrium.   The right atrium is a thin-walled muscular pouch placed between the ends of the two major systemic veins. Superiorly is the *superior vena cava* which drains venous blood from the head, neck, upper limbs, and a portion of the thoracic wall. Approaching from below is the *inferior vena cava* which delivers venous blood from the inferior parts of the body. The wall of the right atrium internally is thrown into a network of ridges (trabeculae) away from the entrance of the venae cavae. The trabeculation is distinctly marked in the small earlike appendage, the *right auricle,* which projects from the atrium. The endocardium presents three openings. These are the opening of the superior vena cava, the opening of the inferior vena cava, and the right atrioventricular opening through which blood passes into the right ventricle. The atrioventricular opening is guarded by the three cusps of the *right atrioventricular* (tricuspid) *valve.* The internal surface about this opening is frequently termed the "floor" of the right atrium. The veins which drain the heart of blood which has nourished its own tissue also open into the floor of the right atrium.

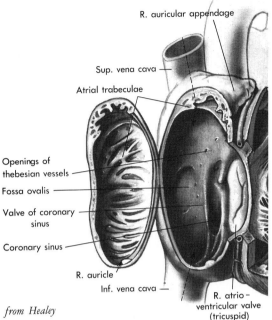

from Healey

Left Atrium. The smaller left atrium is situated opposite the right atrium in the base of the heart. The pulmonary artery and aorta ascend between the two atria as these vessels rise upward from the ventricular chambers. An *interatrial septum* separates the atrial chambers. The septum in embryonic development was closed only incompletely. A foramen in the septum, marked in adult life by the *fossa ovalis* on the median wall of the right atrium, permitted blood from the right atrium to pass into the left atrium. (The reasons are described in the section on the fetal circulation, p. 358.) Four *pulmonary veins* open into the left atrium to bring oxygenated blood from the lungs for recirculation to the tissues of the body. The *pulmonary openings* are located posteromedially. On the floor of the left atrium is found the *left atrioventricular opening* through which the blood passes into the left ventricle; this opening is guarded by two cusps of the *left atrioventricular* (bicuspid, mitral) *valve.* (See illustration at top of right column.)

Right Ventricle. The right ventricle is smaller than the left. The two ventricles share a common wall, the interventricular septum. This is marked externally by *anterior* and *posterior interventricular grooves* in which the main vessels of the heart itself descend from the atrioventricular groove. The right ventricular chamber is cleft-like, and the internal surface is raised into muscular ridges called the *trabeculae carneae.* Several of the muscular ridges are more prominent and are detached superiorly from the myocardial mass. These are the *papillary muscles* from the tips of which collagenous threads, the *chordae tendineae,* extend superiorly like minute parachute cords to the tricuspid valve. One muscle bar crosses the lower portion of the ventricular cavity from the interventricular septum to the opposite wall. Early anatomists thought that this muscular ribbon restrained the right ventricle from overexpanding when it filled with blood, and they named it the *moderator band.* It is mainly a bridge for the tissue which carries the impulse to cardiac contraction.

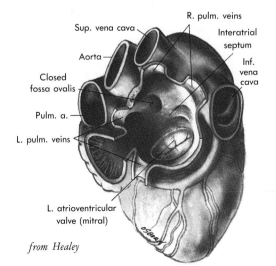

from Healey

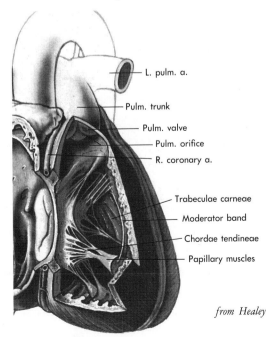

from Healey

The right atrioventricular opening and the under-surface of the tricuspid valve appear centrally in the upper part of the right ventricular chamber. The *pulmonary orifice,* which opens into the *pulmonary trunk* artery, is located considerably anterior to the tricuspid opening. (See illustration at top of following page.)

Left Ventricle. In cross section the left ventricle is more oval and more massive, and its walls are two to three times thicker than the right ventricle. Since the left ventricle receives oxygenated blood destined for all the tissues of the body, and since it must forcefully pump this blood out into the major conducting and distributing arteries, it is obvious that it will dominate the smaller right ventricle which pumps venous blood only a short distance to the lungs.

The thicker myocardium is raised into prominent trabeculae carneae, projects into stronger papillary muscle bars, but lacks a moderator band. The interventricular wall, at one time patent during embryonic life, is strongly muscular in its lower two thirds but distinctly fibrous in its upper third near the "roof" of the chamber. The fibrous portion of the interventricular septum is part of the *fibrous skeleton of the heart.* The fibrous skeleton is a heavy collagenic plate which roughly corresponds to the plane of the atrioventricular groove. This fibrous plate is perforated by the atrioventricular openings and the ostia of both the pulmonary trunk and the aorta. It provides the attachment for the spirally wrapped myocardial muscle sheets. The openings of the left ventricle are the left atrioventricular opening,

guarded by the mitral (bicuspid) valve, and the funnel shaped ostium of the ascending aorta located at the right of the atrioventricular opening.

Valves of the Heart

The atrioventricular openings and the ostia of the two great arterial vessels are but openings in the fibrous skeleton of the heart. They are closed by reflections of the endocardium which take the form of leaflike or cuplike flaps. The valves prevent blood from returning to the chamber from which it just came when the heart contracts. In so doing the blood is diverted to the proper ostium by the atrioventricular valves and prevented from returning to the ventricles by the pulmonary and aortic valves.

Tricuspid Valve. The right atrioventricular valve is formed by three triangular flaps of endocardium which, when stretched out by the pressure of blood within the right ventricle, meet to close the aperture between the right atrium and the right ventricle. The three cusps are named, according to their positions, *anterior, posteroinferior, and septal.*

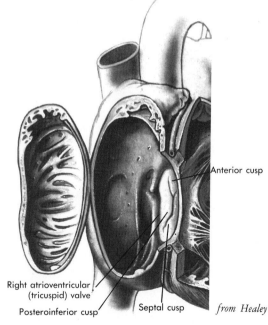

Anterior cusp

Right atrioventricular (tricuspid) valve

Posteroinferior cusp

Septal cusp

from Healey

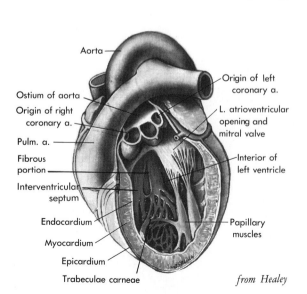

Aorta

Ostium of aorta

Origin of right coronary a.

Pulm. a.

Fibrous portion

Interventricular septum

Endocardium

Myocardium

Epicardium

Trabeculae carneae

Origin of left coronary a.

L. atrioventricular opening and mitral valve

Interior of left ventricle

Papillary muscles

from Healey

Each cusp is connected by the white collagenous strands of the chordae tendineae to the papillary muscles of the right ventricular myocardium. The valve opens passively when the right atrium propels its blood into the ventricle. Closure of the valve is accomplished mainly by ventricular blood pressure. The blood could pass just as easily back into the atrium as into the pulmonary trunk when the ventricle contracts. But the blood forces the cusps "upward" into the orifice. Then contraction of the papillary muscles prevents the cusps of a healthy valve from inverting. The blood has only one way to go—into the pulmonary circulation.

Bicuspid Valve. The left atrioventricular valve has only two cusps, the *anterior* and the *posterior,* and only two sets of chordae tendineae and papillary muscles.

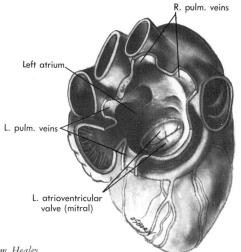

from Healey

Otherwise its construction and its action is the same as the tricuspid valve. Its function is to prevent the reflux of blood into the left atrium to ensure the propulsion of the oxygenated blood into the aorta.

Semilunar Valves. The ostia of both the pulmonary trunk and the aorta are closed by pocket-like valves, which, except for their larger size, resemble the valves of veins. As with venous valves, the semilunar valves are constructed to permit the flow of blood in only one direction. Blood is expelled from the ventricular chambers in jetlike spurts. When the friction and hydraulic resistance of the more distal vessels are encountered, these spurts are "ironed out" into pulselike waves of blood through these vessels. The aorta and pulmonary trunk are highly elastic to accommodate to the high pressure of the blood propelled into them. As the ventricular contraction ceases, arterial blood fills the pockets of the

semilunar cusps. The valve cusps close upon each other to close the ostia effectively, thus preventing dilatation of the heart because of reverse flow.

The *right semilunar valve,* better known as the *pulmonary valve,* is situated in the orifice of the pulmonary trunk. It is oval and is made up of three cusps, the *posterior, left anterior,* and *right anterior.*

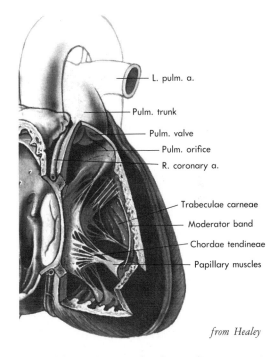

from Healey

The *left semilunar valve,* better known as the *aortic valve,* occupies the more circular aortic orifice. The three cusps are named differently. They are the *anterior, right posterior,* and *left posterior cusps.*

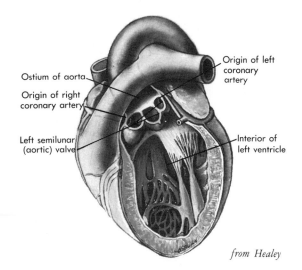

from Healey

THE CARDIAC CONDUCTION SYSTEM

The heart beat is independent of the central nervous system. A network of clumps and strands of primitive cardiac muscle permeates most of the heart. The connections and ramifications of this tissue form the *cardiac conduction system* which initiates and distributes the impulses stimulating cardiac contraction.

Tissue of the Cardiac Conduction System

Certain fibers of the embryonic heart do not specialize into the adult form of cardiac muscle, but retain an embryonic appearance and interconnect in a diffuse network. Some of the fibers are gathered into tangled knots in several locations. The rest are gathered into strands and bundles which spread throughout the myocardium.

Nature of Conduction Tissue.   Whether in the form of knotted nodes or ramifying bundles, the tissue is made up of fibers of primitive cardiac muscle. The fibers of the bundles are broader than adult cardiac muscle and lighter in appearance, whereas those of the nodes are narrower. The broad fibers are characterized by oval double nuclei surrounded by a large clear zone. The myofibrillae are sparser than in the adult form and swing wide around the clear zone to enhance the appearance of a "halo zone" about the double nuclei. Although the fibers branch, with the myofibrillae passing into one branch or another, the distinctive intercalated disks of the adult cardiac muscle are absent.

Electron microscopic observations of cardiac conduction tissue confirm the looser organization of fewer myofilaments in the sparser myofibrillae. Numerous mitochondria are distributed in a random manner amidst the myofibrils.

Disposition of the Conduction Tissue.   The *sinuatrial node* (of Keith and Flack) is a subendocardial knot of conduction tissue located high on the posteromedial wall of the right atrium adjacent to the opening of the superior vena cava. The *atrioventricular node* (of His and Tawara) is a similar mass of tissue located in the floor of the right atrium. There is no morphological connection between the two nodes. Extending medially from the atrioventricular node is a thick bundle of conduction tissue which crosses the rough floor of the right atrium behind the posterior cusp of the tricuspid valve. This is the *common atrioventricular bundle* (of His and Kent). This bulky strand enters the fibrous (membranous) upper portion of the interventricular septum where it divides into *right* and *left limbs* or bundles. At the point where the atrioventricular bundle enters the septum it is the only muscular connection between the atria and the ventricles. Otherwise the atria are completely separated from the ventricles by the fibrous skeleton of the heart.

The *right limb* forms a cylindrical strand which descends in the myocardium of the right side of the interventricular septum. As it does, it distributes branches to the septum and the lower myo-

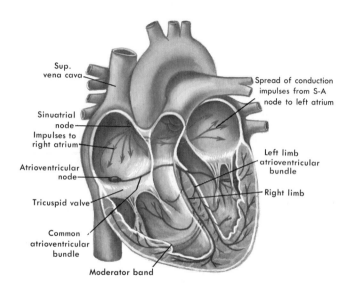

Sup. vena cava

Spread of conduction impulses from S-A node to left atrium

Sinuatrial node

Impulses to right atrium

Left limb atrioventricular bundle

Atrioventricular node

Right limb

Tricuspid valve

Common atrioventricular bundle

Moderator band

cardium of the right ventricle. It sends a thick column across to the lateral wall of myocardium through the *moderator band.* The right limb then ascends through the lateral wall, giving off branches in its course, to terminate in the myocardium up under the fibrous skeleton.

The *left limb* is in the form of one or several ribbon-like strands that, different from its counterpart on the right, lie under the endocardium. It descends to the apex, which it supplies, giving off branches to the left side of the septal myocardium on the way. There is no moderator band. The distal portion of the left limb ascends from the apex through the lateral wall of the left ventricle. It gives off branches in its ascent and terminates as does the right limb. (See illustration at bottom of opposite page.)

The Purkinje Fibers. The branches of the distributing strands of the cardiac conduction system ramify, as do nerves, and enter the microscopic zone while giving off sub-branches to all portions of the ventricular myocardium. An actual morphological connection is made between the finest conduction system fibers and an adult cardiac muscle fiber. The primitive conducting fiber tapers down to become continuous with an adult cardiac muscle fiber. The first portion of the conduction system to be described was this connection, and earlier anatomists ascribed the name "Purkinje fiber" to it in recognition of the discoverer. The term, however, has come to be loosely applied to the whole system.

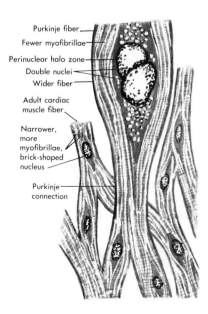

Purkinje fiber
Fewer myofibrillae
Perinuclear halo zone
Double nuclei
Wider fiber
Adult cardiac muscle fiber
Narrower, more myofibrillae, brick-shaped nucleus
Purkinje connection

Function of the Cardiac Conduction System

The impulse that stimulates cardiac contraction is generated in the sinuatrial node. It apparently spreads as an electrical wave over the surface of the right atrial cardiac muscle fibers and across the interatrial septum to the myocardium of the left atrium. These two chambers contract in response almost simultaneously to move the blood contained within them through the atrioventricular orifices into the ventricles.

The impulses spreading across the right atrial musculature are gathered in by the atrioventricular node. This node can be a subsidiary "pacemaking center" but usually passes sinuatrial impulses on through the common ventricular bundle toward the ventricles. Further distribution of the impulse is made via the common bundle, its limbs, and its ramifications.

The impulse to contraction reaches the interventricular septum first. This causes the septum to contract early in order to shorten as the size and volume of the ventricular myocardium are reduced with contraction of these chambers. The impulse stimulates contraction then in the apex, then in the lower body myocardium with the last contraction being of the myocardium up near the atrioventricular groove and, hence, nearest the point of egress of the blood from the ventricles into the great arteries. The muscular action of the heart, because of the timing of the arrival of the impulse and the spiral nature of the myocardial sheets, is more of a wringing action than like the shortening of a skeletal muscle. The contraction of the heart can be likened to the wringing out of a wash cloth. The distribution of the conduction system tissue and the sequence of the contractile response become the morphological basis for the "cardiac cycle" and the action of the valves of the heart. The sinuatrial node has a blood supply of its own through a sinus node artery, usually stemming from the right coronary artery. The richness of the arterial blood supply lends support to its autonomous nature in initiating the impulse for cardiac contraction.

Nerves to the Heart

Branches of nerves from the two portions of the autonomic nervous system end within the sinuatrial node or adjacent to its tissue. In addition these nerves send fine branches to parallel the cardiac muscle fibers of the myocardium. (See color plate 14.) It is not believed currently that any of these fibers end upon the surface or within the muscle fibers as is true for smooth and skeletal muscle. It is known, however, that autonomic nerve fibers emanating from blood vessel-regulating centers in the brain are able to influence or to modify the inherent rhythm of the heart. Vagus (parasympathetic) nerve impulses result in slowing of the heart beat whereas thoracolumbar outflow fibers (sympathetic) speed up the rate of cardiac contraction.

BLOOD SUPPLY OF THE HEART

The heart, constantly beating throughout life, never rests in the sense that other organs, such as the stomach, have resting phases. Its metabolic requirements are exceeded only by those of the brain. An abundant supply of oxygenated blood is a necessity to a vital organ such as the heart. It has been mentioned that the capillary network of the myocardium is so profuse that each muscle fiber is paralleled by a capillary.

The Coronary Arteries

Just above the aortic semilunar valve is the origin of the aorta, which is widened into three pockets or *aortic sinuses;* these correspond to the cusps of the aortic valve. The *left coronary artery* branches from the aorta from the left posterior aortic sinus. The *right coronary artery* springs from the anterior sinus. These two arteries are the main blood supply to the heart.

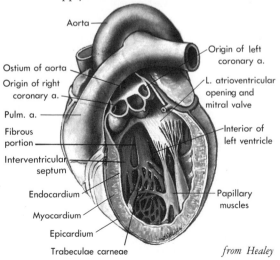

Aorta —

Ostium of aorta —
Origin of right coronary a. —

Pulm. a. —

Fibrous portion —

Interventricular septum —

Endocardium —

Myocardium —

Epicardium —

Trabeculae carneae

— Origin of left coronary a.

— L. atrioventricular opening and mitral valve

— Interior of left ventricle

— Papillary muscles

from Healey

Left Coronary Artery. The pulmonary trunk is thrust forward by the aorta. The left coronary artery winds between the trunk and the left atrium to reach the atrioventricular groove. The vessel courses to the left in the epicardial fat of this groove, giving off the following branches:

1. The *anterior interventricular artery* descends in the anterior interventricular groove along the plane of the interventricular septum to the apex. It provides the main supply to the septum and gives off branches to the myocardium of both ventricles. It passes posteriorly around the apex to ascend in the posterior interventricular groove where it anastomoses with a branch of the right coronary artery.

2. The *circumflex artery* is a continuation of the left coronary artery along the atrioventricular groove. Myocardial branches are provided to the upper part of the left ventricle and to the adjacent left atrium. The circumflex artery ends as it meets and anastomoses in the groove with the terminal part of the right coronary artery.

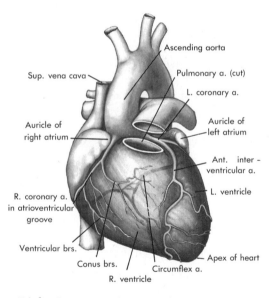

— Ascending aorta

Sup. vena cava —

— Pulmonary a. (cut)

— L. coronary a.

Auricle of right atrium —

— Auricle of left atrium

— Ant. inter-ventricular a.

— L. ventricle

R. coronary a. in atrioventricular groove

Ventricular brs.

Conus brs.

— Apex of heart

Circumflex a.

R. ventricle

Right Coronary Artery. The right coronary artery proceeds to the right in front of the aorta. The vessel winds under the edge of the right atrium to enter the atrioventricular groove. It courses to the right and to the posterior surface of the heart. It terminates by anastomosing with the circumflex branch of the left coronary artery. It gives off the following branches:

1. *Atrial branches:* Several branches are given off to the right atrium. Special *nodal arteries* are sent to the sinuatrial node, atrioventricular node,

and common atrioventricular bundle. Any of these may receive blood from branches of the left coronary artery instead.

2. *Conus branches:* These branches are sent to the origin of the aorta (often called the conus arteriosum) and the adjacent ventricular myocardium.

3. *Ventricular branches:* A *right marginal artery* originates in the groove and descends along the right border of the heart, supplying the right ventricular myocardium. The *posterior interventricular artery* descends in the posterior interventricular groove along the posterior cardiac surface. It supplies the upper part of the interventricular septum and adjacent portions of the myocardium of both ventricles.

Intracardiac Blood Supply

The branches of the coronary arteries are essentially *"end arteries"* in that they supply demarcated portions of the myocardium without an overlapping supply from other branches. The pathological consequence of such a morphologic arrangement is the common coronary artery infarction. This occurs when an artery is occluded by a disease, such as arteriosclerosis. A mass of myocardial tissue may be suddenly deprived of its needed blood when the lumen becomes too small, with or without spasm of the smooth muscle of the arterial wall, to supply sufficient blood. There are many small artery-to-artery anastomoses between adjacent vessels of arteriolar size. These are insufficient to provide blood when a main artery is suddenly occluded, but the anastomotic vessels may dilate to become vessels of considerable size if the disease process develops slowly.

Veins of the Heart

The heart's own capillary blood returns to the right atrium. Several vessels are involved:

1. Veins from the musculature of the right atrium and right ventricle which open directly into the right atrium.

2. Major *cardiac veins* roughly paralleling the main coronary arteries. These veins empty into a widened venous channel, the *coronary sinus,* which pursues a short course in the atrioventricular groove. It empties into the right atrium between the opening of the inferior vena cava and

the tricuspid orifice. A flap of endocardium, the *valve of the coronary sinus,* deflects the blood toward the tricuspid opening. The main veins which contribute to the formation of the coronary sinus are the

a. *great cardiac vein,* which parallels the anterior interventricular artery,

b. *left marginal vein,* which drains the left ventricular margin,

c. *middle cardiac vein,* which accompanies the posterior interventricular artery,

d. *small cardiac vein,* which drains venous blood from the border of the right ventricle, and the

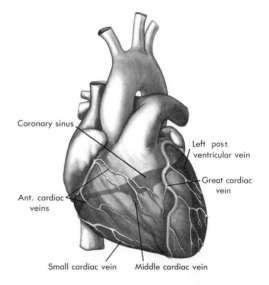

e. *oblique vein of the left atrium,* which conveys blood from the left atrium to the coronary sinus.

3. *The venae cordis minimae* (smallest cardiac veins) are minute veins (of Thebesius) which collect blood from small areas near the endocardial surface of any chamber. They open by minute pores into the crevices between the trabeculated interior surface of their own chamber.

THE CARDIAC CYCLE AND PATHS OF CIRCULATION

The circulation of blood is through a continuous, closed tubular system. The heart is inserted into the system during development, and the two great circulatory paths stem from it. The heart, therefore, is a starting point in considering the paths of circulation, since all blood is pumped from it and through it and returns to it.

The Cardiac Cycle

Except for the venae cordis minimae of other chambers, all blood returned to the heart from the tissues and organs of the body (except the lungs) enters the right atrium. The venous blood arrives via the superior or inferior venae cavae, the direct veins of the right atrium, or the heart's own coronary sinus.

The *first phase of cardiac contraction* occurs when impulses from the sinuatrial node cause a quick contraction of the atrial musculature. Although both atria contract almost simultaneously, the right atrium leads because the impulse must travel farther to reach the left atrium. In studying the blood flow through the circulatory system, however, the left atrium will be ignored for the moment. At the end of the first phase, blood has passed into the right ventricle to fill this chamber.

The *second phase of cardiac contraction* occurs when the *right ventricle* contracts to force the venous blood out into the pulmonary trunk through which it is distributed to the lungs for oxygenation and for the removal of volatile wastes.

The *third phase of cardiac contraction* begins with the return of oxygenated blood from the lungs via the four pulmonary veins to the *left atrium.* This chamber fills with oxygenated blood at the same time that the right atrium is receiving venous blood. With contraction of the left atrium the limp cusps of the bicuspid valve are pushed aside as the blood flows through the left atrioventricular orifice into the left ventricle.

The *fourth phase of cardiac contraction* begins with the powerful contraction of the thick myocardium of the *left ventricle.* The blood is compressed by the wringing action of ventricular contraction. This pressure forces the blood toward the base of the heart. The mitral valve is closed by the onrush of the blood which is then diverted into the aortic orifice. The arterial blood is then conducted through the aorta into the coronary arteries and into the great systemic arterial vessels of the superior mediastinum. (See color plate 5.)

A. During diastole chambers fill and expand to dashed line dimensions.

B. During systole atrial and ventricular musculature contract to solid line dimensions to pump the blood as shown.

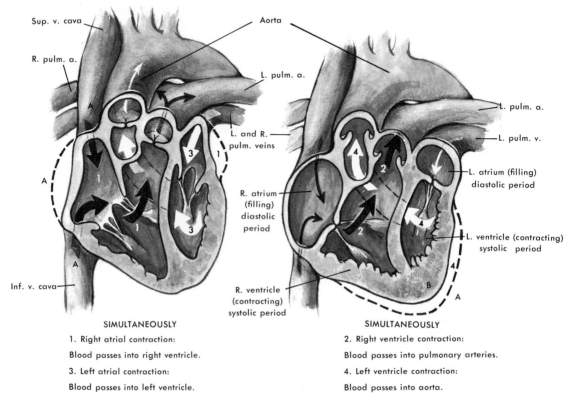

SIMULTANEOUSLY

1. Right atrial contraction:

Blood passes into right ventricle.

3. Left atrial contraction:

Blood passes into left ventricle.

SIMULTANEOUSLY

2. Right ventricle contraction:

Blood passes into pulmonary arteries.

4. Left ventricle contraction:

Blood passes into aorta.

Vascular Circuits in Relation to the Heart

There are two circuits within the circulatory system, the *pulmonary* and the *systemic,* which originate from the heart.

Pulmonary Circuit. The right atrium and the right ventricle are *functionally* separate from the chambers on the left. The right atrium and right ventricle receive venous blood and pump it to the lungs. They make up a separate system which includes the pulmonary trunk, the pulmonary arteries, the vasculature of the lungs, and the pulmonary veins. This vascular loop is called the *pulmonary circuit* of the circulatory system.

Systemic Circuit. The left atrium and the left ventricle receive oxygenated blood from the pulmonary veins and pump the arterial blood through the great conducting arteries to the tissues and organs of the body. The left chambers, therefore, join with all the major arteries, capillary networks, and systemic veins in forming a larger vascular loop, the *systemic circuit* of the circulatory system.

THE ARTERIES OF THE BODY

The main function of the cardiovascular system is to bring the blood into proximity with the intercellular medium and cells of tissues. In keeping with this concept, all the arteries of the body can be grouped together as conductors and distributors of blood. Blood is conveyed from the ventricles of the heart through the great vessels of the mediastinum which are the beginnings of the distribution system.

THE GREAT ARTERIES OF THE MEDIASTINUM

The pulmonary and systemic arterial circuits originate respectively with the pulmonary trunk and the ascending aorta.

PULMONARY TRUNK AND PULMONARY ARTERIES

The pulmonary trunk extends superiorly and posteriorly from the right ventricle enclosed with the ascending aorta within the pericardium. About 2 inches above its origin, the pulmonary trunk ends by dividing into the *right* and *left pulmonary arteries.* This division occurs at the left of the midline under the arch of the aorta. The *right pulmonary artery* extends to the right, crossing in front of the esophagus and right main bronchus to reach the root of the lung.

The *left pulmonary artery,* with a shorter course, plunges laterally in front of the descending portion of the aorta and the left main bronchus to reach the root of the lung. The *ligamentum arteriosum* connects the artery to the arch of the aorta. Its significance is described with the fetal circulation (p. 358).

Both pulmonary arteries divide within the lungs as companions to the divisions of the main bronchi, first into lobar and then into segmental pulmonary arteries.

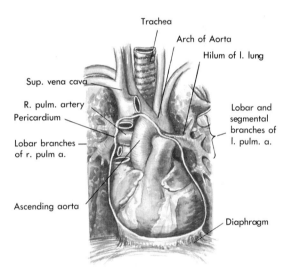

Trachea

Arch of Aorta

Hilum of l. lung

Sup. vena cava

R. pulm. artery

Pericardium

Lobar and segmental branches of l. pulm. a.

Lobar branches of r. pulm a.

Ascending aorta

Diaphragm

THE AORTA

This great systemic arterial conductor can be divided for descriptive purposes into three portions: the *ascending aorta,* the *aortic arch,* and the *descending* (thoracic) *aorta.*

Ascending Aorta

The first portion of the aorta rises from its aortic sinuses to ascend in front of the left atrium. At this point the ascending aorta lies to the left of the right atrium and the superior vena cava. The pulmonary trunk passes over to the left side of the ascending aorta before dividing into the pulmonary arteries. The only branches of the ascending aorta are the coronary arteries.

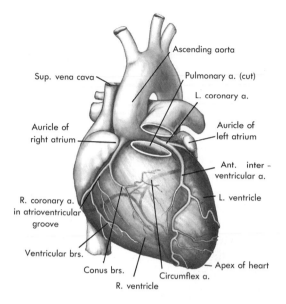

Ascending aorta

Sup. vena cava

Pulmonary a. (cut)

L. coronary a.

Auricle of
right atrium

Auricle of
left atrium

Ant. inter -
ventricular a.

R. coronary a.
in atrioventricular
groove

L. ventricle

Ventricular brs.

Conus brs.

Apex of heart

Circumflex a.

R. ventricle

Arch of the Aorta

As the aorta rises from the pericardial sac it becomes a wide curving arch. The direction of the aortic arch is first superior and posterior, then posterior and to the left, and finally posterior and inferior as it continues as the descending aorta. The arch encloses the bifurcation of the pulmonary trunk and passes in front of the right pulmonary artery but behind the left pulmonary artery and bronchus. Three major arteries are given off from the convex superior surface of the arch. These are the *brachiocephalic* (innominate) artery to the right upper limb, the neck, and the head, the *left common carotid artery* to the neck and the head; and the *left subclavian artery* to the neck and left upper limb.

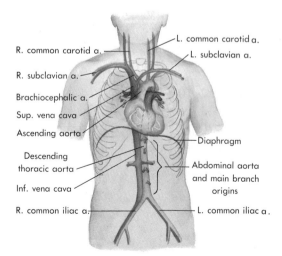

R. common carotid a.

L. common carotid a.

R. subclavian a.

L. subclavian a.

Brachiocephalic a.

Sup. vena cava

Ascending aorta

Diaphragm

Descending
thoracic aorta

Abdominal aorta
and main branch
origins

Inf. vena cava

R. common iliac a.

L. common iliac a.

Descending Aorta

The aorta continues inferiorly through the posterior mediastinum to convey blood to all the regions and organs of the body below the level of the aortic arch. The descending aorta gradually courses medially and comes forward in the mediastinum until it lies in front of the body of the twelfth thoracic vertebra. At this point the esophagus, which has descended through the mediastinum to the right of the aorta, swings over to the left to intervene between the aorta and the pericardium. The esophagus and the aorta then pass into the abdomen through separate apertures in the thoracoabdominal diaphragm. There are many small branches from the descending aorta within the thorax: *Posterior intercostal arteries* pass to the thoracic wall. Other branches are the *bronchial, esophageal, pericardial,* and *mediastinal arteries* which supply arterial blood to the structures for which they are named.

THE GREAT CONDUCTING ARTERIES

Blood reaches the major regions of the body by the following routes:

1. The *brachiocephalic artery* rises through the superior mediastinum, behind the large veins draining the upper part of the body, as the first branch of the aortic arch. It ascends through the superior thoracic aperture in front of the apex of the lung, where it divides behind the sternoclavicular joint into the *right subclavian* and *right common carotid arteries.* The *subclavian* is the artery to the upper limb but it also sends arterial blood to the brain via the *vertebral artery,* to the anterior wall via the *internal thoracic* (internal mammary) *artery,* and to parts of the neck and shoulder via the *thyrocervical* and *costocervical arteries.*

The *right common carotid* artery sends blood to the right side of the neck, face, jaws, deep head structures, scalp and brain by way of its *external* and *internal carotid branches.*

2. The *left common carotid* artery as the second branch of the aortic arch supplies the left side of the same areas of the head and neck as its counterpart on the right.

3. The *left subclavian artery* is the third branch of the aortic arch. Its area of supply on the left side is the same as that of its fellow on the right.

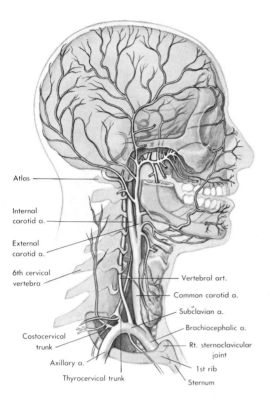

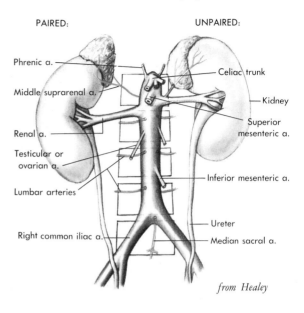

from Healey

branches of the main conducting vessels just described.

4. The thoracic branches of the *descending aorta* and the *pulmonary arteries* supply thoracic structures as described previously.

5. The descending aorta becomes the *abdominal aorta* upon passing through the thoraco-abdominal diaphragm. This vessel not only carries arterial blood to the abdominal organs and portions of the abdominal wall but also conveys blood toward the pelvis, perineum, and lower limbs. The major paired branches of the *abdominal aorta* are the *phrenic* (diaphragmatic) and *lumbar arteries* to the abdominal wall, the *suprarenal arteries* to the *adrenal glands,* the *renal arteries* to the kidneys, and either the *testicular* or *ovarian arteries* to the reproductive glands. Single arterial branches of the aorta are the *celiac trunk* to the upper digestive organs, liver, and spleen; the *superior mesenteric artery* to the upper intestinal tract; and the *inferior mesenteric artery* to the lower portions of the digestive system.

6. At the brim of the pelvis the abdominal aorta branches into the *right* and *left common iliac arteries* which supply the pelvis, pelvic organs, perineum, gluteal and hip regions, lower abdominal wall, and lower limbs.

The specific distributing vessels are smaller

ARTERIES OF THE HEAD AND NECK

The blood supply to the head and neck is from branches of the subclavian and carotid arteries.

SUBCLAVIAN BRANCHES TO THE HEAD AND NECK

The subclavian artery on each side crosses the root of the neck in front of the apex of the lung. The vessel passes behind the scalenus anterior muscle and over the outer border of the first rib to continue as the *axillary artery* to the upper limb. Branches from the subclavian artery to the head and neck are the *vertebral, thyrocervical,* and *costocervical arteries.*

Vertebral Artery

This artery ascends deeply through the root of the neck to enter the foramen of the transverse process of the sixth cervical vertebra. Its upward course carries it successively through the cervical transverse foramina above until it turns medially over the atlas to enter the skull through the foramen magnum. *Vertebral branches* are given to the cervical spinal column and associated muscles, ligaments, and membranes. Within the skull it participates in forming the *arterial circle of the base of the brain* (circle of Willis).

Thyrocervical Trunk

The thyrocervical trunk is a short common stem artery which ascends only a short distance in the root of the neck before dividing into its main branches. These are the *inferior thyroid, transverse cervical,* and *suprascapular arteries.*

Inferior Thyroid Artery.   The *inferior thyroid artery* is a most important visceral and muscular artery in the neck. It ascends amid the neck structures to supply the lower part of the thyroid and parathyroid glands, larynx, trachea and esophagus, and pharynx by means of branches of the same names. The scalenus anterior and longus capitis muscles are supplied by the ascending cervical branch. Vertebral branches are given off to the spinal column and associated structures.

Transverse Cervical Artery.   This vessel crosses posteriorly over the side of the neck to give off muscular branches to the muscles of the posterior scapular area, such as the levator scapulae, trapezius, and rhomboids.

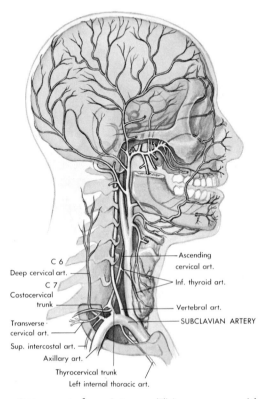

C 6
Deep cervical art.
C 7
Costocervical trunk
Transverse cervical art.
Sup. intercostal art.
Axillary art.
Thyrocervical trunk
Left internal thoracic art.

Ascending cervical art.
Inf. thyroid art.
Vertebral art.
SUBCLAVIAN ARTERY

Suprascapular Artery.   This artery roughly parallels the course of the transverse cervical artery but courses more laterally. It sends small branches to the acromial area, muscles, and ligaments located superiorly about the shoulder joint, and its terminal branches join an anastomotic ring of vessels about the scapula. The *scapular anastomosis* is described in connection with the axillary artery.

Costocervical Trunk

The third main branch of the subclavian artery only incidentally supplies the neck structures through its small *deep cervical artery* to the deep posterior neck muscles. (See the description of the blood supply of the thoracic wall for the superior intercostal branch of the costocervical trunk and for the other subclavian branch, the internal thoracic [internal mammary] artery.)

THE CAROTID ARTERIES TO THE HEAD AND NECK

It has been seen that the subclavian artery supplies structures of the root and lower parts of the neck. The *common carotid arteries* ascend deeply through these regions, therefore, with no branches. These vessels share the tubular *carotid sheath* of fascia with the internal jugular vein and the vagus nerve. These structures lie upon the prevertebral muscles and the transverse processes of the cervical vertebrae. The common carotid artery may be compressed against the transverse processes, especially the more prominent carotid tubercle of the sixth cervical vertebra. This maneuver, the application of deep pressure posteromedially with the thumb about an inch and a half above the sternoclavicular joint, is a first aid pressure point procedure. It is used bilaterally and violently in hand to hand combat to shut off blood to the head, to compress the trachea (asphyxiation) and to abruptly slow the heart (vagus nerve). The common carotid artery divides into its *internal* and *external carotid artery* branches at about the level of the superior border of the larynx, about two finger-breadths below the body of the mandible. A dilatation at the point of bifurcation of the common carotid artery is known as the *carotid sinus.* Nerve endings, mainly from the vagus nerve, in the wall of the dilatation respond to changes in blood pressure. Reflexes through the central nervous system result in changes in heart rate and strength of contraction. Adjacent nerve cells making up the *carotid body* respond similarly to changes in the composition of the blood, particularly the oxygen–carbon dioxide ratio.

Internal Carotid Artery

The *internal carotid artery* continues in the direction of and along the deep course of the common carotid artery. The internal carotid artery courses upward along the pharynx toward the base of the skull without giving off branches in the neck. This is the major artery to the brain and other intracranial and orbital structures. It enters the carotid canal at the base of the skull and turns through the petrous portion of the temporal bone just anterior to the middle ear. The vessel follows a sinuous course, as shown in the diagram, as it rises through the foramen lacerum and crosses the cavernous sinus beside the sphenoid bone and pituitary fossa. It attains the cranial cavity by a sharp upward bend where it gives off the following branches:

1. *Tympanic branches* to the tympanic cavity.
2. *Inferior* and *superior hypophysial arteries* to the pituitary gland.
3. *Ophthalmic artery* to the eye and other orbital structures.
4. *Cerebral branches* to the brain, including the *anterior choroidal, anterior cerebral, middle cerebral,* and *posterior communicating arteries.* These are part of the circle of Willis at the base of the brain (described with the blood supply of the central nervous system, p. 264).

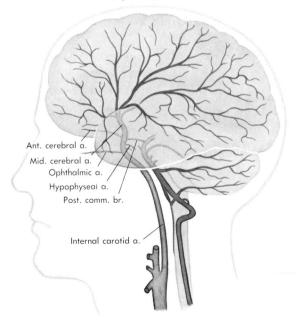

Ant. cerebral a.
Mid. cerebral a.
Ophthalmic a.
Hypophyseal a.
Post. comm. br.

Internal carotid a.

External Carotid Artery

The *external carotid artery* is the important artery to neck structures above the level of the larynx; to the jaws, face and scalp; and to the deep head structures below the floor of the skull. This vessel is directed superficially from its origin at the carotid bifurcation. It ascends to pass through the parotid gland in front of the temporomandibular joint. Its many branches are:

Branches to Neck Structures. The external carotid artery gives off the *superior thyroid artery* early in its course. Any of the following branches to neck structures from the superior thyroid vessel may spring directly from the external carotid artery instead.

1. *Infrahyoid* to the hyoid bone and muscles attaching to its inferior border.
2. *Sternomastoid* to the muscle.
3. *Superior laryngeal* to the larynx.
4. *Cricothyroid* to the cricothyroid membrane and muscle between these two laryngeal cartilages.
5. *Glandular* to the superior pole of the thyroid gland with anastomoses to both the artery of the opposite side and the inferior thyroid artery.

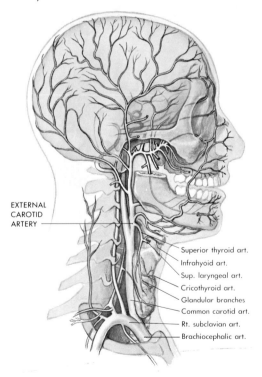

EXTERNAL CAROTID ARTERY

Superior thyroid art.
Infrahyoid art.
Sup. laryngeal art.
Cricothyroid art.
Glandular branches
Common carotid art.
Rt. subclavian art.
Brachiocephalic art.

Lingual Artery. This artery arises opposite the greater cornu of the hyoid bone. It ascends in a sinuous course to the base of the tongue, giving off the following branches which extensively anastomose with their fellows from the opposite side:

1. *Suprahyoid* to the hyoid bone and muscles attaching to its superior border.
2. *Dorsales linguae* branches to the dorsal (superior) surface of the tongue.
3. *Arteria profunda linguae* (deep artery of the tongue) to the muscles of the tongue.
4. *Sublingual artery* to the sublingual salivary gland beneath the tongue.

1. In the deep cervical part of its course, the facial artery gives off the following branches: *ascending palatine* to the upper pharynx and palate, *tonsillar* to the palatine (faucial) tonsil, *glandular* to the submandibular salivary gland, and *submental* to the muscles under the chin.
2. *Inferior* and *superior labial arteries* to the skin, muscles, and glands of the lips.
3. *Lateral nasal artery* to the wing and bridge of the nose.
4. *Angular artery,* which is the termination of the facial artery near the medial angle of the eye. This vessel joins with eyelid arteries that are the termination of the orbital part of the internal carotid system and, thereby, establishes a connection between the two carotid arteries.

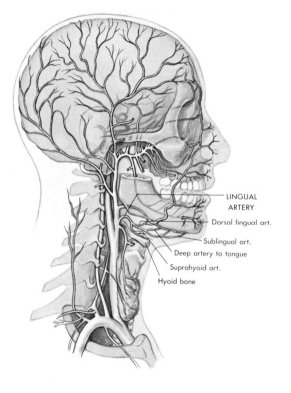

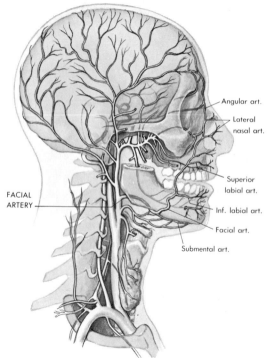

Facial (External Maxillary) Artery. Sometimes this vessel arises from a short stem shared with the lingual artery (linguofacial trunk). The facial artery runs upward at first along the pharynx and then deeply under the lower jaw and the submandibular (submaxillary) salivary gland. It then winds its way tortuously over the border of the mandible where its pulsation can be felt at a first aid pressure point about 1 inch medial to the angle of the jaw. The facial artery gives off a tremendous number of branches to all structures of the face. Only the more important ones will be listed here:

Occipital Artery. The *occipital artery* runs posteriorly and upward to become the important artery to the back of the skull. Its branches are:

1. *Sternomastoid* to the upper part of the muscle.
2. *Mastoid* to the mastoid process.
3. *Meningeal* to the coverings of the brain by passing through minute foramina of the skull.
4. *Muscular* to the posterior cervical muscles attaching to the posterior surface of the occipital bone.
5. *Occipital* to the scalp over the posterior surface of the skull. These branches are many

and tortuous. They bleed profusely when the scalp of this region is lacerated.

Posterior Auricular Artery. This vessel also courses posteriorly. It divides behind the ear into its branches:

1. *Occipital* and
2. *Auricular* which supply the medial and lateral surfaces of the external ear as well as the scalp near the ear.
3. *Stylomastoid* which travels into the stylomastoid foramen along the facial nerve to supply the middle ear and the internal ear by many smaller branches.

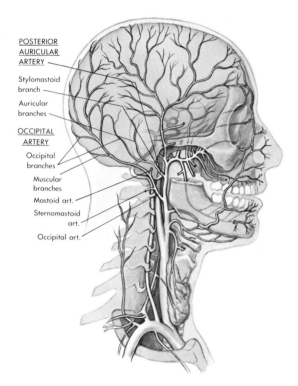

POSTERIOR
AURICULAR
ARTERY

Stylomastoid
branch

Auricular
branches

OCCIPITAL
ARTERY

Occipital
branches

Muscular
branches

Mastoid art.

Sternomastoid
art.

Occipital art.

Maxillary (Internal Maxillary) Artery. The external carotid artery, after giving off the posterior branches, rises in front of the ear surrounded by the deeper portions of the parotid salivary gland. The artery terminates by dividing into the deeper *maxillary artery* and the more superficial *temporal artery*. The maxillary artery is the important artery of the deep infratemporal region or pterygoid region. It is the artery which supplies the teeth, jaws, and muscles of mastication. The vessel pursues a deep course anteriorly from its origin and almost horizontally as it gives off its main branches which are the:

1. *Middle meningeal artery* which ascends to the base of the skull to pass into the cranial cavity through the foramen spinosum. This vessel, from the standpoint of health, is even more important than branches to the jaws or to the masticatory muscles. Its branches fan out to supply the dura mater covering the brain. The branches travel on the internal aspects of the temporal and parietal bones. The branches course in grooves or tunnels in an intimate relationship to the bony calvarium. In cases of head injury, the consequence of bleeding and the pressure of blood clots may cause severe damage to the brain or death.

2. *Inferior alveolar* (dental) *artery,* the artery to the lower jaw, which descends in an anterior direction to enter the mandibular foramen. *Mylohyoid* branches to the floor of the mouth and *lingual* branches to the mucous membrane lining the cheek come off the vessel before it enters the mandibular canal. *Dental* branches pass to the teeth, while a *mental* branch passes forward through the mental foramen to supply tissues of the chin.

3. *Pterygoid arteries* extend to supply these masticatory muscles. Others supply the *masseter* muscle. *Buccal arteries* ramify in the buccinator muscle and also send twigs to the skin and mucous membrane of the cheek.

4. *Posterior, middle,* and *anterior superior alveolar arteries* convey blood to the teeth and gums of the upper jaw.

5. *Infraorbital artery* which runs forward from the pterygopalatine fossa into the floor of the orbit through the inferior orbital fissure. The vessel keeps to the infraorbital groove and leaves the orbit anteriorly through the infraorbital foramen (with the infraorbital nerve) to supply the skin and other tissues of the lower eyelid, adjacent areas of the cheek, and lacrimal sac.

6. Other branches pass as *greater* and *descending palatine arteries* to the roof of the mouth and the palates. A *pharyngeal artery* passes posteriorly through the bony pharyngeal canal to supply the roof of the nasal cavity and the nasopharynx. The maxillary artery terminates as the *sphenopalatine artery* by passing from the pterygopalatine (sphenopalatine) fossa through the sphenopalatine foramen. This arterial branch divides extensively to form *posterior nasal arteries* to the sinuses, nasal turbinates, and the mucous membrane covering the nasal meatuses. A final branching forms a network of *posterior septal arteries.*

Superficial Temporal Artery. The superficial continuation and termination of the external

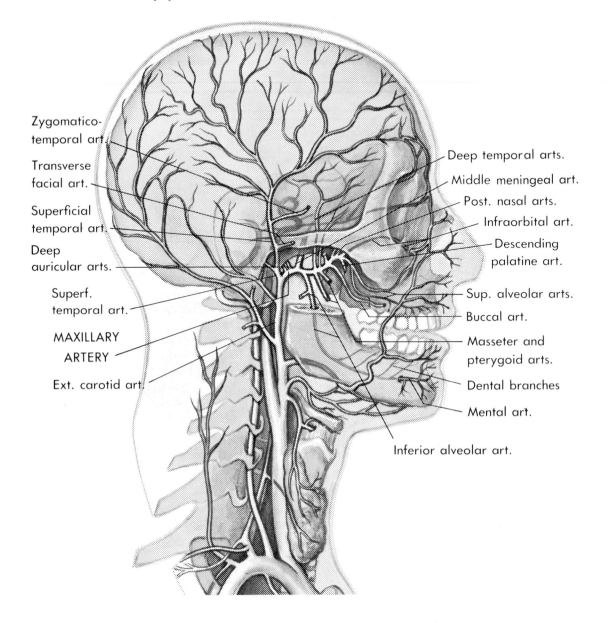

Zygomatico-temporal art.

Transverse facial art.

Superficial temporal art.

Deep auricular arts.

Superf. temporal art.

MAXILLARY ARTERY

Ext. carotid art.

Deep temporal arts.

Middle meningeal art.

Post. nasal arts.

Infraorbital art.

Descending palatine art.

Sup. alveolar arts.

Buccal art.

Masseter and pterygoid arts.

Dental branches

Mental art.

Inferior alveolar art.

carotid artery is the *superficial temporal artery*. This vessel rises from the grasp of the parotid gland to cross over the zygomatic process, another point of first aid pressure and a point for "taking the pulse." Many branches ramify over the *temporal, auricular, zygomatic,* and *facial* regions from which the arteries receive their names.

ARTERIES OF THE AXILLA AND UPPER LIMB

The subclavian artery, after giving off its neck branches, continues as the great arterial stem of the upper limb. At the outer border of the first rib this same arterial trunk is renamed the *axillary artery.*

AXILLARY ARTERY

The *axillary artery* also supplies the structures of the axilla. This region has been described with both the skeletal system and the muscles attaching to the upper end of the humerus.

The Cervicoaxillary Canal

The axilla is a blunt pyramidal space which is filled with important structures: It is the route taken by the arterial stem of the upper limb to reach the arm, by veins returning from the arm, and by the nerve cables entering to innervate the arm. All these structures pass through the root of the neck. The aperture forming the entrance to the axilla is the *cervicoaxillary canal*. The bony boundaries of this canal are the first rib, with the clavicle and coracoid process above, the head and surgical neck of the humerus laterally, and the upper three ribs medially. These osseous boundaries protect the vessels and nerves as they pass through a circumscribed region which is subject to trauma from the front and above.

Course of the Axillary Artery

Beginning at the outer border of the first rib, the *axillary artery* traverses the cervicoaxillary canal in the company of the *axillary vein* and cords of the *brachial plexus* of nerves to the arm. The artery is divided conventionally into three parts for descriptive purposes but its branches may not respect their arbitrary assignment to such divisions.

First Part of Axillary Artery. The proximal division extends from the outer border of the first rib to the point where the vessel is crossed by the medial border of the pectoralis minor muscle. There is but one branch from this part of the artery. It is the *superior thoracic artery* which comes off medially and anteriorly, traverses the axillary adipose tissue, and is distributed to tissues in the upper thoracic interspaces of the anterior chest wall.

Second Part of Axillary Artery. This part lies under the pectoralis minor as this muscle reaches for the coracoid process of the scapula for insertion. The boundaries of the middle division of the artery, therefore, are the medial and lateral borders of the muscle. Two branches spring from this part:

1. *Thoracoacromial artery* which sends four branches to the tissues of the pectoral and infraclavicular regions. The *pectoral branch*(es) are distributed to the pectoral muscles and fasciae; the *deltoid branch* nourishes the anterior portion of the deltoid muscle; the *clavicular branch* passes to the clavicle and subclavius muscle; and the *acromial branch* supplies the tela subcutanea and skin over the acromion.
2. *Lateral thoracic artery* which sends branches to the lateral thoracic wall and, in the female,

supplies the outer quadrants of the mammary gland.

Third Part of the Axillary Artery. The third part has three branches. It gives off a large branch, the *subscapular artery,* which is the major artery to the posterior axillary muscles, an *anterior humeral circumflex artery,* and a *posterior humeral circumflex artery.*

1. The *subscapular artery* courses postero-inferiorly as the artery to the latissimus dorsi muscle and to the muscles of the posterior wall of the axilla. Its *circumflex scapular* branch passes to muscles of the infraspinous fossa of the scapula and enters into the scapular anastomosis.
2. The *anterior humeral circumflex artery* is a small vessel which passes laterally to anastomose with its fellow.
3. The *posterior humeral circumflex artery* winds laterally and posteriorly around the upper end of the humeral shaft. By anastomosis, both humeral circumflex vessels nourish the periosteum and bone of the upper end of the humerus. One may be lacking or both may spring from a short common trunk.

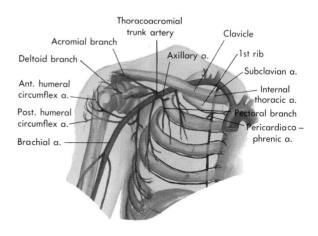

BRACHIAL ARTERY

The distal boundary of both the axilla and its artery is formed by the lower border of the teres major as it approaches the upper end of the shaft of the humerus for insertion. From this point to the elbow, the arterial stem to the upper limb takes the name *brachial artery*. This is the artery of the brachium (upper arm) which also enters into an anastomotic network about the elbow joint. Branches of the brachial artery are the following:

Muscular Branches

Many direct branches come from the brachial artery to supply the anterior muscles of the anterior compartment of the brachium. They are short, direct in course, and frequently multiple. The vessels are named for the muscle supplied, e.g., "biceps brachii branch" or "artery to the brachialis muscle." Similarly, a *nutrient artery* passes to the humerus.

Deep Brachial Artery

This artery (in older terminology the *profunda brachii*) comes from the upper portion of the brachial artery and courses along the spiral groove of the humeral shaft in company with the radial nerve. This artery supplies the muscles of the posterior compartment of the brachium.

Collateral Arteries

Vessels from the brachial artery descend on each side of the lower part of the brachium to enter into an anastomosis about the elbow joint through which blood could reach the forearm if the brachial artery were occluded at the elbow. The *radial collateral artery* passes around the lateral side of the lower end of the humerus. The *ulnar collateral artery* is distributed to the medial side. These vessels supply the periarticular tissues of the elbow joint in company with more distal partners soon to be described.

The Scapular Anastomosis

The borders of the scapula are ringed by vessels whose ramifications permeate the muscles and fasciae of the scapular fossae. Branches from two of the arteries stemming from the thyrocervical trunk, the *suprascapular* and the *transverse cervical arteries,* approach the scapula from the neck. Branches of the *subscapular artery* and its *circumflex scapular* branch wind toward the lower parts of the scapula from the axilla. The anastomosis of terminal branches of these vessels effects a union between the subclavian artery and the third part of the axillary artery. A potential pathway of collateral circulation exists in which blood can

reach the distal parts of the axilla and, therefore, the upper limb if it is necessary to tie off the first or second part of the axillary artery because of injury or disease.

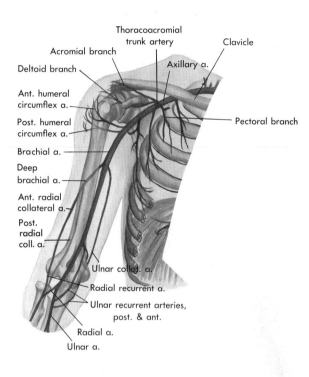

ARTERIES OF THE FOREARM

The brachial artery usually divides into two terminal branches, the *ulnar* and the *radial arteries,* in the depths of the antecubital fossa. These arteries send recurrent branches to the anastomosis about the elbow joint, serve the forearm and wrist, and supply blood to the hand.

Ulnar Artery

As the *ulnar artery* leaves the antecubital fossa, it angles close to the medial muscle mass of the forearm and follows a common course with the ulnar nerve toward the wrist. The branches are:

Recurrent Arteries. *Anterior* and *posterior ulnar recurrent arteries* pass upward respectively in front of and behind the medial epicondyle of the humerus. They supply adjacent muscles of the medial mass and supply blood to the capsule and ligaments of the elbow joint. They end by anastomosing with the corresponding ulnar collateral arteries.

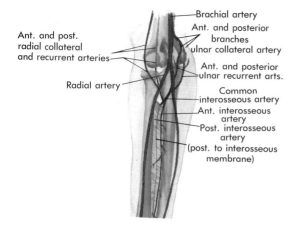

Ant. and post. radial collateral and recurrent arteries

Radial artery

Brachial artery
Ant. and posterior branches ulnar collateral artery
Ant. and posterior ulnar recurrent arts.
Common interosseous artery
Ant. interosseous artery
Post. interosseous artery (post. to interosseous membrane)

Common Interosseous Artery. This artery is as important to the forearm as are the ulnar and radial vessels. The *common interosseous artery* branches from the ulnar at the lower border of the antecubital fossa, passes posteriorly to the radioulnar interosseous membrane, and divides into *anterior* and *posterior interosseous arteries*.

1. The *anterior interosseous artery* courses on the anterior aspect of the interosseous membrane. It provides *muscular branches* to the flexor muscles, which cover it, and *nutrient branches* to the radius and ulna and sends a *median* (interosseous) *artery* distally along the membrane as far as the wrist.

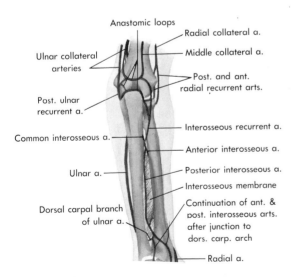

Anastomic loops

Ulnar collateral arteries

Post. ulnar recurrent a.

Common interosseous a.

Ulnar a.

Dorsal carpal branch of ulnar a.

Radial collateral a.
Middle collateral a.
Post. and ant. radial recurrent arts.
Interosseous recurrent a.
Anterior interosseous a.
Posterior interosseous a.
Interosseous membrane
Continuation of ant. & post. interosseous arts. after junction to dors. carp. arch
Radial a.

2. The *posterior interosseous artery* passes through the interosseous membrane to the posterior compartment of the forearm. This vessel sends an *interosseous recurrent artery* proximally to anastomose with a posterior descending branch

of the deep brachial artery. The posterior interosseous artery then extends irregularly downward amid the extensor muscles to which it gives off many *muscular branches*. The artery terminates at the wrist by anastomosing with terminal branches of the anterior interosseous artery.

Carpal and Palmar Arteries.   See arteries of the hand below.

Radial Artery

The *radial artery* more directly continues the course of the parent brachial artery. In the antecubital fossa the radial vessel crosses the bicipital tendon but passes under the bicipital aponeurosis. The artery retains a superficial course as it proceeds distally along the lateral side of the anterior compartment of the forearm. The radial artery is quite superficial as it nears the wrist. It lies between the tendon of the flexor carpi radialis muscle and the radius. The superficial location and easy compressibility against the radius at this point make the distal radial artery the prime vessel for "taking the pulse." Its branches are:

Radial Recurrent Artery. This branch ascends to the lateral part of the arterial anastomosis about the elbow where it joins the radial collateral artery. The structures supplied are identical with those nourished by ulnar recurrent arteries.

Muscular Branches. These are provided to the lateral forearm muscles.

Carpal and Palmar Arteries. These are described with the arteries of the wrist and hand.

ARTERIES OF THE WRIST AND HAND

The blood supply to the wrist and hand comes from interconnecting networks of the ulnar and radial arteries.

Blood Supply of the Wrist

Small *carpal arteries* form a network on both surfaces of the wrist. The *anterior* and *posterior carpal arches* receive branches contributed from the ulnar and radial arteries, from the anterior and posterior interosseous arteries, and from the median interosseous artery, which in some persons may be a sizable vessel. Both arches supply the ligaments, tendon sheaths, and joints of the wrist. In addition, the posterior carpal arch sends the second, third, and fourth *dorsal metacarpal arteries* between the second and third, the third and fourth, and the fourth and fifth metacarpal bones. These vessels nourish the bones, periostea, and muscles attaching to these metacarpal bones. Short terminal twigs cross the knuckles and divide to form short posterior digital arteries to the proximal part of the corresponding fingers.

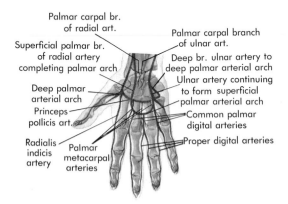

between deep muscles of the base of the thumb, it gives off immediately the *princeps pollicis artery* to the muscles of the thumb and the *radialis indicis artery,* which is the digital artery to the lateral side of the index finger. The termination of the radial artery becomes the *deep palmar arch,* which is completed by the deep branch of the ulnar artery. Three *palmar metacarpal arteries* are given off from the deep arch to adjacent surfaces of the second to fifth metacarpal bones. These palmar metacarpal arteries, at the webs of the fingers, anastomose with the digital arteries from the superficial palmar arch.

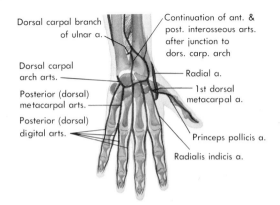

Blood Supply of the Hand

The terminal parts of the radial and ulnar arteries supply the hand. The *ulnar artery* divides near the hamate bone to form *superficial* and *deep palmar branches.*

Superficial Palmar Arch. The superficial branch of the ulnar artery crosses the palm convexly at about the level of the middle skin crease of the palm. It lies superficial to the flexor tendons. This arterial arch is completed laterally by the palmar cutaneous branch of the radial artery. The superficial palmar arch gives off three *common palmar digital arteries* which divide at the webs of skin between the second and third, third and fourth, and fourth and fifth fingers. The divisions form the *digital arteries* which supply adjacent surfaces of the fingers mentioned. An additional digital branch of the arch passes to the medial (ulnar) side of the fifth finger.

Deep Palmar Arch. The *radial artery,* at the distal limits of the forearm, turns laterally and posteriorly through the "anatomical snuff box." In its course toward the deeper parts of the palm, the radial artery provides two *dorsal digital arteries to the thumb* and a *dorsal digital artery to the index finger.* When the radial artery enters the palm

There are extensive cross connections between the digital arteries on each side of the fingers, particularly at the interphalangeal joints, which ensure a rich arterial supply to the tendons, joints, ligaments, and skin of the fingers. This is important to the function and the healing after injuries of these valuable manual appendages.

ARTERIAL SUPPLY OF THE THORAX

The blood vessels of the thorax, other than those supplying the lungs and heart, are those of the thoracic wall and mediastinum.

ARTERIES TO THE MEDIASTINUM

The connective tissue and serous membranes of the mediastinum on the one hand, and structures traversing the partition on the other hand, are supplied with arterial blood from the subclavian artery, the descending aorta, and the axillary artery.

Internal Thoracic Artery

The *internal thoracic* (internal mammary) artery has been mentioned as a branch of the subclavian artery. Stemming from the parent vessel in the root of the neck, the internal thoracic artery courses inferiorly through the superior thoracic aperture as a major artery of the mediastinum and anterior thoracic wall. Its branches are extensive and include the following:

1. *Anterior intercostal arteries:* These vessels branch from the main stem to the medial end of the upper six thoracic interspaces. Branches supply the costal pleura and endothoracic fascia, the anterior intercostal membrane, and the intercostal muscles. The terminal portions perforate the anterior intercostal membrane just lateral to the sternum. These *perforating branches* supply the sternal fibers of the pectoralis major and the skin of the parasternal area. The vessels of the second to fourth interspaces in the female are incorporated into the arterial supply of the mammary gland. Anastomosis is made with terminal twigs from the posterior intercostal arteries.

2. *Pericardiacophrenic artery:* This vessel comes from the internal thoracic artery at the level of the first costochondral articulation. Its high origin indicates the longitudinal growth of the heart since the vessel must descend between the mediastinal pleura and the pericardium to supply the contiguous surfaces of both structures. This artery ends in the diaphragm as a companion to the phrenic nerve.

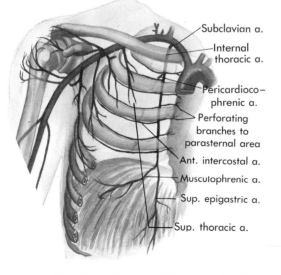

Subclavian a.
Internal thoracic a.
Pericardiocophrenic a.
Perforating branches to parasternal area
Ant. intercostal a.
Musculophrenic a.
Sup. epigastric a.
Sup. thoracic a.

3. *Musculophrenic artery:* The name implies the distribution of this vessel from the lower part of the vertical course of the internal thoracic artery. Branches pass to the endothoracic fascia between the vaultlike diaphragm and the anterior thoracic wall. The *anterior intercostal arteries* of the seventh to ninth interspaces come from this vessel, which finally passes through the diaphragm to end in lower intercostal branches.

4. *Superior epigastric artery:* After giving off the musculophrenic artery, the internal thoracic artery terminates as the *superior epigastric artery,* which passes through the attachments of the anterior diaphragm to become the superior artery of the anterior abdominal wall. This artery supplies the rectus abdominis muscle and its fascial sheath, the anteromedial portions of the other abdominal muscles, and overlying areas of the skin and tela subcutanea. An anastomosis between the superior epigastric and the inferior epigastric arteries links the subclavian and iliac arteries.

Mediastinal Arteries from the Aorta

The descending aorta gives off paired visceral and parietal arteries in its course through the thorax. Parietal branches are described with the blood supply to the thoracic wall. The visceral branches pass to organs of the mediastinum:

1. *Bronchial arteries* pass along the bronchi, supplying them, the lymph nodes of the tracheobronchial system, and the non-air bearing connective tissue framework (parenchyma) of the central two-thirds of the lung. Anastomoses occur between the bronchial arteries and the pulmonary arteries. There is variation in the number and exact origin of the bronchial arteries. The *right bronchial artery* frequently arises from the first posterior intercostal artery instead of from the aorta. There are often two *left bronchial arteries* from the aorta.

2. *Pericardial branches,* varying in number, supplement the arterial blood supply to the pericardium.

3. Four to six *esophageal branches* supply the thoracic portion of the esophagus.

4. Many minute *mediastinal branches* pass to the connective tissue of the partition and to the serous mediastinal pleural membrane which covers it.

BLOOD SUPPLY OF THE THORACIC WALL

The tissues of the curving surfaces of the thoracic wall receive arterial blood via many arteries which, in general, come from the subclavian artery, the axillary artery, or from parietal branches of the aorta.

Arteries of Subclavian Origin

The subclavian artery sends arterial blood to both the anterior thoracic wall and to the highest part of the posterior thoracic wall. Its supply of the anterior chest wall has already been described on page 343 by its contribution of anterior intercostal arteries from the internal thoracic artery.

The main source of blood for the internal aspect of the posterior thoracic wall, the descending thoracic aorta, does not give off thoracic wall branches until the level of the third thoracic interspace posteriorly. The gap between this level and the superior thoracic aperture above is filled in by the subclavian artery, which sends a descending branch to this area. The branch, coming from the subclavian's costocervical trunk, is the *superior (highest) intercostal artery.* This vessel drops backward and downward in the root of the neck over the apex of the lung. It passes to the neck of the first rib, where it divides to form *posterior intercostal arteries* for the first and second thoracic interspaces. The course of the superior intercostal artery is further evidence that thoracic structures were located far cephalically in the embryo. This vessel—like the heart, the cardiac, vagus, and phrenic nerves, and the tracheobronchial system—grew inferiorly in postnatal life to keep pace with the longitudinal growth of the trunk.

Descending branches of the transverse cervical and suprascapular arteries, as well as vessels of the scapular anastomosis, contribute arterial twigs to the *external* aspect of the posterior thoracic wall. The subscapular and circumflex scapular arteries from the axillary artery also enter into this pattern.

Arteries of Aortic Origin

Posterior intercostal arteries course from the aorta to each of the third to eleventh thoracic interspaces. Each artery has two branches:

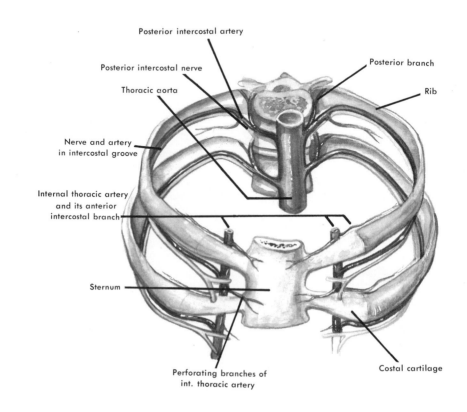

Posterior intercostal artery

Posterior intercostal nerve

Posterior branch

Thoracic aorta

Rib

Nerve and artery
in intercostal groove

Internal thoracic artery
and its anterior
intercostal branch

Sternum

Costal cartilage

Perforating branches of
int. thoracic artery

1. *Posterior branch* to the deep longitudinal muscles of the back.

2. *Spinal branch* to the vertebrae and spinal cord via the intervertebral foramen.

The intercostal artery courses around the vertebral body to the costal groove along the lower border of the rib, sending branches to the muscles, bone, fasciae, and skin of the interspace. (See illustration on following page.) Its location along the undersurface of the rib causes the physician, when inserting needles through the thoracic wall, to direct the point over the superior surface of a rib. The cutaneous branches of the second to fourth posterior intercostal arteries send *mammary arteries* to the mammary gland. The artery terminates by anastomosing with the corresponding anterior intercostal artery.

ARTERIAL SUPPLY OF THE ABDOMEN

The abdominal viscera and the abdominal wall receive arterial blood from the abdominal aorta, from terminations of thoracic arteries, and from blood vessels which ascend from the inguinal region. Some vessels are paired but others are unpaired.

BLOOD SUPPLY OF THE ABDOMINAL WALL

The blood supply to the anterior abdominal wall is derived from sources different from those supplying the posterolateral regions since the anterior abdominal wall closes only late in embryonic life. The posterolateral abdominal wall is developmentally older and receives blood more in accordance with the segmental body plan of vertebrates.

Arteries of the Anterior Abdominal Wall

The arteries supplying the anterior abdominal wall grow inferiorly from the thorax and superiorly from the inguinal region to anastomose in and about the rectus abdominis muscle and its fascial sheath.

Superior Arteries. The upper part of the anterior abdominal wall is supplied by the *superior epigastric* and *musculophrenic* branches from the internal thoracic artery.

Inferior Arteries. The superior vessels enter into a great *ventral arterial anastomosis* of the anterior abdominal wall with branches of the *inferior epigastric artery,* a branch of the external iliac artery. The muscles of the lower part of the anterior abdominal wall are supplied by the inferior epigastric vessel as its ramifications ascend from the inguinal region.

Arteries of the Posterior Abdominal Wall

Arteries to the posterolateral abdominal wall spring from the abdominal aorta as a series of paired, segmentally organized vessels. The vessels are the:

1. *Subcostal arteries* from the thoracic aorta which descend between the posterior attachments of the diaphragm to enter the transversus abdominis. The subcostal artery supplies the uppermost parts of the posterolateral abdominal muscles and then approaches the anterior midline.

2. *Phrenic arteries* which arise from each side of the aorta just inferior to the diaphragm. The phrenic arteries ramify on the inferior surface of the diaphragm, nourishing it and anastomosing with phrenic branches from the thoracic aorta which enter the superior surface. A *superior suprarenal* (adrenal) *artery* branches from each phrenic artery as part of the blood supply of this endocrine gland.

3. *Lumbar arteries* continue a regular parietal series of arteries which include the superior intercostal, posterior intercostal, subcostal, and phrenic arteries. There are four pairs of lumbar arteries which course through and supply the posterolateral abdominal muscles below the level supplied by the subcostal artery. (See illustration at top of page 346.)

BLOOD SUPPLY OF THE ABDOMINAL ORGANS

The abdominal organs of the alimentary system are supplied by three unpaired visceral arteries: the *celiac trunk,* the *superior mesenteric artery,* and the *inferior mesenteric artery.* Paired renal arteries provide arterial blood to the kidneys. Either *testicular* or *ovarian arteries* spring from the abdominal aorta to supply the reproductive glands.

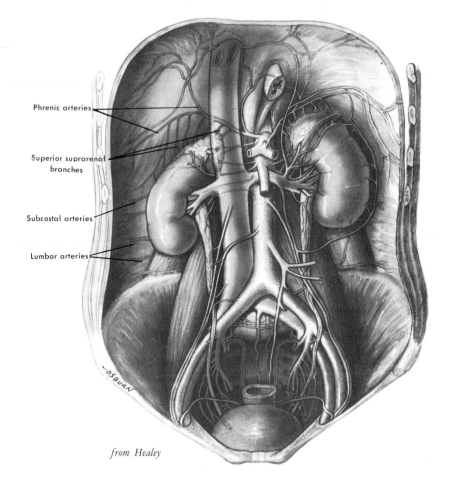

Phrenic arteries

Superior suprarenal
branches

Subcostal arteries

Lumbar arteries

from Healey

Blood Supply to the Upper Abdominal Organs

The *celiac trunk* is a short, common stem artery arising from the front of the abdominal aorta. This distributing trunk (in older terminology called the *coeliac axis*) pursues only a short course before dividing into main branches to the upper abdominal organs. These branches are the *hepatic, left gastric,* and *splenic arteries.*

Hepatic Artery.   The *hepatic artery* passes to the right toward the duodenum and the inferior surface of the liver. It gives off an inferiorly coursing main branch, the *gastroduodenal artery,* and the *right gastric artery* before turning superiorly under the liver. (See illustration on following page.)

1. The *right gastric artery* passes to the lesser curvature of the stomach. It supplies both surfaces of the stomach adjacent to the lesser curva-

ture and also sends a small branch to the first portion of the duodenum.

2. The *gastroduodenal artery* is the main distributing artery to the head and body of the pancreas and to the duodenum. The *right gastroepiploic artery* is given off to the greater curvature of the stomach and to the great omentum. The gastroduodenal artery then forms arcades of vessels which encircle the duodenum as the *anterior* and *posterior pancreaticoduodenal arteries.* These branches supply the duodenum and head of the pancreas. The vessels on the posterior surface may merge to form a sizable vessel called the *retroduodenal artery.* These arteries anastomose with inferior pancreaticoduodenal branches from the superior mesenteric artery.

3. The *hepatic artery* continues as the hepatic artery proper which gives off the *cystic artery* to the gallbladder and then divides into *right* and *left hepatic branches.* Considerable variation exists in the number, course, and relations of the cystic and hepatic arteries with accessory vessels occurring quite commonly.

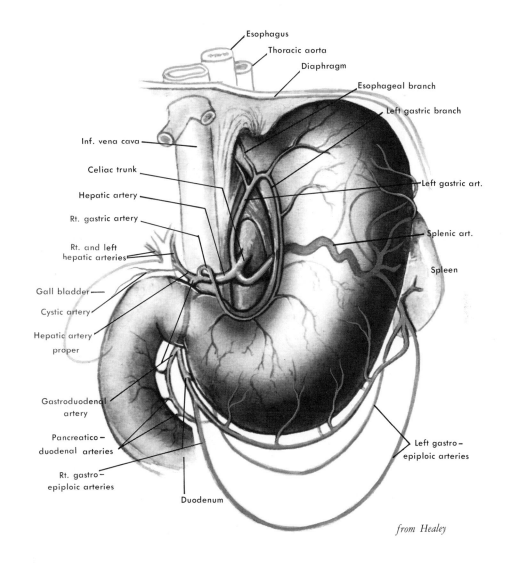

Esophagus
Thoracic aorta
Diaphragm
Esophageal branch
Left gastric branch
Inf. vena cava
Left gastric art.
Celiac trunk
Hepatic artery
Rt. gastric artery
Splenic art.
Rt. and left
hepatic arteries
Spleen
Gall bladder
Cystic artery
Hepatic artery
proper
Gastroduodenal
artery
Pancreatico -
duodenal arteries
Left gastro -
epiploic arteries
Rt. gastro -
epiploic arteries
Duodenum

from Healey

Left Gastric Artery. The *left gastric artery* curves posteriorly from the celiac trunk to provide *left gastric branches* to the lesser curvature of the stomach and *esophageal branches* to the terminal part of the esophagus.

Splenic Artery. The splenic artery undulates from the celiac trunk to the left and along the superior border of the pancreas. This tortuous vessel supplies in its course *pancreatic branches* to the body and tail of the pancreas, *short gastric branches* to the cardiac portion of the stomach, and the *left gastroepiploic artery* to the greater curvature. The splenic artery ends in several *splenic*

branches which plunge into the hilum of the spleen to bring arterial blood to this organ.

Blood Supply to the Gastrointestinal System

The blood supply to the stomach is encompassed by the anastomoses of the gastric, gastroepiploic, and short gastric arteries which receive arterial blood via the celiac trunk. The small and large intestines are supplied by the unpaired *superior* and *inferior mesenteric arteries* which arise from the ventral surface of the abdominal aorta inferior to the celiac trunk.

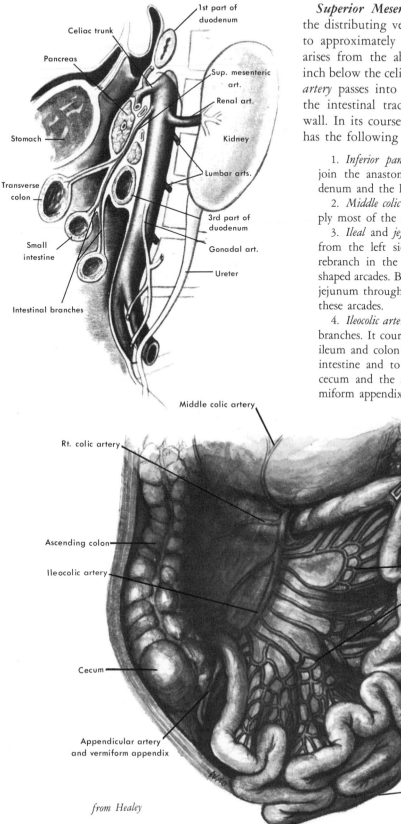

Celiac trunk
1st part of duodenum
Pancreas
Sup. mesenteric art.
Renal art.
Stomach
Kidney
Transverse colon
Lumbar arts.
Small intestine
3rd part of duodenum
Gonadal art.
Ureter
Intestinal branches

from Healey

Middle colic artery
Rt. colic artery
Jejunal portion of small intestine
Ascending colon
Ileocolic artery
Arcades formed by ileal and jejunal arteries
Cecum
Appendicular artery and vermiform appendix
Ileal portion of small intestine

Superior Mesenteric Artery. This vessel is the distributing vessel to the small intestine and to approximately half of the large intestine. It arises from the abdominal aorta about half an inch below the celiac trunk. The *superior mesenteric artery* passes into the mesentery which attaches the intestinal tract to the posterior abdominal wall. In its course the superior mesenteric vessel has the following branches:

1. *Inferior pancreaticoduodenal arteries* which join the anastomotic arcades about the duodenum and the head of the pancreas.

2. *Middle colic artery* which descends to supply most of the transverse colon.

3. *Ileal* and *jejunal* branches which, coming from the left side of the artery, branch and rebranch in the mesentery as a series of fan-shaped arcades. Blood reaches the ileum and the jejunum through smaller intestinal branches of these arcades.

4. *Ileocolic artery* which is one of the terminal branches. It courses toward the junction of the ileum and colon to supply this portion of the intestine and to give off *cecal branches* to the cecum and the *appendicular artery* to the vermiform appendix.

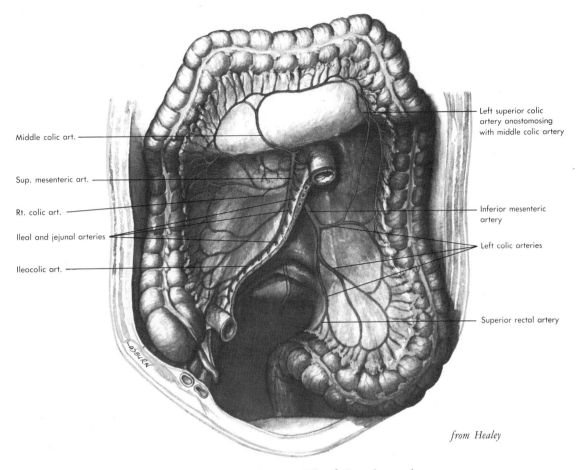

Middle colic art.

Sup. mesenteric art.

Rt. colic art.

Ileal and jejunal arteries

Ileocolic art.

Left superior colic artery anastomosing with middle colic artery

Inferior mesenteric artery

Left colic arteries

Superior rectal artery

from Healey

5. *Right colic artery* which, as the other terminal branch of the superior mesenteric artery, supplies the junction of the cecum and ascending colon, the ascending colon, and the hepatic flexure of the colon. The right part of the transverse colon is also supplied in common with branches from the middle colic artery.

Inferior Mesenteric Artery. This artery arises from the abdominal aorta below the origins of the superior mesenteric and renal arteries. The *inferior mesenteric artery* descends toward the left, giving off branches to the lower part of the intestinal tract. The branches are the:

1. *Left colic arteries* which are the single *left superior colic* and three to four *left inferior colic arteries.* These vessels supply the descending colon and sigmoid colon.

2. *Superior rectal artery* which descends to supply arterial blood to the upper part of the rectum. This artery, representing a terminal branch of an abdominal vessel from the aorta, anastomoses with middle and inferior rectal arteries which have pelvic origins.

Blood Supply to the Genitourinary System

The genitourinary organs are nourished by a number of different vessels because the genitourinary system develops in several stages in which the earlier organs are replaced by more definitive forms. Precursors of the kidney rise out of the pelvis and the reproductive glands descend into it. Therefore, some vessels persist in postembryonic life to supply adult organs which occupy a different location than their forebears.

Blood Supply of the Kidneys. A pair of *renal arteries* arise from the sides of the abdominal aorta. Since the kidneys lie behind the peritoneal (abdominal) cavity, the renal arteries remain close to the posterior abdominal wall. Each artery pursues a horizontal course toward the hilum of the kidney of its side where it breaks up into *renal branches* to the subdivisions of the kidney. An *inferior suprarenal* (adrenal) *artery* is given off to the inferior portion of the suprarenal gland as well as small *ureteric arteries* to the upper portion of the ureter. (See illustration on next page.)

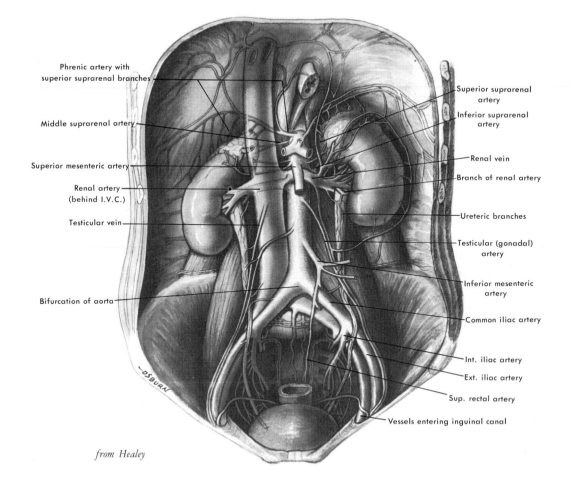

Phrenic artery with
superior suprarenal branches

Middle suprarenal artery

Superior mesenteric artery

Renal artery
(behind I.V.C.)

Testicular vein

Bifurcation of aorta

Superior suprarenal
artery

Inferior suprarenal
artery

Renal vein

Branch of renal artery

Ureteric branches

Testicular (gonadal)
artery

Inferior mesenteric
artery

Common iliac artery

Int. iliac artery

Ext. iliac artery

Sup. rectal artery

Vessels entering inguinal canal

from Healey

Occasionally small *accessory renal arteries* spring directly from the abdominal aorta to pass to either the upper or lower poles of the kidney. In some cases no true renal artery may be present. Its place is taken by multiple, small renal arteries.

Blood Supply to the Reproductive Glands. The sex of the individual determines whether the gonadal arteries are termed the *testicular* or *ovarian arteries*. The abdominal course of these vessels is similar. They arise from the aorta just below the renal arteries to follow a descending course behind the peritoneum rather closely paralleling the ureters which they supply. At the brim of the pelvis, the *testicular artery* enters the inguinal canal on its way to the scrotum. In the female, the *ovarian artery* is directed through the suspensory ligament of the ovary to the ovary and remains a pelvic vessel.

Blood Supply to the Adrenal Glands. Mention has been made of *superior suprarenal arteries* as branches of the parietal phrenic arteries and of the *inferior suprarenal arteries* as branches of

the renal vessels. *Middle suprarenal arteries* arise from each side of the abdominal aorta and pass to the adrenal gland. There are many variations in the actual pattern of suprarenal blood supply and in the number of vessels.

ARTERIAL SUPPLY OF THE PELVIS

The abdominal aorta bifurcates above the level of the pelvic brim to form the two *common iliac arteries*. The common iliac vessels conduct and distribute blood to the pelvic walls, pelvic organs, perineum, and gluteal region. The entire arterial blood supply of the lower limbs is derived from one of the main iliac branches.

ILIAC ARTERIES

The *common iliac arteries* are short conducting arterial trunks. A small midline continuation of the aortic bifurcation, the *median sacral artery,* runs downward between the diverging common iliac vessels to supply fascial and muscular structures of the anterior surface of the sacrum. The branches of the common iliac arteries are the *internal* and *external iliac arteries.*

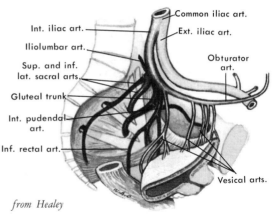

from Healey

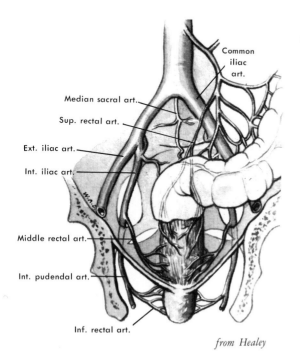

from Healey

INTERNAL ILIAC ARTERY

The internal iliac artery is the artery of the pelvis, perineum, and gluteal region. It gives off parietal and visceral branches.

Parietal Branches

Parietal branches of the internal iliac artery in part pass to the pelvic wall to continue the series of trunk parietal vessels which have been noted from the superior thoracic aperture caudally. Parietal branches also serve the gluteal region.

Pelvic Wall Branches. A short common stem branch from the internal iliac artery continues as the superior gluteal artery after giving off the:

1. *Iliolumbar artery* which curves upward to supply the *iliacus branch* to the iliacus muscle and the iliac bone of the pelvic wall. It sends *lumbar branches* to the muscles of the lower abdominal wall and the flank region. These vessels freely enter into the great ventral anastomosis.

2. *Superior* and *inferior lateral sacral arteries* which supply structures of and about the sacrum including *spinal branches* to the contents of the sacral canal. These vessels anastomose with the median sacral artery.

Gluteal Branches. A gluteal trunk, continuing the common stem from the internal iliac artery after the pelvic wall branches have been given off, gives origin to many vessels which nourish the gluteal muscles. It continues to supply the muscular and fascial structures of the perineum which closes the pelvic outlet. The arteries are the:

1. *Superior gluteal artery* which is the artery to the gluteus maximus, gluteus medius, and gluteus minimus muscles and their related fascial structures.

2. *Inferior gluteal artery* which continues the gluteal stem downward and backward to supply muscles of the pelvic diaphragm such as the levator ani. It also nourishes muscles located proximally in the thigh about the hip joint and terminates as a cutaneous vessel of the skin of the buttocks.

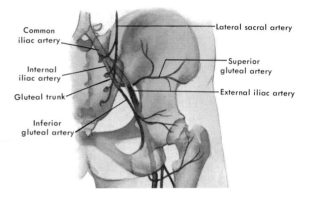

3. *Internal pudendal artery* which is the terminal part of the gluteal trunk. It leaves the pelvis through the greater sciatic foramen but enters the perineum through the lesser sciatic foramen. This is the arterial vessel to the muscle and fascial layers of the urogenital diaphragm, to the alimentary and genitourinary organs which pass through it, and to the external genitalia.

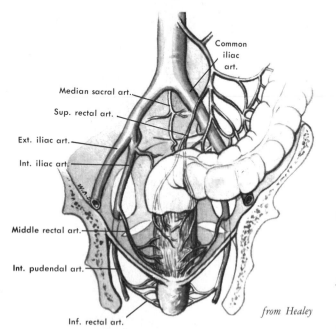

from Healey

The more important branches are the:

a. *Inferior rectal artery* which supplies the lower anal canal, the anal sphincter muscles, and the skin of the anus. This artery anastomoses with the middle and superior rectal arteries from the inferior mesenteric artery and with the middle rectal artery which is a *visceral* branch of the internal iliac artery.

b. *Transverse perineal, scrotal,* and *penile arteries* in the male.

c. *Transverse perineal, clitorine,* and *labial arteries* in the female.

d. *Obturator artery* which does not supply perineal structures but courses in the direction of the inguinal canal to nourish the obturator muscles, the pubic soft tissues, and the acetabulum. This vessel anastomoses freely with the inferior epigastric artery from the external iliac vessel.

Visceral Branches

The internal iliac artery supplies the pelvic organs through the following branches:

1. *Superior vesical artery* to the upper domed surface of the urinary bladder.

2. *Inferior vesical artery,* mainly in the male to the lower part of the bladder and its associated seminal vesicles and the prostate gland.

3. Middle rectal artery to the rectum in both sexes and, in the male, to the prostate gland and seminal vesicles.

4. *Uterine artery,* in the female, to the uterus and, through its prominent *vaginal artery* branch, to the vagina, bladder, and rectum.

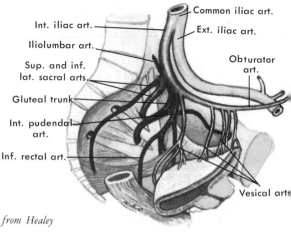

from Healey

Both the ureters and the urinary bladder receive arterial vessels from almost any visceral or parietal branch of the internal iliac artery that courses near them. Further information about the blood supply of the genital organs will be found with the description of those organs. The fate of the *umbilical artery* is described with the discussion of the fetal circulation on page 358.

EXTERNAL ILIAC ARTERY

The external iliac artery becomes the major arterial trunk to the lower limb. The *external iliac artery* curves anteriorly and downward along the brim of the pelvis upon the psoas muscle. The following branches are given off as the vessel approaches the inguinal ligament:

1. *Inferior epigastric artery* which ascends from behind the inguinal ligament to supply the rectus abdominis, its fasciae, and the skin of the anterior abdominal wall. The terminations of this artery enter into the great ventral arterial anastomosis, thereby linking superficially the subclavian and external iliac arteries. Minor *pubic branches* are given off in the pelvis in both sexes, *cremasteric*

branches to the spermatic cord in the male, and an *artery to the round ligament* in the female.

2. *Deep circumflex iliac artery* which parallels the inguinal ligament in a lateral course to enter and supply the lower portions of the transverse abdominis and internal oblique muscles of the lower abdominal wall.

As it passes under the inguinal ligament the external iliac arterial conducting trunk to the lower extremity is renamed the *femoral artery.*

ARTERIES OF THE LOWER LIMB

The muscular, tendinous, and osteoarticular structures of the lower limb form a large tissue mass. Many arterial blood vessels are required to supply these structures. It is helpful to recognize major arterial trunks in the lower limb from which all the structures of the limb derive their blood supply.

1. The main stem artery is the *femoral artery* which supplies the thigh by many branches. Its important *profunda femoris* branch is analogous to the deep brachial artery in the upper limb.

2. The femoral artery becomes the *popliteal artery* just above the knee. This is the artery of the knee region in combination with terminal branches of the femoral artery from the thigh.

3. At the lower border of the popliteal fossa the popliteal artery divides into the *anterior* and *posterior tibial arteries* which supply the leg and foot in combination with the important *peroneal artery,* a branch of the posterior tibial artery. (See illustration on page 354.)

The blood supply of the structures of the lower limb may be traced to the arteries listed. In the following sections the blood supply is given by the regions of the limb rather than by listing every branch to every structure. The reader's knowledge of the skeletal, articular, and muscular organization of the limb will suffice to supply the details if he remembers that every structure in a region will have an arterial branch. (See illustration on page 355.)

BLOOD SUPPLY TO THE THIGH

Branches of the femoral artery supply the thigh.

Course of Femoral Artery

The *femoral artery* enters the *femoral triangle* as it emerges from under the inguinal ligament. This triangle, it will be recalled, is a region of closely related arteries, veins, nerves, and lymphatic structures at the proximal boundary of the thigh. The femoral artery and vein are enclosed within a tubular extension of the transversalis fascia from the pelvis known as the *femoral sheath.* The artery is in front of the hip joint at this point. The vessel descends first in front of the tendon of the psoas muscle and then over the pectineus and adductor longus muscles. It then enters the *adductor* (subsartorial) *canal* where it is roofed over by the muscles of the anteromedial aspect of the thigh. The femoral artery beomes the popliteal artery as the arterial stem emerges from the distal end of the adductor canal.

Branches in the Femoral Triangle

The femoral artery can be compressed against the pubic bone, at its superficial location in the femoral triangle, as a first aid maneuver and to detect the arterial pulse. Four small arterial branches arise at the proximal boundary (base) of the femoral triangle.

1. *Superficial circumflex iliac artery* which winds laterally toward the anterior superior iliac spine.

2. *Superficial epigastric artery* which ascends in a medial direction to superficial structures of the lower abdominal wall and enters into the ventral arterial anastomosis.

3. *Superficial external pudendal artery* which courses upward to supply superficial tissues over the pubic tubercle.

4. *Deep external pudendal artery* which also ascends medially. It supplies *anterior scrotal branches* in the male and *anterior labial branches* to the labium major of the female as well as supplying the inguinal lymph nodes.

Blood Supply of the Thigh

The bulk of the muscles, fasciae, and superficial tissues of the thigh is supplied with arterial blood by the major branch of the femoral artery, the *profunda femoris artery.* The *femoral artery* below the femoral triangle is a conducting arterial stem to more distal parts of the limb, except for its direct *muscular branches* to muscles adjacent to its course.

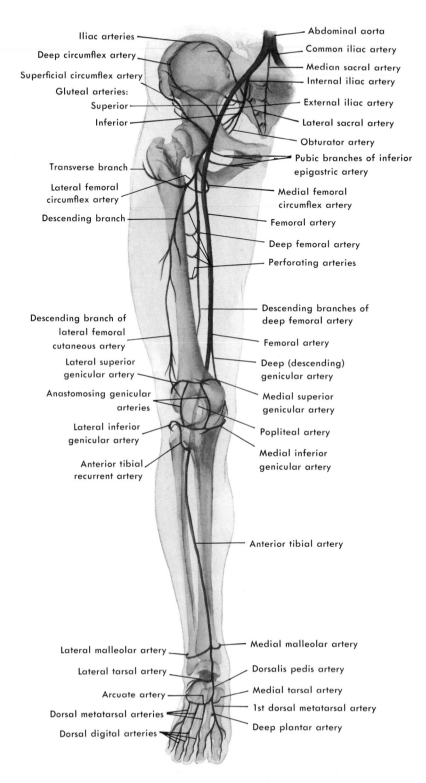

Iliac arteries

Deep circumflex artery

Superficial circumflex artery

Gluteal arteries:

Superior

Inferior

Transverse branch

Lateral femoral
circumflex artery

Descending branch

Descending branch of
lateral femoral
cutaneous artery

Lateral superior
genicular artery

Anastomosing genicular
arteries

Lateral inferior
genicular artery

Anterior tibial
recurrent artery

Lateral malleolar artery

Lateral tarsal artery

Arcuate artery

Dorsal metatarsal arteries

Dorsal digital arteries

Abdominal aorta

Common iliac artery

Median sacral artery

Internal iliac artery

External iliac artery

Lateral sacral artery

Obturator artery

Pubic branches of inferior
epigastric artery

Medial femoral
circumflex artery

Femoral artery

Deep femoral artery

Perforating arteries

Descending branches of
deep femoral artery

Femoral artery

Deep (descending)
genicular artery

Medial superior
genicular artery

Popliteal artery

Medial inferior
genicular artery

Anterior tibial artery

Medial malleolar artery

Dorsalis pedis artery

Medial tarsal artery

1st dorsal metatarsal artery

Deep plantar artery

Anterior View

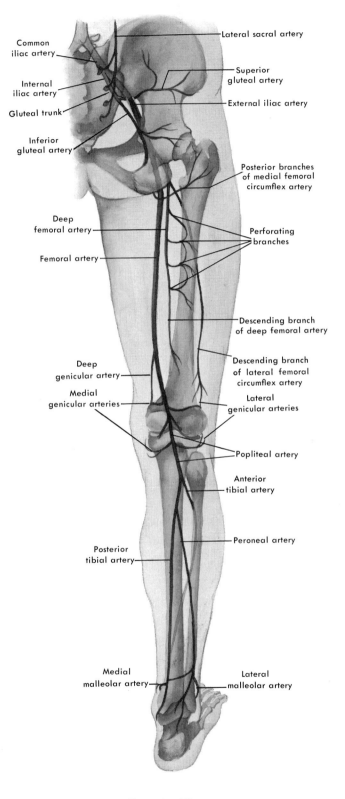

Common iliac artery

Internal iliac artery

Gluteal trunk

Inferior gluteal artery

Lateral sacral artery

Superior gluteal artery

External iliac artery

Posterior branches of medial femoral circumflex artery

Deep femoral artery

Femoral artery

Perforating branches

Descending branch of deep femoral artery

Deep genicular artery

Medial genicular arteries

Descending branch of lateral femoral circumflex artery

Lateral genicular arteries

Popliteal artery

Anterior tibial artery

Peroneal artery

Posterior tibial artery

Medial malleolar artery

Lateral malleolar artery

Posterior View

The profunda femoris artery, therefore, although structurally analogous to the deep brachial artery, supplies structures on both the anteromedial and posterolateral aspects of the thigh.

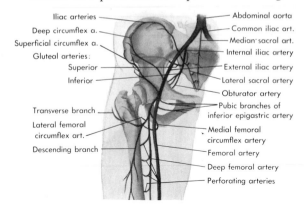

Arterial Supply to Proximal Structures. The *medial* and *lateral circumflex femoral arteries* arise from the profunda femoris vessel to course, as implied by their names, to muscles of the proximal thigh region. The two vessels loop about the upper thigh, participate in an anastomotic circle with branches of the gluteal arteries about the hip joint, and supply both the femoral head and the acetabulum.

Arterial Supply to Distal Structures. The mid and lower thigh muscles, their associated fasciae, and the superficial soft tissues are supplied by *direct muscular branches* from the profunda femoris artery or from the femoral vessel and by four *perforating arteries* from the profunda femoris artery. The perforating arteries arise in a numbered series proceeding distally. Each perforating vessel winds in a posterior direction supplying whatever muscle, or part of a long muscle, extends within its province. The femoral periosteum may be similarly supplied although the *nutrient artery* most often comes from one of the upper two perforating vessels.

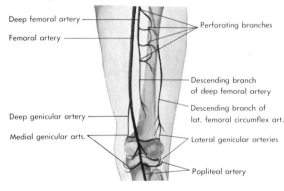

Deep Genicular Artery. The *femoral artery* trunk gives off a *deep genicular artery* just before emerging from the adductor canal. The deep genicular vessel divides into *muscular branches* to the distal parts of the long thigh muscles and into *articular branches* which enter into an arterial anastomosis around the knee joint.

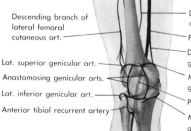

BLOOD SUPPLY OF THE KNEE REGION

The popliteal artery continues the arterial stem of the lower limb into the leg and also provides arteries to supply the knee region.

Course of the Popliteal Artery

The *popliteal artery* begins as the femoral artery emerges from the adductor canal. The popliteal vessel courses in the popliteal fossa posterior to the knee joint. It is surrounded by tendons crossing the back of the knee joint and covered by the popliteal fascia which roofs over the fossa.

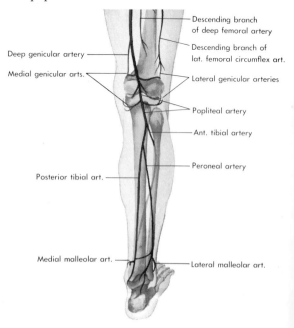

Detection of arterial pulsation in this vessel is quite difficult. The course of the popliteal artery is generally lateral. Its relationships to other popliteal structures are depicted in the adjoining diagram.

Popliteal Blood Supply

The *popliteal artery* gives off *genicular* and *muscular branches.*

Genicular Arteries. *Medial* and *lateral superior genicular arteries* pass horizontally in these respective directions. These vessels anastomose with the deep genicular artery and with a long descending branch from the lateral circumflex femoral artery. *Medial* and *lateral inferior genicular arteries* arise from the more distal part of the popliteal artery. Branches of all these vessels, coupled with twigs from muscular branches above and below the knee joint, join in an *arterial anastomosis of the knee joint*. This anastomosis is effective in providing collateral circulation about the knee joint if the popliteal artery is occluded.

A *middle genicular artery* penetrates deeply from the popliteal artery to supply the intrinsic ligaments and articular membranes of the knee joint.

Muscular Arteries. *Upper muscular branches* from the popliteal artery supply the lower parts of the hamstring muscles. *Lower muscular branches* are the important arterial supply to the calf muscles. These vessels, also called the *sural arteries*, descend to nourish the gastrocnemius, soleus, and plantaris muscles. These are the arteries which, when occluded or narrowed by vascular diseases, are unable to provide sufficient blood to the legs. Progressive pain in the calf muscles, known as "intermittent claudication," makes walking difficult and impairs locomotion. Different diseases affect young persons as well as the elderly.

BLOOD SUPPLY TO THE LEG AND FOOT

The popliteal artery divides at the lower border of the popliteal fossa into the *anterior* and *posterior tibial arteries*. This bifurcation is analogous to the formation of ulnar and radial arteries in the upper limb. The tibial arteries, as the main terminal branches of the arterial stem of the lower limb, provide arterial blood to the leg and foot. Diseases may also affect these arteries to impair nutrition of the distal parts of the limb or impede locomotion.

Anterolateral Blood Supply of Leg and Foot

The *anterior tibial artery* is the smaller of the terminal divisions of the popliteal vessel. The anterior tibial artery is the artery of supply to the muscles and superficial soft tissues of the anterolateral aspect of the leg and ankle and to the dorsum of the foot. It sends an *anterior tibial recurrent artery* proximally to join the genicular anastomosis, provides *muscular branches* in its region, and supplies *medial* and *lateral malleolar arteries* to a network of vessels about the ankle.

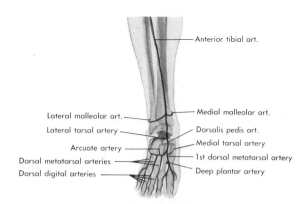

The anterior tibial artery continues onto the dorsal aspect of the foot by passing superficially in front of the two malleoli to become the *dorsalis pedis artery* of the foot. The dorsalis pedis artery is utilized clinically as one of two points of arterial pulsation at the foot and ankle. This vessel gives off an *arcuate artery* and *lateral tarsal branches* over the metatarsal bones. It then sends a *first metatarsal artery* toward the great toe before passing deeply to the sole of the foot to complete the plantar arterial arch. *Second* to *fourth metatarsal arteries* arise from the arcuate artery. All the metatarsal vessels supply muscles, tendons, ligaments, and articular structures of the metatarsal region. *Dorsal digital arteries* proceed from the metatarsal vessels to adjacent sides of the second to fifth toes with an extra branch from the fourth metatarsal artery passing to the lateral aspect of the little toe as shown in the drawing.

Posteromedial Blood Supply to the Leg

The *posterior tibial artery* is the vessel of arterial supply to the posteromedial aspect of the leg through its own trunk and through its important branch, the *peroneal artery*. The *posterior tibial artery* serves the structures of the more medial portion of the posterior aspect of the leg and continues behind the medial malleolus at the ankle (another pulsation point) into the sole of the foot.

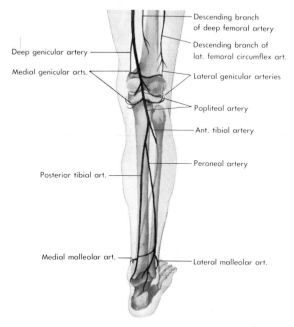

Descending branch of deep femoral artery

Descending branch of lat. femoral circumflex art.

Deep genicular artery

Medial genicular arts.

Lateral genicular arteries

Popliteal artery

Ant. tibial artery

Peroneal artery

Posterior tibial art.

Medial malleolar art.

Lateral malleolar art.

The *peroneal artery* branch inclines more laterally as it descends along the same course as the fibula. The peroneal artery ends behind the inferior tibiofibular joint and, therefore, does not pass into the foot, but it does send a *communicating branch* over to the posterior tibial vessel. Both arteries contribute *malleolar branches* to the arterial anastomosis about the ankle joint.

Blood Supply to the Sole of the Foot

The plantar surface of the foot is served by terminal branches of the *posterior tibial artery* which enters the foot between the medial malleolus and the medial tubercle of the calcaneal bone. The smaller of the two terminal branches of the posterior tibial artery is the *medial plantar artery* which courses along the medial aspect of the longitudinal arch of the foot, supplying muscles and superficial tissues. The major terminal branch of the posterior tibial vessel is the *lateral plantar artery*. This vessel provides muscular and cutaneous branches to the lateral aspect of the sole of the foot and then curves deeply under the arches of the foot to form the *plantar arch*. Metatarsal and plantar digital arteries form from this arch as described for the arcuate arterial arch on the dorsal surface. *Plantar perforating* and *communicating* arteries link the two arches and supply the tarsal and metatarsal joints.

THE FETAL CIRCULATION

The embryo is dependent upon the placenta for nutrition and the oxygenation of its tissues. The lungs are not functional and, therefore, an operating pulmonary circuit is not necessary. Circulation is needed only in the bronchial arteries to nourish the growing but noninflating lungs. The embryo's alimentary tract does not absorb and digest food during the prenatal period. Nor does it convert food to simpler diffusible substances for tissue building and nutrition. The mother's supply of such substances is drawn

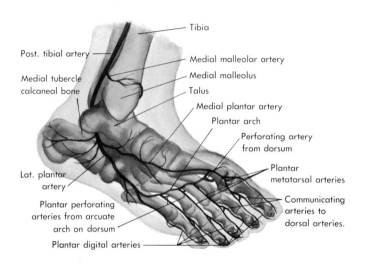

Tibia

Post. tibial artery

Medial tubercle calcaneal bone

Medial malleolar artery

Medial malleolus

Talus

Medial plantar artery

Plantar arch

Perforating artery from dorsum

Lat. plantar artery

Plantar metatarsal arteries

Communicating arteries to dorsal arteries.

Plantar perforating arteries from arcuate arch on dorsum

Plantar digital arteries

upon through the placenta. Therefore, much of the blood in the developing vascular system is diverted from the fetal gastrointestinal tract and liver.

Prenatal mechanisms exist to provide for major arterial and venous connections to the placenta, to shunt blood around the liver, and to divert blood from the pulmonary arterial circuit.

CONNECTIONS TO THE PLACENTA AND THE ALIMENTARY SHUNT

Excretion of wastes and the absorption of oxygen and nutritive substances occur in the placenta. A lake of maternal blood surrounds the capillaries of the fetal part of the circulation in the placenta. Blood passes from the fetal placental veins into the *umbilical vein.* This vessel carries oxygenated arterial blood as do the adult pulmonary veins, but it is like the portal vein of the adult digestive tract also in that the blood is rich in nutritive substances. The umbilical vein pursues a spiraling course along the umbilical cord, enters the fetal body at the umbilicus, and ascends along the *falciform ligament.* This ligament is a fold of peritoneum between the abdominal wall and the liver. The umbilical vein joins with the

portal system of veins from the gastrointestinal tract. Some of the blood enters the liver via radicles of the portal vein to nourish this important organ. Most of the umbilical vein blood, however, is shunted to the inferior vena cava by an embryonic vein, the *ductus venosus.* At this point the rich arterial blood of placental origin blends with venous blood from the inferior regions of the fetal body. The blood from both sources then enters the heart.

Blood which has circulated through the upper parts of the fetal body is a "gray" mixture of oxygenated arterial blood and venous blood. This mixture of blood reaches the common iliac artery and is distributed through its branches (internal and external iliac vessels) to the pelvis and lower limbs. There is a major fetal branch of the internal iliac artery on each side in the fetal pelvis. In the adult these branches are marked only by the superior vesical arteries and by fibrous strands leading toward the umbilicus. The arterial branches are the *umbilical arteries* which, after birth, atrophy to form the *lateral umbilical ligaments.* During the prenatal period, however, the umbilical arteries are major conducting vessels which convey most of the aortic blood through the umbilical cord to the placenta.

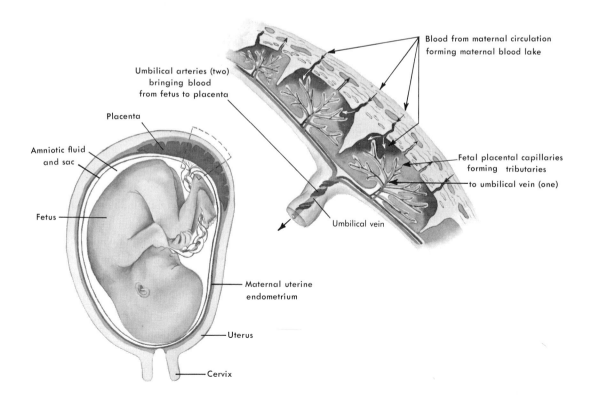

Blood from maternal circulation forming maternal blood lake

Umbilical arteries (two) bringing blood from fetus to placenta

Placenta

Amniotic fluid and sac

Fetus

Fetal placental capillaries forming tributaries
to umbilical vein (one)

Umbilical vein

Maternal uterine endometrium

Uterus

Cervix

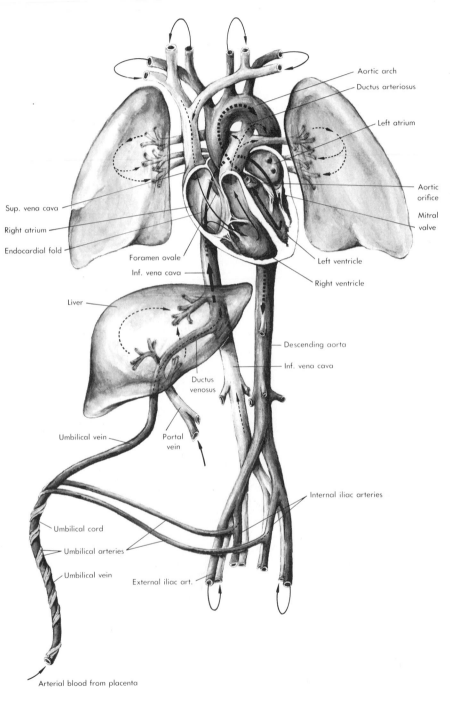

Aortic arch

Ductus arteriosus

Left atrium

Aortic orifice

Mitral valve

Sup. vena cava

Right atrium

Endocardial fold

Foramen ovale

Inf. vena cava

Left ventricle

Right ventricle

Liver

Descending aorta

Inf. vena cava

Ductus venosus

Umbilical vein

Portal vein

Internal iliac arteries

Umbilical cord

Umbilical arteries

Umbilical vein

External iliac art.

Arterial blood from placenta

The Fetal Circulation

PULMONARY BY-PASS

As blood enters the right atrium from the inferior vena cava, it is deflected by an endocardial fold so that most of the blood is directed toward the interatrial septum. This partition has not closed completely in the fetus. The *foramen ovale* is a large opening in the interatrial septum through which blood is shunted into the left atrium to mix with the small amount of blood that returns through the pulmonary veins. The blood is then pumped from the left ventricle through the aorta, as in the adult, with some of the blood passing to the head region from which it returns via the superior vena cava to the right atrium. (See illustration on opposite page.)

Blood returning from the superior vena cava, plus the smaller amount of inferior vena caval blood which was not deflected into the foramen ovale, flows into the pulmonary arterial trunk.

A fetal by-pass, the *ductus arteriosus,* diverts most of the pulmonary circuit blood into the aortic arch where it joins the stream of blood emanating from the left ventricle. The ductus arteriosus closes at birth and a full volume of blood reaches the now-functioning lungs through the pulmonary arteries. This fetal vascular shunt persists in the adult only as a fibrous strand, the *ligamentum arteriosum,* extending from the left pulmonary artery to the aortic arch.

Closure of the foramen ovale after birth further obliterates the pulmonary bypass mechanism by preventing interatrial passage of blood. A persistent foramen ovale or a patent ductus arteriosus or both present simultaneously are types of congenital heart defects found in "blue babies." In such cases too much blood continues to by-pass the lungs and inadequate oxygenation of tissues results.

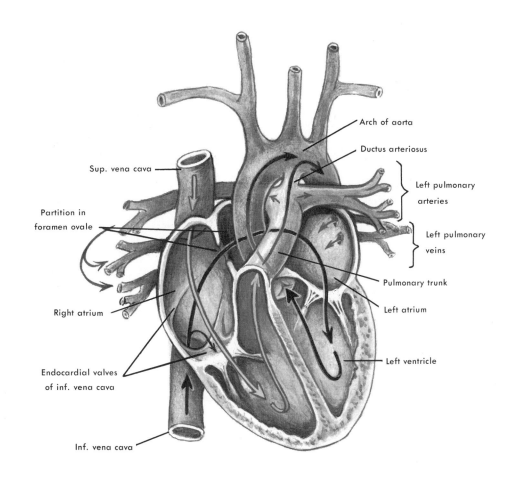

Arch of aorta

Ductus arteriosus

Sup. vena cava

Left pulmonary arteries

Partition in foramen ovale

Left pulmonary veins

Pulmonary trunk

Right atrium

Left atrium

Left ventricle

Endocardial valves of inf. vena cava

Inf. vena cava

VEINS OF THE HUMAN BODY

It is not necessary to memorize descriptions or lists of the venous vessels of the body if the following basic premises are kept in mind.

1. The veins are a blood return system originating in peripheral or distal parts, regions, or organs and leading toward the heart. The venous system begins by the reassembly of capillaries into venules after the diffusion-exchange of gases, metabolites, and wastes has occurred.

2. In contrast to arteries and arterioles, venules and veins are frequently multiple, form irregular networks, and often communicate with each other. Variations are very common in the venous system. Except for the major collecting trunks, standard patterns are difficult to detect or describe.

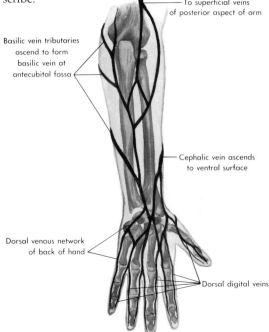

3. Veins of comparable size join with each other at acute angles on converging courses as they extend centrally toward the heart. The larger venous vessels are formed as the result of *tributary convergence.* The veins differ in this regard from arteries in which larger vessels proceeding peripherally branch to form smaller vessels by *arterial ramification.* Remember, therefore, the arteries have branches but veins have tributaries. It is a mistake to speak of a branch of a vein.

4. The convergence of many small, unnamed venous tributaries from the tissues results in the formation of large veins which parallel the course of named arteries. It will be remembered, however, that the blood flow in the parallel vessels is opposite: in the veins toward the heart and in the arteries toward the periphery. Veins accompanying arteries are known as the *venae comitantes* or *accompanying veins.* If one knows the name and course of a given artery to a region, part, muscle, or organ, one knows that in most cases the name and course of the vein will be the same. A few major exceptions are noted in the summary of veins.

GENERAL PLAN OF THE VENOUS SYSTEM

All the blood of the body except that from the lungs and the heart itself returns to the heart by way of the two final venous drainage pathways, the *superior* and *inferior venae cavae.*

Blood from the head and the deeper structures of the neck is drained by the *internal jugular vein.* Venous blood from the upper limb reaches the root of the neck by way of the *subclavian vein* which is joined by the *external jugular vein* draining more superficial neck structures.

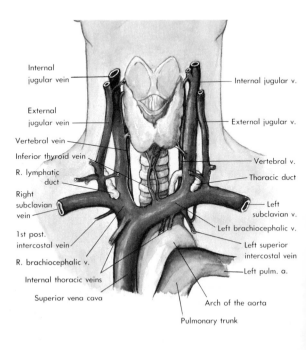

The juncture of the subclavian and internal jugular veins forms the *brachiocephalic* (innominate) vein behind the sternoclavicular joint. The two brachiocephalic veins unite in the superior mediastinum to form the *superior vena cava.*

Blood from the lower limbs is collected into the common venous pathway of the *femoral vein* on each side. Tributaries corresponding to the branches of the femoral and external iliac arteries to the lower abdominal wall also join the femoral vein. The collecting trunk passes centrally into the pelvis as the *external iliac vein* where it is joined by the *internal iliac vein* from the pelvic and gluteal regions to form the *common iliac vein.*

The union of the two common iliac veins gives origin to the *inferior vena cava* which ascends through the abdomen receiving the *renal* and *hepatic veins.* Other visceral tributaries, corresponding to the branches of the abdominal aorta, are not received because of the special alimentary venous collecting system of the *portal vein.*

The *portal vein* drains the abdominal viscera of the alimentary tract and the spleen. The *splenic vein* receives the *inferior mesenteric vein* which drains the large intestine. The splenic vein joins with the *superior mesenteric vein* which receives blood from the small intestine, stomach, and lower esophagus. The union of the splenic and superior mesenteric veins results in the formation of the portal vein which passes into the liver. (See illustration on following page.) *Portal blood* is rich in the products of digestion, and blood flow in the portal vein may be expected to increase markedly during the period of digestion after eating. It may be seen, therefore, that the hepatic veins from the liver are significant tributaries to the inferior vena cava which ascends through the diaphragm to enter the pericardium.

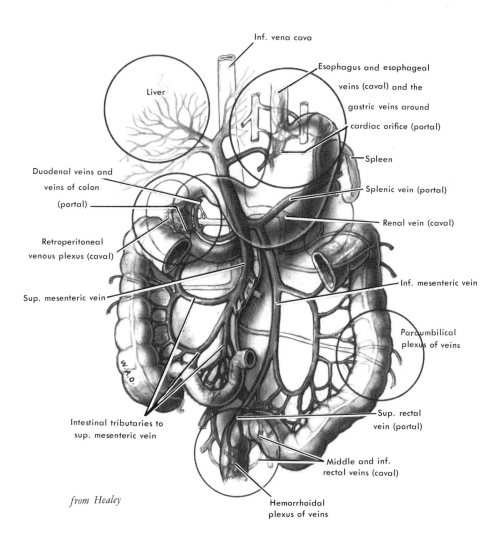

Inf. vena cava

Liver

Esophagus and esophageal veins (caval) and the gastric veins around cardiac orifice (portal)

Duodenal veins and veins of colon (portal)

Spleen

Splenic vein (portal)

Renal vein (caval)

Retroperitoneal venous plexus (caval)

Inf. mesenteric vein

Sup. mesenteric vein

Paraumbilical plexus of veins

Intestinal tributaries to sup. mesenteric vein

Sup. rectal vein (portal)

Middle and inf. rectal veins (caval)

from Healey

Hemorrhoidal plexus of veins

VEINS OF THE BODY REGIONS

The figures which follow show the major veins of the various regions of the body. These are the tributaries to the main venous collecting trunks described in the preceding section. The venous drainage of the brain and other structures within the skull is described with the nervous system and organs of special sense. Notes appended to certain of the regional diagrams explain special arrangements (e.g., the azygos veins of the thorax) or veins of considerable practical importance (e.g., veins of the antecubital fossa and the saphenous system).

The *basilic* and *cephalic* veins are the main superficial veins of the upper limb. They arise from venous arches and networks on the dorsal and palmar aspects of the hand and wrist. A *median antebrachial vein* follows the course of the interosseous arteries by traversing the central area of the anterior surface of the forearm. It may

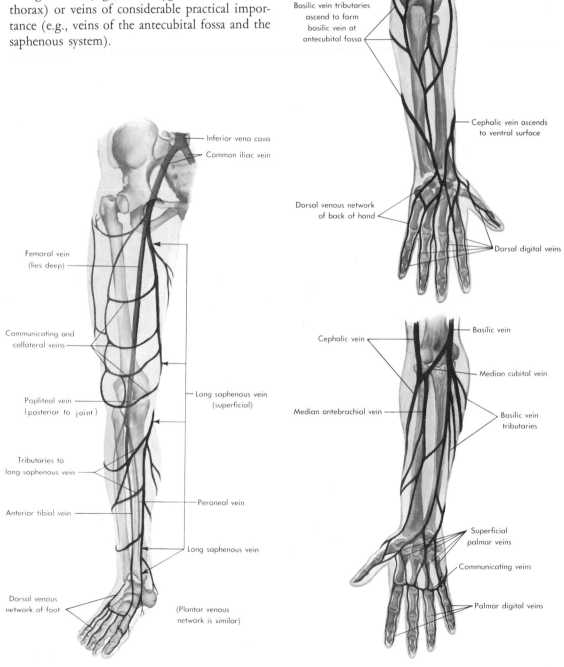

To superficial veins of posterior aspect of arm

Basilic vein tributaries ascend to form basilic vein at antecubital fossa

Cephalic vein ascends to ventral surface

Dorsal venous network of back of hand

Dorsal digital veins

Inferior vena cava

Common iliac vein

Femoral vein (lies deep)

Communicating and collateral veins

Popliteal vein (posterior to joint)

Tributaries to long saphenous vein

Anterior tibial vein

Dorsal venous network of foot

Long saphenous vein (superficial)

Peroneal vein

Long saphenous vein

(Plantar venous network is similar)

Cephalic vein

Median antebrachial vein

Basilic vein

Median cubital vein

Basilic vein tributaries

Superficial palmar veins

Communicating veins

Palmar digital veins

join the cephalic, basilic, or *median cubital vein.* All are superficial veins, coursing in the tela sub-cutanea. They drain the skin and superficial connective tissues of the upper limbs.

The dorsal venous network on the back of the hand is sometimes used as a site for obtaining blood for examination. Their more important application is in offering the surgeon easily accessible vessels when it is necessary, as in severe burn cases, to "cut down" to insert a semi-permanent cannula into a vein through which plasma, blood, or isotonic fluids may be administered.

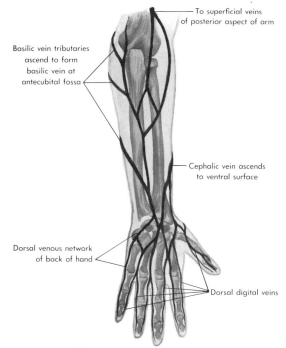

The *antecubital veins,* similarly accessible at the bend of the elbow, are the veins most used by physicians, nurses, or paramedical technicians and corpsmen as a site for *venipuncture.* The *median cubital vein,* ascending from the laterally placed cephalic vein to the medially coursing basilic vein, may be punctured in order to withdraw blood for examination or to introduce fluids or medications. The veins it connects may be used also but these are more mobile. There are many variations in the arrangements of the veins at the antecubital fossa which are learned by those who must use the veins as they learn the techniques involved.

The diagram shows the tributaries to the brachiocephalic vein from the thoracic and upper abdominal walls: the *internal thoracic vein,* the

right first posterior intercostal vein, and the *left superior intercostal vein.* The upper abdominal veins of the trunk drain through the *superior epigastric vein* to the internal thoracic vein.

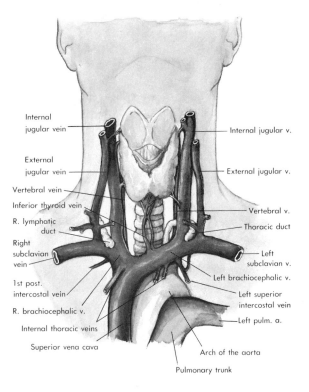

The rest of the parietal veins which accompany the posterior intercostal arterial branches of the aorta do not drain directly and individually into either the superior or the inferior vena cava. Instead these veins on the *right side* are tributaries to the *azygos vein* which commences below the diaphragm as the subcostal vein.

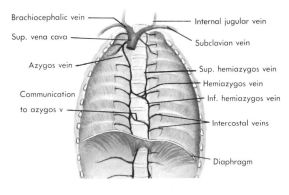

The azygos trunk ascends to the right of the vertebral column through the posterior mediastinum. After receiving its intercostal tributaries, the azygos vein arches over the right lung root to enter the superior vena cava.

On the *left side* are the origins of the *inferior hemiazygos vein* and the azygos vein. A *superior hemiazygos vein* receives the left fourth and sixth intercostal veins. The two either join to form a common venous trunk or cross separately to the right side to drain into the azygos vein.

A superficial system of veins in the lower limb, the *saphenous veins,* forms from venous networks on the foot. The *dorsal* and *plantar venous networks* are similar to those described for the hand. Most of the superficial veins drain to the margins of the foot as may easily be seen in the living adult with the foot dependent. Two saphenous veins commence from the foot margins.

The *long saphenous vein,* as illustrated, proceeds upward from the *medial* margin of the foot by passing in front of the medial malleolus of the ankle. It is very accessible at this point as an alternative point for inserting a needle or a canula into the venous system. The ascent of the long saphenous vein carries it along the medial surface of the leg and knee. This long vessel continues upward along the medial aspect of the thigh to the saphenous opening in the cribriform fascia of the thigh. At this point, below the inguinal ligament, the vein passes deep to join the femoral vein. The long saphenous vein receives many tributaries, near its termination, from the upper thigh and lower abdominal wall. These are seen in the drawing. Many of these tributaries are the veins accompanying branches of the femoral artery.

The *short saphenous vein* begins at the *lateral* margin of the foot. It extends *behind* the lateral malleolus to ascend along the back of the calf of the leg. It passes deep into the popliteal fossa as a tributary to the popliteal vein or the lower part of the femoral vein. See facing page.

The saphenous veins communicate with each other but, more important, they send *communicating veins* deep to connect with the deep veins of the leg and thigh. The erect position of man has placed great pressures upon the saphenous veins, their deep communications, and their valves. The saphenous veins are superficial in the tela subcutanea and are separated from the action of muscles by heavy deep fasciae. The "milking action" of muscle contraction cannot help significantly in promoting the upward flow of blood

in the saphenous system. The effect of gravity tends to impede the venous flow despite the presence of valves. If the valves become incompetent, because of long-standing back pressure or weakening of the vein walls, *varicose veins* result.

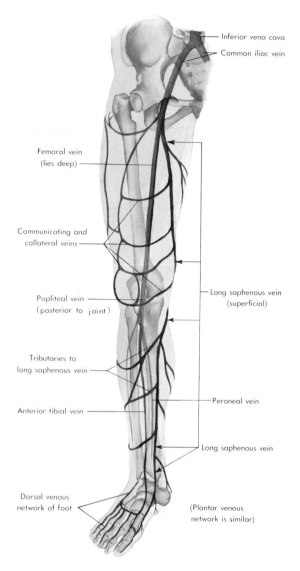

Inferior vena cava

Common iliac vein

Femoral vein (lies deep)

Communicating and collateral veins

Long saphenous vein (superficial)

Popliteal vein (posterior to joint)

Tributaries to long saphenous vein

Peroneal vein

Anterior tibial vein

Long saphenous vein

Dorsal venous network of foot

(Plantar venous network is similar)

In this syndrome, which is partly hereditary and is also linked to occupations requiring long periods of standing, back pressure upon the faulty veins enlarges them. The dilated veins become tortuous and knotted. The venous stagnation may be quite disabling and cosmetically disfiguring and interfere ultimately with the nutrition of superficial tissues. The result of chronic untreated *varicosities* is the breakdown and ulceration of the skin of the more dependent parts of the lower limb. Surgical treatment, prior to this latter stage, is effective in the correction of a patient's problems.

By way of the *deep plantar venous arch,* digital and metatarsal veins of the foot contribute to the formation of *posterior tibial veins.* These deep veins ascend as venae comitantes to the posterior tibial artery which receives the *peroneal veins.* The *anterior tibial veins* join the deep veins at the lower border of the popliteal fossa to form the *popliteal vein* which continues upward as the deep vein of the thigh, the *femoral vein.* The saphenous veins are received by the deep veins as shown previously.

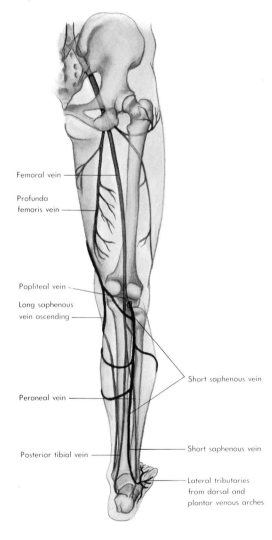

Femoral vein

Profunda
femoris vein

Popliteal vein

Long saphenous
vein ascending

Short saphenous vein

Peroneal vein

Posterior tibial vein

Short saphenous vein

Lateral tributaries
from dorsal and
plantar venous arches

The femoral vein is a large vein which also receives the large *profunda femoris* vein from the large muscles of the back of the thigh. The two large veins represent a potentially large cross sec-

tional area for venous stagnation. When a person has been ill and confined to the inactive recumbent position for a long time, or in cases of inflammation of the femoral veins, there is a great danger of the formation of *thrombi.* In such situations, blood cells clump together and become attached to the endothelium; a form of clotting occurs *within* the large veins. A greater danger is that fragments of the clot may break loose and, as *emboli,* float through the inferior vena caval system to the heart, eventually to lodge in an arterial vessel, often in the lungs, with serious consequences.

The pelvic organs of the reproductive, urinary, and lower alimentary systems are drained by veins accompanying the branches of the internal iliac artery. They are shown in the diagram. These veins are thin-walled but profuse and have many interconnections. The *internal iliac vein* or direct visceral tributaries from venous plexuses about the rectum, prostate gland and seminal vesicles, penis, urinary bladder, uterus, and vagina join with the external iliac vein to form the *common iliac vein.* The testicular or ovarian veins travel upward from the pelvis on a course corresponding to their arteries. On the right side such veins join the inferior vena cava; on the left they are tributaries to the left renal vein. (See illustration on page 368.)

Several important practical points relate to the pelvic veins. *First,* the same principles of venous stagnation, inflammation of veins, and the possibility of thrombosis apply to the pelvic veins as described for the femoral veins. *Second,* many of the major lymphatic vessels from pelvic and perineal organs course upon these veins. (See illustration on page 369.) The spread of malignant cells from these organs may, therefore, be expected to follow the course of the veins. Cancers also may spread by the introduction of malignant cells directly into the venules of the pelvic organs. Clumps of tumor cells may be circulated to other organs to seed secondary tumors distant from the site of the primary tumor in the pelvis. *Third,* the pelvic veins, especially those of the prostatic venous plexus in the male, have connections with the lower vertebral veins. The latter connect with the more cephalic vertebral veins via irregular networks about the vertebral column and within the vertebral canal. It may be seen, therefore, that a brain tumor may be secondary to a primary pelvic tumor through the insidious

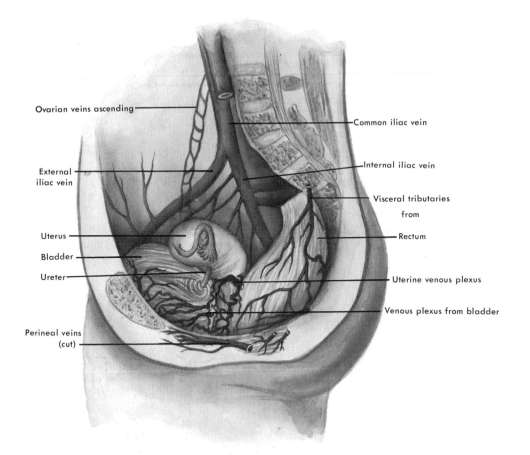

Ovarian veins ascending

External
iliac vein

Uterus

Bladder

Ureter

Perineal veins
(cut)

Common iliac vein

Internal iliac vein

Visceral tributaries
from

Rectum

Uterine venous plexus

Venous plexus from bladder

spread of cancer along far-flung, indirect pathways.

The diagram on this page shows the tributaries of the *portal vein,* which collects the venous blood of the alimentary tract below the diaphragm, the spleen, the pancreas, and the gallbladder. It conveys blood from these organs to the liver.

The portal venous system has numerous connections with the vena caval system. These occur between the superior rectal veins (portal) and the middle and inferior rectal veins (caval); between gastric veins surrounding the cardiac orifice of the stomach (portal) and tributaries to the esophageal and hemiazygos veins (caval); between veins of the duodenum and the colon (portal) which lie against the posterior abdominal wall and the retroperitoneal plexus of veins (caval); and between paraumbilical veins (portal) which traverse the round ligament of the liver to the umbilicus and the periumbilical venous anastomosis around the umbilicus. The periumbilical plexus of veins provides a connection between the superior and

inferior epigastric veins medially and the *thoracoepigastric* venous channel laterally.

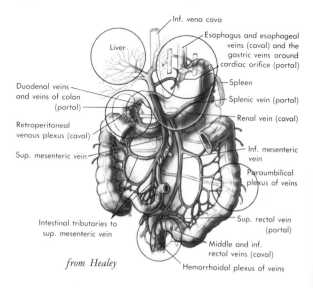

Inf. vena cava

Liver

Esophagus and esophageal
veins (caval) and the
gastric veins around
cardiac orifice (portal)

Spleen

Splenic vein (portal)

Renal vein (caval)

Inf. mesenteric
vein

Paraumbilical
plexus of veins

Sup. rectal vein
(portal)

Middle and inf.
rectal veins (caval)

Hemorrhoidal plexus of veins

Duodenal veins
and veins of colon
(portal)

Retroperitoneal
venous plexus (caval)

Sup. mesenteric vein

Intestinal tributaries to
sup. mesenteric vein

from Healey

When disease of the liver blocks the entrance of the portal vein and its blood to the liver, back pressure and stagnation in the portal system dilates all the connections to the caval system tremendously in an attempt to form an adequate collateral circulation around the block. The vessels become enlarged and tortuous. A process similar to varicose veins develops in the lower esophagus and rectum to form "esophageal varicosities" and "hemorrhoids" respectively. The veins of the anterior abdominal wall form an interlacing tortuous network around the umbilicus which has been given the fanciful mythological name of the *caput medusae.*

The *thoracoepigastric veins* are an irregular network connecting the epigastric veins to the axillary vein. These channels ascend along the lateral thoracic wall to the axilla, frequently receiving mammary veins. The thoracoepigastric veins share in the processes forming the caput medusae.

LYMPHATIC SYSTEM AND SPLEEN

The lymphatic system is a separate system of vessels which assists the venous system in the return of fluids from the tissues. There is no counterpart in this system to the arteries, and it does not open into the heart but into the venous system in the base of the neck. Lymphatic organs, the lymph nodes, are interposed in the course of lymphatic vessels. An abdominal organ, the spleen, is closely associated with the blood and with the lymphatic system.

LYMPH AND LYMPHATIC CAPILLARIES

The lymphatic system begins as a network of blind *lymphatic capillaries* which permeate the tissues of the body. Except for a larger diameter and their origin within the tissues, the structure of these endothelial vessels is similar to that of the blood capillaries. *Lymph* is the fluid which fills the vessels of the lymphatic system. It is a clear colorless fluid except in the lymph vessels of the intestinal tract where the lymphatic system carries absorbed fats in a milky suspension. The composition of lymph is similar to that of blood plasma. (See illustration below.) Components of the tissue fluid which are in excess of its needs and do not pass into blood capillaries pass into the lymph vessels. Proteins, particularly those whose molecular size does not permit passage through the endothelium of vascular capillaries, are received into the lymph. Cells do not appear in lymph vessels of capillary size. (See also color plate 8.)

LYMPHATIC VESSELS AND LYMPH NODES

Lymphatic capillaries join to form a series of collecting vessels of increasing size as they course centrally. The lymphatic vessels or *lymphatics* course in close relationship with the veins of any body part. Lymphatics, although thinner walled, have the tissue structure of veins and, like them, possess valves.

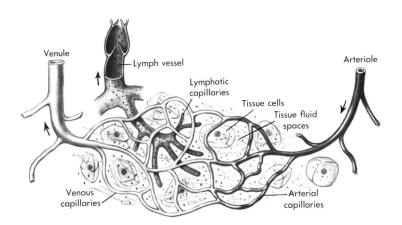

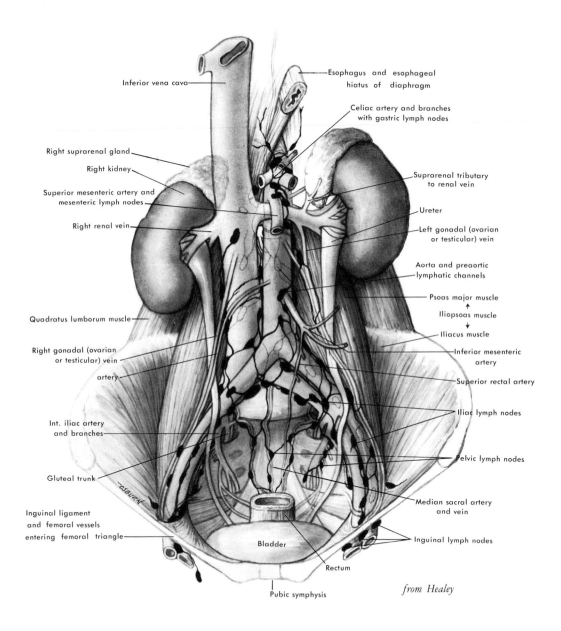

Inferior vena cava

Esophagus and esophageal hiatus of diaphragm

Celiac artery and branches with gastric lymph nodes

Right suprarenal gland

Right kidney

Suprarenal tributary to renal vein

Superior mesenteric artery and mesenteric lymph nodes

Ureter

Right renal vein

Left gonadal (ovarian or testicular) vein

Aorta and preaortic lymphatic channels

Psoas major muscle

Iliopsoas muscle

Quadratus lumborum muscle

Iliacus muscle

Inferior mesenteric artery

Right gonadal (ovarian or testicular) vein

artery

Superior rectal artery

Iliac lymph nodes

Int. iliac artery and branches

Pelvic lymph nodes

Gluteal trunk

Median sacral artery and vein

Inguinal ligament and femoral vessels entering femoral triangle

Inguinal lymph nodes

Bladder

Rectum

Pubic symphysis

from Healey

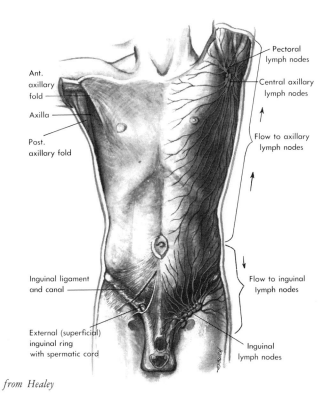

Pectoral
lymph nodes

Central axillary
lymph nodes

Ant.
axillary
fold

Axilla

Post.
axillary fold

Flow to axillary
lymph nodes

Inguinal ligament
and canal

Flow to inguinal
lymph nodes

External (superficial)
inguinal ring
with spermatic cord

Inguinal
lymph nodes

from Healey

At intervals along the course of the lymphatics, *lymph nodes* are interposed. These organs have a dual function: the filtering of the lymph and the addition of lymphocytes which are produced within the node. Lymph nodes are encapsulated organs formed from lymphatic tissue and channels through which the lymph flows. *Afferent lymphatics* from more distal parts enter through the capsule of the lymph node. After passing through the node, lymph leaves in centrally bound *efferent lymphatics* from an indented hilum. *Lymphatic* (lymphoid) *tissue* consists of collections of lymphocytes supported by a framework of delicate reticular connective tissue. Within the lymph node is an outer cortex and an inner medulla. The *cortex* is made up largely of dense collections of lymphoid tissue called lymphoid follicles. The *medulla* is formed by irregular cords of lymphocytes. The lymph flows between the follicles and cords in irregular sinusoids. In its passage the lymph is filtered of foreign matter and bacteria, and detoxification may take place. New lymphocytes, destined for the blood, are added to the lymph.

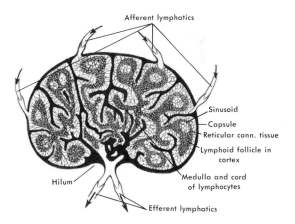

Afferent lymphatics

Sinusoid

Capsule
Reticular conn. tissue

Lymphoid follicle in
cortex

Medulla and cord
of lymphocytes

Hilum

Efferent lymphatics

LYMPHATIC DRAINAGE PATHWAYS

The course taken by lymphatics and the location of lymph nodes are described as the lymphatic drainage of the body or its parts.

The lymph from the left side of the head, neck, and thorax; from the abdomen and pelvis; and from the left upper and both lower limbs flows ultimately into a large lymphatic trunk, the *thoracic duct*. Lymph from the right side of the head,

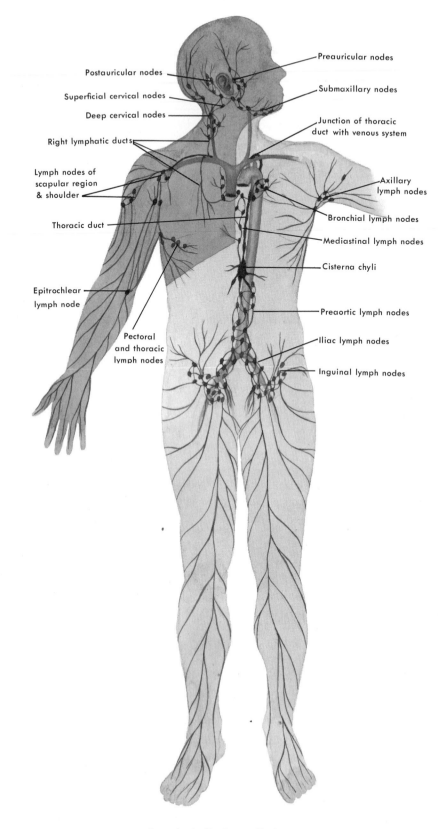

Preauricular nodes

Postauricular nodes

Submaxillary nodes

Superficial cervical nodes

Deep cervical nodes

Junction of thoracic
duct with venous system

Right lymphatic ducts

Lymph nodes of
scapular region
& shoulder

Axillary
lymph nodes

Bronchial lymph nodes

Thoracic duct

Mediastinal lymph nodes

Cisterna chyli

Epitrochlear
lymph node

Preaortic lymph nodes

Iliac lymph nodes

Pectoral
and thoracic
lymph nodes

Inguinal lymph nodes

Lymphatic Drainage Pathways

neck, and thorax and from the right upper limb is collected by the *right lymphatic duct*. These main ducts empty into the venous system by entering the juncture of the internal jugular and subclavian veins where the brachiocephalic vein of their side is formed. The thoracic duct begins along the posterior abdominal wall in a saccular dilatation, the *cisterna chyli*, into which a number of collecting trunks empty. It pursues an upward course through the abdomen and thorax close to the vertebral column, receiving other lymphatics, and finally crosses to the left side of the thorax and base of the neck. The right lymphatic duct originates from the confluence of a number of collecting trunks on the right side of the thorax.

The major lymphatic pathways are listed below.

1. Lymph from each side of the head flows downward through submaxillary, preauricular, and postauricular nodes. This lymph and that from neck structures pass through superficial or deep cervical nodes whose efferent trunks enter either the thoracic or right lymphatic ducts just before their termination.

2. Lymphatics of the upper limb follow veins centrally to the axilla. Axillary lymph nodes also drain the pectoral region, mammary gland, lateral thoracic wall, and scapular region.

3. Lower limb lymphatics drain to the inguinal lymph nodes which cluster above and below the inguinal ligament. Lymph from the external genitalia and from the abdominal wall below the umbilicus also passes to the inguinal nodes.

4. Lymph leaving the inguinal lymph nodes passes through nodes along the iliac blood vessels which also receive lymphatics from the pelvic viscera. A chain of lymph nodes located in front of the abdominal aorta passes the lymph on to the cisterna chyli.

5. Lymphatics from the abdominal organs pass through chains of mesenteric and gastric lymph nodes which are located along the main arteries in mesenteries and at points of vascular branching from the aorta. The ultimate destination of the lymph is the thoracic duct.

6. Mediastinal and tracheobronchial lymph nodes receive lymph from the internal aspect of the thoracic wall, the thoracic viscera, and the mediastinum. Collecting trunks pass to either the right lymphatic duct or thoracic duct.

NONNODAL LYMPHATIC TISSUE

Lymphoid tissue is found in the walls of the pharynx and intestinal tract. Such aggregations of lymphocytes appear at locations which may be subject to bacterial invasion from inspired or ingested substances. These lymphoid collections, of which the tonsils are an example, are not on lymphatic pathways but instead reinforce the blood leukocytes in areas of potential bacterial activity.

SPLEEN

The *spleen* is a fist sized, intensely vascular organ located high in the left side of the abdomen under the shelter of the diaphragm and lower ribs. Its costodiaphragmatic surface is convex but its medial surface is molded to the contours of other abdominal organs. A prominent indented hilum provides entry and exit for the splenic vessels. These vessels reach the spleen in folds of peritoneum which then invest the spleen. The vascular pulp of the interior of the organ is surrounded by a fibrous capsule into which smooth muscle is intermixed. Partitioning trabeculae spring inward from the capsule to irregularly subdivide the spleen and form a framework for the splenic tissue.

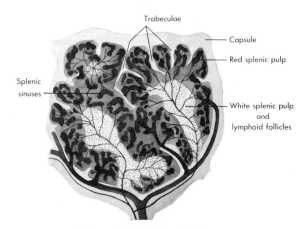

Schematic Section of Spleen

The *splenic tissue* is divided into the white pulp and the red pulp which are mixed together. The *white pulp* is formed by lymphoid follicles similar to those of lymph nodes. The more diffuse *red pulp* is composed of cords of red blood cells, lymphocytes, and phagocytic cells. Blood flows through the spleen in sinuses which provide for intimate contact with cells of the cords. During fetal life the spleen produces both red and white blood cells. In adult life the blood cell-forming function of the spleen is restricted to lymphocytes and monocytes. The spleen is active in the destruction of aging erythrocytes during which the iron fraction of hemoglobin is conserved. It is believed that the spleen has a role in antibody production. By means of its many blood passageways the spleen acts as a reservoir for blood during periods when circulation is slowed.

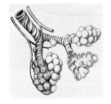

7

The Respiratory System

The respiratory system is formed by a series of continuous passageways in the head, neck, and thorax that provide conduits through which atmospheric air can come into intimate contact with the blood so that gaseous exchange may take place. See color Plate 5. The following parts will be considered:

1. Air inlet and exit: the external nose and the nostrils, with the secondary channels of the oral cavity and the mouth.

2. Head passageways: the nasal cavity and nasopharynx where the air is initially warmed and humidified as it moves inward.

3. Neck and thorax conduit: the trachea with the vibrating chamber for sound production, the larynx.

4. Extrapulmonary distributors: the tracheal bifurcation and the extrapulmonary bronchi.

5. Intrapulmonary distributors and the mechanism for gaseous exchange: the lungs and their bronchovascular patterns.

6. Mechanism for breathing: the coverings of the lung, pleural potential space, expansibility of the lung, intrathoracic relationships, thoracic wall, and the respiratory muscles.

THE UPPER RESPIRATORY SYSTEM

EXTERNAL NOSE

The bony nasal fossa is just a triangular opening into the deeper recesses of the head. The living person has an additional feature of facial anatomy, the *external nose,* which is fastened to the boundaries of the bony nasal fossa. Much of the identity and appearance of a person is represented by his external nose. Similarly, deformity or destruction of the nose by disease or trauma is a threat to his personality, equanimity, and social relationships. The external nose provides a more protected opening to the respiratory passages, increases the surface area of their warming and humidifying membranes, and serves as a coarse filter against foreign matter. It also acts with other head structures to modify sounds produced by the larynx in the production of speech.

Construction

The external nose is basically a pyramidal framework of cartilage and fibrous tissue covered by facial skin and lined by a mucous membrane.

375

Framework. Separate *nasal cartilages* of varying size and shape are bound together by dense fibrous tissue. This tissue fastens them to the margins of the nasal fossa, molds them into the characteristic shape of the nose, and fills in the apertures between the cartilages. The nose is divided at the midline by a *nasal septum* which increases the mucous membrane surface area and lends further support to the pyramid. Each side of the nose is open anteriorly to the external environment through the *nostril.* Posteriorly the external nose opens into the *nasal cavity,* which is subdivided by the nasal septum into right and left sides. The *nasal cartilages* are as shown below.

Shape. Familial, racial, and national genetic factors influence the shape of a person's nose. It changes gradually from infancy to assume its final adult form, characteristic of each individual, during the latter part of the second decade. Injuries, invasive disease processes, and degenerative conditions in the later years of life modify its shape.

In general, the pyramidal external nose slopes downward from the *root* (nasion) as the bridge or *dorsum* to the *apex.* The sides also slope laterally to join the facial contour while the *base* of the nose expands into the wings or *alae.* The alae surround the nostrils and are the only portion of the nose with considerable mobility. The skin is thin and mobile over the root, but at the base it becomes thicker, adherent to underlying tissues, and more vascular. Large sebaceous glands, here and in the *nasolabial fold* of facial skin adjoining the alae laterally, are responsible for the oiliness of this part of the nose, comedones ("blackheads"), and the disturbing blemishes of facial acne. Although the subcutaneous tissue of the alae is thin, its vascular mass may increase in later years to markedly alter a person's appearance.

Interior

The cavity of the external nose follows the shape of the external features and is divided by

CARTILAGE	LOCATION	SHAPE	SUPPORT
Septal	Interior midline	Vertically disposed, thin, flat plate	Bony nasal septum posteriorly; hard palate inferiorly
Lateral nasal	Bridge below nasal bone	Molded to individual shape of nose	Frontal process of maxilla; meets its fellow at midline
Greater alar	Lower part of bridge, tip, wing, borders of nostril except laterally	Irregular plate; molded to individual shape of nose	Fibrous tissue attached to nasal fossa margins
Lesser alar	Completes nostril; part of wing (rest is fibrous)	Small irregular plate or several fragments	Fibrous tissue joining it to greater alar cartilage; facial deep fasciae

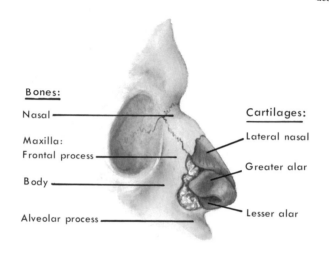

Bones:
Nasal
Maxilla:
Frontal process
Body
Alveolar process

Cartilages:
Lateral nasal
Greater alar
Lesser alar

the midline septum. The most anterior part of the cavity is expanded to form the *vestibule* into which the index finger may be inserted through the nostril. The facial skin extends around the margins of the nostril to line the vestibule before becoming continuous with the mucous membrane of the nasal passageways. Coarse bristle-like hairs grow from this skin, particularly in the male. These *vibrissae* act to strain coarse particulate matter and small insects from the air that enters the respiratory system.

NASAL CAVITY

Most of the nasal passageways are hidden from view but represent a large mucous membrane surface area over which the environmental air passes into the depths of the head. The *nasal cavity* includes the interior of the external nose and the nasal passages within the head. It is divided by the nasal septum into right and left *nasal fossae.* The nasal cavity opens posteriorly through an oval opening on each side, the *posterior choanae,* into the nasal part of the pharynx.

Surfaces of the Nasal Cavity

The nasal cavity is roughly an isosceles triangle in cross section before being divided by the nasal septum into the nasal fossae. Each nasal fossa is roughly a right triangle. Projections of its lateral wall narrow the triangle into an irregular cleft which extends from the nostril to the posterior choana. Each nasal fossa has medial (septal) and lateral walls, a roof, and a floor.

Floor of the Nasal Fossa. The floor is formed by a horizontal shelf of the maxilla anteriorly, by the palatine bone of the hard palate centrally, and by the soft palate posteriorly. The floor is wider than the apex or roof. There is a hump in the floor, where the external nose joins the base of the bony nasal fossa, from which the floor of the external nose slopes downward and posteriorly into the head.

Medial Wall of the Nasal Fossa. The medial wall is the nasal septum which is shared by both nasal fossae. The septum is partly cartilaginous and partly bony in the nasal fossa. The *cartilaginous nasal septum* forms the anterior and anteroinferior part of the septum. It is slightly movable and possesses a rich arterial blood supply which

becomes obvious in some types of acute nosebleeds due to nose injuries, habitual nose picking, high blood pressure, or congenital knots of fragile blood vessels. The *bony nasal septum* is the fixed posterior and posteroinferior part of the septum which also extends up under the nasal bones. It is made by a perpendicular plate hanging downward from the ethmoid bone of the base of the skull to meet the separate vomer bone above the hard palate. The bony nasal septum is seldom a perfect midline partition. It often displays angulations or deviations which in their more severe forms interfere with passage of air, drainage of the sinuses, and flow of nasal mucus, especially when infections of the sinuses, or allergies coexist.

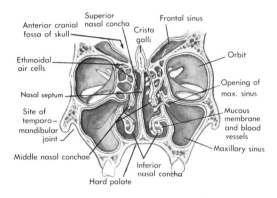

Roof of the Nasal Fossa. The roof is the narrowest part of the nasal cavity. It is a groove related to the cribriform plate of the ethmoid bone, the incurving orbital portion of the frontal bone, and the nasal bones. The fibers of the olfactory nerve commence in this area.

Lateral Wall of the Nasal Fossa. The lateral wall is hidden by its own projections. Three scroll-like shelves from bones forming the lateral wall project across toward the nasal septum. These are the *nasal conchae.* Their free margins turn downward to obscure the lateral wall itself, the intervals between the conchae which are called *meatuses,* and the openings of cavities and ducts which convey secretions from outside the nasal cavity. The nasal conchae are bony shelves which are padded with a highly vascular tissue which may become congested and swollen in allergic conditions to reduce the dimensions of the nasal passageways. The nasal mucous membrane overlies the vascular tissue. When the conchae are clothed with these soft tissues, the term nasal *turbinate* is applied. The three nasal turbinates are the:

1. SUPERIOR: This is the smallest and is formed by a ledge of the ethmoid bone high on the posterolateral wall.

2. MIDDLE: Also formed by a scroll-like shelf of the ethmoid bone, this turbinate comes farther forward and presents a scalloped, turned down, free edge.

3. INFERIOR: A separate *inferior nasal conchal bone* forms the framework of this largest and lowest turbinate which extends from the posterior choana almost to the nostril. See color plate 5.

The interval of nasal cavity that lies immediately under a turbinate has already been termed a *meatus*. Each meatus is named for the turbinate which lies above it. (See illustration on opposite page and also color plate 16.)

Nasal Mucous Membrane

A thin epithelial membrane, with the highly vascular bed of connective tissue underlying it, smoothly lines the entire nasal cavity and covers all its projections and irregularities. This important membrane contains many mucus-secreting glands which produce the film of mucus that coats the nasal lining. The combination of moist mucus and the warmth of the nasal cavity resulting from the rich blood supply has the following functions:

1. Warms and humidifies the incoming air.

2. Traps fine foreign particulate matter in the air so that many impurities are removed.

Microscopic cilia of the epithelial cells wave ceaselessly toward the external nose to produce a slow current of mucus across the surfaces of the nasal cavity.

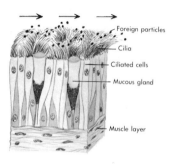

Foreign particles
Cilia
Ciliated cells
Mucous gland
Muscle layer

This film dries slowly as it moves toward the nostrils unless the secretion is copious as in colds and allergies. In this way the particulate matter is expelled from the nasal passages.

THE PARANASAL SINUSES

During the development of the nasal mucous membrane before birth, outpouchings of the epithelium occur which grow and gradually move away from the margins of the nasal cavity. The bones of the skull are forming at this time and they surround the series of glandular outpouchings which, however, maintain their connections with the nasal cavity by short ducts or openings. These sacs, varying from interconnected bubbly air cells in the bones to single irregular chambers within them, are called, collectively, the *paranasal sinuses*. (See color plate 16.) They possess the same type of glandular lining, vascular supply, and nervous innervation as do the portions of the nasal mucous membrane from which they grew and with which they are continuous.

Location of the Sinuses

The sinuses are named for the bones within which they lie.

Frontal Sinus.   The frontal bone is hollowed by an irregular chamber which extends across the midline of the forehead at the region of the glabella and above the medial supraorbital margin. It is connected by the *frontonasal duct* to the *frontal recess* of the nasal cavity which lies high on the lateral wall under the anterior end of the middle turbinate.

Ethmoid Sinuses.   Many air cells, clustered in anterior, middle, and posterior groups, form excavations of the ethmoid bone in the roof of the nasal cavity. The *anterior group* of ethmoid air cells drain through tiny ducts into the middle meatus; the *middle group* have similar connections just behind them. The *posterior group* drains to the superior meatus under the short superior turbinate. The ethmoid air cells, collectively, may be termed the ethmoid sinuses.

Sphenoid Sinus.   A single chamber hollows a portion of the sphenoid bone at the base of the skull above the posterior extremity of the nasal cavity and the nasopharynx behind it. This sinus opens into the *sphenoethmoidal recess* which is located above and behind the superior turbinate.

Maxillary Sinuses.   The body of each maxilla holds a large, air-filled chamber, the maxillary sinus. This sac within the bone abuts the lateral wall of the nasal cavity to which it is connected by an opening, the *maxillary ostium,* which may be found posteriorly in the middle meatus hidden

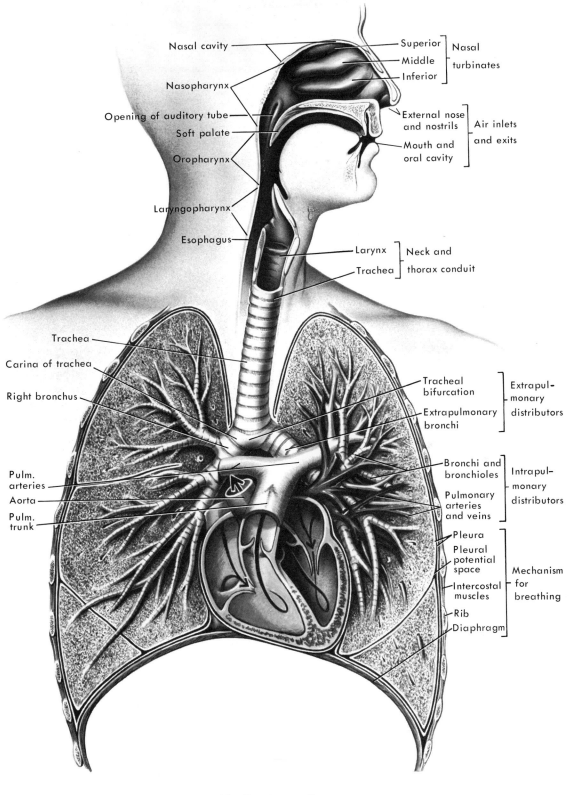

Nasal cavity

Superior ⎫
Middle   ⎬ Nasal turbinates
Inferior ⎭

Nasopharynx

Opening of auditory tube

Soft palate

Oropharynx

External nose
and nostrils ⎫
⎬ Air inlets and exits
Mouth and
oral cavity ⎭

Laryngopharynx

Esophagus

Larynx ⎫
⎬ Neck and thorax conduit
Trachea ⎭

Trachea

Carina of trachea

Right bronchus

Tracheal
bifurcation ⎫
⎬ Extrapulmonary distributors
Extrapulmonary
bronchi ⎭

Pulm.
arteries

Aorta

Pulm.
trunk

Bronchi and
bronchioles ⎫
⎬ Intrapulmonary distributors
Pulmonary
arteries
and veins ⎭

Pleura ⎫
Pleural
potential
space
Intercostal
muscles ⎬ Mechanism for breathing
Rib
Diaphragm ⎭

The Respiratory System

by the wavy free margin of the middle turbinate. (See color plate 16.)

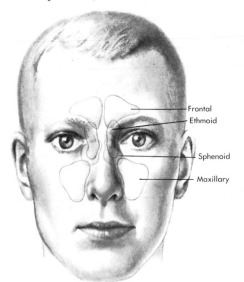

Significance of the Paranasal Sinuses

The exact role of the paranasal sinuses is unclear. A number of functions have been proposed which include:

1. Lightening of the bones of the head by the presence of these air spaces.

2. Adding further surface area to the nasal passages for warming and humidifying the incoming air.

3. Producing mucus to add to the nasal mucous membrane.

4. Assisting in speech production by acting as resonating chambers.

The first function may be debated; the second is doubted because the small diameter of the ostia and ducts precludes much interchange of air between the nasal cavity and the sinuses; the third is believable, and the fourth is reasonable although debated by many authorities.

The greater significance of the sinuses is in their diseases rather than in their normal function. Pneumatization of the bones of the skull, which includes the mastoid process of the temporal bone, seems to recall the four-footed position of mammals. The direction of the ducts and the position of the sinus ostia on the lateral wall and roof of the nasal cavity are not very well suited for the erect position of man. Consider the following:

1. The duct of the sphenoid sinus is considerably higher than its floor. It does not drain well in the erect position.

2. The maxillary ostium is also located high

on the medial wall of the sinus (lateral wall of the nasal cavity).

3. The drainage of mucus from the frontal sinus gravitates along the groove made by the junction of the middle turbinate with the lateral wall. If the sinus is infected, the purulent material from the frontal sinus may enter the openings of the anterior ethmoid and maxillary sinuses to extend the infection to them.

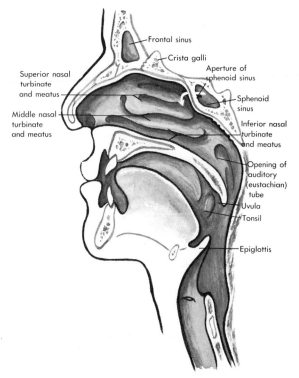

Since there is continuity of the nasal mucous membrane with that of the paranasal sinuses, infections of the nose may spread into these bone-locked chambers. The large amounts of mucus resulting from overproduction by the inflamed sinuses may not be able to escape if the membranes of the ducts and ostia of the sinuses also are swollen.

Relations of the Paranasal Sinuses

The close relations of the sinuses to other structures of the head are important to the understanding of these troublesome chambers. The *frontal sinus* is close to the orbit and the roof of the nasal cavity. The *ethmoidal air cells* are but bubbles in the thin roof of the nasal cavity and, therefore, the thin floor of the skull. Only a thin shell of bone separates them from the olfactory nerve and the frontal lobe of the brain above. The *sphenoid sinus* is close to the optic nerves and pituitary gland just above and it is separated by only a fraction of an inch from the main artery to the brain, the internal carotid, and from the nerves entering the orbit by way of the superior orbital fissure. The *maxillary sinus* is surrounded by the soft tissues of the face in front, by the orbit above, the upper teeth below, and the nasal cavity medially. When, in addition to the closeness of important structures, it is realized that not only the sinus membranes but most of the related structures are provided with pain sensation by the same (fifth) cranial nerve, it can be seen that the misery of sinus disease can be widespread.

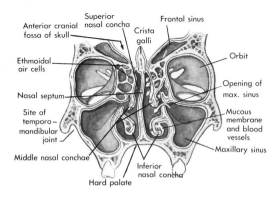

Living Anatomy of Nose, Nasal Cavity, and Paranasal Sinuses

The external nose is easily outlined. The smooth transition from the nasal bones to the lateral nasal cartilages along the dorsum is not apparent unless an injury has occurred. At the tip the nose is softer and more movable because of the greater flexibility and resilience of the greater alar cartilage and the mobility and thinness of the lower part of the septal cartilage. The alae are soft and compressible because the lesser alar cartilage is thin and frequently broken up into smaller pieces which are combined with fibrous tissue. Such factors produce the soft margins of the nostrils and make possible the sphincter-dilator action of the nasal muscles. The firm buttress formed by the frontal process of the maxillary bone can be felt between the nose and the medial corner of the eye.

Little of the nasal cavity can be seen without the assistance of special illuminated instruments. The hump in the floor of the nasal cavity produces problems in seeing even the lower parts of the septum and the lateral wall. By thrusting a partner's head back by upward pressure on the

chin, and elevation of the tip of the nose, one can see into the vestibule through the nostrils with flashlight illumination. The vibrissae will be seen and the most anterior portion of the septum. The limited diameter of the nostrils and the encroachment of the inferior turbinate upon the cleftlike nasal fossa preclude further visualization.

The area of the frontal sinus may be approximated as described previously. The maxillary sinus can be outlined by pressing the fingers beside the nose from the inferior orbital margin downward until the scalloped alveolar margin of the maxilla and the upper teeth are felt through the soft tissues. The other sinuses are hidden deep in the head. If a pen light is placed in the mouth between closed lips, and no sinus disease is present, faint red areas of *transillumination* can be seen over the frontal and maxillary sinus regions in a dark room.

PHARYNX AND NASOPHARYNX

The *pharynx* is a musculofibrous tube suspended from the base of the skull in front of the cervical vertebrae. Its predecessor in development was the foregut of the primitive alimentary tract. Later, an outpouching from the anterior wall of the foregut became the forerunner of the lower respiratory system. In the head region and in the upper parts of the neck the original digestive passages became modified to serve both the alimentary and the respiratory systems. The part of the pharynx which is posterior to the nasal and oral cavities serves both the respiratory and digestive systems. The pharynx is arbitrarily divided into nasal, oral, and laryngeal portions. The *nasopharynx* is the subdivision that lies posterior to the nasal cavity.

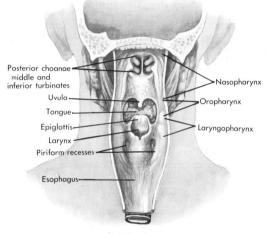

Posterior view

Boundaries of the Nasopharynx

The nasopharynx is in close relation to the base of the skull. Its *roof* is the body of the sphenoid bone. The sphenoid sinus is, therefore, above and anterior to the nasopharynx. The important pituitary gland, the crossing of optic nerve fibers (optic chiasm), and the cavernous venous sinus are just superior and immediately posterior to the nasopharynx. *Anterior* to the nasopharynx are the two posterior choanae of the nasal cavity.

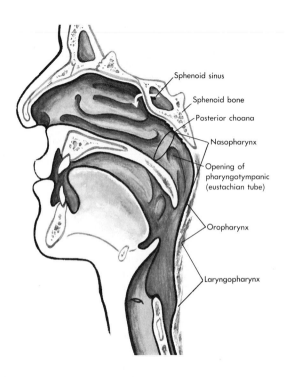

Sphenoid sinus

Sphenoid bone

Posterior choana

Nasopharynx

Opening of pharyngotympanic (eustachian tube)

Oropharynx

Laryngopharynx

Through these tall oval apertures, which are separated by the posterior extremity of the nasal septum, air passes in breathing. On the *lateral* walls of the chamber are the openings of the pharyngotympanic (eustachian) tube, which connects the middle ear cavity with the pharynx. There is no *medial* wall because the chamber is situated in the midline of the head posterior to the nasal septum. The *posterior* wall is separated only by fascia from the bodies of the upper cervical vertebrae. *Inferiorly* the nasopharynx is continuous with the oropharynx which is posterior to the soft palate and to its dangling projection, the uvula.

Mucous Membrane

The nasal lining continues posteriorly as the mucous membrane of the nasopharynx. Around the opening of each pharyngotympanic tube, the membrane is thickened by very vascular connective tissue to form the *eustachian cushion.* Under the lining of the roof and on the posterior wall are collections of lymphatic system tissue which cause the membrane to bulge. When chronic infections produce enlargement of these *nasopharyngeal tonsils* they are known as the "adenoids." In such situations the membrane enlarges and droops downward in fringes which may block the nasopharyngeal airway.

Living Anatomy

The nasopharynx is hidden from view unless illuminated instruments are used. The physician inserts a warmed angled mirror through the mouth in order to see the boundaries, walls, and contents of the nasopharynx as already described. He can also see the posterior end of the nasal septum, the posterior ends of the middle and inferior nasal turbinates, and the respective nasal meatuses.

PARTS OF THE RESPIRATORY SYSTEM IN THE NECK

Both the respiratory and alimentary systems in the head are composed of irregular passageways and chambers. The head is rigid and bony. Therefore such passages are relatively inflexible and have only a thin mucous membrane overlying bony or cartilaginous walls. In the neck these systems function under different circumstances. The neck is cylindrical and is capable of a variety of movements. Systems for transferring air and food to the thorax have a flexible, tubular structure which is reminiscent of the primitive alimentary tube from which they developed. The two systems share a common musculofibrous chamber in the upper cervical region of which the nasopharynx is the subdivision immediately posterior to the nasal cavity. It lies superior to the *oropharynx,* which is the subdivision immediately posterior to the oral cavity of the head.

Oropharynx

The oropharynx is the central subdivision of the pharynx. It communicates with the nasopharynx superiorly, the oral cavity anteriorly, and the laryngeal portion of the pharynx inferiorly. Some of the oropharynx can be seen in the mirror if one opens his mouth widely, depresses his tongue on the floor of the oral cavity, and says "aah." It is the region one can see at "the back of the throat" when a sore throat is suspected. The pharynx as a whole and its oral and laryngeal portions are considered with the digestive system but the following are essential to understanding the respiratory system:

1. The common musculofibrous chamber is suspended from the base of the skull in front of the cervical vertebrae.

2. The pharynx is flexible and capable of changes in dimension in response to the contraction of its musculature. Such contractions, coordinated with similar muscular movements of the soft palate and tongue, are of concern in the swallowing of food but not in the passage of air to the lower respiratory system.

3. The oropharynx is bounded anteriorly by the *faucial* or *tonsillar region,* where muscular folds of the tonsillar pillars laterally, and the soft palate superiorly, bound the opening from the oral cavity.

4. The base of the tongue rises from the anterior wall of the oropharynx as a bulky muscular fold of the mucous membrane which projects upward and forward along the floor of the oral cavity.

Laryngopharynx

In the pharynx there is a crossing of the stream of air passing to and from the lower respiratory system and the food destined for the lower parts of the alimentary system. Since the entire lower respiratory system lies anterior to the digestive tract, food must pass across the airway in a posterior direction to gain entrance to the esophagus. Similarly, the incoming air must pass anteriorly across the axis of swallowing to gain access to the lower respiratory system. Anyone who has "breathed water" while swimming or "swallowed down the wrong throat" in eating or coughing knows the discomfort that ensues if the crossing mechanism functions improperly.

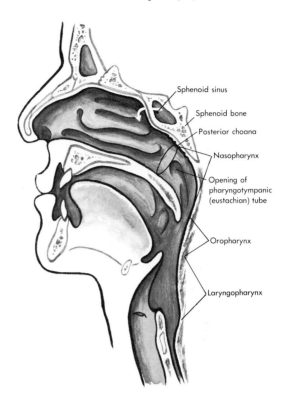

The *laryngopharynx* is the entrance to the esophagus which lies posterior to the vastly more important larynx. The laryngopharynx is described further with the alimentary system.

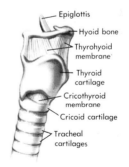

(See also color plate 5.)

Larynx

The entrance to the lower respiratory system is guarded by a boxlike chamber of cartilage, muscle, and fibrous tissue. The larynx:

1. Transfers air from the common pharyngeal chamber into the trachea.

2. Prevents the entrance of food or water into the lower respiratory system.

3. Produces vibrations in the air column within it. These are modified by the tongue, palates, oral cavity, and teeth (and also by the nasal cavity and paranasal sinuses) to become the sounds perceived by others as speech, song, or a variety of grunts, groans, moans, or laughs. The vibrations of coughs and hiccups, while passing through this chamber, have origins lower in the respiratory system.

Development.   The *tracheal bud* is an epithelial outgrowth which grew inferiorly in front of the primitive alimentary tract from which it originates. It became surrounded by totipotential mesenchymal cells which specialized to form:

1. A *cartilaginous framework* whose component *laryngeal cartilages* maintain an open airway, protect the superior aperture of the larynx, and are moved by muscles to change the dimensions of the chamber and to exert traction on folds of the mucous membrane (the *vocal cords.*)

2. The *connective tissue* which supports the mucous membrane, fills in the openings between the laryngeal cartilages, ensheathes the whole organ, and separates it from other structures of the neck.

3. The *muscles* of the larynx.

Framework.   The larynx is a somewhat squared tubular box which is longer than it is broad. It tapers from a broader triangular *inlet* to a narrower tubular *outlet* which is the continuation of the airway into the trachea. Nine *laryngeal cartilages* perform the general functions already noted and have the specific functions listed in the table on page 385.

The *epiglottis* is a flap of mucous membrane which is strengthened by the epiglottic cartilage. It is normally erect to provide an open airway. Muscle action narrows the laryngeal inlet but does not draw the epiglottis downward and backward to cover the constricted aperture. It is pressure of the base of the tongue and the impact of the bolus (ball) of food in swallowing that actually closes the epiglottis.

There are many apertures in the laryngeal framework. These openings between the cartilages are closed by fibrous membranes and the laryngeal muscles. (See table on opposite page.)

Form and Major Features of the Larynx.
Immediately inferior to the inlet of the larynx is a widened superior portion of the *cavity of the larynx* called, appropriately, the *vestibule.* The vestibule is limited inferiorly by the *ventricular folds* which are also termed the *false vocal cords.* A short

distance below these are the true *vocal folds* (*vocal cords*). Between the false and true folds on each side is a shallow indentation in the lateral walls of the larynx, the *ventricles* of the larynx. The cavity of the larynx tapers below the true vocal folds to assume the tubular form that is characteristic of the trachea.

The ventricular folds are given substance by ligamentous bands. The aperture between the two ventricular ligaments, as seen when the physician looks down the larynx during *laryngoscopy* with special illuminated instruments, is the *rima vestibuli.* The true vocal folds are formed by the highly elastic *vocal ligaments* which stretch between the thyroid cartilage anteriorly and the arytenoid cartilages posteriorly. The opening between the two vocal folds is triangular when the vocal cords are relaxed and is then called the *glottis;* it is narrower than the rima vestibuli above it. When the vocal folds become tense during phonation, the aperture between them narrows, even to a slit, and is then termed the *rima glottidis.* The *mucous membrane* of the larynx continues from the pharynx to the trachea and closely invests the folds and crevices of the organ. It is thick over the ventricular folds and thin over the vocal folds and contains not only many mucus-secreting glands but, strangely enough, a few taste buds. The latter are especially prominent over the epiglottis.

Muscles of the Larynx.   The *laryngeal muscles* are striated voluntary muscles, which, however, function at a reflex level under the control of the tenth cranial nerve. These muscles change the dimensions of the larynx and regulate the tension of the vocal folds in maintaining an open airway and in controlling phonation. They also act to narrow the inlet as part of the protective mechanism in swallowing. The names of the laryngeal muscles are a clue to their attachments:

The actions of the muscles are listed in the table on page 386. Their role in *phonation* is an outgrowth of these actions. The production of vibrations in the air column of the respiratory system depends in part upon the changes in laryngeal volume, in part upon the changes in position of the vocal folds, and in part upon the tension of these folds. As adduction of the vocal cords occurs, the glottis narrows to change the character of the sounds produced. As abduction occurs, the aperture widens with a consequent alteration in the sounds. Tension changes of the

CARTILAGES	NO.	SHAPE	LOCATION	MAIN FUNCTION
Epiglottic	1	Duck-billed; free superior border; attached inferiorly	Erect: behind base of tongue anteriorly at inlet	Bent backward during swallowing to close the laryngeal inlet
Thyroid	1	Like opened book laid with covers up	Midline and lateral	Main upper and anterior support of larynx; muscle attachments
Cricoid	1	Ring-shaped, wide part is posterior	Surrounds lower border and outlet	Main lower and posterior support of larynx; muscle attachments
Arytenoid	2	Pyramidal	Posteriorly on either side of wide part of cricoid	Muscle attachments; moved to produce tension upon vocal cords; moved to narrow larynx during swallowing
Corniculate	2	Conical	On tips of arytenoid cartilages	Same as arytenoid cartilages; extends them backward and upward
Cuneiform	2	Small cylinders	Anterior to corniculate cartilages in tissue folds between them and the epiglottis	Stiffens important aryoepiglottic folds

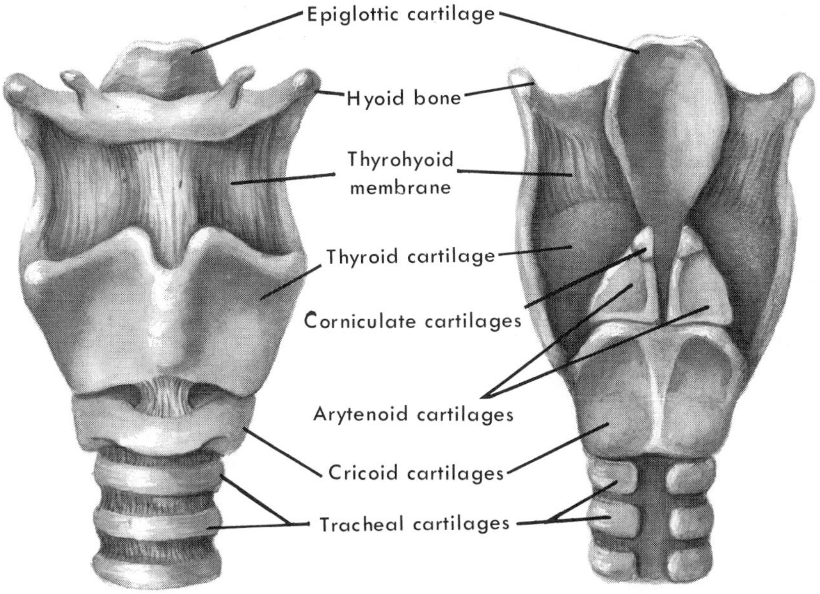

Epiglottic cartilage

Hyoid bone

Thyrohyoid membrane

Thyroid cartilage

Corniculate cartilages

Arytenoid cartilages

Cricoid cartilages

Tracheal cartilages

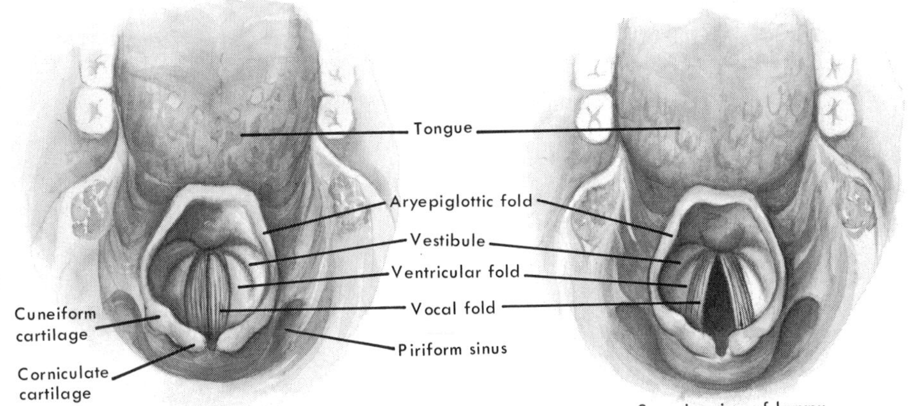

Tongue

Aryepiglottic fold

Vestibule

Ventricular fold

Vocal fold

Piriform sinus

Cuneiform cartilage

Corniculate cartilage

Superior view of larynx, vocal folds closed

Superior view of larynx, vocal folds open

vocal folds influence the sounds produced by both movements. A danger in certain clinical conditions is *laryngospasm* in which excessive volleys of nerve impulses produce an uncontrolled adduction and extreme tension of the vocal folds to obstruct the airway. This may occur in athletes who are subject to allergic attacks, such as asthma, although the bronchi are more apt to be involved.

Living Anatomy of the Larynx.   The larynx lies in the anterior midline of the neck. It is located at the level of the fourth to sixth cervical vertebrae. If the index finger is stroked inferiorly from the chin in the midline of the neck, it will encounter the deep indentation of the *superior thyroid notch* of the thyroid cartilage. Spreading of the thumb on one side and the middle finger on the other will outline the *laryngeal prominence* of the neck commonly known as the "Adam's apple." The larynx can be moved from side to

MUSCLE	ORIGIN	COURSE	INSERTION	ACTION
Cricothyroid (paired)	Anterolateral surface of cricoid cartilage	Posteriorly and superiorly	Inferomedial surface of thyroid cartilage	Draws thyroid cartilage downward to increase vocal fold tension
Posterior cricoarytenoid (paired)	Posterior surface of wide posterior part of cricoid cartilage	Laterally and superiorly	Posterior surface of arytenoid cartilage	Turns arytenoid cartilage backward which draws vocal folds laterally (abduction)
Lateral cricoarytenoid (paired)	Posterolateral arch of cricoid cartilage	Posteriorly and superiorly	Anterior surface of arytenoid cartilage	Draws arytenoid cartilage forward which directs vocal folds medially (adduction)
Transverse arytenoid (single)	Posterolateral border of one arytenoid cartilage	Transversely across back of larynx	Posterolateral surface of other arytenoid cartilage	Pulls arytenoid cartilages together to draw vocal folds together (adduction)
Oblique arytenoid (paired)	Posterior surface of one arytenoid cartilage	Obliquely and superiorly across back of larynx	Superior tip of other arytenoid cartilage	Contracts diameters of larynx to reduce its volume
Thyroarytenoid (paired)	Inner surface of lateral side of thyroid cartilage	Follow vocal folds	Lateral border of arytenoid cartilage	Pulls arytenoid forward to reduce vocal fold tension; reduces size of glottis

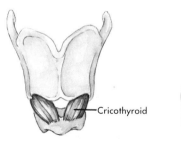

—Cricothyroid

Anterior view

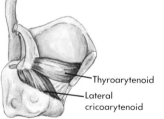

—Thyroarytenoid
—Lateral cricoarytenoid

Lateral view

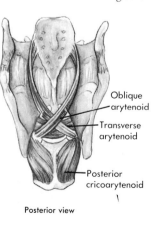

Oblique arytenoid

Transverse arytenoid

Posterior cricoarytenoid

Posterior view

side in outlining this prominence which is made up largely by the *thyroid cartilage*. While the fingers are still in position note how the larynx is drawn upward in swallowing and then returns to its former position. At the lowermost part of the laryngeal prominence, the *cricoid* cartilage may be felt as a flattened arc separated from the inferior border of the thyroid cartilage by a narrow horizontal crevice, the *cricothyroid groove*. Below the cricoid cartilage a well defined indentation can be felt which represents the interval between this cartilage and the *first tracheal cartilage*. Several tracheal cartilages can be felt as transverse bands below this level by running the fingers downward between the converging sternocleidomastoid muscles into the suprasternal fossa.

THE LOWER RESPIRATORY SYSTEM

The lower respiratory system really commences at the inlet of the larynx which was described with the upper respiratory system because of its intimate connection with the pharynx and its role in speech production. The trachea and bronchi (known together as the trachobronchial system) and the respiratory organs, the lungs, make up the balance of the lower respiratory system.

THE TRACHEA

The manner in which the tracheobronchial system developed explains many of its complex features.

Development

The *tracheal bud* is an anterior outpouching from the embryonic alimentary tube. The anterior position of the tracheal bud forecasts the adult relationship of structures in the neck and in the thorax: the respiratory components (larynx, trachea, bronchi, and lungs) are always located *anterior* to the alimentary counterparts (laryngopharynx, cervical esophagus, and thoracic esophagus). The tracheal bud grows downward in front of the embryonic alimentary tract as a solid cord. The cord hollows into the primitive trachea as the tracheal bud grows slowly in a caudal direction through the mesenchyme that invests it. The epithelial tracheal bud becomes coated with primitive mesenchymal cells which add connective tissue, cartilages, and smooth muscle to the epithelial portion of the tracheal lining and its glands.

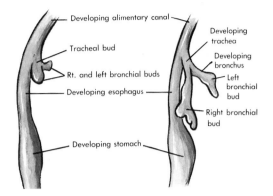

Structure of the Trachea

The tracheal tube in cross section is seen to consist of well defined layers. The mucous membrane is formed by ciliated pseudostratified columnar epithelium with mucus-secreting goblet cells and small accessory *tracheal glands*. The epithelium is underlaid by areolar connective tissue which supports it, holds the glands, provides passage for blood vessels and nerves, and unites the epithelium to deeper layers. Cartilaginous C-shaped bands, the *tracheal cartilages*, are spaced horizontally at regular intervals around the tracheal wall. These maintain the patency of the airway and support the organ. Their open posterior ends are completed by bands of smooth muscle. The apertures in the cartilaginous framework are filled with highly elastic connective tissue.

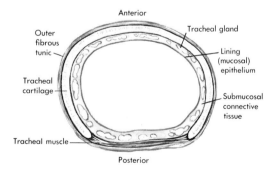

The trachea is resilient enough so that as food is swallowed from the pharynx, the soft, flat, empty esophagus can expand to receive it without being compressed by the air tube which is immediately anterior.

Form and Relations of the Cervical Trachea

The trachea, though rigid enough to preserve its patency, is also elastic enough to stretch and then to reassume its normal position with changes in breathing and swallowing. The distensibility and elasticity of the trachea permit it to help suspend the heart because of the intertwining of the tracheal branches (the bronchi) with the pulmonary arteries which reach upward from the heart.

The tracheal tube maintains a midline position in the neck in front of the esophagus. The trachea is closely related to the thyroid gland, the thoracic duct, and the laryngeal and vagus nerves in the neck. (See illustration below.) The trachea passes through the superior aperture of the thorax to continue its course as a thoracic organ in the mediastinum.

THE TRACHEOBRONCHIAL SYSTEM

The trachea descends through the superior aperture of the thorax in company with many of the structures to which it is related in the neck. These, including the esophagus, traverse the mediastinum, the median connective tissue partition of the thorax. The *tracheobronchial system* consists of the thoracic trachea, its division into the right and left bronchi, and their ramifications throughout the lungs.

Development

The tracheal bud divides into right and left *bronchial buds* amid the mesenchyme of the developing thorax. At this time the thoracic wall is not closed and there is no cavity within the thorax. Each bronchial bud continues to grow through the mesenchyme, receiving investments of totipotential cells. The *right bronchial bud* is

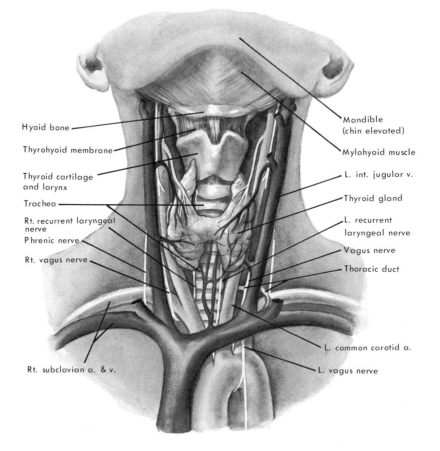

Hyoid bone

Thyrohyoid membrane

Thyroid cartilage and larynx

Trachea

Rt. recurrent laryngeal nerve

Phrenic nerve

Rt. vagus nerve

Rt. subclavian a. & v.

Mandible (chin elevated)

Mylohyoid muscle

L. int. jugular v.

Thyroid gland

L. recurrent laryngeal nerve

Vagus nerve

Thoracic duct

L. common carotid a.

L. vagus nerve

more the direct continuation of the trachea and descends at a more acute angle with the longitudinal axis of the trachea. The *left bronchial bud* grows at a more obtuse angle and is less a continuation of the tracheal axis. These facts influence the adult form and course of the bronchi.

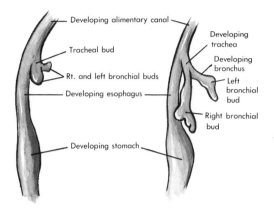

The bronchial buds, growing laterally and inferiorly, reach a zone which corresponds embryologically to the lateral borders of the adult mediastinum. At this point the bronchial bud begins to divide. From the original tracheal bifurcation there are 33 suborders of division in the tracheobronchial system before gas-exchange air spaces are formed. As each division takes place, the new and smaller caliber bronchi acquire an investment of totipotential mesenchymal cells. The budding process forms only the epithelial lining of the tracheobronchial system. The mesenchymal cells provide cartilage, connective tissue layers, and smooth muscle. The totipotential cells also form the outer mesothelial covering of the lung (pulmonary pleura). As the tracheobronchial system develops, pulmonary and bronchial arterial blood vessels are being formed. They grow into the embryonic lung and acquire an intimate association with the new bronchi.

Tracheal Bifurcation

At about the level of the junction between the manubrium and body of the sternum, the trachea bifurcates into the right and left extrapulmonary bronchi. Several features are characteristic of the *tracheal bifurcation:*

1. Position slightly to the right of the midline owing to pressure from the arch of the aorta. (See illustration below.)

2. Modification of the lowest tracheal ring to form the *carina* (keel) of trachea. This is an irregular keellike plate of hyaline cartilage which strengthens the branching of the main air tube.

3. The right extrapulmonary bronchus is larger, more directly the continuation of the trachea, and springs at a steeper, more acute angle from the tracheal axis. Inspired foreign objects which pass through the larynx are more apt to enter and lodge in the right bronchial system. The left extrapulmonary bronchus is smaller and

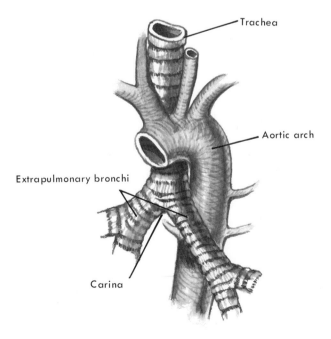

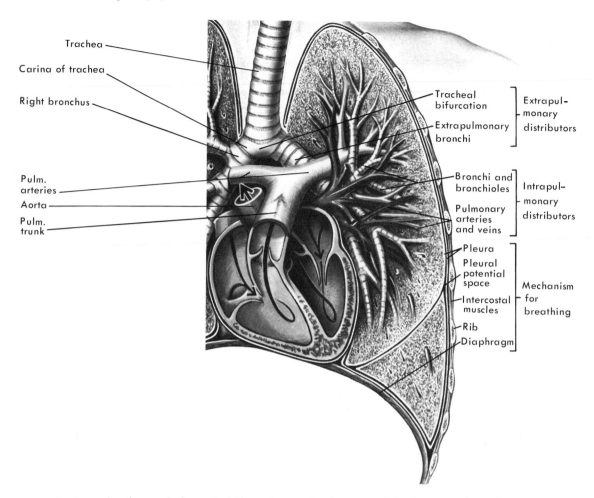

more horizontal and extends from the bifurcation at a more obtuse angle. It is also longer because of the position of the tracheal bifurcation to the right of the midline. (See also color plate 5.)

Extrapulmonary Bronchi and Root of the Lung

The extrapulmonary bronchi pursue a short course between the tracheal bifurcation and their point of entry into the lungs. The extrapulmonary bronchi are much like the trachea in structure. C-shaped bands of hyaline cartilage preserve the patency of the airway, although their irregularity forecasts a change in the wall of the smaller major bronchi. The extrapulmonary bronchus is the key component of the bridge of structures which cross the potential space between the lateral surface of the mediastinum and the mediastinal (medial) surface of the lung. These structures together form the *root of the lung.*

Associated with the extrapulmonary bronchus

in the root of the lung are the pulmonary artery enroute to the lung, the pulmonary veins leaving the lung, the bronchial arteries entering the lung, the pulmonary lymphatic vessels leaving the lung, the tracheobronchial lymph nodes, and the nerves of the lung. These structures are wrapped by connective tissue and covered by the pleural membranes of the lung.

The following structures are adjacent to the root of either lung:

1. *Anteriorly,* the phrenic nerve en route from its origin in the neck to its destination, the diaphragm; arteries and veins supplying the pericardium and diaphragm and the anterior part of the pulmonary nerve plexus which innervates the lungs.

2. *Posteriorly,* the vagus nerve enroute from the neck to thoracic and abdominal organs and the posterior part of the pulmonary nerve plexus.

The right and left lungs differ in other relationships of their roots:

1. As the *right* extrapulmonary bronchus nears

the lung surface, it commences to divide into subsidiary major bronchi. The first division yields a *right upper lobe bronchus,* sometimes called the *epiarterial bronchus,* and a continuing stem bronchus. The epiarterial bronchus assumes a position superior to the pulmonary artery in the lateral part of the root. Other divisions of the parent stem, called the right *middle* and *lower lobe bronchi,* occur just at the entry of the root structures into the lung.

The superior vena cava, descending through the superior mediastinum, lies just anterior to the right lung root, whereas the right atrium of the heart nudges posterosuperiorly against the root. The azygos vein arches over the root just at the lung surface.

2. The *left* major extrapulmonary bronchus, in traversing its root, is first superior to the pulmonary artery but shortly passes below it. Although it may divide in the lateral part of the root, no epiarterial bronchus occurs on this side. The aorta rises, arches, and descends through the mediastinum on the left side. Root structures, therefore, pass first under the aortic arch and then lie anterior to the thoracic aorta. The esophagus, veering to the left of the midline of the mediastinum, comes to lie behind the left lung root as it descends toward the diaphragm enroute to the abdominal cavity.

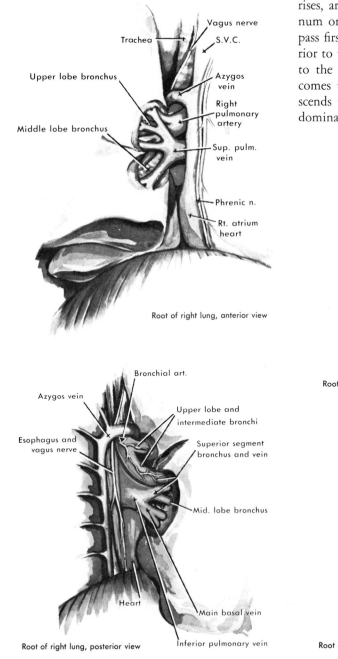

Root of right lung, anterior view

Root of right lung, posterior view

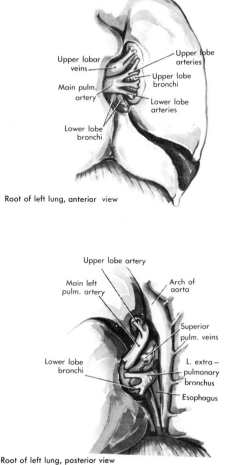

Root of left lung, anterior view

Root of left lung, posterior view

Intrapulmonary Bronchi

The root structures, of which the newly formed major bronchi are the key components, either plunge into the substance of the lungs or leave the organ at the tear-drop shaped *hilum* of the lung.

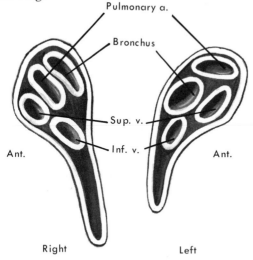

Relations at root of lung and hilum

All bronchi within the lung are termed *intrapulmonary bronchi*. As they branch, the intrapulmonary bronchi become progressively smaller in diameter. Rings of hyaline cartilage become replaced by irregular plaques. Smooth muscle bands crisscross around the tube. The ciliated epithelium shows a great increase in mucus-secreting cells. The smaller bronchi are called *bronchioles*. In these small air tubes the cartilage steadily diminishes to small pieces and then disappears. Smooth muscle and fibrous tissue increase in proportion in the walls. Small bronchi and bronchioles are flexible; they are subject to changes in caliber, as their smooth muscle contracts or relaxes, and these changes affect the amount of air passing into and out of the lungs. (See illustration below and also color plate 6.)

THE LUNGS

The *lungs* represent the branching of the tracheobronchial system to form a system for gaseous exchange in association with the pulmonary vascular circuit. The lungs are the functional organs of the respiratory system. All structures

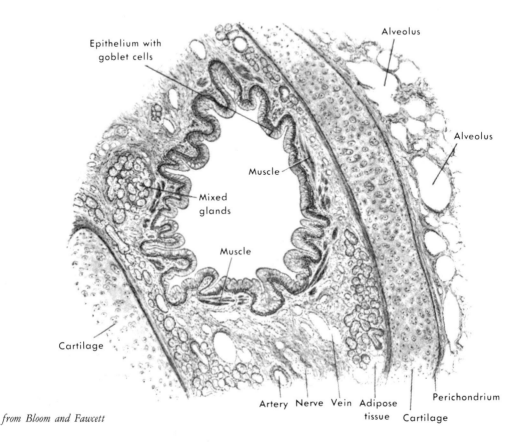

from Bloom and Fawcett

previously described are passageways and conduits which serve primarily to convey air from the external environment to the lungs and secondarily to warm, humidify, and remove foreign particulate matter from it.

Development of the Lungs

After forming the many suborders of bronchi and bronchioles, the bronchial buds complete the growth of the bronchial tree with the formation of bubble-like gaseous exchange compartments called *pulmonary alveoli.* (See right column illustration.) The pulmonary vascular system keeps pace with these processes so that an intimate association develops between the pulmonary capillaries and the alveoli. Nerve fibers grow in from the autonomic portion of the nervous system and rudimentary lymphatic channels form.

External Morphology of the Lungs

The lungs are soft spongy organs which are covered by a glistening mesothelial covering, the

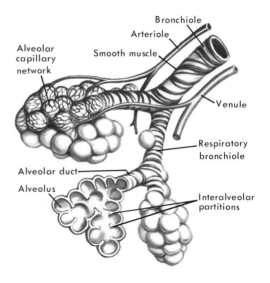

pleura. These organs are expanded to whatever volume the thoracic wall permits. They surge with blood delivered by the pulmonary circulation and return it via the pulmonary veins to the left atrium of the heart. The lungs are irregular, pyramid shaped, air-filled organs molded to the

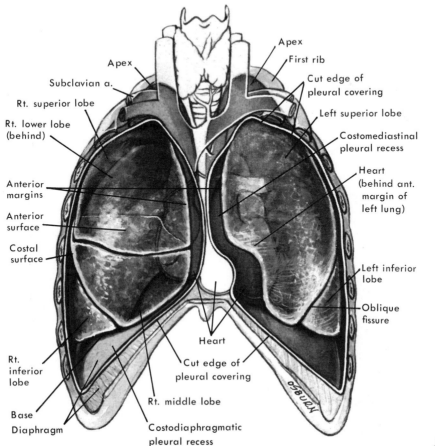

from Healey

interior surfaces of the thorax which contains them. The thorax is closed superiorly by structures of the base of the neck against which the superior part of the lung (apex) presses through the thoracic inlet. Inferiorly, the thorax is closed by the thoracoabdominal diaphragm. The lung has *lateral* (costal), *medial* (mediastinal), *inferior* (diaphragmatic), and *anterior* (sternocostal) *surfaces* as well as a sharp *anterior margin,* a rounded *posterior margin,* an *apex,* and a *base.* Factors related to these are summarized in the table that follows. The lung perpetually expands against the firmer structures surrounding it. It becomes marked by the structures against which it presses, transiently in life, permanently in death. These are termed *surface landmarks* of the lung which are also summarized in the table on page 395.

Coverings of the Lungs

Totipotential cells are carried along with the branching tracheobronchial system of the developing lung. As the lung completes its development prior to birth, it is a lobulated ball of bronchovascular and epithelial elements. It has yet to be fully inflated. Some of the mesenchymal cells form a continuous membrane of mesothelium which covers the external surface of the adult lung. It is called the *pulmonary* or *visceral pleura* and is affixed to the lung by cobweb-like subpleural connective tissue.

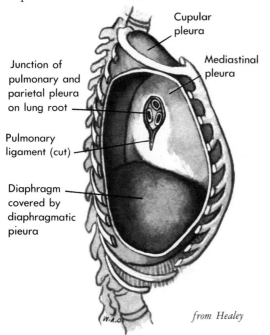

Cupular pleura

Mediastinal pleura

Junction of pulmonary and parietal pleura on lung root

Pulmonary ligament (cut)

Diaphragm covered by diaphragmatic pleura

from Healey

A sheet of mesothelium also forms external to the lung, on the internal surface of the thoracic wall, on the lateral surface of the mediastinum, over the superior surface of the diaphragm, and about the apex. (See opposite page.) The mesothelium is supported by a cobweb-like areolar tissue called the *endothoracic fascia* which attaches it to the internal surfaces of the thorax. The membrane is termed the *parietal pleura.* The surfaces on which the parietal pleura lies give their names to subdivisions of this membrane: the *cupular* (apical) *pleura* in the base of the neck, the *costal pleura* over the internal surfaces of the thoracic wall, the *diaphragmatic pleura,* and the *mediastinal pleura.* The pulmonary and parietal pleurae become continuous with each other as they blend on the root of the lung.

The Pleural Potential Space (Pleural Cavity)

A purely potential space is found between the pulmonary and parietal pleurae. It contains only a thin film of *pleural fluid* formed by the pleural membranes. The fluid lubricates the pleural surfaces as they move against each other when the lung changes its size during breathing. Only when the lungs or pleurae are diseased does a cavity of any size develop. Therefore, the term "pleural cavity" is a misnomer.

Pleural recesses are formed where the parietal pleura crosses from one geographical surface to another. The *costodiaphragmatic pleural recess* is a long curved gutter between the edge of the diaphragm and the internal costal surfaces of the thorax.

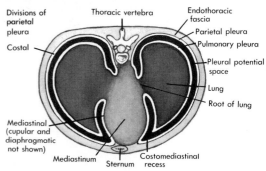

Divisions of parietal pleura

Thoracic vertebra

Endothoracic fascia

Costal

Parietal pleura

Pulmonary pleura

Pleural potential space

Lung

Root of lung

Mediastinal (cupular and diaphragmatic not shown)

Mediastinum

Sternum

Costomediastinal recess

The *anterior costomediastinal pleural recess* is a vertical gutter located on each side, where the costal parietal pleura turns onto the lateral surface of the mediastinum. The lung slides in and out of these recesses during breathing.

SURFACE LANDMARKS OF THE LUNG

SURFACE OR MARGIN	LOCATION ON LUNG	PRESSES AGAINST	SURFACE LANDMARKS
Apex	Superior	Thoracic inlet structures	Subclavian artery groove
Base or diaphragmatic surface	Inferior	Diaphragm and costodiaphragmatic recess	Diaphragmatic impression
Costal surface	Lateral	Lateral thoracic wall	Costal grooves, linear intercostal bulges
Medial surface	Medial	Mediastinum	*Right lung:* azygos groove arching immediately superior to hilum, superior vena caval groove vertically above hilum
			Left lung: aortic groove vertically above hilum anterior to esophageal groove; deeper cardiac notch below hilum anterior to esophageal groove
Anterior surface	Anterior, sternocostal	Parasternal part of sternum, sternochondral articulations, anterior ends of ribs	Relatively smooth
Anterior margin	Sharp anterior border	Between anterior thoracic wall and anterior mediastinum	Smooth, less prominent on left side owing to presence of heart
Posterior margin	Rounded posteromedial surface	Origin of ribs, costovertebral articulations, vertebral bodies	Relatively smooth except for costal markings

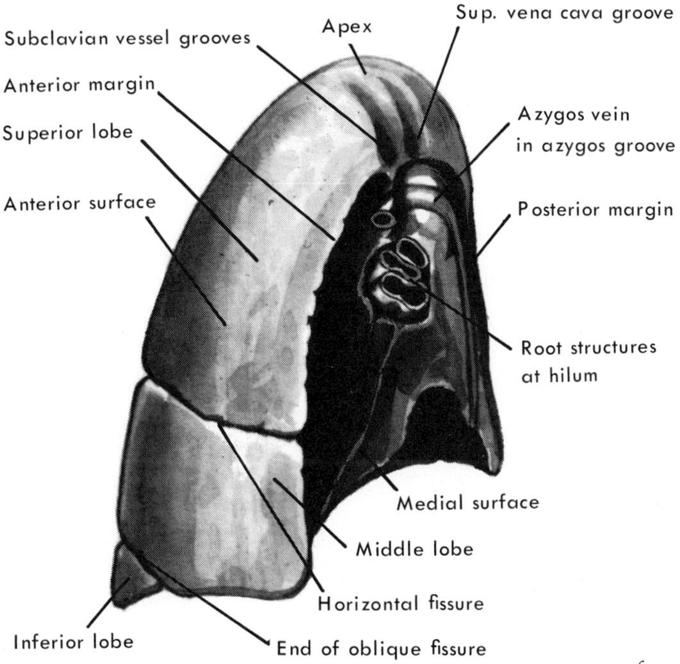

RIGHT LUNG, MEDIOLATERAL VIEW

from Healey

Lobes of the Lungs

Each major bronchus in development forms a block of air-bearing tissue by repeated branching. Each block, which is called a *lobe,* is structurally independent of all the others. When all the lobes are wrapped in connective tissue and invested by pulmonary pleura, an organ, the lung, results. Since lobes are independent units, it is not surprising to find that they are separated from each other by deep indentations. These clefts, which extend from the external surface of the lung almost to the hilum, are the *fissures* of the lung. They externally mark off the lung into its main *lobar* divisions. Three lobes are described for the *right* lung: the *right superior* (upper), *middle,* and *inferior* (lower) *lobes.*

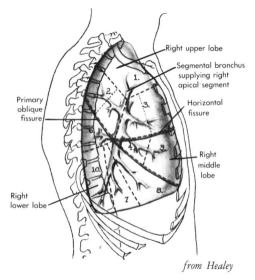

from Healey

The external division of the *left* lung is incomplete; it displays only *left superior* and *inferior lobes.*

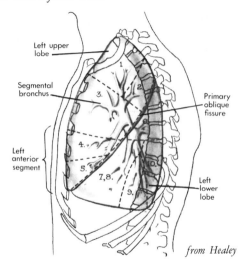

from Healey

Fissures of the Lungs

The *primary oblique fissure* of each lung commences high on the posterior margin of the lung. It slants laterally and inferiorly across the posterior margin and costal surface. Then it curves anteriorly and anteromedially to ascend finally across the anterior surface. It divides both lungs into upper and lower divisions which, on the *left side,* are the corresponding lobes. On the *right side,* there is a secondary or *horizontal fissure* which commences from the oblique fissure along the costal surface and courses medially along the line of the fourth rib or its interspace to reach the anterior margin. This fissure subdivides the upper division of the right lung into right superior and middle lobes. The inferior lobe on the right is no different than that on the left side.

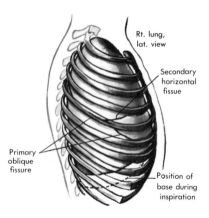

The pulmonary pleura usually follows the external surface of the lobes of the lungs very closely. It dips into the fissures to cover the opposing surfaces of the lobes which are separated by those fissures. In this sense, the pulmonary pleura can be said to "line" the fissures, although it is really investing the surfaces of the lobes. Sometimes the pleura skips over the fissures or a fissure extends only a short distance between lobes. In these cases the lobes are separated only by subpleural connective tissue.

Internal Structure of the Lungs

The developmental processes which have been described continue the development of the tracheobronchial system to the functioning microscopic level. This is more familiarly known as the "branching of the bronchial tree." There is no conflict between the *lungs,* as intrathoracic organs, and the *tracheobronchial system* which makes up the lungs. It must be realized by now that the internal structure of the lung is a combination of both the respiratory and vascular systems. Not only are the bronchi and the pulmonary artery branches intimately associated structurally, but they are inseparable functionally in order for the lungs to rid the blood of carbon dioxide and other wastes that can be volatilized. The constant relationship of the branches of the bronchi and the pulmonary artery is the basic principle of the internal structure of the lung. It is embodied in the modern concept of the organization of the lung according to *bronchopulmonary segmentation.*

Bronchopulmonary Segmentation

The continuation of the bronchovascular subdivisions through 33 suborders of branching demonstrates that there is a logical internal structure in the smaller units of the lung. Each lobar bronchus supplies air, and its corresponding branch of the pulmonary artery supplies blood from the right ventricle of the heart, to the independent block of air-bearing tissue of a lobe. The bronchus and the arterial branch enter the lobe near the hilum. They immediately subdivide, also near the hilum, into two or more *segmental* bronchi or the accompanying *segmental branches* of the *pulmonary artery.* Bronchial arteries and nerves accompany these divisions. (See color plate 5.)

A *bronchopulmonary segment* is a portion of air-bearing tissue of the lobe of a lung, which is supplied with air by one of the segmental bronchi. The same block of air-bearing tissue is also supplied, through the segmental pulmonary arterial branch, with blood which requires oxygenation and the removal of carbon dioxide. The separate bronchial arteries, conveying oxygenated arterial blood from the aorta, supply the bronchial walls and the connective tissue framework of the lung.

Bronchopulmonary Concept of the Lung. The lung is constructed of a series of independent blocks of air-bearing tissue, each of which has developed from and is supplied with air by a segmental bronchus. Two or more bronchopulmonary segments are wrapped together by an investment of reticular connective tissue to form a lobe. The lobes are separated from each other by the fissures of the lungs and are joined ex-

ternally only near the hilum. The lobes are invested by subpleural connective tissue and pulmonary pleura. A bronchopulmonary segment is covered by pulmonary pleura only where its periphery forms a part of the external surface of a lobe.

Names of the Bronchopulmonary Segments. The segments are named for the position they occupy in the *lobe* of the lung of which they are a part. In the list which follows, the name of the segment is a clue to its lobar position.

RIGHT LUNG	LEFT LUNG
Superior lobe	Superior lobe
Apical	Apicoposterior
Posterior	Anterior
Anterior	Superior lingular
Middle lobe	Inferior lingular
Lateral	Inferior lobe
Medial	Superior (apical)
Inferior lobe	Medial basal
Superior (apical)	Anterior basal
Medial basal	Lateral basal
Anterior basal	Posterior basal
Lateral basal	
Posterior basal	

(See illustration at top of following page.)

Subsegmental Structure of the Lung. The clue to the internal structure of any bronchopulmonary segment is the segmental bronchus which is constantly accompanied, division by division and branch for branch, by corresponding bifurcations of the segmental branch of the pulmonary artery. In accord with the acceptance of several bronchopulmonary segments as the basis of organization in the lobe, it is reasonable to include "subsegments" and "sub-subsegments" within a bronchopulmonary segment, for the same principle of organization is characteristic of smaller units within the segment.

Respiratory bronchioles continue the bronchial tree as the very smallest branches of the air conduit system. The respiratory bronchiole is modified by bubble-like outpouchings from its wall.

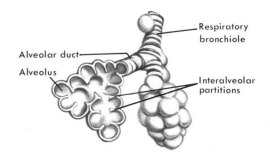

Respiratory bronchiole

Alveolar duct

Alveolus

Interalveolar partitions

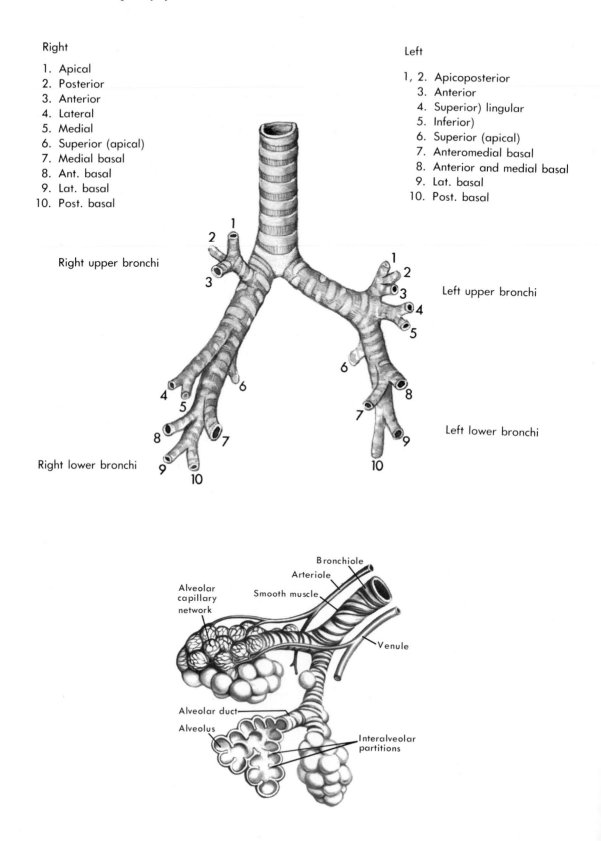

Right

1. Apical
2. Posterior
3. Anterior
4. Lateral
5. Medial
6. Superior (apical)
7. Medial basal
8. Ant. basal
9. Lat. basal
10. Post. basal

Left

1, 2. Apicoposterior
3. Anterior
4. Superior) lingular
5. Inferior)
6. Superior (apical)
7. Anteromedial basal
8. Anterior and medial basal
9. Lat. basal
10. Post. basal

Right upper bronchi

Left upper bronchi

Right lower bronchi

Left lower bronchi

Bronchiole
Arteriole
Smooth muscle
Alveolar capillary network
Venule
Alveolar duct
Alveolus
Interalveolar partitions

These are the first of the functioning air spaces of the lung to be encountered as the bronchial system is followed out to the periphery of the lung. The appearance of alveoli along the respiratory bronchioles is secondary to the function of the respiratory bronchioles as tiny air conduits leading toward the functioning air spaces of the main pulmonary alveoli. The continuation of the respiratory bronchiole is the *alveolar duct.* At the peripheral end of this duct the smooth muscle of the respiratory system ends. The smooth muscle, which has appeared in a crisscrossing pattern around the larger air tubes is at this point thrown

around the termination of the alveolar duct as a muscular sphincter. This sphincter can contract to regulate the amount of air entering or leaving the air spaces.

The *primary functioning lobule* of the lung consists of all the pulmonary alveoli at the distal end of the alveolar duct. These spongy air spaces expand from the end of each of the myriad of alveolar ducts as delicate bubbles in what has been called the "blossoming of the bronchial tree." Although each alveolus is shaped as a dome, they are clustered together so compactly that their abutting surfaces form common *interalveolar par-*

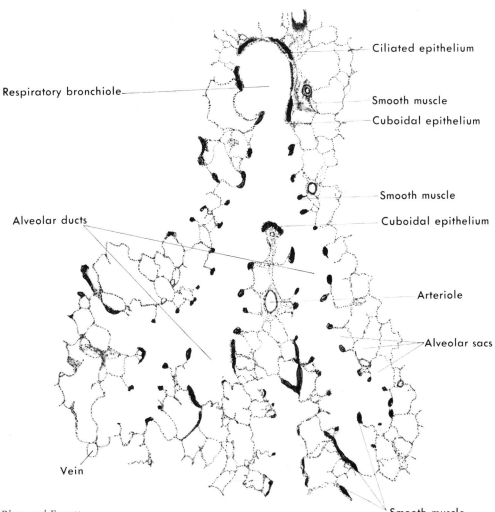

Respiratory bronchiole

Ciliated epithelium

Smooth muscle
Cuboidal epithelium

Smooth muscle

Alveolar ducts

Cuboidal epithelium

Arteriole

Alveolar sacs

Vein

from Bloom and Fawcett

Smooth muscle

titions. The organizational pattern of broncho-pulmonary structure is manifested by an associated pulmonary arteriole. This microscopic vessel forms a capillary network on the domed surfaces of the pulmonary alveoli and in the interalveolar partitions. The vessels are supported by reticular connective tissue interlaced by elastic fibers and are in close association with monocytic tissue cells. The connective tissue is the framework of the alveoli. The elastic fibers provide for elasticity and recoil of the air spaces during expiration. The monocytes are the first line of body defense against microorganisms or foreign particulate matter which escape the cilia and mucus of the preceding parts of the lower respiratory system.

Air reaches its destination as it passes into the alveoli where it contacts the almost transparent inner surface of the air spaces. The rich capillary network is practically exposed to the air with hardly more than the thickness of endothelial cells and possibly squamous epithelial cells of the air space intervening.

It has taken the electron microscope to resolve one of the oldest disputes among microscopists: is there a complete pulmonary alveolar epithelial lining? It is now believed that a very thin squamous cellular lining may exist, which is interrupted at intervals by *alveolar pores,* providing direct communication from one alveolus to another. In addition, phagocytic *alveolar septal cells* bulge out the thin alveolar wall, adding to an impression of discontinuity in the lining. Free phagocytic cells are found in the lumen of the alveolus as a first line of defense against invading microorganisms or inspired foreign particles.

Blood in the pulmonary capillary network gives up carbon dioxide and other volatile wastes to the air and receives oxygen in return. Gaseous exchange, which is the primary function of the entire respiratory system, occurs in the primary functioning lobule of the lung, which carries to a functional termination the general pattern of bronchopulmonary segmentation.

MECHANICS OF NORMAL BREATHING

In respiration the following steps take place:
1. The thorax expands by the action of the muscles of respiration and the diaphragm. The intrathoracic volume increases in all diameters because of the outward movement of the ribs and the descent of the diaphragm.

2. The lungs follow and expand into the base of the neck, outward with the thoracic cage, and inferiorly with the diaphragm. Margins of the lungs slide into the pleural recesses.

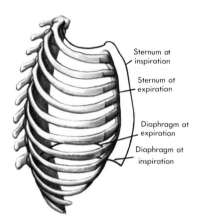

Sternum at inspiration
Sternum at expiration
Diaphragm at expiration
Diaphragm at inspiration

3. Air moves through the upper and lower respiratory system, mixing in the lower parts with a certain amount of residual air. The air is warmed, humidified, and filtered of foreign particulate matter, but some microorganisms and viruses may escape the protective devices which have been described.

4. Gaseous exchange occurs in the primary functioning lobule.

5. As thoracic expansion ceases at the end of inspiration, the diaphragm rises and the elastic tissue of the lungs aids in the decrease of thoracic volume by its inherent recoil.

A demand for greater oxygenation of the blood and for faster removal of volatile wastes accompanies exercise, violent physical activities, and the emotions of anxiety and fear. Such situations result in the more rapid and forceful action of the respiratory muscles and diaphragm and the action of other muscles, not used in ordinary breathing, which are called the *accessory muscles of respiration.* These are the abdominal wall muscles, neck muscles such as the scalene and sternocleidomastoid muscles, and the dilator muscles of the nasal aperture.

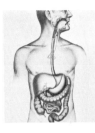

8
The
Digestive
System

The digestive or alimentary system is a continuous series of interconnecting cavities and tubes which receive food into the body, utilize the nutritive components of the food, and evacuate the residues. Since many of these processes go on within the stomach and intestines, the digestive system is also called the *gastrointestinal tract,* although the stomach and intestines undertake only a portion of the digestive functions.

FUNCTIONS OF THE DIGESTIVE SYSTEM

The anatomic structures of the digestive system can be related to the following functions which are carried out in the order of listing.

Oral Cavity

1. Reception and testing of food to determine its suitability for consumption (temperature, taste, odor, texture).
2. Chewing and grinding of food to reduce its particle size to aid in swallowing and to increase the particle surface area for enzyme action.
3. Addition of enzymes during the masticatory process to initiate the chemical processes of digestion.

4. Addition of mucus and moisture to aid in shaping the masticated food into a bolus, or lump, for swallowing.

Pharynx and Esophagus

5. Conveyance of the bolus or liquids from the mouth to the stomach.

Stomach

6. Churning of food mixtures to facilitate the breakdown of food substances as other enzymes and digestive substances are added.

Small Intestine

7. Addition of further enzymes to complete digestion, followed by the absorption of the digested chemical food substances and liquids through the walls of the intestine into the blood.

Large Intestine

8. Reabsorption of water and electrolytes from the digestive residue to conserve them and to preserve the fluid-electrolyte balance of the body.
9. Evacuation of the residues of digestion.
10. Excretion of those waste substances of

body metabolism which are eliminated by the gastrointestinal tract or its associated organs.

DEVELOPMENT OF THE DIGESTIVE SYSTEM

The digestive system is the product of development and specialization within a long tube which appeared in the embryo as the *gut*. The superior end of this tube established continuity with an invagination of the head which became the oral cavity.

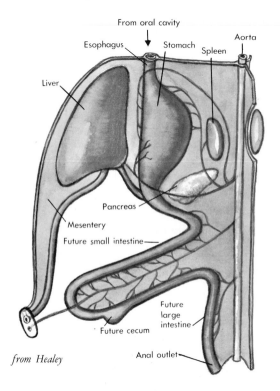

from Healey

A similar connection was established with a caudal invagination to form the anal outlet of the digestive system. This system, therefore, is found in the head and neck and in all portions of the trunk of the body.

RECEPTION OF FOOD AND PREPARATION FOR SWALLOWING

Food is received through the *mouth,* the aperture of which is controlled in part by the musculature of the lips and in part by the mechanism for opening and closing of the jaws. (See color plate 9.) The *oral cavity* is the chamber of varia-

ble size between the hard palate and soft palate above and the tissues comprising the floor of the mouth below. Its lateral limits are formed by the teeth and the tissues of the cheeks. The major part of the chamber, the *oral cavity proper,* is limited anteriorly and laterally by the teeth. A narrow area between the lips or cheeks and the teeth is designated as the *vestibule of the mouth.* This is but a narrow crevice which is usually considered part of the oral cavity. Within the oral cavity are the teeth and gums, the tongue, the palates, and the openings of the salivary glands. The oral cavity is bounded posteriorly by the opening of the oropharynx which is formed by the posterior edge of the soft palate with its downward projecting uvula, the base of the tongue, and the tonsils.

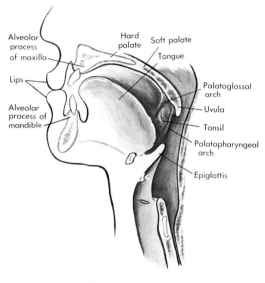

STRUCTURES FOR TESTING OF FOOD

Food is tested visually before it approaches the mouth. Then, and as it nears the mouth, the odor of food is perceived by the olfactory apparatus. Such visual and olfactory stimuli, coupled with anticipatory thoughts or visceral sensations of hunger, set a number of autonomic nerve impulses in motion to produce a readiness for digestion such as an increase in salivary secretion or moistening of the lips. By training, a person tends to reject foodstuffs which present an unpleasant appearance or smell.

General Sensory Testing

As food approaches the skin of the face and of the lips, temperature receptors detect its degree of warmth or cold so that unusually hot or cold foods are rejected primarily upon a reflex basis. As the food enters the lips and goes into the oral cavity all the receptors for general sensation are stimulated to apprise the person of its nature. Coarse, irregular, sharp, or indigestible substances (metal, glass, earth) are detected by pain, touch, and pressure receptors in the skin of the lips; in the mucous membrane of the gums, cheeks, floor, and roof of the mouth; and in the tongue. Temperature of the food is again tested by receptors in the mucous membrane of all the surfaces mentioned. An infant or young child may disregard such sensory information until training, disagreeable results, or parental admonitions establish criteria for judgment. If food is deemed unsuitable or unpleasant, it is rejected.

Special Sensory Testing

Olfaction and taste are closely associated in the mouth. In fact, many substances which one believes he tastes are actually perceived by the olfactory sense as food is initially received. The *tongue* is the main organ for special sensory testing within the mouth, although taste receptors are scattered through the oral mucous membrane as well.

The Tongue

The tongue is a thick musculofibrous organ, capable of many changes of shape, which projects upward and forward from the floor of the mouth and anterior wall of the pharynx. When the mouth is closed, the occlusion (approximation) of the upper and lower teeth prevents the roof of the mouth, formed by the palates, and the floor of the mouth from touching. The oral cavity then is a wide slitlike chamber. The tongue almost completely fills the closed oral cavity. When the mouth is opened, the tongue is seen covering the floor of the mouth. Only the tip, dorsum (upper surface), and margins (sides) of the anterior two-thirds of the tongue are ordinarily visible. The posterior portion, known as the base of the tongue, sinks below the line of vision to the anterior wall of the pharynx. The inferior surface or root of the tongue is V-shaped and is attached to the floor of the mouth by a midline fold of mucous membrane, the *frenulum*. A long crevice glistening with a vascular mucous membrane runs along the floor of the mouth on each side, forming a gutter for saliva and food.

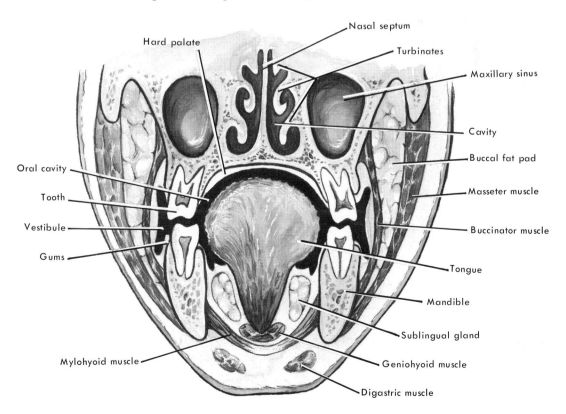

The highly mobile tongue is capable of twisting and turning into many positions. It aids in general sensory testing of food, in mastication by pressure and moving the food about, and in initiating swallowing.

The *special sense receptors* of the tongue are called the *taste buds.* The surface epithelium and fibrous tissue of the tongue mucous membrane are thrown up into many projections called *papillae* which give a rough appearance. Papillae vary in shape between *filiform* (threadlike), *fungiform* (mushroom shaped), and *vallate* (raised islands surrounded by moats). The filiform and fungiform papillae are scattered over the anterior two-thirds of the tongue. The vallate papillae form into a V where the posterior third of the tongue meets the anterior portion. Both fungiform and vallate papillae contain taste buds. The taste buds are neuroepithelial modifications of the mucous membrane which represent the specialized end organs of special visceral sensory neurons which carry the special sensory component of taste.

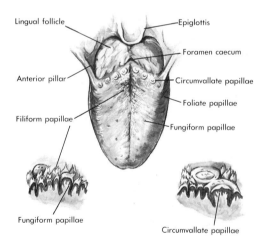

Lingual follicle — Epiglottis
— Foramen caecum
Anterior pillar — — Circumvallate papillae
— Foliate papillae
Filiform papillae — — Fungiform papillae
Fungiform papillae
Circumvallate papillae

The cranial nerves which carry these components have been identified in Chapter 5 (pp. 288–89, 92). Taste buds are not highly selective to stimuli, being sensitive only to substances which are sweet, sour, bitter, or salty. Taste sensations are combinations of these or combine with the olfactory sense. Dryness of the tongue or a heavy coating impede taste.

The *musculature* of the tongue is responsible for its form and mobility. The main substance of the visible portion of the tongue is made up of bands of striated muscle fibers which form the intrinsic *muscles* of the tongue. These are unnamed masses and bands which interlace tridimensionally as they course longitudinally, transversely, and vertically. Their action changes the form of the tongue. The *extrinsic muscles* of the tongue arise outside the organ and course to it from three main skeletal points on the mandible, hyoid bone, and styloid process.

1. The *genioglossus* is a vertically disposed sheet of muscle fibers which fans out into the posterior and inferior aspect of the tongue from a narrow origin on the genial tubercle behind the symphysis of the mandible.

2. The *hyoglossus* is a flat sheet of fibers which rises from the hyoid bone to the side of the tongue behind the mylohyoid muscle of the floor of the mouth.

3. The *styloglossus muscle* descends into the side of the tongue from the styloid process of the temporal bone. (See illustration on facing page.)

These muscles receive motor innervation from the twelfth cranial nerve. The pull of the genioglossus depresses the tongue. If its anterior fibers are active, the tip of the tongue is retracted. The posterior fibers protract (protrude) the tongue. The main body of the tongue is retracted by the styloglossus and hyoglossus muscles.

STRUCTURES FOR MASTICATION

Chewing of the food is generally termed mastication. This process is one of many simultaneous movements which include incising and tearing apart of pieces of food. The resulting smaller masses are ground into a pulpy mass by the pressure and lateral movements of the teeth against each other. The motive power for these movements is provided by the muscles of mastication. The tongue is important in mastication, as are the buccinator muscles of the cheeks and the musculature of the lips. The food masses are shaped by these muscular structures and are continually rearranged or positioned by the teeth or pressed against the hard palate, teeth, and gums for further disintegration.

The Teeth and Gums

The *teeth* are organs, modified from connective tissues, which project from sockets in the alveolar processes of the maxillae and mandible. The teeth

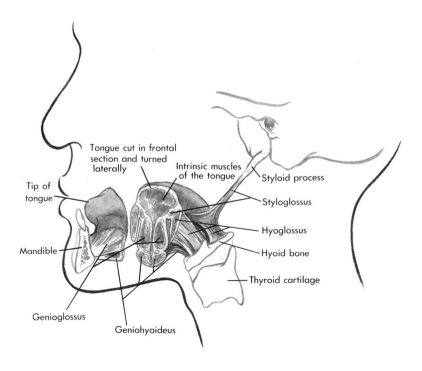

vary in shape in accordance with their position and general function. A *tooth* consists of a central connective tissue *pulp* which conveys blood vessels and nerves, the *dentin* which is a calcareous substance forming the body of the tooth, and the *enamel* which is a highly calcified connective tissue covering the grinding surfaces. The alveolar socket of a tooth is a depression in 'the bone of the jaw lined by a connective tissue, the *periodontal membrane*. The portion of the tooth which actually fits into the socket is formed into one or more *roots*. The root is joined to the periodontal membrane and held into the socket by another calcified connective tissue, the *cementum*. The projecting portion of the tooth bearing the enamel is called the *crown*. The *gums* are a very vascular fibrous connective tissue, covered by a continuation of the oral mucous membrane, which shrouds the alveolar processes. The gingival (gum) tissue also covers the base of the tooth crown and projects between adjacent surfaces of the teeth. (See color plate 9.)

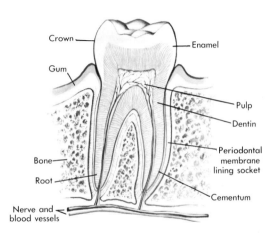

Dental Patterns

The teeth are borne in *dental arches* which follow the posterolateral curvature of each jaw. The upper and lower dental arches are divided between right and left sides so that there are four *dental quadrants.*

There are two sets of teeth. The primary or *deciduous dentition* forms before birth in the jaws but does not appear superficially through the gums (erupt) until the "teething period" after birth. Twenty deciduous teeth form the primary dentition with five appearing in each quadrant in the order and at the approximate times listed:

1. Central incisor: 6–8 months
2. Lateral incisor: 15–20 months
3. Canine (cuspid): 16–20 months
4. First molar: 12–20 months
5. Second molar: 20–24 months

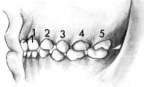

DECIDUOUS DENTITION

The deciduous teeth are quite variable in time of eruption. The teeth of the lower quadrants frequently precede the upper ones. By the age of 2½ years all deciduous teeth should be present. The roots of deciduous teeth are smaller than the roots of permanent teeth. At intervals before the twelfth year the roots are resorbed, the tooth gradually loosens, and the deciduous tooth is shed, as a permanent tooth grows and erupts in its place.

The *secondary dentition* begins with the eruption of a sixth tooth in each quadrant beyond the deciduous teeth. This is the first permanent or adult molar. There are 32 *permanent teeth* with eight in each quadrant. These teeth and their approximate times of eruption are listed below:

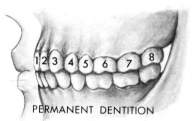

PERMANENT DENTITION

1. Central incisor: 6–7 years
2. Lateral incisor: 7–8 years
3. Canines (cuspids): 10–11 years
4. First premolar: 9–11 years
5. Second premolars: 10–11 years
6. First molar: 6th year
7. Second molar: 12–13 years
8. Third molar: 17–21 years (wisdom tooth)

ORGANS OF SALIVARY SECRETION

The mucous membrane of the oral cavity, tongue, and palates is studded with individual glandular cells which moisten the surfaces and lubricate them by secreting a film of mucus. In the development of the body, some of these cells divided to form glandular groups in the submucosal connective tissue. Some groups grew large enough to be connected with the surface membrane by short ducts through which their secretions pass. The growth of three groups on each side becomes so extensive in development that they move away from the oral cavity to form the *salivary glands.* In the adult, only the openings of their ducts mark the location from which the salivary glands developed. The salivary secretion or *saliva* varies in fluidity, depending on the presence or absence of food in the oral cavity. Between meals, saliva flow is scanty and tends toward a higher percentage of mucus. During mastication, autonomic impulses stimulate a profuse flow of watery saliva which, though containing mucus, is high in the amount of *ptyalin* contained. Ptyalin is an amylase enzyme which begins the breakdown of complex starch and other polysaccharides into simpler sugar compounds. The physical presence of saliva moistens food particles during mastication and helps bind food into a lubricated bolus for swallowing.

Structure of Salivary Glands

Other than the glands associated with the trachea and bronchi, the first major gland types encountered so far in the discussion are the glands associated with the oral cavity. It is suggested that reference be made to the description of membranes and glands (pages 15 and 16) in the first chapter.

Glands of the oral cavity may secrete mucus or the enzyme, ptyalin. The salivary glands may specialize in one secretion or the other, or have cellular units for both secretions. *Mucus-secreting cells* (mucous cells), previously called "goblet cells," are of a columnar type which are lined up along a gland duct. The individual cells may also act as unicellular glands when they appear in the mucous membrane lining of the oral cavity.

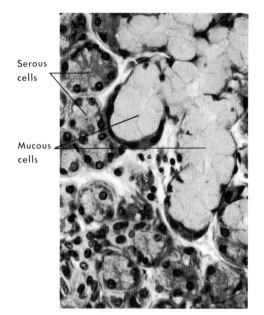

Serous cells

Mucous cells

from Bloom and Fawcett

Mucous cells appear to be very light or pale under the light microscope and may be either engorged with their product, resting after discharge, or in the process of filling again. Along with the mitochondria around the nucleus, the cell product is the main constituent of the cytoplasm. *Mucigen* droplets (granules), the product of manufacture, are polysaccharides which become converted to *mucin* when secreted from the cell. The mucin material within a duct or on the surface of a membrane is commonly called mucus.

Serous-secreting cells form into basketlike collections around the origin of salivary gland ducts.

These squared columnar or cuboidal cells appear very dark and granular under the light microscope because of the staining reaction of the contained *zymogen* droplets (granules). Serous cells appear to have a vertically oriented RNA cytoplasmic structure, along which the zymogen droplets seem to be oriented when the cell is examined by the electron microscope. The Golgi apparatus (see pages 7 to 10) is near the nucleus as may be expected in a secretory cell. Abundant rodlike mitochondria lie between the vertical plates of RNA. Zymogen, when secreted, is converted to the active enzyme *ptyalin.*

Parotid Gland

The largest salivary gland, the *parotid gland,* migrates in development to a position below the zygomatic arch, in front of the mastoid process and behind the ramus of the mandible. Its *deep portion* is a wedge which fits into the fossa between the mandibular ramus and the mastoid process, below the external acoustic meatus, and points inward toward the wall of the pharynx. The *superficial portion* stretches out anteriorly over the ramus and the masseter muscle. The two portions of the gland are connected by glandular isthmuses between which course the first main branches of the facial nerve. The superficial temporal artery ascends, and the retromandibular vein descends, within its substance. The secretion of the parotid gland is serous or zymogenic in type, having a high percentage of water and ptyalin.

The *parotid duct* (Stensen's duct) leaves the anterosuperior angle of the gland, crosses the masseter muscle, and turns medially through the buccinator muscle. It then turns anteriorly under the buccal mucous membrane through which it opens opposite the upper second molar tooth.

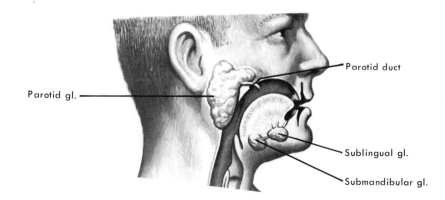

Parotid gl.

Parotid duct

Sublingual gl.

Submandibular gl.

Submandibular Gland

The *submandibular* (submaxillary) *gland* is a predominantly serous type of salivary gland. It lies sheltered by the mandible, folded partly above and partly below the posterior border of the mylohyoid muscle of the floor of the mouth. Its duct passes forward and medially in the transversely disposed sublingual fold of the floor of the mouth to open near the frenulum of the tongue.

Sublingual Gland

The sublingual gland is the smallest salivary gland. This predominantly mucous gland lies beneath the gutter of the floor of the mouth. Its salivary secretion flows through many separate sublingual ducts which open into the sublingual fold.

STRUCTURES RELATED TO SWALLOWING

When food has been chewed to particles of proper size, mixed with salivary enzymes and mucus, and shaped into a lubricated bolus, it is ready for swallowing. During the initiation of the act of swallowing (deglutition) the tip of the tongue is raised by its intrinsic muscles to the front teeth and hard palate. The dorsum of the tongue is converted to a trough by other intrinsic muscles. The base of the tongue is depressed and the bolus of food or liquid is forced backward through the mouth and downward into the oropharynx.

OROPHARYNX

The *pharynx* has been described previously as a musculofibrous box suspended in front of the cervical vertebrae from the base of the skull. One of its three portions, the *nasopharynx*, was described with the respiratory system. It was mentioned that the nasopharynx opens inferiorly into the oropharynx where there is a crossing of respiratory and alimentary streams.

The *oropharynx*, therefore, opens superiorly into the nasopharynx and anteriorly to the oral cavity. Food or liquid entering the oropharynx must pass through the *oropharyngeal* (faucial) *isthmus* whose boundary structures demarcate oral cavity from oropharynx. The base of the tongue forms the floor of the isthmus, along which the bolus is shunted in swallowing. The *soft palate*, with its dependent tag, the *uvula*, forms the superior boundary. A group of muscles, collectively called the palatal muscles, form the main substance of the soft palate and, by their contraction, can draw up the soft palate to close off the nasopharynx during swallowing.

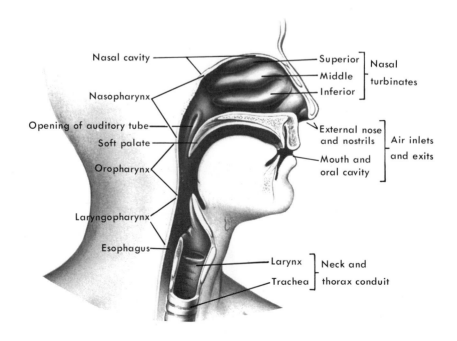

The lateral walls of the oropharyngeal isthmus are formed by two folds of mucous membrane which are raised into the *anterior* and *posterior faucial pillars* by muscles within them. The anterior pillar (palatoglossal arch or fold) contains the *palatoglossal muscle* which runs from the soft palate to the side of the base of the tongue. The muscles of each side act together to narrow or close the aperture between the oral cavity and the pharynx. The posterior pillar (palatopharyngeal arch or fold) contains the *palatopharyngeus muscle* which extends between the posterior edge of the hard palate, through the soft palate, to course downward to attachments on the thyroid cartilage and into the pharyngeal and esophageal walls. A depressed area occurs between the two faucial pillars on each side, which is termed the *tonsillar fossa.* Masses of lymphoid tissue covered by mucous membrane are located in the tonsillar fossa. These masses are called the *faucial* or *palatine tonsils,* or just *the tonsils.* They are part of a lymphatic ring which lies around the oropharynx and includes the adenoids above and the lingual tonsils below in the base of the tongue. The tonsils are related to the lymphatic vessels of the pharynx and assist in guarding this region against infection. Tonsils frequently enlarge and become inflamed during childhood and adolescence. They diminish in size and undergo involution later.

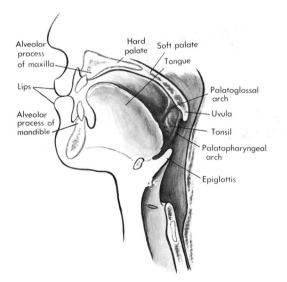

The posterior wall of the oropharynx continues the downward slope of the nasopharynx directly downward to the esophagus. The musculofibrous wall lies along the prevertebral muscles and the bodies of the cervical vertebra.

LARYNGOPHARYNX

The oropharynx opens below into the *laryngopharynx* with which it is continuous. Anteriorly the inlet to the larynx, guarded by the epiglottis, is the prominent feature. Laterally the laryngopharynx broadens on each side into the *piriform recess* which extends partly around the back of the larynx. Food, in the process of being swallowed, is diverted through the piriform recesses. An unusually large food bolus or a foreign object may be temporarily trapped in the piriform recess until coughing or convulsive swallowing movements return it to the oral cavity. The esophagus directly continues the laryngopharynx downward and represents the first appearance of the truly tubular alimentary tract.

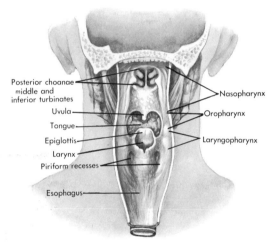

MUSCLES OF THE PHARYNX

The striated skeletal muscles of the pharynx are arranged into outer circular groups and inner longitudinal sheets. Although the muscle is of striated nature it is not under voluntary control. The circular muscles are called the constrictor muscles of the pharynx. The *superior constrictor muscle* drapes downward from bony points on the base of the skull and mandible to curve backward around the upper part of the pharynx. There it meets its fellow of the opposite side in a median raphe on the posterior aspect of the pharynx. The *middle constrictor muscle* arises from the greater and lesser horns of the hyoid bone. It fans posteriorly around the middle part of the pharynx, its upper portion overlapping the superior muscle and its lower portion being overlapped by the inferior muscle as they course into the median raphe. The *inferior constrictor muscle* originates from portions of the laryngeal cartilages and passes around the lower part of the pharynx to the posterior raphe.

The inner longitudinal muscles of the pharynx form longitudinal sheets. One, the *palatopharyngeus,* has been mentioned previously. The other, the *stylopharyngeus,* descends from the styloid process onto the side of the pharynx.

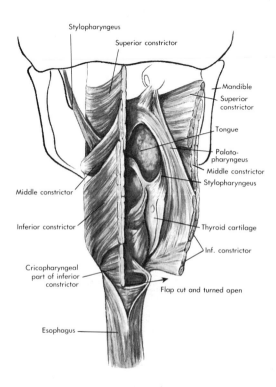

Stylopharyngeus

Superior constrictor

Mandible
Superior constrictor

Tongue

Palato-pharyngeus

Middle constrictor

Stylopharyngeus

Middle constrictor

Inferior constrictor

Thyroid cartilage

Inf. constrictor

Cricopharyngeal part of inferior constrictor

Flap cut and turned open

Esophagus

The pharyngeal muscles are innervated by branches of the vagus nerve which originate in the cranial part of the accessory nerve, as do direct accessory nerve fibers to the palatal muscles.

PHARYNGEAL COMPONENT OF SWALLOWING

It has been seen that in the first of three stages of swallowing, the *oral component,* the food bolus is voluntarily passed through the oropharyngeal isthmus into the oropharynx, and the nasopharynx is sealed off by palatal action. The second or *pharyngeal component* of swallowing is automatic. The longitudinal muscles of the pharynx draw the upper part of the pharynx superiorly to receive and accommodate the bolus. At the

same time the larynx is elevated under the base of the tongue and its aperture closed by the epiglottis so that food or liquid does not enter the respiratory passageway. Suprahyoid and laryngeal muscles draw the hyoid bone and larynx upward next to receive the bolus into lower portions of the pharynx. (See illustration on facing page.) This movement is discernible externally in the neck, both in quiet swallowing and in "gulping" as the laryngeal prominence moves upward. Throughout these actions of longitudinal nature the constrictor muscles act in succession to narrow the pharynx so that the food or liquid is passed along to the lower pharynx and then into the esophagus.

The lower part of the inferior constrictor is active even when swallowing is not being performed. It acts as a pharyngeal sphincter to keep air from entering the esophagus during breathing.

ESOPHAGEAL COMPONENT OF SWALLOWING

The tubular structure of the alimentary tract is not revealed in the digestive cavities and chambers of the head or pharynx. Beginning with the esophagus a basic tubular pattern is set for the rest of the alimentary system.

Basic Pattern of the Digestive Organs

A central *lumen* of varying size and patency is enclosed by the structures of the wall of the digestive tube. The *mucosa* faces the central

Connective tissue between muscle layers, site of myenteric plexuses

Lumen

Outer longitudinal muscle layer

Inner circular muscle layer

Mucosa

Submucosa, site of submucosal plexuses

lumen. Food and secretions pass over the epithelial surface of the mucosa. Frequently their passage is facilitated by a film of mucus which lubri-

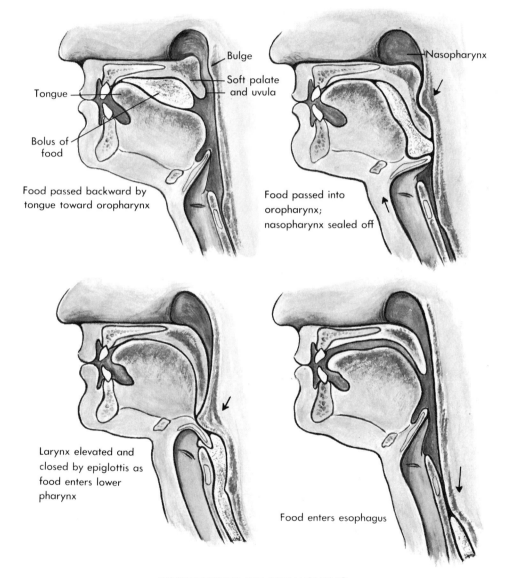

Bulge

Soft palate
and uvula

Tongue

Bolus of
food

Food passed backward by
tongue toward oropharynx

Nasopharynx

Food passed into
oropharynx;
nasopharynx sealed off

Larynx elevated and
closed by epiglottis as
food enters lower
pharynx

Food enters esophagus

COMPONENTS OF SWALLOWING

cates and protects the epithelium of the mucous membrane. The epithelium varies in its cellular composition in accordance with the function of each part or organ of the digestive tube. A strengthening and binding layer of connective tissue, the *submucosa,* lies under the epithelium. This layer frequently contains the bases of the more complex glands of the epithelium or is the site of collections of lymphoid tissue or of the cell bodies of parasympathetic neurons to the glands (*submucosal plexuses of Meissner*). Two muscular layers lie outside the submucosa. Although these smooth muscle layers are called the *inner circular* and *outer longitudinal muscle layers,* the actual disposition of these layers is respectively a close spiral and a loose (elongated) spiral around the tube. Parasympathetic neurons and their postganglionic fibers form the *myenteric plexuses of Auerbach* at intervals between the muscle layers. A connective tissue investment surrounds the entire digestive tube and merges with the regional connective tissue or fasciae through which the tube travels. In the abdominal region a layer of mesothelium combines with this connective tissue as a reflection of the peritoneal lining to form the outer serous membrane or *serosa.*

Esophagus

The *esophagus* is a flattened musculofibrous tube which runs a vertical course 10 inches in length to connect the lower end of the pharynx with the stomach. Since the pharynx descends behind the larynx to the level of the sixth cervical vertebra, only about 2 inches of the esophagus lies within the neck. The balance of the esophagus is located within the thorax and upper abdomen.

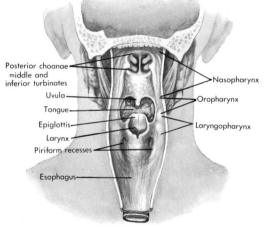

Posterior choanae middle and inferior turbinates
Uvula
Tongue
Epiglottis
Larynx
Piriform recesses
Esophagus
Nasopharynx
Oropharynx
Laryngopharynx

The esophagus is narrowed at three points, the first constriction being at its very beginning where it is attached to the back of the cricoid cartilage by the cricoesophageal tendon. The cervical esophageal tube, collapsed except during passage of food, descends to the superior thoracic aperture with the vertebral bodies behind it, and covered along its right margin by the trachea. (See illustration on facing page.) The lobes of the thyroid gland wrap around the sides of the esophagus. In its thoracic course in the mediastinum the esophagus is covered anteriorly by the trachea until the latter ends in the tracheal bifurcation. A second constriction is found at this point where the left bronchus crosses the esophagus. Lower in the mediastinum, the esophagus passes behind the left atrium of the heart and close to the medial surface of the right lung. Below the heart the descending thoracic aorta comes to lie to the left of the esophagus. Just above the diaphragm the esophagus swings to the right and in front of the aorta to pass through the esophageal hiatus of the diaphragm where a third constriction occurs. The abdominal course of the tube brings it to the upper end of the stomach which it joins at the gastroesophageal junction. (See also color plate 9.)

The mucous membrane of the esophagus is greatly folded because the alimentary tube is collapsed unlike the trachea which is kept open by its cartilaginous rings. A thick stratified squamous epithelium makes a tough smooth lining. The muscular layers of the esophagus are of striated muscle in the upper part but of smooth muscle below. (See illustration on page 414.)

The *esophageal component of swallowing* is automatic. Food or liquids are conveyed through the tube in short, rapid, peristaltic spurts until the gastroesophageal junction is reached. Here the movement is slowed until a muscular sphincter relaxes to allow the bolus to enter the stomach.

PARTS AND DISPOSITION OF THE GASTROINTESTINAL TRACT

The *gastrointestinal tract* is a convenient term to use to group the organs of the alimentary system located in the abdomen. The *abdomen* is the portion of the trunk below the diaphragm which is surrounded by the abdominal muscles

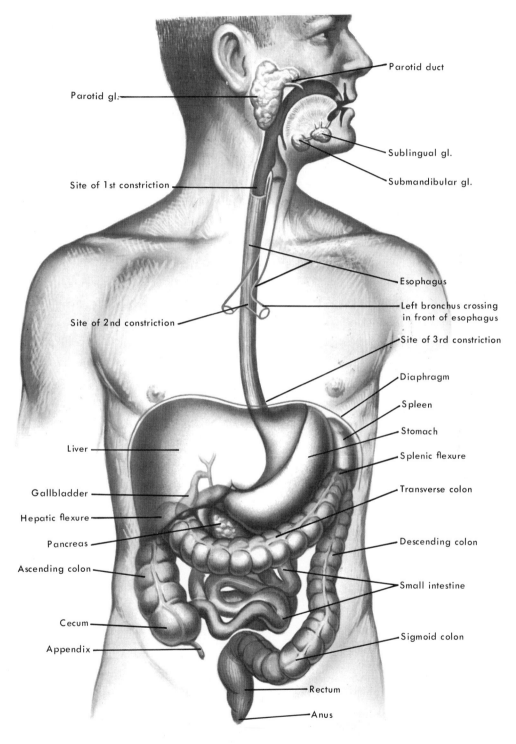

Parotid duct

Parotid gl.

Sublingual gl.

Submandibular gl.

Site of 1st constriction

Esophagus

Left bronchus crossing
in front of esophagus

Site of 2nd constriction

Site of 3rd constriction

Diaphragm

Spleen

Stomach

Liver

Splenic flexure

Transverse colon

Gallbladder

Hepatic flexure

Pancreas

Descending colon

Ascending colon

Small intestine

Cecum

Appendix

Sigmoid colon

Rectum

Anus

DIGESTIVE SYSTEM

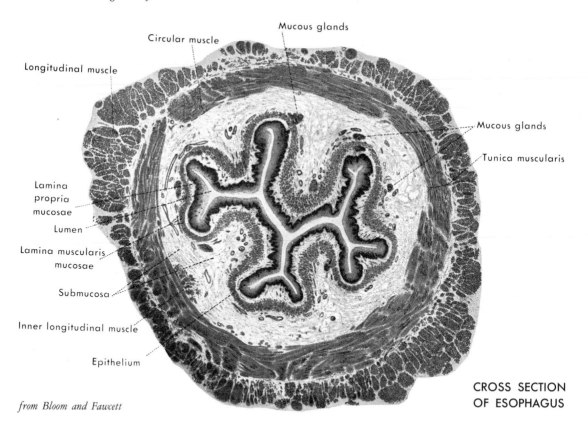

Longitudinal muscle

Circular muscle

Mucous glands

Mucous glands

Tunica muscularis

Lamina propria mucosae

Lumen

Lamina muscularis mucosae

Submucosa

Inner longitudinal muscle

Epithelium

from Bloom and Fawcett

CROSS SECTION OF ESOPHAGUS

and limited inferiorly by the muscles which close the pelvic outlet. Within the abdomen is the *abdominal cavity proper,* between the undersurface of the diaphragm and the pelvic inlet, and the *pelvic cavity* which lies between the walls of the true pelvis. Most of the organs of the digestive tract lie within the abdominal cavity proper, although in the absence of a partition such as that found between the thorax and the abdomen some of the alimentary organs may come to lie partly in the pelvic cavity. (See illustration on opposite page.)

ORGANS OF THE GASTROINTESTINAL TRACT

The abdominal portion of the esophagus, the stomach, the small intestine, and the large intestine are the primary alimentary organs. Two large glands, the liver and the pancreas, are associated with the digestive tube. The spleen, primarily a blood-forming and lymphatic organ, is also situated in the abdomen. The kidneys of the urinary system are located in the abdomen, and their tubes, the ureters, pass along the posterior abdominal wall. The urinary bladder, which is a

reservoir for urine, and most of the reproductive organs lie within the pelvic cavity with the terminal part of the large intestine, the rectum.

DISPOSITION OF ABDOMINAL ORGANS

The *liver* is a large gland, which occupies the upper right portion of the abdominal cavity under the shelter of the diaphragm and the lower right costal margin. Its left portion extends across the midline of the upper abdomen. The *spleen* is tucked under the left side of the diaphragm under the cover of the lower left ribs. The main portion of the stomach fills the upper left portion of the abdominal cavity and extends to the right across the midline of the upper abdomen under the liver. The *small intestine* fills the central and lower parts of the abdominal cavity with its loops and coils. The *large intestine* commences in the lower right portion of the abdominal cavity where the cecum is located. The *ascending colon* passes upward along the right side of the abdomen to the undersurface of the liver where it turns transversely at the *hepatic flexure.* The *transverse colon* crosses the mid-abdominal region, frequently sagging downward into the lower abdominal cavity or pelvis before reaching upward toward the spleen in the upper

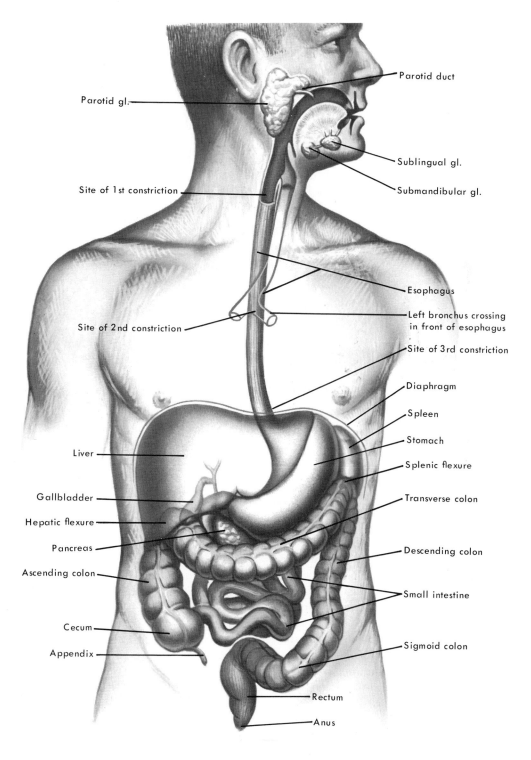

Parotid duct

Parotid gl.

Sublingual gl.

Submandibular gl.

Site of 1st constriction

Esophagus

Left bronchus crossing
in front of esophagus

Site of 2nd constriction

Site of 3rd constriction

Diaphragm

Spleen

Stomach

Splenic flexure

Liver

Transverse colon

Gallbladder

Hepatic flexure

Pancreas

Descending colon

Ascending colon

Small intestine

Cecum

Appendix

Sigmoid colon

Rectum

Anus

DIGESTIVE SYSTEM

left portion of the abdomen. At the *splenic flexure* the large intestine descends along the left side of the abdomen as the *descending colon.* The terminal part of the colon is named the *sigmoid* or *pelvic* colon because it turns transversely across the pelvic brim toward the midline to become the *rectum.* The rectum is a pelvic organ which descends amid genitourinary organs to the anal opening.

PERITONEUM

The *peritoneum* is the thin, glistening internal lining of the abdominal cavity. It is a sheet of mesothelial cells that is strengthened by connective tissue fibers. The peritoneum is loosely united by extraperitoneal connective tissue to the transversalis fascia which surrounds the abdominal cavity under the abdominal muscles. The part of the peritoneal membrane which lines the anterolateral and posterior abdominal walls is the *parietal peritoneum.* Most of the abdominal organs are covered by peritoneum to a varying degree, because in development they project from the posterior wall of the abdomen or have grown out to become independent organs. The peritoneum which covers the surface of abdominal viscera is called the *visceral peritoneum.* Organs like the kidneys and the pancreas, which project only partly into the abdominal cavity, are covered on only one surface. These are termed *retroperitoneal organs* because they lie behind the parietal peritoneum. Other organs, particularly those of the alimentary system, are almost completely covered by peritoneum. Their only attachment to the posterior abdominal wall is by double-layered folds of peritoneum, termed *mesenteries,* through which blood vessels reach each organ. These viscera are called *peritoneal* or *intraperitoneal organs* because of their situation as organs within the abdominal cavity. The projection of intraperitoneal viscera fills the abdominal cavity and reduces the *peritoneal cavity* to a slit between the parietal and visceral membranes. The peritoneal cavity contains only peritoneal fluid which lubricates the peritoneal membranes. The peritoneum as a whole serves to provide a lubricated surface for the movement of abdominal organs and as a barrier to infection. Pelvic organs are outside the peritoneum and are, therefore, termed *extraperitoneal organs.* Some pelvic organs, such as the bladder, rectum, and uterus, receive a partial peritoneal investment on surfaces which rise up under the peritoneal sac.

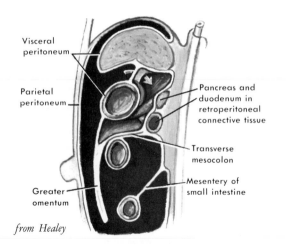

from Healey

The disposition of the peritoneum into folds and "ligaments" as it invests the abdominal organs is complicated. These factors will be summarized after the organs have been described.

ORGANS RELATED TO DIGESTION

The alimentary organs can be grouped according to their principal function. Digestion of the swallowed food occurs mainly in the stomach and duodenum. Absorption of food materials, reduced to simple, utilizable form by digestion, occurs in the small intestine. Reabsorption of water and electrolytes takes place in the large intestine as food residues and body wastes are moved to the rectum for elimination.

STOMACH

The *stomach* is a J-shaped, mobile, dilated portion of the alimentary tube which receives swallowed food from the esophagus, continues the digestive process, performs limited absorption, and moves the food elements to the small intestine. The shape and position of this part of the gastrointestinal tract are highly variable, depending upon the degree of distention and the posture of the person. (See illustration at top of facing page.) When the stomach is moderately distended, the J-shape changes to a thick crescent, and when full the stomach becomes a distended sac. Its position varies from an oblique presentation downward and to the right across the upper abdomen to an elongated dependent position with the limb of the J hanging downward toward the brim of the pelvis. The position is affected by body form. The vertical elongated stomach is seen particularly in a tall slender person with a long trunk; the oblique position is found more often in an individual with a short but broad and capacious abdomen.

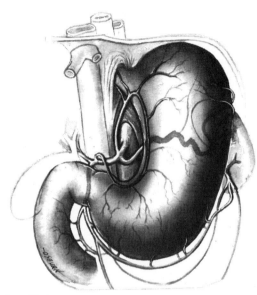

from Healey

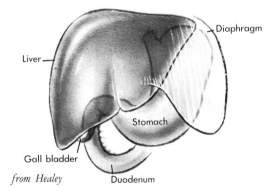

Liver

Diaphragm

Stomach

Gall bladder

Duodenum

from Healey

Parts of the Stomach

The stomach presents two *curvatures.* The *lesser curvature* is short. Its concave surface looks upward and to the right. The longer *greater curvature* is convex. It presents anterolaterally to the left. Between these two curvatures are the posterior and anterior *walls.* The *gastroesophageal junction* occurs at the upper end of the lesser curvature. This is the more fixed end of the stomach. The walls of the stomach around this junction are termed the *cardiac part* of the stomach. To the left of the gastroesophageal junction the

stomach balloons upward above the level of the entrance of the esophagus. This portion of the organ, frequently filled with swallowed air, is termed the *fundus.*

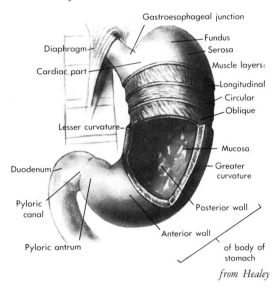

Gastroesophageal junction
Diaphragm
Cardiac part
Lesser curvature
Duodenum
Pyloric canal
Pyloric antrum
Fundus
Serosa
Muscle layers:
Longitudinal
Circular
Oblique
Mucosa
Greater curvature
Posterior wall
Anterior wall
of body of stomach

from Healey

The greater part of the walls of the stomach between the curvatures expands into the more dilated *body* of the stomach. The lower end of the stomach narrows abruptly, at a point about three-fifths of the distance along the lesser curvature, into the *pyloric part* of the organ. This is divided into the *pyloric antrum* adjacent to the body and a muscular constricted part, the *pyloric canal,* which surrounds the *pyloric opening* into the small intestine. The cardiac, fundic, and pyloric parts of the stomach are not set apart from the body by any demarcations. These are general geographic areas which are indicated generally by the features mentioned and identified specifically only by the type of glands in their mucosae.

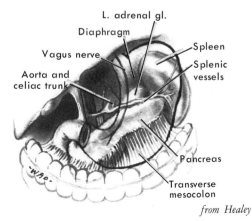

L. adrenal gl.
Diaphragm
Vagus nerve
Aorta and celiac trunk
Spleen
Splenic vessels
Pancreas
Transverse mesocolon

from Healey

Structure of the Stomach

The characteristic layers of the alimentary tube are present in the stomach. The *mucosa* is thick and thrown up into prominent longitudinal folds, the *rugae,* bearing upon and between them the *gastric pits* (foveolae) into which the gastric glands open. The mucosal gastric glands are of several types. *Mucous glands* are scattered throughout the stomach in the mucosa of all parts. They are the only glands of the cardiac region. The branched tubular glands of the fundic and body regions contain mucous cells in the neck portions near the gland openings but, in the lower portions, other cell types are present. These are the acid-secreting *parietal cells* and the enzyme-secreting *chief* or *zymogenic cells.* Mucous cells are found in the pyloric glands in addition to those which elaborate a gastric secretion hormone.

The enzyme-secreting *zymogenic cells* have the typical ultrastructural appearance of a protein-making cell, such as has already been described for the fibroblast and the serous salivary cell.

The rather cuboidal zymogenic cells present stubby microvilli from the surface membrane bordering the gastric gland lumen. (See illustration below.) The submicroscopic framework of the cell is that of the channels of a granular endoplasmic reticulum studded with ribosomes. The latter are also found free within the cytoplasm, along with large oval granules which represent the cell product, *pepsinogen,* which is the precursor of the enzyme *pepsin.*

The acid-secreting *parietal cells* have no secretory granules, but the cytoplasm is loaded with mitochondria which lie amidst a framework of smooth, endoplasmic reticulum. (See illustration at top of following page.) The main feature of the acid-secreting cell is an extensive invagination of the surface membrane to form an intricate intracellular canal system termed the *secretory canaliculus.* The intricate folding of the lining of the secretory canal is made by many microvilli.

Epithelial cells, which form the actual surface mucosal lining of the stomach, are a type of mucus-secreting cell. The unit membrane making the free surface of such cells is formed into short microvilli and, in addition, is covered by a filamentous polysaccharide material. The

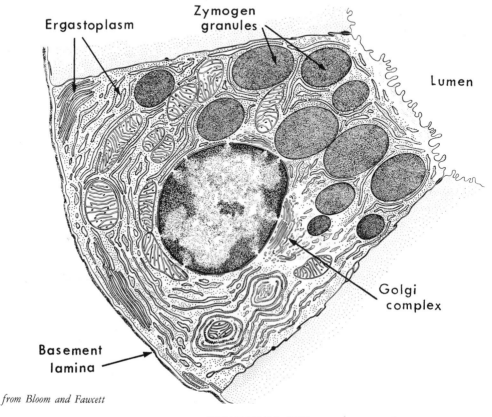

from Bloom and Fawcett

ZYMOGENIC CELL

PARIETAL CELL

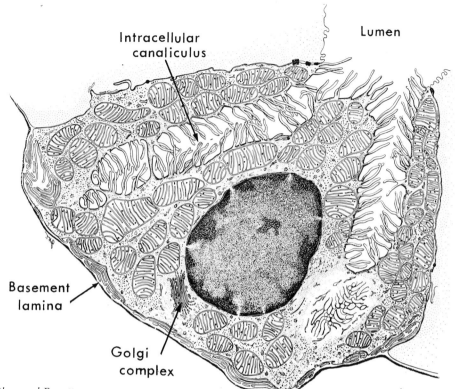

Lumen

Intracellular canaliculus

Basement lamina

Golgi complex

from Bloom and Fawcett

mucigen secreted by these cells and the polysaccharide surface layer are believed to provide protection from the enzymes and hydrochloric acid formed by the gastric glands.

The connective tissue of the *submucosa* conveys blood vessels and nerves, contains autonomic ganglion cells, and loosely binds the mucosa to the muscular layer. The *muscularis* of the stomach

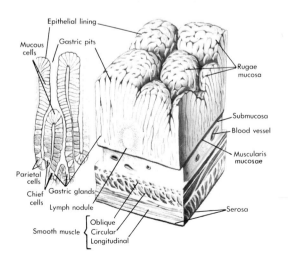

is composed of layers of smooth muscle. Inner circular and outer longitudinal muscle layers continue from the esophagus in the cardiac region. The fundic and body portions are characterized by the addition of an innermost oblique layer of smooth muscle. The musculature of the pyloric canal is dense and is formed into the *pyloric sphincter* which guards the pyloric opening and regulates the flow of gastric contents. The *serosa* consists of fibrous adventitial connective tissue which binds the mesothelial membrane of visceral peritoneum to the surface of the organ. (See also color plate 10.)

Gastric Functions

Food entering the stomach usually progresses along the lesser curvature to the pyloric antrum where it is stopped by the closed pyloric sphincter. The presence of undigested food stimulates parasympathetic nerve endings in the mucosa of the pyloric antrum. These endings are further stimulated as the antrum and body of the stomach distend with additional increments of swallowed food and liquids. Afferent impulses of the parasympathetic system produce reflexes in the brain stem which initiate parasympathetic motor impulses to the glandular cells of the pyloric mucosa and the secretion of a hormone, *gastrin,* into the blood. When gastrin is circulated to other parts of the stomach, the parietal cells secrete hydrochloric acid and the chief cells elaborate enzymes such as pepsin. These substances pour out into the gastric contents as components of the *gastric juice.* The acid environment is necessary to assist in the breakdown of meat fibers and to activate the pepsin, which requires an acid medium to carry out its proteolytic action. Mucus forms a protective film on the surface of the mucosa to prevent digestion of the gastric lining by its own enzymes.

Waves of muscular contraction (peristalsis) progress successively from the cardiac region to the pylorus. These peristaltic waves temporarily constrict the walls of the segments of the stomach as they pass. The alternate constriction and dilation of the walls brings pressure upon the gastric contents to thoroughly mix food and enzymes together. The successive processes, occurring over a period of one to two hours, reduce the food to a liquid called *chyme.* Only then does the pyloric sphincter relax to permit passage of the chyme into the small intestine. Liquids such as water are passed very quickly through the pylorus, and there is evidence that this may occur selectively even when more solid food is present. Limited absorption, mainly of water and some drugs, may occur through the gastric mucosa.

DUODENUM

The small intestine begins with the duodenum which is its shortest part and the portion more closely related to the digestive function of the stomach. Of the approximately 20 feet of small intestine, only the first 12 inches are formed by the duodenum.

Parts of the Duodenum

The *duodenum* is relatively firmly fixed in position in contrast to the other portions of the small intestine. Four successive parts of its tube are described. The *first* or *upper part* initiates a C-shaped curve of the whole organ by rising posteriorly and to the right from the pylorus to come into close relation with the gallbladder on the undersurface of the liver. This part is freely movable to follow the positional changes of the

pyloric end of the stomach. At the height of its rise this part turns downward. This area is visualized radiologically as the *duodenal cap.* Its mucosa and that of the stomach are subject to the occurrence of ulcers.

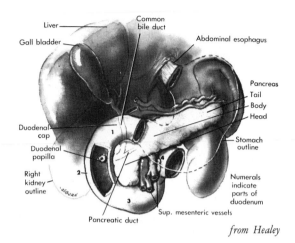

from Healey

The *second* or *descending part* of the duodenum descends in front of the right kidney. This part is abutted to the left by the head of the pancreas. The pancreatic and common bile ducts enter the posteromedial aspect of the wall of this part to empty their conveyed secretions through the *greater duodenal papilla* into the lumen. The *third* or *horizontal part* turns to the left across the midline over psoas muscles, inferior vena cava, and aorta. The first three parts of the duodenum form the C-shaped bend, concave to the left, into which fits the head of the pancreas. The root of the peritoneal mesentery of the small intestine issues forward, as do the superior mesenteric blood vessels, between the inferior border of the pancreas above and the third part of the duodenum below. The *fourth* or *ascending part* rises to the left of the aorta and then turns forward at the duodenojejunal flexure. Only the mobile first part of the duodenum is invested more or less completely by peritoneum. The other parts, like the pancreas, are retroperitoneal.

Structure of the Duodenum

When the duodenum is opened its mucosa presents a fringed velvety appearance owing to the many *villi* which project into the lumen. These increase the epithelial surface and bring the mucosa into greater contact with duodenal contents. Surface area is also increased by *circular folds* of both mucosa and submucosa. Tubular glands, also known as the *crypts of Lieberkühn,* run downward in the mucosa between the bases of the villi. Other glands (of Brunner) are found in the submucosal connective tissue. Isolated collections of lymphoid tissue appear in the submucosa as do autonomic plexuses and ganglion cells. Inner circular and outer smooth muscle coats provide a vigorous churning type of peristalsis.

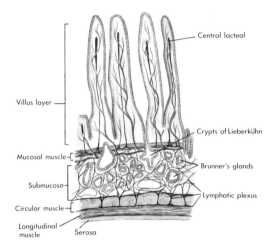

Central lacteal

Villus layer

Crypts of Lieberkühn

Mucosal muscle

Brunner's glands

Submucosa

Lymphatic plexus

Circular muscle

Longitudinal muscle   *Serosa*

The tall, columnar epithelial cells of the duodenal villus have a striated border. The electron microscope shows this border to be made up of myriad straight *microvilli* thrusting into the lumen of the organ. A typical trilaminar *unit membrane* covers the microvilli, which are also clothed by a filamentous feltwork which lies upon the surface membrane, as is seen in the illustration below. The filamentous coat is made of a mucopolysaccharide material which attaches to the other layer of the unit membrane. Extending inward into the cytoplasm are microfilaments which lie amidst a smooth endoplasmic reticulum. Mitochondria and free ribosomes are plentiful. As in most secretory cells, a well-developed Golgi complex is found.

The tubular epithelial glands making up the *crypts of Lieberkühn* are similar to the villous epithelial cell. The cells of the submucosal *Brunner's glands* are like the serous salivary cells or gastric zymogenic cells in ultrastructural appearance.

Duodenal Functions

The chyme does not remain long in the duodenum but an amazing number of digestive processes occur in a short time. The acidity of the chyme is neutralized and altered to an alkaline reaction necessary for the activity of pancreatic and intestinal enzymes. This function is performed by the submucosal Brunner glands. The alkaline mucus secreted by these glands also protects the epithelial surface. The intestinal glands of the crypts of Lieberkühn produce several enzymes which are combined into the *intestinal juice.* These enzymes are waved into and

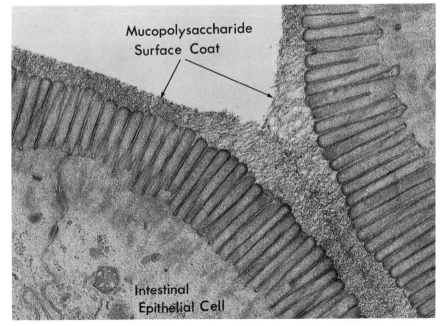

Mucopolysaccharide Surface Coat

Intestinal Epithelial Cell

DUODENUM

from Bloom and Fawcett

intermixed with the chyme by movements of the villi and peristaltic contractions of the wall. Proteins, broken down to polypeptides in the stomach, are reduced to assimilable simple amino acids by peptidases from the intestinal glands and trypsin received in pancreatic secretions. Fats are broken down into fatty acids and glycerol by intestinal lipases aided by bile and pancreatic lipase arriving through the greater duodenal papilla. A number of specific intestinal enzymes, aided by pancreatic amylase, convert the various polysaccharides derived from carbohydrate foods into simple sugars of which glucose predominates.

Many of these digestive functions are only initiated in the duodenum where many of the enzymes are formed or arrive from the pancreas and biliary system. After a short time for intermixture with the chyme the duodenal contents are moved into the jejunum where the digestive processes are completed. The duodenal villi serve to initiate the absorption of the various simple products of digestion. (The process of absorption is described with the jejunum.)

The duodenum and jejunum have common functions of digestion and absorption with digestion predominating in the duodenum and absorption becoming the more important role of the small intestine.

ORGANS RELATED TO ABSORPTION

The small intestine is concerned with the absorption of food materials into the blood although in its upper part it continues the digestive process. The large intestine conserves water and electrolytes by reabsorbing these elements from the residual intestinal components before they are evacuated through the rectum and anal opening.

JEJUNUM AND ILEUM

The duodenum has been described as the first 12 inches of the small intestine. The proximal two-fifths of the balance of the small intestine is formed by the *jejunum;* the distal three-fifths is termed the *ileum.* There are no absolute differentiating characteristics between these two coiled parts of the small intestine. They are both intraperitoneal organs and share an extensive peritoneal fold, the *mesentery,* which connects them to the posterior abdominal wall. The mesentery arises from a narrow strip which crosses obliquely downward and to the right over the retroperitoneal structures. This double-layered fold of peritoneum containing the blood vessels, lymphatic vessels, and nerves to the small intestine has been seen to issue forth between the pancreas and lower parts of the duodenum. The mesentery fans out from narrow origins to connect to the small intestine along the inner margin of all of the loops and coils. Although differences in the jejunum and ileum occur only gradually along the entire length of the small intestine, certain differentiating characteristics are described.

Differences Between Jejunum and Ileum

The characteristics of these organs are summarized in the table on page 423.

The coils of the small intestine are packed tightly together, but, because of their lubricated peritoneal surfaces, can change volume, diameter, and position by sliding upon each other. The surgeon utilizes the tabulated differentiating characteristics to determine the portion of the small intestine that appears in the relatively limited area of an incision of the abdominal wall.

Structure of the Small Intestine

The internal appearance of the small intestine and the structure of its wall are quite similar to those of the duodenum. Exceptions are the absence of the greater duodenal papilla and the submucosal glands. The submucosa does contain increasingly larger collections of lymphoid tissue. As the distal portions of the small intestine are reached, the lymphoid tissue is organized into prominent nodules termed *Peyer's patches.* (See also color plate 10.)

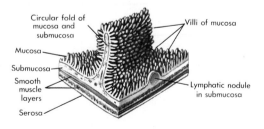

The *villi* of the mucosa of the small intestine function as minute organs for absorption. These

CHARACTERISTICS OF THE JEJUNUM AND ILEUM

CHARACTERISTIC	JEJUNUM	ILEUM
Location	Central upper abdomen	Lower abdomen and pelvis
Diameter	Wider	Narrower
Wall thickness	Thicker	Thinner
Vascularity	More vascular; redder	Less vascular
Emptying	Rapid by more vigorous peristalsis	Slower
Fat in mesentery	Little; many translucent windows in peritoneum	More; thick fatty mesentery with few windows
Arcade arrangement of vessels in mesentery	Only one or two tiers of arcades	Four or five tiers of arterial arcades

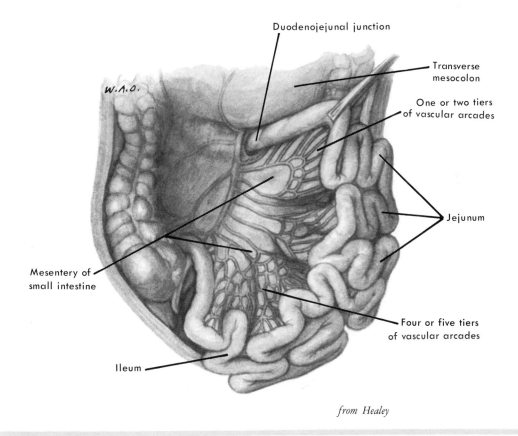

from Healey

finger-like processes are comprised of a central core of connective tissue which is covered by columnar epithelium much like the glandular epithelium of the crypts.

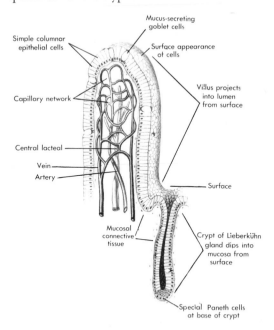

Some of these cells secrete mucus, but most are principally concerned with the absorption of food substances into the capillary network and venules of the central core. A blind lymphatic tubule, the *lacteal,* is the central occupant of the villus. Fats absorbed into the villus do not enter the blood but pass into the lacteal. Simple forms of fats, absorbed by the epithelial cells of the villus, form minute intracellular particles called *chylomicrons.* (See illustration below.) These move to the borders of the cells and pass through the unit membrane into the intercellular space, as shown in the electron photomicrograph. The chylomicrons filter through the connective tissue of the villus to enter the lacteal. Lacteals open into lymphatic vessels of the wall of the small intestine which join mesenteric lymphatic vessels. After a fatty meal the milky fats display the route of the lymphatics through the mesenteries to the thoracic duct. Although other foods are carried to the liver by way of tributaries to the portal vein, the fats bypass this organ and reach the systemic venous circulation via the lymphatic system.

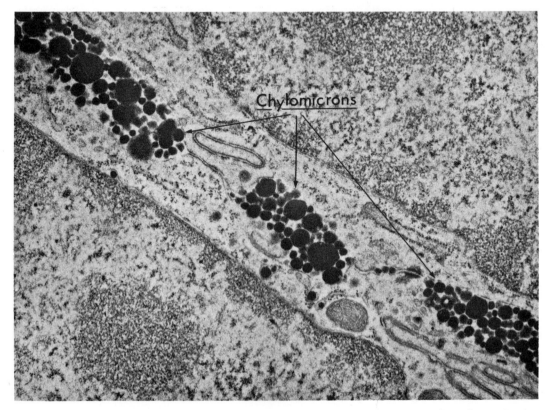

CHYLOMICRONS IN LACTEAL

from Bloom and Fawcett

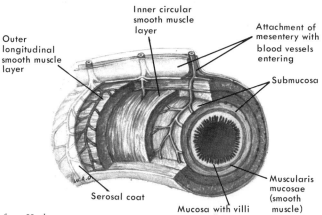

Outer longitudinal smooth muscle layer

Inner circular smooth muscle layer

Attachment of mesentery with blood vessels entering

Submucosa

Muscularis mucosae (smooth muscle)

Mucosa with villi

Serosal coat

from Healey

The microscopic structure of the various parts of the small intestine are similar to the descriptions already given. Only the gradual increase in nodular lymphoid tissue (Peyer's patches) distinguishes distal from proximal small intestine.

Muscular Activity of the Small Intestine

The inner circular and outer longitudinal layers of smooth muscle are richly supplied with autonomic motor fibers of which the parasympathetic components are most important. The musculature is functionally divided into a series of short segments which contract successively in peristaltic movements to move the liquid contents along the long coiled tube. Local reflexes, apparently mediated in the wall of the coils themselves by the cells of the myenteric plexuses, adjust the rate of peristalsis. As the more proximal of two adjacent segments distends, the more distal segment contracts to move its increment of contents along so that it will be ready to fill again. Afferent fibers carry impulses to the central nervous system along the autonomic pathways which relate to sensations of distention. If the intestine is overactive, these impulses may be interpreted as "cramps" or colicky pain.

LARGE INTESTINE

The *large intestine* is the larger caliber continuation of the gastrointestinal tract which receives the liquid residues of digestion and alimentary absorption.

Parts of the Large Intestine

The ileum crosses the midline of the lower abdomen to open into the large intestine at the

ileocecal opening, which is guarded by two folds which form the *ileocecal valve.* The ileum enters the medial side of the blind end of the large intestine which is dilated to form a sac called the *cecum.*

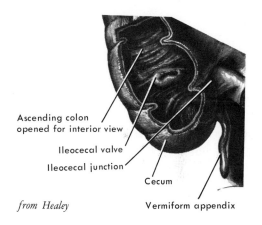

Ascending colon opened for interior view

Ileocecal valve

Ileocecal junction

Cecum

from Healey

Vermiform appendix

The cecum rests in the right iliac fossa with much of its sac below the level of the ileocecal opening. A vestigial narrow prolongation of the cecum, the *vermiform appendix,* projects from its point of apparent attachment to the apex of the cecum below the terminal portion of the ileum.

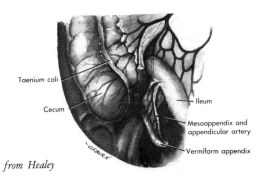

Taenium coli

Cecum

Ileum

Mesoappendix and appendicular artery

Vermiform appendix

from Healey

APPENDIX

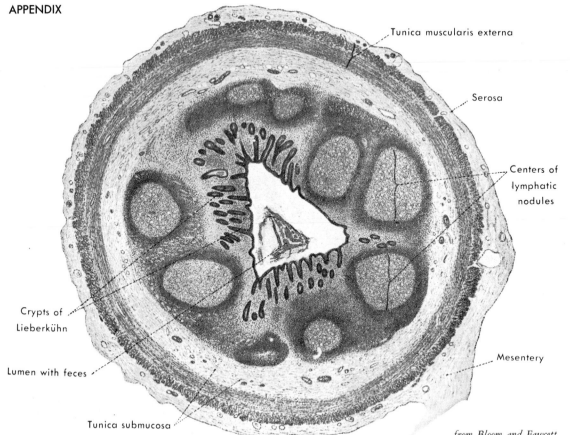

Tunica muscularis externa

Serosa

Centers of lymphatic nodules

Crypts of Lieberkühn

Lumen with feces

Mesentery

Tunica submucosa

from Bloom and Fawcett

The appendix is not active in the functions of the large intestine. Its opening into the cecum is constricted but its hollow lumen may receive and trap some food residues. The wall of the appendix contains large collections of lymphoid tissue which become active and enlarged if the appendix becomes inflamed. The position of the appendix is quite variable. It is usually directed inferomedially toward the pelvis but may occupy a position behind the cecum and colon or between ileal loops along an arc swung from the point of attachment of the appendix. (See illustration at top of opposite page.)

The cecum is continued upward along the right posterolateral abdominal wall as the *ascending colon*. The ascending colon is fixed posteriorly to the fascia of the abdominal wall with peritoneum covering only its anterior and lateral surfaces. The large intestine turns medially under the liver at the *right colic* or *hepatic flexure*. (See illustration at bottom of opposite page.) The *transverse colon* crosses the upper abdomen. This part of the large intestine acquires a peritoneal mesentery, the *transverse mesocolon,* which provides mo-

bility not possessed by the ascending colon. The position of the transverse colon is, therefore, variable. It only rarely crosses the upper abdomen in a direct horizontal course. More often the central part of the transverse colon droops downward toward the pelvis in a U or V shape before ascending beneath the spleen. The *descending colon* begins in the *left colic* or *splenic flexure* and descends along the left posterolateral abdominal wall. This part of the large intestine is similar to the ascending colon in having no mesentery and being fixed to the abdominal wall. The descending colon ends at about the level of the left iliac crest where the large intestine swings medially to become the *pelvic* or *sigmoid colon.* The sigmoid colon forms an S-shaped loop at first passing forward to the right toward the midline and then coursing posteriorly around the brim of the pelvis. (See illustration on page 428.)

Structure of the Large Intestine

The mucosa of the colon has no villi, but the surface epithelium dips into the crypts of Lieber-

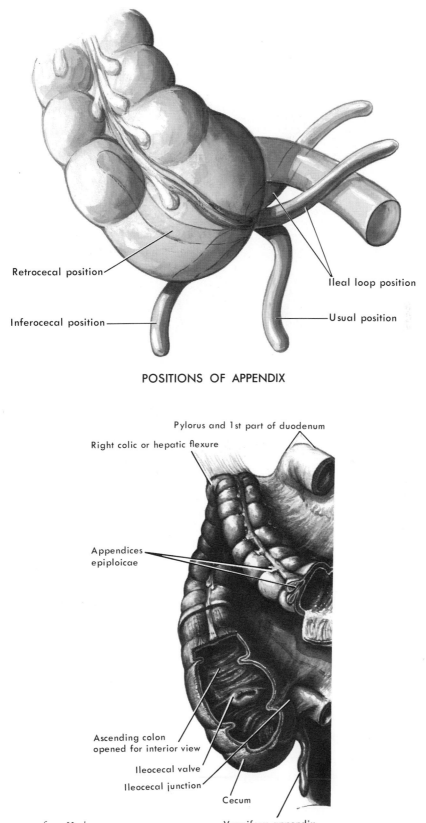

Retrocecal position

Ileal loop position

Inferocecal position

Usual position

POSITIONS OF APPENDIX

Pylorus and 1st part of duodenum

Right colic or hepatic flexure

Appendices epiploicae

Ascending colon opened for interior view

Ileocecal valve

Ileocecal junction

Cecum

Vermiform appendix

from Healey

COLON STRUCTURE

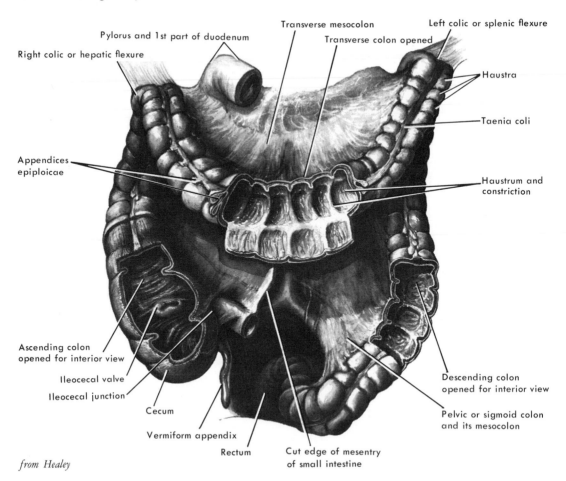

Pylorus and 1st part of duodenum

Right colic or hepatic flexure

Transverse mesocolon

Transverse colon opened

Left colic or splenic flexure

Haustra

Taenia coli

Appendices epiploicae

Haustrum and constriction

Ascending colon opened for interior view

Ileocecal valve

Ileocecal junction

Cecum

Vermiform appendix

Rectum

Cut edge of mesentry of small intestine

Descending colon opened for interior view

Pelvic or sigmoid colon and its mesocolon

from Healey

kühn which are lined by columnar and mucus-secreting epithelial cells. The goblet cells of mucus secretion increase in number and percentage toward the distal end of the colon. Lymphoid collections are present in both the mucosa and submucosa. The circular smooth muscle layer is disposed as elsewhere in the gastrointestinal tract, but the longitudinal layer is formed into three discrete bands, the *taeniae coli*. (See also color plate 10.)

These muscular bands pull upon the large intestine in the manner of a purse string so that it is drawn longitudinally into a series of alternating dilated sacculations (*haustra*) and constrictions. Localized collections of fat in the serosal layer form fatty tags covered by peritoneum which are termed the *appendices epiploicae*.

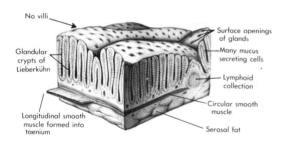

No villi

Glandular crypts of Lieberkühn

Longitudinal smooth muscle formed into taenium

Surface openings of glands

Many mucus secreting cells

Lymphoid collection

Circular smooth muscle

Serosal fat

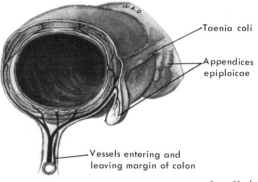

Taenia coli

Appendices epiploicae

Vessels entering and leaving margin of colon

from Healey

Functions of the Colon

Peristaltic movement slows in the colon. The cecum and ascending colon fill passively from the ileum. Movement of the contents is by way of large segments of the musculature of the transverse and descending colons in contrast to the short peristaltic contractions of the small intestine. During the passage of the contents, large amounts of fluid are absorbed so that semisolid fecal masses are formed. With the absorption of water more and more mucus is added toward the end of the colon. The feces is usually retained in the sigmoid colon and moved into the rectum only prior to or during evacuation from the body.

ORGANS RELATED TO EVACUATION

Alimentary wastes consist of the residues of digestion, undigestible food elements, and the mucus added in the lower intestinal tract. Also included are excretory products of tissue metabolism which have been eliminated by the liver and enter the alimentary system in the bile.

RECTUM

The rectum, as its name implies, is straight only in that its course approximates the midline of the body. From the rectosigmoid junction the organ follows the curvature of the sacrum and coccyx, having these structures behind it and the bladder and prostate gland (of the male) or the uterus (of the female) in front of it. The rectum bends forward and then backward as it passes through the pelvic diaphragm which is formed by the muscles closing the pelvic outlet. Just above the diaphragm the tube expands into the *ampulla* of the rectum. Peritoneum drapes over the front and sides of the upper part of the rectum but only over the front of the middle part as it dips down between the rectum and the organs in front of it to form the *rectovesical pouch* (male) or the *rectouterine pouch* (female). The lower rectum has no peritoneal investment. Its fibrosal covering blends with the fasciae of the other extraperitoneal pelvic organs.

The mucosa of the rectum when empty is ridged into many longitudinal folds. Several *transverse rectal folds* incorporate also the submucosa and inner circular muscle layers.

ANAL CANAL AND ANUS

The *anal canal* continues the rectum below the pelvic diaphragm to open at the surface of the skin of the perineum in the anal opening or *anus*. The interior of the anal canal is thrown up into longitudinal *anal columns* which are joined together inferiorly by transverse ridges called the *anal valves*. Below this level the mucosa smooths and gradually undergoes a transition to the skin type of epithelium of the anal opening. The inner circular smooth muscle forms the *sphincter ani internus* muscle about the anal canal. The more effective *sphincter ani externus* muscle is formed of voluntary striated muscle.

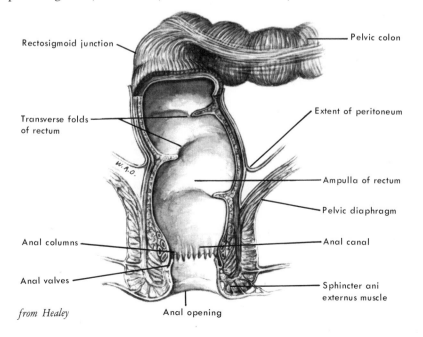

from Healey

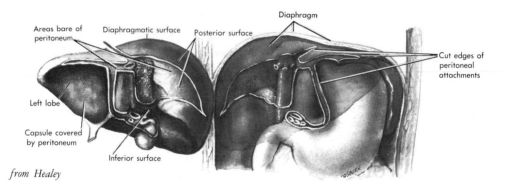

from Healey

ACCESSORY DIGESTIVE ORGANS

The liver and pancreas are glandular organs which have important relationships to the alimentary system because of their exocrine secretions which contribute to digestion. The pancreas also functions as an endocrine gland, and the liver has a number of vital functions which make it one of the more important glands of the body.

THE LIVER

The liver is the largest gland and one of the largest organs of the body. It is located in the right upper portion of the abdomen under the right leaf of the diaphragm and the right costal margin. It presents an extensive *diaphragmatic surface* which is smoothly convex to fit the arch-

ing undersurface of the diaphragm. The inferior or *visceral surface* of the liver faces downward and to the left. This surface is related to the stomach, the first part of the duodenum, right kidney, and right colic flexure of the colon. The visceral surface is shaped to receive the gallbladder. The liver is generally shaped into a large *right lobe,* which makes up the bulk of the organ, and a smaller *left lobe,* which crosses the midline of the abdomen above the stomach. Smaller *quadrate* and *caudate lobes* on the inferior surface are separated by the *porta hepatis,* which is the point of entrance or exit of the main vessels and hepatic ducts. The smaller lobes are really surface projections of the right lobe. Division of the liver into lobes is of no functional significance as all parts of the liver function as one organ. Liver segmentation is found, however. It is based upon the distribution patterns of the hepatic ducts and accompanying blood vessels; this is of surgical rather than functional importance.

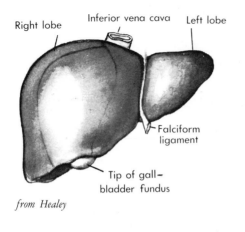

from Healey

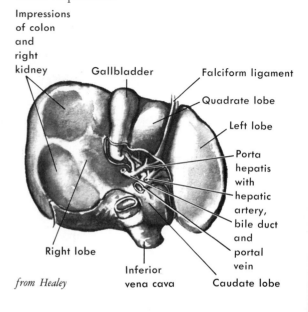

from Healey

Lobular Structure

Within the fibrous capsule which surrounds the whole organ the liver has a remarkably homogeneous structure to the eye. Connective tissue septa are found mainly along the distribution paths of vessels and ducts. Between these the liver appears as a reddish brown, intensely vascular mass. The liver does have an organizational structure, being composed of myriads of microscopic *hepatic lobules*. The hepatic lobule is a hexagonal structure made up of anastomosing plates of *hepatic cells* which radiate outward from a *central vein*. The plates are separated by vascular channels, the *hepatic sinusoids,* whose walls are formed by the liver cells occasionally covered by phagocytic cells. At the angles of the hexagons are strands of connective tissue through which course branches of the portal vein and the hepatic artery. These vessels open into the peripheral ends of sinusoidal channels of a number of adjacent lobules. Blood from the alimentary tract (portal vein) joins freshly oxygenated blood (hepatic artery) to seep through the sinusoids toward the central vein of the lobule. During its passage the blood nourishes the liver cells and the contained food substances are withdrawn or acted upon by the hepatic cells. Many metabolic actions are performed simultaneously within the cytoplasm of the hepatic cells.

Fine *biliary ductules* run between adjacent hepatic cells to receive their exocrine secretion. The bile traverses biliary tributaries to reach biliary ducts which run in the connective tissue strands at the angles of the hexagons making up the *portal triad* with the portal vein and the hepatic artery. (See illustration below.)

The central veins of adjacent lobules form larger and larger tributaries to the hepatic veins which leave the liver at the porta hepatis to join the inferior vena cava.

Biliary System

The *right* and *left hepatic ducts* drain bile from the right and left portions of the liver. These ducts unite to form the *common hepatic duct*. As this duct descends in the gastrohepatic ligament of peritoneum it is paralleled and then joined by the *cystic duct* from the gallbladder to form the *bile duct,* often called the common bile duct. The common hepatic duct above and the bile duct below are paralleled by the portal vein and hepatic artery on their way to the porta hepatis. Bile in the common hepatic duct flows through

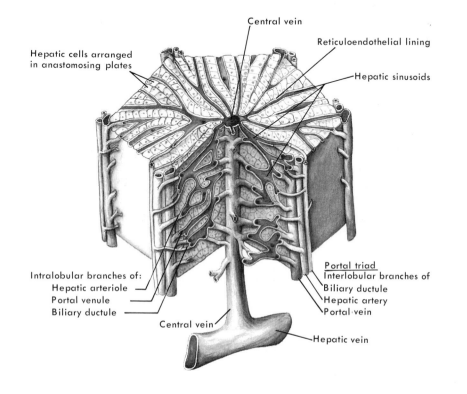

Central vein

Reticuloendothelial lining

Hepatic cells arranged in anastomosing plates

Hepatic sinusoids

Intralobular branches of:
Hepatic arteriole
Portal venule
Biliary ductule

Central vein

Portal triad
Interlobular branches of
Biliary ductule
Hepatic artery
Portal vein

Hepatic vein

the cystic duct into a saclike terminal expansion, the *gallbladder.*

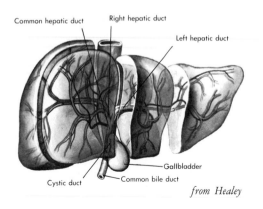

Common hepatic duct

Right hepatic duct

Left hepatic duct

Gallbladder

Cystic duct

Common bile duct

from Healey

The gallbladder hangs downward from the cystic duct in the fossa of the gallbladder on the visceral surface of the liver. Its main part or *body* ends blindly in the expanded *fundus* which reaches or bulges below the inferior margin of the liver. The gallbladder serves as a reservoir because bile is needed in the duodenum only when fatty foods are to be digested. While stored in the gallbladder the bile is concentrated by the absorption of water and salts across the mucosal columnar epithelium. When fats enter the duodenum a substance, *cholecystokinin,* is secreted into the blood. This substance stimulates contraction of smooth muscle in the thin fibromuscular wall of the gallbladder. Bile flows through the cystic duct into the bile duct and is conveyed to the second part of the duodenum.

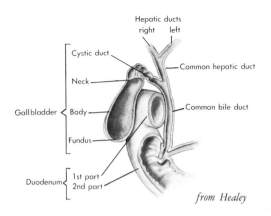

Hepatic ducts
right   left

Cystic duct

Common hepatic duct

Neck

Gallbladder

Common bile duct

Body

Fundus

Duodenum
1st part
2nd part

from Healey

Bile has no enzymes but by its emulsifying action breaks down fat globules into fine droplets which are reduced to fatty acids and glycerol by the intestinal and pancreatic lipases.

The Liver Cell

The hepatic cell was pictured in the first chapter as an example of cells in general because it has so many features in common with many cells of the body. Review of that description will be helpful now. (See illustration on opposite page.)

The large spherical nucleus is located near the center of the cell and contains typical nuclear chromatin, as well as having common nucleolar and membrane structure. The tubular cisterns of both *smooth* and *rough endoplasmic reticulum* make up the submicroscopic framework of the cytoplasm. *Ribosomes* stud the surface of the framework and also are found free in the cytoplasm. Macromolecules of *glycogen* appear in particulate form. The narrow sacs of the *Golgi complex* are located near the tiny radicles of the bile canaliculi at the periphery of the cell. *Mitochondria* abound throughout the cell. Dense vacuoles of *lysosomes,* indicating enzyme activity within the complex cell, are scattered through the cytoplasm. Other puzzling membrane-limited vacuoles, called *microbodies,* which contain crystalline substances are often found near the lysosomes and the mitochondria. Microbody function is not known. Also yet to be clarified by research is the question of how such a generalized cell with almost routine constituents can carry on so many functions.

Hepatic Functions

The liver has many functions other than its exocrine secretion of bile. Some of these are:

1. Reabsorption from the portal venous blood of bile salts after they play their role in the digestion of fats. These are used again in the production of new bile.

2. Conservation of iron-bearing pigments resulting from the breakdown of old red blood cells in the spleen. Since iron is very poorly absorbed from food its conservation and storage by the liver until needed for new red blood cells is very important.

3. Blood plasma proteins, such as globulin and albumin, are synthesized from amino acids. Other amino acids may be stored as part of the complex glycogen molecule.

4. Carbohydrates are stored by conversion of blood glucose to glycogen. When the level of blood glucose falls because of the demand of peripheral tissues, glycogen is broken down and glucose is released into the blood. This is described as an endocrine function of the liver.

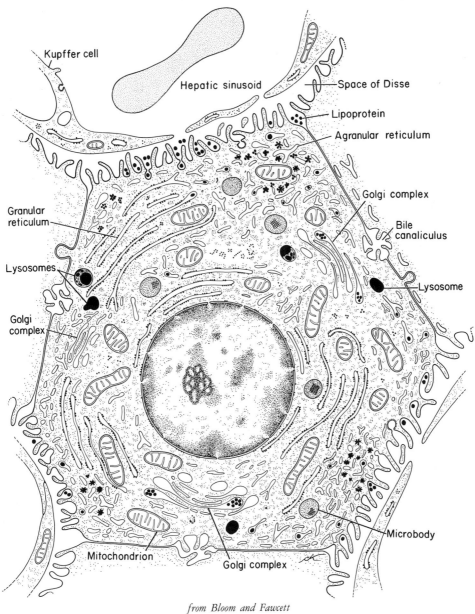

Kupffer cell

Hepatic sinusoid

Space of Disse

Lipoprotein

Agranular reticulum

Golgi complex

Granular reticulum

Bile canaliculus

Lysosomes

Lysosome

Golgi complex

Microbody

Mitochondrion

Golgi complex

from Bloom and Fawcett

ELECTRON MICROSCOPIC APPEARANCE OF LIVER CELL

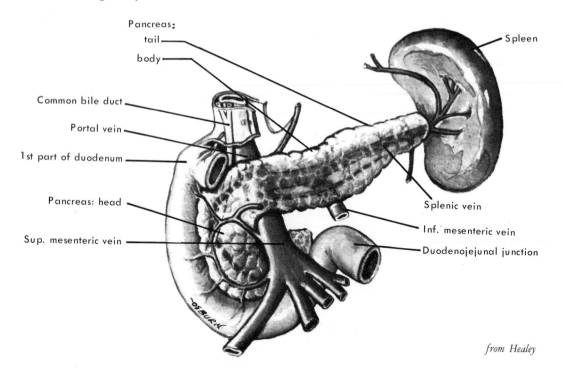

Pancreas:
tail
body
Common bile duct
Portal vein
1st part of duodenum
Pancreas: head
Sup. mesenteric vein

Spleen
Splenic vein
Inf. mesenteric vein
Duodenojejunal junction

from Healey

5. Fats from the circulating blood, entering via the hepatic artery, are stored.

6. Waste products within the blood may be converted into a form suitable for excretion by other organs such as the kidney. The formation of urea from nitrogenous wastes is an example. Other substances are combined with the bile salts for biliary system excretion.

7. Some substances, ingested by the body through the alimentary or respiratory system into the blood or formed by disease processes, are actually injurious to the body tissues. These are detoxified by the liver cells into noninjurious substances by combination with other materials and are excreted in the bile.

8. The synthesis or storage of vitamins and of proteins needed for blood clotting is accomplished.

9. Blood cells are formed during embryonic development.

THE PANCREAS

The pancreas is a thin, flat, elongated organ which lies retroperitoneally against the posterior abdominal wall behind the stomach. The bulkier *head* of the pancreas fills the concavity formed by the first three parts of the duodenum. The *body* extends to the left across the vertebral column, finally tapering into the *tail* of the organ

which reaches across to the spleen. The pancreas presents many flat lobulations of glandular tissue which are related to both exocrine and endocrine functions. The lobules are formed of many secretory acini which are composed of exocrine cells which surround an excretory duct. Secretions containing the pancreatic enzymes for the digestive breakdown of fats, protein polypeptides, and intermediate carbohydrates pass into intralobular and interlobular ducts. Most of the ducts are tributaries to the *pancreatic duct* which runs the length of the gland to the head. This duct joins the bile duct to form the *hepatopancreatic papilla* in the wall of the posteromedial aspect of the second part of the duodenum. The ampulla perforates the duodenal mucosa at the greater duodenal papilla. An *accessory pancreatic duct* is often located in the head of the pancreas.

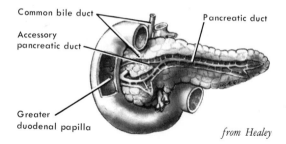

Common bile duct
Accessory pancreatic duct
Pancreatic duct
Greater duodenal papilla

from Healey

Although it may connect with the main duct within the head, the accessory duct usually opens separately into the duodenum. The endocrine part of pancreatic structure and function is described with the endocrine glands as is the microscopic structure of the exocrine portion, in order to avoid repetition in description of the closely related cells.

THE PERITONEUM

The smooth sweep of the peritoneal lining about the anterior and lateral walls of the abdominal cavity is interrupted posteriorly. The intraperitoneal organs of the adult developed behind the peritoneum in the embryo. These organs pushed the peritoneum in front of and around them as they projected forward to gain room and to acquire mobility. Their visceral peritoneal coat was acquired in the process, and they left behind them a trail of reflected peritoneum which, to some extent, positions them and conveys vessels, nerves, and lymphatics from the retroperitoneal area.

A view of the abdomen, cut in the sagittal plane, as illustrated, shows the disposition of the major peritoneal folds and reflections.

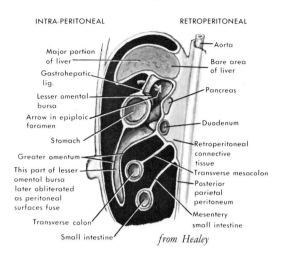

from Healey

Peritoneum from the posterior wall of the abdomen is reflected over the diaphragmatic surface and visceral surface of the liver. These folds form the ligaments of the liver which diverge to enclose a *bare area* where the liver is in direct contact with retroperitoneal connective tissues. The liver is also connected to the anterior abdominal wall by a sickle-shaped peritoneal fold, the *falciform ligament,* which carries the *ligamentum teres* in its free border. The latter is the

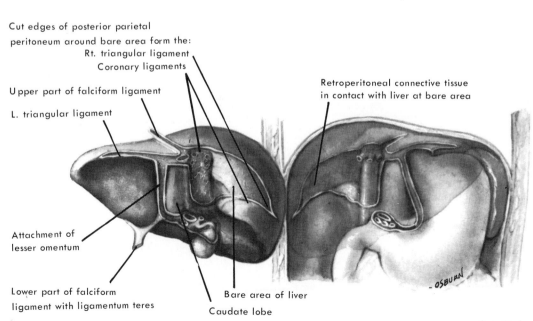

from Healey

obliterated umbilical vein, which carried blood in the embryo from the placenta to the liver. The peritoneal coverings of the liver fuse on its visceral surface to form a sheet, the *lesser omentum,* which extends downward to the lesser curvature of the stomach. The lesser omentum (also known as the gastrohepatic ligament) splits to surround the stomach as its visceral peritoneum. The two layers of peritoneum fuse again along the greater curvature of the stomach to form the anterior layers of an apron of peritoneum and fat called the *greater omentum.*

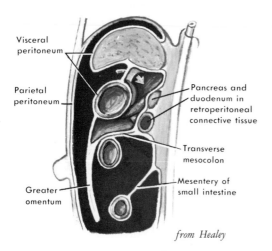

Visceral peritoneum

Parietal peritoneum

Pancreas and duodenum in retroperitoneal connective tissue

Transverse mesocolon

Greater omentum

Mesentery of small intestine

from Healey

The greater omentum hangs downward over the loops and coils of the small intestine. This structure possesses considerable mobility. It is able to gravitate to one area or another within the abdomen and frequently surrounds an inflamed organ. The two anterior layers of peritoneum of the greater omentum turn backward at the border of the apron. They ascend, fused into the posterior surface of the fatty sheet, up to a point below the transverse colon. The posterior layers split away during the ascent, surround the transverse colon, and then ascend to the posterior abdominal wall as the *transverse mesocolon.* A cavity, the *lesser sac* (lesser omental bursa), is formed by the divergence of the posterior layers of peritoneum to surround the transverse colon and form its mesentery. This cavity communicates with the peritoneal cavity (greater sac) at the *epiploic foramen* (foramen of Winslow). The lesser sac is limited in front by the gastrohepatic ligament (lesser omentum) enroute downward to the first part of the duodenum and stomach. The posterior parts of the duodenum and the stomach, therefore, face the lesser sac and form part of its anterior boundary. The anterior layers of the greater omentum form the lower part of the anterior boundary of the lesser sac until fused with the posterior layers. The posterior boundary is made

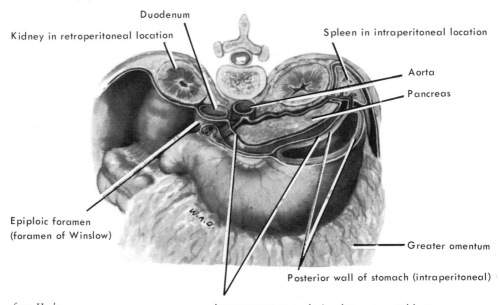

Duodenum

Kidney in retroperitoneal location

Spleen in intraperitoneal location

Aorta

Pancreas

Epiploic foramen (foramen of Winslow)

Greater omentum

Posterior wall of stomach (intraperitoneal)

from Healey

Lesser omentum enclosing lesser omental bursa

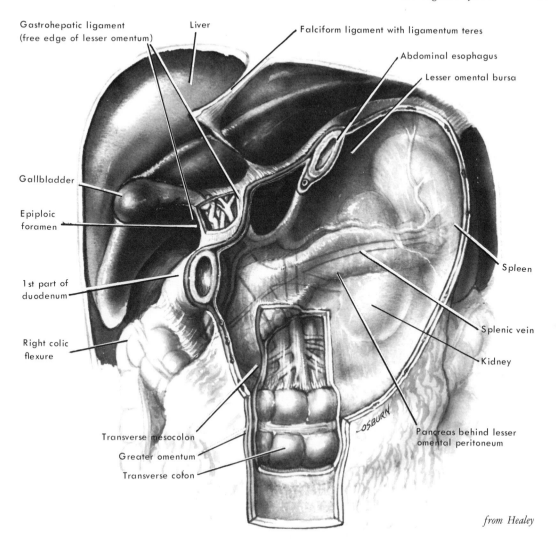

Gastrohepatic ligament
(free edge of lesser omentum)

Liver

Falciform ligament with ligamentum teres

Abdominal esophagus

Lesser omental bursa

Gallbladder

Epiploic foramen

Spleen

1st part of duodenum

Splenic vein

Right colic flexure

Kidney

Transverse mesocolon

Greater omentum

Transverse colon

Pancreas behind lesser omental peritoneum

-OSBURN

from Healey

by the posterior layers of the greater omentum, transverse colon, and transverse mesocolon. The attachment to the posterior abdominal wall covers the pancreas. If the stomach were to be removed, its "bed" would be the posterior wall of the lesser sac.

The posterior parietal peritoneum is again re-flected to form the two layers of the *mesentery of the small intestine,* which with the ileum and jejunum project forward to fill most of the peritoneal cavity. Similar reflections of the peritoneum form the *sigmoid mesocolon* and turn forward over the pelvic organs, dipping between them to form the peritoneal pouches.

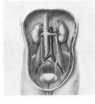

9

The Genitourinary System

Purification of the circulating blood and reproduction of the species are of importance to the continuance of life. It has been seen that one of the functions of the circulatory system is to remove the waste products of metabolism from the tissue fluid. The blood in turn must be relieved of such substances lest they accumulate in amounts which would be toxic to the body or interfere with its other functions. The *urinary system* is a group of organs which serve to excrete from the body those waste products from the blood which are not volatile and cannot be expired through the respiratory system. These organs, like those of the *reproductive system,* develop from cells which are set apart early in the development of the body. The organs of both systems are located in the pelvis and, to some extent, in the abdomen. Both systems have outlets through the perineum. In the male the lower passageways of the two systems are combined. The close structural and functional association of the urinary and reproductive systems justifies their description as parts of the larger *genitourinary system.*

THE URINARY SYSTEM

The *urinary system* consists of a group of organs and passageways by which the waste products

dissolved in the plasma of circulating blood pass through an epithelial membrane and are drained from the body. The *kidneys* are the organs of excretion. (See color plate 11.) The *ureters* are the main ducts which bear the urinary fluid or *urine* away from the kidneys and into the *urinary bladder,* a reservoir which temporarily holds the urine until a sufficient volume is collected. The urine is expelled at intervals under voluntary control through a ductlike passageway, the *urethra,* which runs through the pelvis and perineum to open on the surface of the external genital organs at the *urethral meatus.* The kidneys, ureters, and urinary bladder of the male and female are similar, but the urethra is subject to sexual differences in its course and in its relationships to the organs of the reproductive system and the perineum.

THE KIDNEYS

The *kidneys* are a pair of bean-shaped organs which lie on each side of the vertebral column nestled against the psoas major. These organs are located high on the posterior abdominal wall behind the peritoneum of the upper abdominal cavity. Although the kidneys are retroperitoneal, they are not covered directly by the peritoneum of the posterior abdominal wall. Extraperitoneal

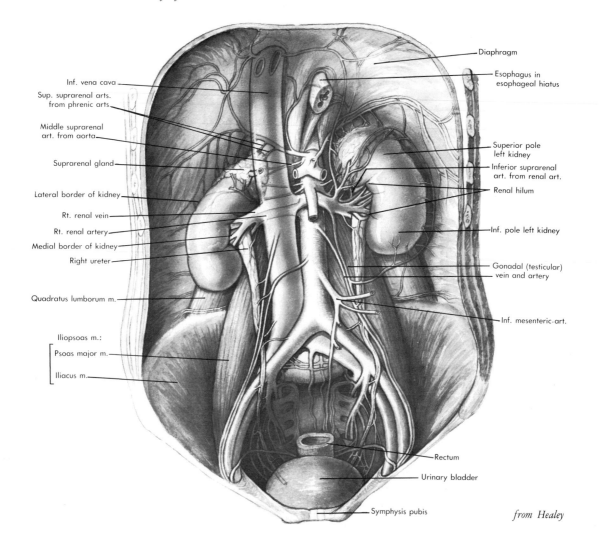

Inf. vena cava

Sup. suprarenal arts. from phrenic arts.

Middle suprarenal art. from aorta.

Suprarenal gland

Lateral border of kidney

Rt. renal vein

Rt. renal artery

Medial border of kidney

Right ureter

Quadratus lumborum m.

Iliopsoas m.:

Psoas major m.

Iliacus m.

Diaphragm

Esophagus in esophageal hiatus

Superior pole left kidney

Inferior suprarenal art. from renal art.

Renal hilum

Inf. pole left kidney

Gonadal (testicular) vein and artery

Inf. mesenteric art.

Rectum

Urinary bladder

Symphysis pubis

from Healey

connective tissue splits at the lateral border of each kidney to form anterior and posterior layers of *renal fascia*. These layers in turn enclose a large collection of adipose tissue called the *perirenal fat* which cushions the kidney and forms the bed in which it lies.

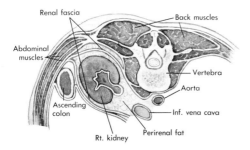

Renal fascia

Back muscles

Abdominal muscles

Vertebra

Aorta

Ascending colon

Inf. vena cava

Rt. kidney

Perirenal fat

Appearance and Relations

The kidney has a long convex lateral border and a shorter indented medial border. These borders meet at the rounded upper and lower poles. The upper pole is capped by the *suprarenal (adrenal) gland* which is tilted medially. The renal artery enters the kidney, and the renal vein and ureter leave it at the *hilum,* which is the deep indentation of the medial border. The relationships of the kidneys differ in accordance with the structures which surround them on the right and left sides of the abdomen. The upper half of the *right kidney* faces toward the liver. The lower half is overlapped anteriorly by the hepatic flexure of the colon laterally, the second part of the duodenum medially, and the

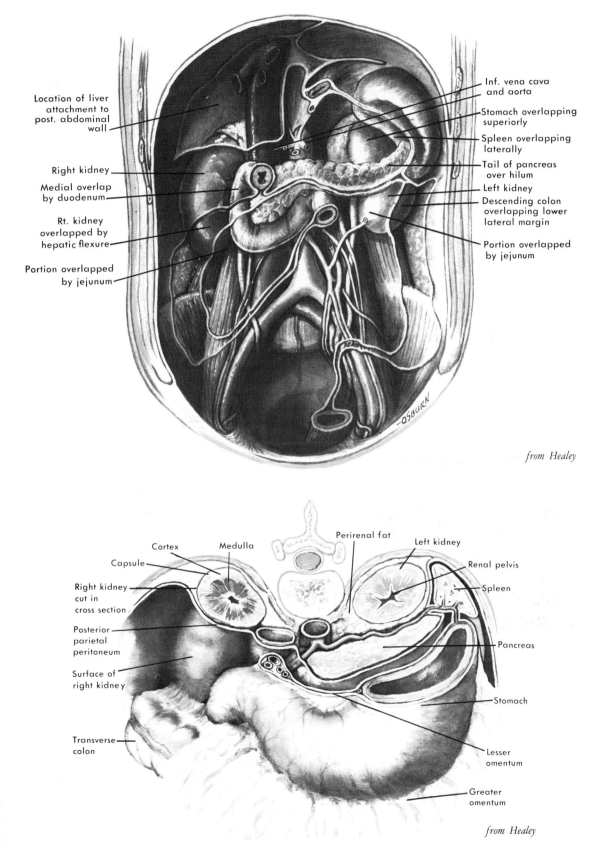

Location of liver attachment to post. abdominal wall

Right kidney

Medial overlap by duodenum

Rt. kidney overlapped by hepatic flexure

Portion overlapped by jejunum

Inf. vena cava and aorta

Stomach overlapping superiorly

Spleen overlapping laterally

Tail of pancreas over hilum

Left kidney

Descending colon overlapping lower lateral margin

Portion overlapped by jejunum

from Healey

Cortex

Medulla

Capsule

Perirenal fat

Left kidney

Renal pelvis

Spleen

Right kidney cut in cross section

Posterior parietal peritoneum

Surface of right kidney

Pancreas

Transverse colon

Stomach

Lesser omentum

Greater omentum

from Healey

RELATIONSHIPS OF THE KIDNEYS

jejunum at the inferior pole. The upper half of the *left kidney* is covered by the stomach medially and the spleen laterally. The tail of the pancreas crosses the hilum. The jejunum lies in front of the lower half of the organ with the descending colon along the lateral margin. The abdominal aorta and the inferior vena cava course longitudinally between the two kidneys. The medial border of the right kidney faces the inferior vena cava; that of the left kidney faces the aorta. These relations produce differences in length of the renal vessels which pass transversely to each hilum. The left renal vein is considerably longer than the right.

Gross Structure

The external surface of the kidney is smooth and covered by a fibrous capsule. The kidney tissue bulges over the hilum from all sides, but when the organ is opened to show its internal structure, it is apparent that the hilum reaches deep into the center. The cavity thus formed is the *renal sinus,* a pocket filled with fat in which the renal vessels and nerves course on their way into or out of the kidney. The *ureter* also lies within the renal sinus. The ureter drains a funnel-shaped expansion, the *renal pelvis,* from which tubular projections, the *major calyces,* reach toward the glandular part of the kidney. Each major calyx divides into several *minor calyces.* The calyces and renal pelvis are parts of an internal drainage envelope by which urine, secreted by the glandular part of the kidney, is collected and conveyed into the ureter.

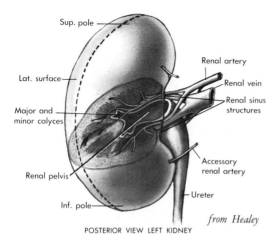

POSTERIOR VIEW LEFT KIDNEY

from Healey

The glandular portion of the kidney which covers and surrounds the structures of the renal sinus is divided into medullary and cortical portions. The *medulla* is composed of a series of bluntly conical masses called the *renal pyramids.* The rounded apex of each pyramid, perforated by the openings of ducts, projects into a minor calyx. A number of longitudinal striations which converge upon the tip of the pyramid indicate the presence of tubules and accompanying blood vessels. The *cortex* of the kidney is located between the bases of the pyramids and the capsule at the surface. Cortical substance, composed of tangled knots of blood vessels and tortuous tubules, also extends between adjacent pyramids to form the *renal columns.*

Mechanisms of Urine Formation

The *nephron* is the structural and functional unit of the kidney which actually forms the urine. Millions of nephrons connect with collecting tubules which course through the pyramids to transfer the urine to the calyces.

The Nephron. The blind end of each nephron, located in the cortex, is the beginning of the tubular structures concerned with urine formation. Here an epithelial sac, the *glomerular capsule* (Bowman's capsule), is invaginated by a knot of capillaries to form the *renal corpuscle.* The tangled capillary loops, representing the termination of an afferent arteriole from a branch of the renal artery, are termed the *renal glomerulus.* The invagination of the glomerular capillary tufts into the epithelial sac causes the blood to be separated from the cavity of the sac by only a double layer of flat epithelial cells. One layer is the capillary wall and the other is the epithelium of the sac itself whose cavity is reduced to a cleftlike space. The capillaries re-form into an efferent arteriole which leaves the renal corpuscle.

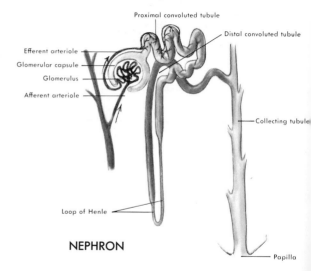

NEPHRON

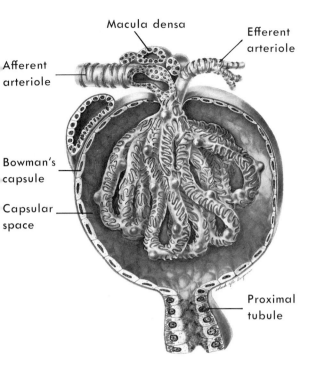

Macula densa

Efferent arteriole

Afferent arteriole

Bowman's capsule

Capsular space

Proximal tubule

from Bloom and Fawcett

RENAL GLOMERULUS AND JUXTAGLOMERULAR APPARATUS

Microscopic Structure. The electron microscope shows the relationship of the epithelial membrane and the endothelium of the glomerular capillary to be much more complicated than that of two abutting flat epithelial layers.

The epithelial cells investing the capillary are actually small bits of central cytoplasm, containing a nucleus, from which a series of *foot processes* radiate. The cells, termed *podocytes*, line up along

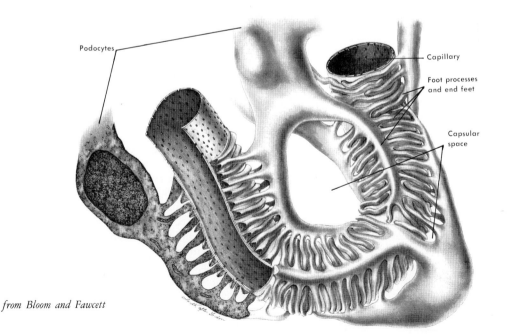

Podocytes

Capillary

Foot processes and end feet

Capsular space

from Bloom and Fawcett

ENLARGED VIEW OF GLOMERULUS

the capillary and surround it. Secondary *pedicles* (*end feet*) extend from the process to touch the capillary. The attachment of many secondary end feet to the capillary creates many small openings of the capsular space (Bowman's capsule) between them.

At the attachment of the podocytic (epithelial) end feet to the capillary endothelium, a *basement lamina* layer intervenes between the endothelial cell and the epithelial cell. The endothelial lining of the capillary is broken at intervals by circular *pores.*

Without becoming exceedingly complex, one may state that all of the structures mentioned combine to create a filtering mechanism which permits electrolytes and small molecular substances to escape from the capillary blood through the endothelial pores and basement lamina, between podocytic end feet, into Bowman's capsule.

Each glomerular capsule leads into a complicated and tortuous tubule, the successive parts of which are separately named and have characteristic locations. The *proximal convolution* (proximal convoluted tubule) twists and turns about in the cortex near the renal corpuscle and then straightens on a course into the medulla. Here it tapers into the U-shaped *loop of Henle.* The caliber of the loop in its *descending limb* is narrow but thickens as the *ascending limb* climbs back up toward the renal corpuscle. When the tube reaches its own glomerulus it becomes the *distal convolution* (distal convoluted tubule). This part of the tube pursues a tortuous path in the cortex adjacent to and above its renal corpuscle. The efferent arteriole breaks up into secondary capillary plexuses about both the proximal and distal convolutions before joining tributaries to the renal vein.

The epithelial wall of the tubular portion of the nephron varies from thick cuboidal to very thin squamous as the gross description changes from thick to thin. Under the light microscope, the cells of the proximal convoluted tubule are characterized by a dense *brush border* next to the lumen. With the electron microscope, the brush border is found to consist of closely packed *microvilli.* Many minute *canaliculi* invaginate the cell surface and lead inward to vacuoles in the cytoplasm. It is believed that this appearance represents a system for the selective reabsorption of protein molecules from the filtrate of the glomerulus.

Collecting Tubules.   The nephron ends as the distal convolution straightens into a link which connects to the first of a series of *collecting tubules.* These are the excretory ducts of the kidney which can be compared to those of any gland since they convey away the products of the secretory portion. Collecting tubules leave the cortex to course in the medullary pyramid, joining with others and receiving the terminations of deeper nephrons. Larger and larger collecting tubules converge upon the apex of the pyramid. *Papillary ducts* are formed by final junctions of the largest collecting tubules. These ducts open into a minor calyx.

Renal Function

The millions of renal glomeruli form, in composite, a tremendous blood filtering mechanism. All of an individual's blood passes through the kidneys about 15 times an hour. The filtering process, across the double epithelial membrane of the renal glomerulus, is selective in that blood cells, fatty substances, and the large molecular plasma proteins are retained within the capillaries. The glomerular filtrate within the lumen of Bowman's capsule is not yet urine. It is a fluid containing in watery solution all the constituents of the blood, other than those just mentioned, in about the same concentration as before filtration. Water may be present in excess but a certain amount must be restored to the blood. Electrolytes and nutritional molecules, in amounts necessary to preserve the composition of the internal environment and to supply the needs of tissue metabolism, must be reclaimed from the filtrate. Metabolic wastes in the form of compounds such as urea have been removed from the blood by filtration into the glomerular fluid. Many drugs, as well as substances toxic to the body, are removed from the blood by this route.

The glomerular filtrate passes along the various tubular portions of the nephron, all of which have an intimate relationship to capillary vessels. During the passage of the glomerular fluid the epithelial cells lining the nephron are active in selectively reabsorbing elements needed by the body and in rejecting wastes. In the early phases of passage the proximal convolution absorbs glucose and chemical ions, such as potassium, sodium, hydrogen, chlorides, and bicarbonates. Further absorption of these elements along the more distal portions of the nephron restores the blood in the accompanying capillaries to a precise balance of fluid and electrolytes. Nutritional substances are reabsorbed but not in excess of normal blood levels; this accounts for the appearance of glucose in the urine when blood sugar levels are high. The tubules may also secrete substances not filtered by the glomerulus. Certain dyes, handled only by the tubules, may be given to patients to test renal function by measurement of the amount appearing in the urine.

The last reabsorption of water is carried out both in the distal convolution and in the smaller collecting tubules. The residual water bearing excreted wastes, now correctly termed urine, is conveyed to the renal pelvis. An occasional epithelial cell is swept along in the urine. The appearance of large numbers of epithelial cells, blood cells, or cellular debris is an indication of disease of the urinary system. In microscopic and chemical examinations of the urine (*urinalysis*) these elements are watched for. The appearance of large molecular proteins, such as albumin, is an indication that damage to the renal filter by disease is permitting them to escape from the blood.

Two structural features, one in the afferent arteriole and one in the distal convoluted tubule beside it are given the joint name of the *juxtaglomerular apparatus*. It has been found that certain smooth muscle cells of the arteriole which abut the distal tubule are modified by the presence of a well-defined *Golgi complex* and by specific granules which appear from the sacs of the Golgi complex and affix themselves to a granular endoplasmic reticulum. The tubular cells beside the arteriole are thickened to form the *macula densa*. It is believed that some sort of enzyme activity in the tubule's macula densa cells influences a secretory process in the juxtaglomerular cells of the arteriole. The secretory product, *renin*, is known to have the effect of raising blood pressure (vasopressor effect).

URETERS

The *ureter* is the excretory duct of the kidney which runs a retroperitoneal course to the urinary bladder. At the hilum the renal pelvis narrows to form a funnel which leads into the tubular ureter. The highly muscular wall of the ureter is composed of inner longitudinal and outer circular coats of smooth muscle. The mucosa, lined by a transitional type of stratified epithelium, is thrown into longitudinal folds by the contraction

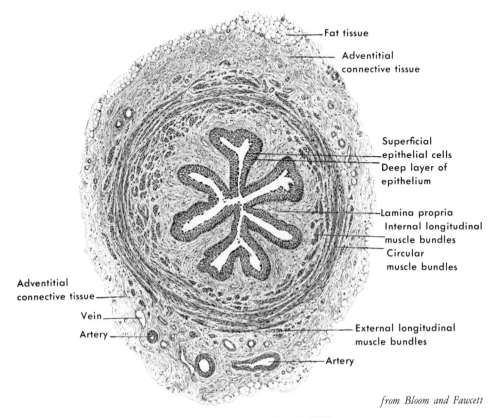

Fat tissue

Adventitial connective tissue

Superficial epithelial cells

Deep layer of epithelium

Lamina propria

Internal longitudinal muscle bundles

Circular muscle bundles

Adventitial connective tissue

Vein

Artery

External longitudinal muscle bundles

Artery

from Bloom and Fawcett

CROSS SECTION OF URETER

of the musculature. As urine flows from the renal pelvis it is carried along the ureter by waves of muscular contraction of the walls. The strength of this contraction causes the urine to enter the bladder in spurts rather than as a steady flow.

In its 10 inches of length the ureter lies in both the abdomen and the pelvis. As it leaves the hilum of the kidney the ureter is behind the renal vein and artery. The tube then arches downward on the psoas major close to the lower pole of the kidney. The *right ureter* starts its descent behind the second part of the duodenum and then passes under the root of the mesentery of the small intestine. The *left ureter* descends behind the sigmoid colon and its mesocolon. Both ureters are crossed by the reproductive artery (testicular or ovarian) of their side. At the brim of the pelvis the ureters cross over the common iliac artery and then run along the lateral wall of the pelvis before turning medially to enter the urinary bladder.

URINARY BLADDER

The *urinary bladder* is a muscular sac which lies in the pelvis behind the pubic bones from which it is separated by the potential connective tissue-filled *retropubic space*. The bladder is characterized by its distensibility. When it is empty its thick muscular walls collapse upon each other and throw the mucous membrane into many folds. As the organ fills with urine its walls separate, the superior surface expands into a dome, and the mucous membrane becomes smooth. Then the bladder rises upward from the pelvis, is draped

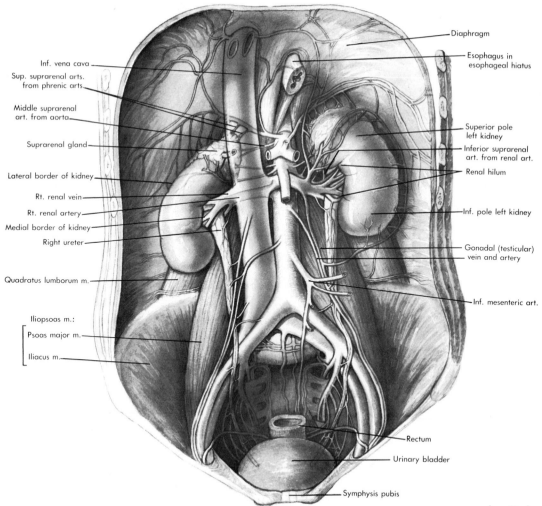

from Healey

superiorly and posteriorly by peritoneum, and comes into relation with coils of the small intestine. In the male the bladder presses posteriorly against the rectum from which it is separated by the *rectovesical pouch*.

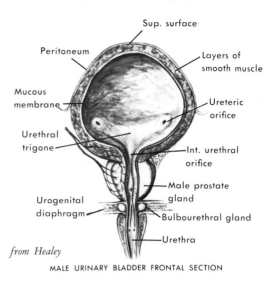

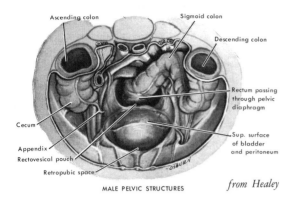

from Healey

MALE PELVIC STRUCTURES

from Healey

MALE URINARY BLADDER FRONTAL SECTION

In the female the bladder is related to the anterior wall of the uterus with the *vesicouterine pouch* intervening between them.

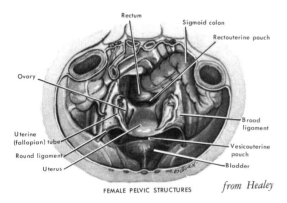

FEMALE PELVIC STRUCTURES   *from Healey*

The internal surface of the floor of the bladder presents a smooth triangular area, the urethral *trigone*. Openings appear at each of the three angles of the trigone. In front is the *internal urethral orifice* through which urine leaves the bladder. The ureters strike obliquely through the muscular wall to open at the posterolateral angles of the trigone through the elliptical *ureteric orifices*. The muscular wall is composed of interlacing bundles of smooth muscle which, because of interweaving, are only incompletely separated into inner longitudinal, middle circular, and outer longitudinal layers. All the musculature is called, in composite, the *detrusor urinae* muscle because the contraction of the bladder wall provides the motive power for the expulsion of the urine.

The bladder musculature is supplied by parasympathetic nerve fibers. The smooth muscle relaxes to permit the bladder to fill. The stretching of the muscle stimulates sensory nerve endings which give rise to sensations of fullness. Continued filling initiates other impulses which, by spinal reflexes, induce rhythmic contractions of the musculature and sensations of urgency. Although the expulsion of urine is under voluntary control, this control is exerted over the sphincter mechanism of the urethra. When the urethral sphincter is relaxed, involuntary reflex stimuli cause powerful contractions of the smooth muscle of the bladder to empty the organ.

Urine passes from the bladder to the external surface of the body through the *urethra*. The course and relationships of this tube differ in the male and the female. Understanding of the morphological arrangements is achieved more easily after knowledge of the structures closing the pelvic outlet and of the reproductive system has been gained.

STRUCTURES CLOSING THE PELVIC OUTLET

The pelvic outlet is closed by several fibromuscular membranes. Alimentary, urinary, and reproductive organs pass through these membranes to reach the surface of the body.

PELVIC DIAPHRAGM

A muscular sheet, covered by fascia on its superior and inferior surfaces, extends transversely

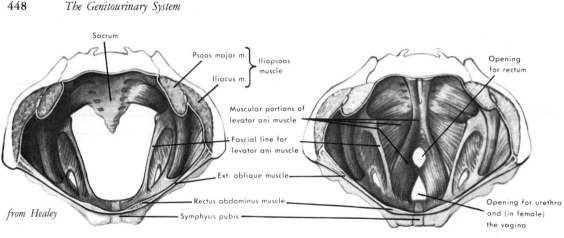

from Healey

PELVIC OUTLET WITHOUT LEVATOR ANI   PELVIC OUTLET CLOSED BY PELVIC DIAPHRAGM

across the floor of the pelvis to form the *pelvic diaphragm.* The major components of this diaphragm are the *levator ani muscles.* The levator ani of each side arises from the posterior surface of the body of the pubic bone and from a fascial line along the lateral wall of the pelvis extending to the ischial spine. From this wide origin the fibers pass in hammock fashion, downward and backward toward the midline as far posteriorly as the coccyx. Although some of the more posterior fibers reach the coccyx for insertion, most of the fibers insert into a tough fibrous structure, the *perineal body,* which lies in the midline between the urethra and the rectum in the male and between the vagina and rectum in the female. Other fibers insert about the walls of the rectum or between it and the coccyx into the anococcygeal ligament. Innervated by the sacral nerves, the levator ani muscles form a sling which assists in closing the pelvic outlet, supports the pelvic viscera, and participates in the regulation of intra-abdominal pressure.

PERINEUM

The curving inferior end of the trunk of the body between the lower limits of the abdomen anterosuperiorly, the inner aspects of the thighs and buttocks laterally, and the coccyx behind is known as the *perineum.* When the thighs are spread the perineum is revealed as a diamond-shaped area. The underlying boundaries of the perineum are the symphysis pubis in front, the ischiopubic ramus of the hip bone on each side anterolaterally, the ischial tuberosity laterally, and the sacrotuberous ligament on each side running

posterolaterally. The coccyx forms the posterior landmark. If a line is drawn transversely between the ischial tuberosities the diamond-shaped perineum becomes divided into two triangular areas. (See illustration at top of following page.) The anterior half is known as the *urogenital triangle* because of its relationship to the urinary and reproductive organs. The posterior half is termed the *anal triangle.*

Urogenital Triangle of the Male

The *urogenital triangle* lies external to the anterior half of the pelvic diaphragm. Just below (external to) the inferior fascia of the pelvic diaphragm is a space called the *deep perineal space (pouch).* This is not a space at all, for it is filled with important structures which together form the strong *urogenital diaphragm.* A delicate superior fascia of the urogenital diaphragm bounds the deep perineal pouch above and a strong inferior fascia (perineal membrane) limits it externally. Between these two fascial layers are the muscles which give substance to the diaphragm. (See illustration at bottom of following page.) The *deep transverse perinei muscle* extends broadly from the ramus of the ischium on each side to the perineal body which it supports and fixes in position by its contraction. The *sphincter urethrae muscle* is a narrower band of muscle fibers which passes within the deep pouch from the pubic ramus of each side toward the midline. The fibers from each side interdigitate with each other to form a sphincter about the urethra.

As the urethra leaves the floor of the bladder, it descends through the *prostate gland* which,

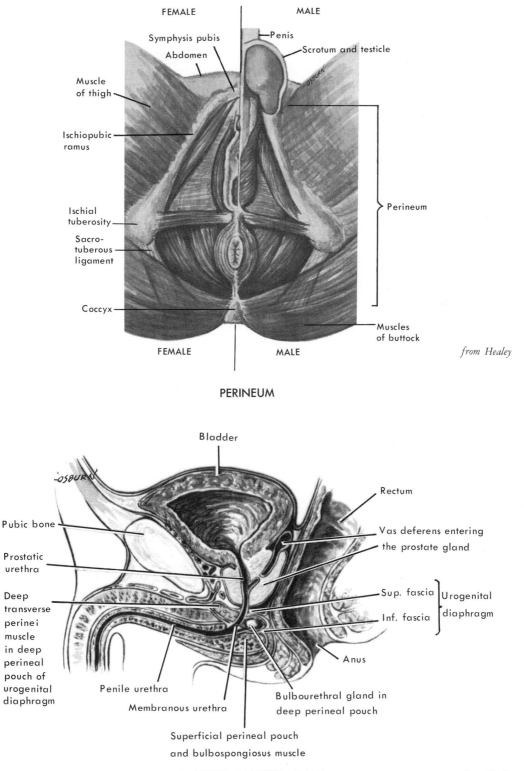

FEMALE   MALE

Symphysis pubis

Penis

Abdomen

Scrotum and testicle

Muscle of thigh

Ischiopubic ramus

Perineum

Ischial tuberosity

Sacro-tuberous ligament

Coccyx

Muscles of buttock

FEMALE   MALE

from Healey

PERINEUM

Bladder

Rectum

Pubic bone

Vas deferens entering the prostate gland

Prostatic urethra

Sup. fascia

Inf. fascia

Urogenital diaphragm

Deep transverse perinei muscle in deep perineal pouch of urogenital diaphragm

Anus

Penile urethra

Bulbourethral gland in deep perineal pouch

Membranous urethra

Superficial perineal pouch and bulbospongiosus muscle

SAGITTAL SECTION: MALE   *from Healey*

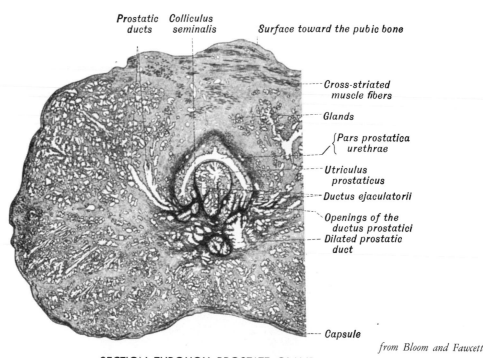

Prostatic ducts   Colliculus seminalis   *Surface toward the pubic bone*

----Cross-striated muscle fibers

---Glands

{Pars prostatica urethrae

----Utriculus prostaticus

---Ductus ejaculatorii

Openings of the ductus prostatici

---Dilated prostatic duct

----Capsule

from Bloom and Fawcett

SECTION THROUGH PROSTATE GLAND

developing as a series of glands in the wall of the urethra, expanded to form a separate organ through which the urethra seems to penetrate. (See above illustration.) It is here that ducts of the male reproductive organs join the *prostatic urethra.* From this point on, the urethra of the male is a combined urinary and reproductive passageway. Below the prostate gland the urethra descends between the medial margins of the levator ani muscles and into the deep perineal pouch. The urethra perforates the urogenital diaphragm and receives the encircling sphincter urethrae muscle fibers. Between the membranous fasciae of the deep pouch the tube is renamed the *membranous urethra.* Other occupants of the deep pouch are the *bulbourethral glands* whose ducts empty below the urogenital diaphragm into the *penile urethra.*

Below (external to) the urogenital diaphragm is the *superficial perineal space (pouch).* This pouch is limited by fasciae which attach to the urogenital diaphragm and ischiopubic ramus, blend with fasciae of the male genital organs, and merge with the fasciae of the abdominal wall. (See illustration at top of page 451.) The root of the penis is formed within the superficial pouch. The *bulbospongiosus (bulbocavernosus) muscles* arise from the perineal body and wrap around each side of the

deep part (bulb) of the penis. Contraction of these muscles helps expel urine or reproductive secretions and maintains erection of the penis by slowing the return of blood from this organ. The *ischiocavernosus muscles* extend forward from the ischiopubic ramus to surround the penis. Their function is similar to that of the preceding muscle. A poorly developed transverse band of fibers forms the *superficial transversus perinei muscles* as they extend from the ischial tuberosity on each side to the perineal body.

Urogenital Triangle of the Female

Similar structures and fibromuscular arrangements are found in the urogenital triangle of the female. These are modified, however, by the passage of the vagina through the perineum and by the presence of the external genital organs which are described with the reproductive organs. (See illustration at bottom of page 451.) The *urogenital diaphragm* is almost divided by the passage of the vagina into whose walls many of the fibers of the deep transversus perinei muscles insert as well as into the perineal body. The fibers of the sphincter urethrae do not encircle the urethra. (See illustration at top of page 449.) In the superficial pouch the bulbospongiosus muscles are sep-

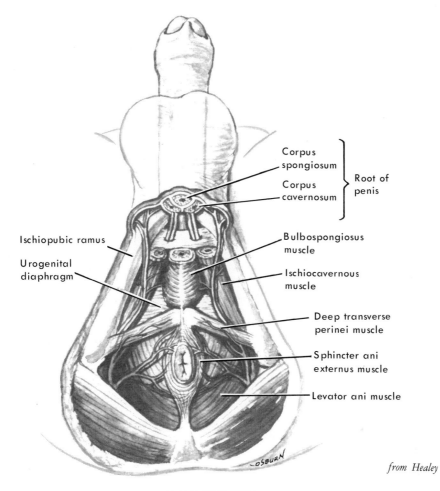

Corpus
spongiosum

Corpus
cavernosum

} Root of
penis

Ischiopubic ramus

Urogenital
diaphragm

Bulbospongiosus
muscle

Ischiocavernous
muscle

Deep transverse
perinei muscle

Sphincter ani
externus muscle

Levator ani muscle

from Healey

MALE PERINEUM

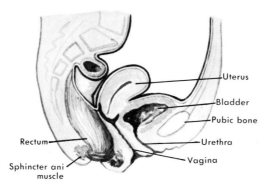

Uterus

Bladder

Pubic bone

Rectum

Urethra

Vagina

Sphincter ani
muscle

SAGITTAL SECTION OF FEMALE PELVIS

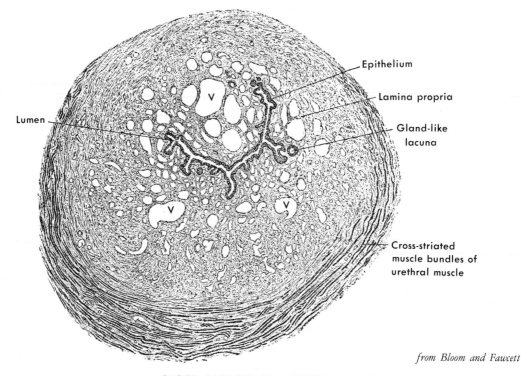

from Bloom and Fawcett

CROSS SECTION OF URETHRA

arated by the vagina around which they pass to insertions upon the clitoris and the pubic arch. Only limited constriction of the vagina results from their contraction.

Anal Triangle

In neither sex is the *anal triangle* as complicated as the urogenital triangle. The *anus* externally marks the location of the central occupant, the *anal canal,* which is surrounded by the fibers of the *sphincter ani externus.* On each side of the anal canal is the deep fat-filled *ischiorectal fossa.* Its boundaries are the inferior layer of fascia of the pelvic diaphragm superiorly, the obturator fascia covering the obturator internus muscle of the pelvic wall laterally, and the anal canal medially. The two fossae communicate behind the anal canal and around the anal sphincter muscle. In addition to containing the ischiorectal fat pad the fossa transmits internal pudendal vessels and nerves through a fascial canal (pudendal, Alcock's canal) on the lateral wall. Inferior rectal vessels and nerves reach the anal canal by traversing the fat of the ischiorectal fossa.

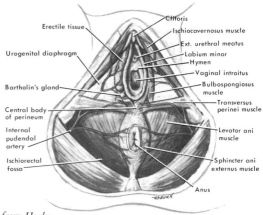

from Healey

THE REPRODUCTIVE SYSTEM

The reproductive system in both sexes is composed of the organs concerned with the continuity of life by the production of offspring. The male reproductive system consists of organs and ducts by which male sex cells are generated, stored, and mixed with glandular secretions which are ejaculated into the female vagina during coitus. In addition to the vaginal passageway the female reproductive system consists of organs in which the female sex cells are formed and transported to other organs where fertilization takes place and the new individual develops.

MALE REPRODUCTIVE SYSTEM

The male reproductive organs are described in the order in which the male sex cells, the spermatozoa, are formed and progress to the male copulatory organ, the penis.

Formation of Spermatozoa

The spermatozoa are developed within the male gonads which are the two *testicles.*

Origin and Descent of the Testicle. The *testicle,* containing primitive sex cells which were set apart early in the development of the embryo, originally forms as a retroperitoneal organ high on the posterior abdominal wall. It is connected by a fibromuscular cord, the *gubernaculum testis,* to the tissues of the scrotal swelling on the surface of the perineal region.

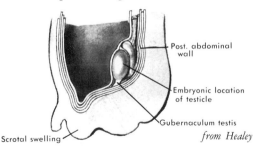

from Healey

Prior to birth the testicle follows or is drawn by the gubernaculum into the pelvis until it reaches the deep inguinal ring. A projection of peritoneum, the *processus vaginalis,* precedes the testicle through the abdominal wall into the scrotal sac which has developed from the skin of perineum to receive the testicle.

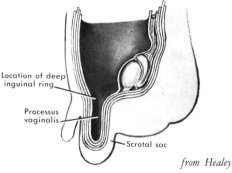

from Healey

The testicle completes its descent by migrating through the inguinal canal, following the processus vaginalis into the scrotum. The testicle invaginates the peritoneal process and thus receives a peritoneal investment although now distant from the abdominal cavity. Usually the continuity of the processus vaginalis with the abdominal peritoneum is obliterated. If it is not, a peritoneal track remains which can be followed by a loop of intestine through the inguinal canal to form a congenital hernia. The testicle draws along from its original location its blood vessels and nerves. These structures become, with the duct of the testicle, constituents of the *spermatic cord* which reaches back through the inguinal canal to the pelvis.

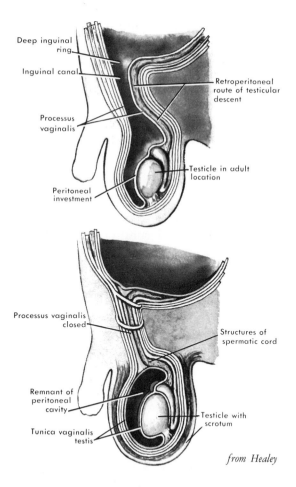

from Healey

Structure of the Testicle and Scrotum. The *scrotum* is a sac of skin and tela subcutanea which hangs downward from the junction of the lower abdominal region and the perineum. Its function is to hold the paired testicles outside the body

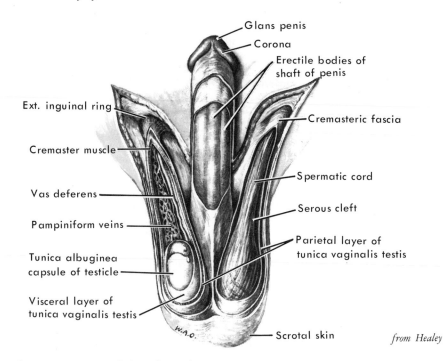

Glans penis
Corona
Erectile bodies of shaft of penis
Ext. inguinal ring
Cremasteric fascia
Cremaster muscle
Spermatic cord
Vas deferens
Serous cleft
Pampiniform veins
Parietal layer of tunica vaginalis testis
Tunica albuginea capsule of testicle
Visceral layer of tunica vaginalis testis
Scrotal skin

from Healey

whose internal temperature is believed to be higher than the optimum for sperm maturation. Smooth muscle fibers, composing the *dartos muscle,* within the tela subcutanea contract to produce a corrugated appearance in the scrotal skin. The interior of the scrotum is divided into a compartment for each testicle. Each compartment is lined by the portion of the processus vaginalis which remained within the scrotum when its connection to the abdominal cavity was obliterated. This serous peritoneal membrane, now called the *tunica vaginalis testis,* reflects over the testicle to provide a serous outer covering. A narrow cleft moistened by fluid intervenes between the parietal layer of the tunica vaginalis lining the scrotum and the visceral layer over the

testicle. Smooth movement of the testicle within the scrotum is thus insured.

Each *testicle* is a smooth ovoid body suspended within the scrotum by the spermatic cord. Under the tunica vaginalis is the tough collagenous *tunica albuginea* which is the capsule of the testicle. The interior of the testicle is divided incompletely into lobules by fibrous septa which extend from the tunica albuginea toward the posterior border of the organ upon a fibrous mass termed the *mediastinum testis.* Each lobule is filled with the tortuous *convoluted seminiferous tubules* in which the spermatozoa are formed. Each tubule straightens and unites with others as they pass into the mediastinum where a network of canals, the *rete testis,* is formed.

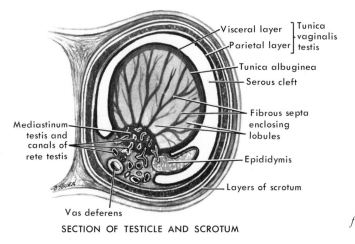

Visceral layer   ⎤ Tunica
Parietal layer   ⎦ vaginalis testis
Tunica albuginea
Serous cleft
Fibrous septa enclosing lobules
Mediastinum testis and canals of rete testis
Epididymis
Layers of scrotum
Vas deferens

from Healey

SECTION OF TESTICLE AND SCROTUM

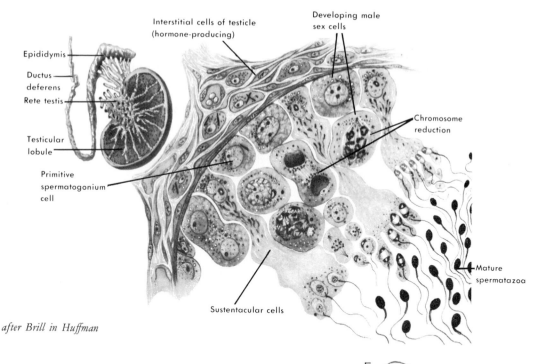

Interstitial cells of testicle (hormone-producing)

Developing male sex cells

Epididymis

Ductus deferens

Rete testis

Chromosome reduction

Testicular lobule

Primitive spermatogonium cell

Mature spermatazoa

Sustentacular cells

after Brill in Huffman

Efferent ductules lead out of the rete testis to enter into a coiled C-shaped structure, the *epididymis,* which curves along the posterior margin of the testicle to its inferior pole. The epididymis is a storehouse for sperm cells.

Development of Spermatozoa. The epithelium of the seminiferous tubules is a complex cellular mass in which the male sex cells in all stages of development are supported by pillarlike sustentacular cells. Original primitive sex cells undergo constant division from puberty onward to form the parent male sex cells, spermatogonia, which lie deepest in the epithelium. A complicated process of histological development ensues as cells move toward the lumen of the tubule. Two main processes occur in spermatogenesis: the number of chromosomes in the developing sperm cells is reduced to half and a highly motile cell form is produced. The mature *spermatozoon* is found in the lumen of the tubule and at the surface of the epithelium. It is a minute organism with a flattened oval *head* which is full of chromosomal material. The long filamentous *tail* is joined to the head by the *middle* piece in which a spiral or helical arrangement of mitochondria surrounds a longitudinal core. The spermatozoa swim within secretions of the male genital tract by lashing movements of the tail.

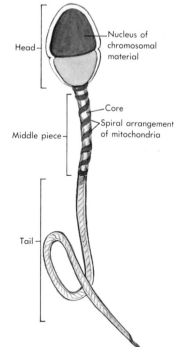

Head

Nucleus of chromosomal material

Core

Spiral arrangement of mitochondria

Middle piece

Tail

Ductus Deferens and Spermatic Cord

The *ductus deferens* (vas deferens) is the excretory duct of the testicle. (See color plate 11.) This thick-walled fibrous tube ascends along the medial side of the testicle to join the bundle of structures which compose the spermatic cord.

After passing through the inguinal canal the ductus deferens pursues an independent course under the peritoneum of the lateral wall of the pelvis. After crossing over iliac vessels and the ureter, the duct turns medially and downward toward the posterior surface of the bladder.

The *spermatic cord,* in addition to the ductus deferens, includes an artery and vein supplying this tube—the testicular artery, lymphatic vessels, and nerves of the testicle—and an extensive *pampiniform plexus* of tortuous veins. Fasciae derived from the transversalis fascia of the abdomen and from muscles of the abdominal wall were carried along during the descent of the testicle to become the connective tissue coverings of the spermatic cord.

Associated Glands and Ducts

As the ductus deferens arches medially over the ureter, it becomes dilated into its *ampulla* which lies along the posterior surface of the bladder also medial to the long pouched *seminal vesicle.* The latter is a glandular outgrowth of the ductus deferens. The seminal vesicle contributes a secretion to the seminal fluid in which the spermatozoa are suspended to form the *semen.* Both the seminal vesicle and the ampulla narrow at the base of the bladder and join to form the *ejaculatory duct* which enters the base of the prostate gland.

The *prostate gland,* located just below the bladder, surrounds the beginning of the urethra from whose walls it originally developed. Many glands, which secrete the bulk of the seminal fluid, open directly into the prostatic urethra. Smooth muscle, originally derived from the urethra, permeates the gland, which it empties by its contraction when the emission of semen is imminent. The ejaculatory ducts converge within the prostate gland to open into the prostatic urethra. The prostate gland is incompletely formed into two lateral lobes and a median lobe. The latter, particularly, may undergo benign enlargement in older men in which its growth presses upon the urethra and base of the bladder to interfere with the flow of urine.

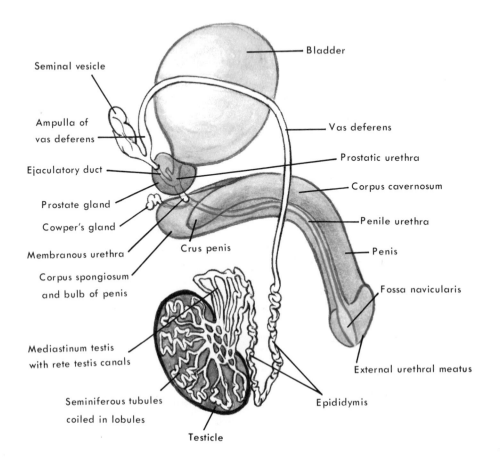

Seminal vesicle

Ampulla of vas deferens

Ejaculatory duct

Prostate gland

Cowper's gland

Membranous urethra

Corpus spongiosum and bulb of penis

Crus penis

Mediastinum testis with rete testis canals

Seminiferous tubules coiled in lobules

Testicle

Bladder

Vas deferens

Prostatic urethra

Corpus cavernosum

Penile urethra

Penis

Fossa navicularis

External urethral meatus

Epididymis

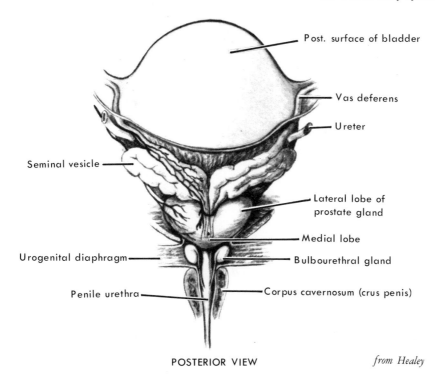

Post. surface of bladder

Vas deferens

Ureter

Seminal vesicle

Lateral lobe of
prostate gland

Medial lobe

Urogenital diaphragm

Bulbourethral gland

Penile urethra

Corpus cavernosum (crus penis)

POSTERIOR VIEW   *from Healey*

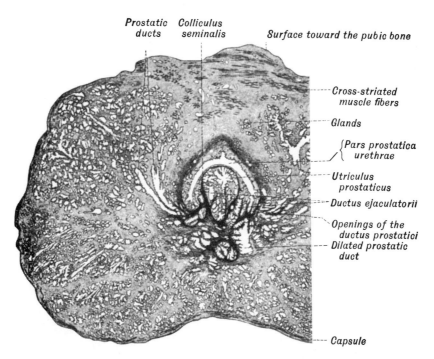

Prostatic
ducts

Colliculus
seminalis

Surface toward the pubic bone

*Cross-striated
muscle fibers*

Glands

*Pars prostatica
urethrae*

*Utriculus
prostaticus*

Ductus ejaculatorii

*Openings of the
ductus prostatici*

*Dilated prostatic
duct*

Capsule

from Bloom and Fawcett

SECTION THROUGH PROSTATE GLAND

Penis and Penile Urethra

The *penis,* the male external genital organ, serves several purposes. It may be stiffened (erection) by the inrush of blood into the cylinders of which it is composed so that the organ may be inserted into the female vagina during coitus. The penis also conveys the penile urethra which, in a flaccid state, conveys urine and, in the state of ejaculation, conveys the semen to the surface of the body.

The penis externally consists of the *body* or shaft which terminates in the *glans.* The glans is molded into an expanded rim, the *corona,* at the junction with the body of the penis. Distal to the corona the glans narrows to a rounded apex which bears the vertical slit of the *external urethral meatus.* The skin of the penis, freely movable upon underlying structures of the body, is bound down and modified to a thin sensitive layer over the glans. A skin fold, the *prepuce* or foreskin, passes forward over the glans unless removed by circumcision shortly after birth.

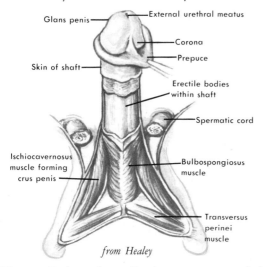

Glans penis — External urethral meatus
— Corona
— Prepuce
Skin of shaft —
Erectile bodies within shaft —
— Spermatic cord
Ischiocavernosus muscle forming crus penis —
Bulbospongiosus muscle —
— Transversus perinei muscle

from Healey

Three cylinders of erectile tissue are surrounded by a tough collagenous tunica albuginea to form the main mass of the penis. Two *corpora cavernosa* are located on the anterolateral surfaces, and the *corpus spongiosum* occupies a median position on the under surface.

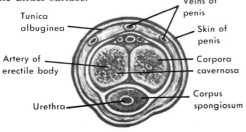

Tunica albuginea —
Veins of penis
Skin of penis
Artery of erectile body —
Corpora cavernosa
Urethra —
Corpus spongiosum

Each corpus cavernosum passes into the perineum as a *crus penis.* These crura diverge as each turns toward its ischial ramus, accompanied by the ischiocavernosus muscle, for attachment. The corpus spongiosum remains in the midline of the perineum where it is covered by the bulbospongiosus muscles and attaches to the urogenital diaphragm. Here it is called the *bulb of the penis.* It receives the membranous urethra which runs in its substance as the penile urethra to the external urethral meatus. The perineal attached portions of the penis constitute its *root.*

The erectile tissue of the corpora consists of a meshwork of trabeculae of fibrous connective tissue and smooth muscle covered by endothelium. In conditions of sexual excitement, parasympathetic nerve fibers produce vasodilatation of branches of arteries entering the penis. As blood enters the cavernous spaces the penis swells into the rigid state of erection during which the blood is prevented from entering the veins. Upon cessation of the stimulus or following ejaculation, sympathetic nerve fibers produce arterial vasoconstriction, blood enters the veins, and the penis returns to the flaccid state.

FEMALE REPRODUCTIVE SYSTEM

The female reproductive organs are described in the order in which the female sex cells, the ova, are formed and are transported to the uterus, from which the vagina leads to the external genital organs. (See color plate 11.)

Formation of Ova

The female sex cells develop within the female gonad or *ovary* which, like the testicle, descends from an original retroperitoneal position high on

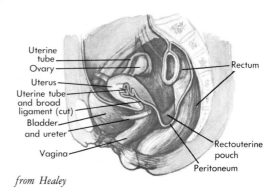

Uterine tube
Ovary —
— Rectum
Uterus —
Uterine tube and broad ligament (cut) —
Bladder and ureter —
Vagina —
Rectouterine pouch
Peritoneum

from Healey

the posterior abdominal wall. The ovary, however, comes to rest within the pelvis where its adult position is upon the lateral wall at about the level of the anterior superior iliac spine.

The Ovary. The *ovary* is a small ovoid body which is lodged in the ovarian fossa of the lateral pelvic wall. It is closely related to the ureter and to the internal iliac artery. The ovary is attached to the shelflike broad ligament of the uterus by a short mesentery, the *mesovarium,* which also conducts blood vessels into the organ. An *ovarian ligament* is a rounded cord which runs across the broad ligament to connect the medial pole of the ovary to the uterus. An ill-defined *suspensory ligament* extends from the ovary over the iliac vessels. Despite its attachments the ovary is relatively mobile. It can rise with the uterus and can be moved about during pelvic examination by a maneuver in which the physician brings pressure against it through the muscles of the lower abdominal wall.

The ovary projects into the abdominal cavity but, instead of a peritoneal investment, presents a surface covering called the *germinal epithelium.* This layer is continuous with the peritoneal mesothelium along the mesovarium. The organ is demarcated into the *cortex,* which lies under the germinal layer and contains the developing ova, and into the medulla, which, containing fibrous tissue, blood vessels, and nerves, merges with the mesovarium.

Ovarian Follicles. At birth the ovary contains all the ova which will be formed during the lifetime of the individual. These, estimated to be 40,000 to 400,000 in number, are far more than will be needed during the childbearing years of a woman. The thousands of immature ova are contained in minute cellular chambers called *primitive follicles* which form a closely packed layer in the cortex under the germinal epithelium. Most of the ova never develop beyond this stage but undergo atrophy (atresia) as other follicles periodically undergo further development. Ordinarily only one ovum and follicle arrive at full maturity at one time, usually alternating between the two ovaries at intervals of about 28 days once the ovulatory cycle becomes established after puberty. A follicle which contains a maturing ovum enlarges and develops specialized layers of follicular cells. As both follicle and ovum enlarge, a clear follicular liquid is secreted by the follicular cells which bathes the ovum and contains one of the ovarian hormones. The *mature (graafian) follicle* occupies considerable space within the ovary and, because of the pressure of the follicular fluid, its distended mass bulges against the germinal epithelium. Just before ovulation the distended follicle appears as a cystic balloon on the surface of the ovary.

Ova and Ovulation. When the ovum first starts to grow it divides until four cells are formed. In this process the number of chromo-

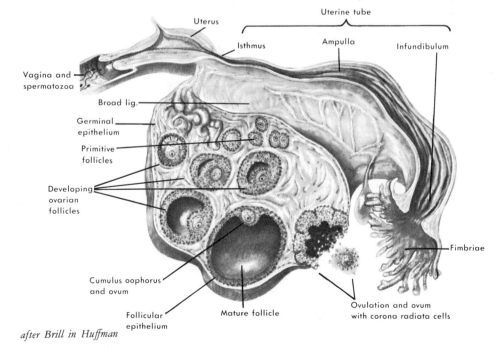

Uterus
Uterine tube
Isthmus
Ampulla
Infundibulum
Vagina and spermatozoa
Broad lig.
Germinal epithelium
Primitive follicles
Developing ovarian follicles
Fimbriae
Cumulus oophorus and ovum
Follicular epithelium
Mature follicle
Ovulation and ovum with corona radiata cells

after Brill in Huffman

somes are reduced by half. Only one of the four cells continues to develop. The *ovum* is a large round cell with a pale nucleus which has scanty chromatin but a large nucleolus. A thick membrane, the *zona pellucida,* surrounds the ovum which is also enveloped by a mantle of follicular cells which form the *corona radiata.* The cellular mass containing the ovum is borne upon a hillock of follicular cells (*cumulus oophorus*) at one side of the distended follicle. *Ovulation* occurs when the distended follicle ruptures and releases the ovum into the peritoneal cavity in a miniature gush of follicular fluid.

After release of the ovum the follicle collapses and by transformation of its cells is converted into the *corpus luteum.* The cells of the thickened lining secrete a second ovarian hormone which is important in preparing the lining of the uterus for reception of the fertilized ovum. (See illustration on page 459.)

Transportation and Fertilization of the Ovum

The uterus is a hollow organ suspended in the midline of the pelvis with its opening presenting into the upper part of the vaginal passageway. The uterus is connected to the abdominal cavity in the vicinity of the ovaries by the *uterine tubes.* (See illustration below.)

Uterine Tubes.   The *uterine* (fallopian) *tubes* extend laterally on each side toward the ovaries from the upper lateral angles of the uterus. These muscular tubes, measuring about 4 inches in length, run in the free margin of the peritoneal shelf formed by the broad ligament of the uterus. Each tube is divided into several parts. The *uterine part* traverses the wall of the uterus from the uterine opening of the tube. The short, thick-walled *isthmus* leads out laterally and opens into the dilated, thin-walled *ampulla* which forms the main part of the uterine tube. This part travels in the free edge of the broad ligament and arches over the ovary. The tube partially encircles the ovary and ends in the trumpet-shaped *infundibulum.* This flared end of the uterine tube bears fringed processes, the *fimbriae,* which drape over and clasp the surface of the ovary.

The interior of the uterine tube is lined by a mucosa which is thrown into longitudinal folds. The epithelial cells bear cilia which establish a current leading toward the uterus. Inner circular and outer longitudinal layers of smooth muscle augment the cilia by producing waves of peristaltic contraction. The outer serosal layer is the peritoneum of the broad ligament which

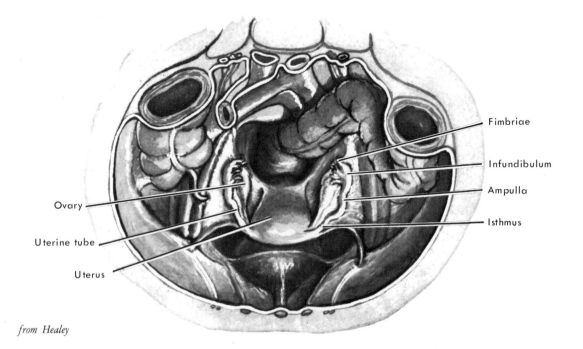

from Healey

Ovary

Uterine tube

Uterus

Fimbriae

Infundibulum

Ampulla

Isthmus

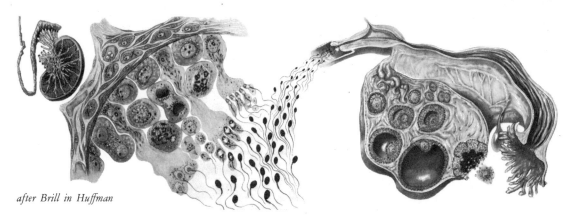

after Brill in Huffman

extends over the fimbriae and down into the funnel of the infundibulum to become continuous with the tubal mucosa.

Path of the Ovum. Although the ovum is expelled into the peritoneal cavity upon ovulation, it is guided by the fimbriae into the infundibulum of the uterine tube. The ovum is steadily propelled toward the uterus by the cilia and by the peristalsis of the tube and bathed in secretions of the epithelium from which it may derive nourishment.

Fertilization usually occurs during the passage of the ovum through the tube. Spermatozoa, deposited within the vagina near the opening of the uterus, swim through the uterine cavity to reach the tubal opening. Sperm cells continue into the uterine tube through tubal secretions against the direction of flow toward the uterus. Of the millions of spermatozoa which are ejaculated, most never reach the outer portion of the uterine tube. Only one spermatozoon is needed for fertilization. Spermatozoa produce an enzyme which softens or dissolves the outer coatings of the ovum. The head of the sperm cell, with the nucleus containing the male component of chromosomes, penetrates the ovum. Fertilization is accomplished by union of the sperm nucleus with that of the ovum after which the ovum becomes unreceptive to the entry of other spermatozoa. The fertilized ovum continues its progress through the uterine tube and undergoes the first cellular divisions of the development of a new individual.

In rare circumstances the fertilized ovum is arrested in its passage. A *tubal pregnancy* occurs in which connections established with the maternal vessels of the tubal wall are insufficient to meet the demands of the growing fetus. Rupture

of the tube, a severe surgical emergency, terminates such an ectopic pregnancy. Rarer still is an *abdominal pregnancy* in which an ovum escapes the fimbria, is fertilized, and becomes implanted upon the wall of an abdominal organ.

Uterus

The *uterus,* in its nonpregnant condition, is a small muscular organ whose thick walls enclose the narrow triangular *uterine cavity*. The shape of the organ is like that of an inverted pear which is positioned above the vagina and bent forward at an angle of 90 degrees to the vaginal axis. In this position of *anteversion* the uterus rises between the bladder in front and the rectum behind. Distention of either of these organs changes the position of the movable uterus. The upper part of the uterus is draped by peritoneum which extends downward between it and the adjacent organs to form the *vesicouterine* and *rectouterine pouches.*

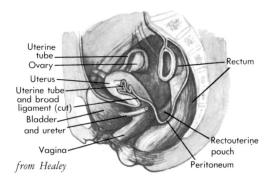

Uterine tube —
Ovary —
Uterus —
Uterine tube and broad ligament (cut) —
Bladder and ureter —
Vagina —
Rectum
Rectouterine pouch
Peritoneum

from Healey

Parts of the Uterus. Above the level of the tubal openings the uterus rounds into its most expanded portion, the *fundus.* The main part of the organ, the *body,* extends downward and backward to a constricted portion, the uterine *isthmus.* The tubular, more fixed *cervix* extends downward from the isthmus and enters the upper portion of the vagina. The cervix is rounded and molded into anterior and posterior lips which surround the *ostium of the uterus* (external os, cervical orifice). In the woman who has not borne children the ostium is a short slit between the lips of the cervix. After pregnancy has ensued the ostium is larger, more open, and irregular.

Uterine Ligaments. The movable uterus is held in position partly by peritoneal folds and ligaments and partly by the pressure of adjacent organs. Its fasciae connect with those of the rectum and bladder and support from below is provided by the vagina and pelvic diaphragm.

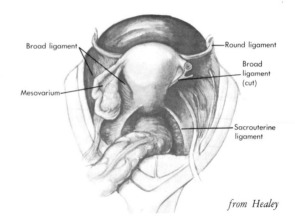

from Healey

These ligaments provide only relative fixation of the uterus in the position of anteversion. The uterus also may be bent forward at the isthmus quite normally to combine *anteflexion* with anteversion.

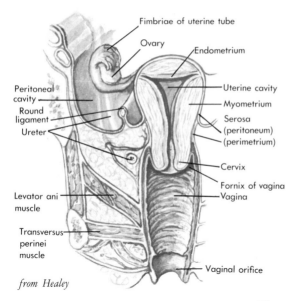

from Healey

The *broad ligament* is a double fold of peritoneum which extends on each side from the lateral margin of the uterus to the lateral wall of the pelvis. The *round ligament* runs through the broad ligament from the lateral uterine wall near the attachment of the uterine tube. This ligament traverses the inguinal canal enroute to the labium majus of the external genital organs. The round ligament splays out into fibrous bands for attachment within the tela subcutanea of the labium. Fascial bands from the cervix pass to the superior fascia of the pelvic diaphragm as the *cardinal ligament.* A *sacrouterine ligament* passes backward to the sacrum.

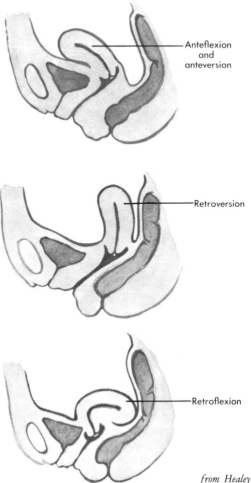

from Healey

| Follicular phase | Progravid phase | Menstrual phase |

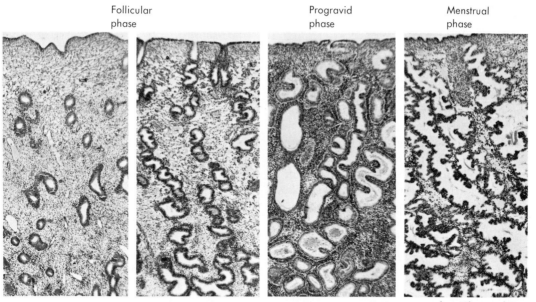

ENDOMETRIAL CHANGES WITH OVULATION   *from Bloom and Fawcett*

The uterus may be turned upward and backward into *retroversion* to which may be added a more pronounced backward tilt if the organ is bent sharply backward at the isthmus into *retroflexion.*

Structure of the Uterus. The uterine cavity is lined by a mucosa of columnar epithelial cells into which open many tubular glands. This layer, termed the *endometrium,* undergoes cyclic changes under the influence of ovarian hormones. The

myometrium or muscular layer is composed of bundles of smooth muscle interlaced with fibrous connective tissue. Definitely demarcated layers of muscle are not present. The muscle fibers are capable of great enlargement and growth in length during pregnancy. A serosal layer is found where peritoneum drapes the uterus. Elsewhere the outer layer is of connective tissue which is called the *perimetrium.*

Cyclic Changes in the Endometrium. The

from Villee

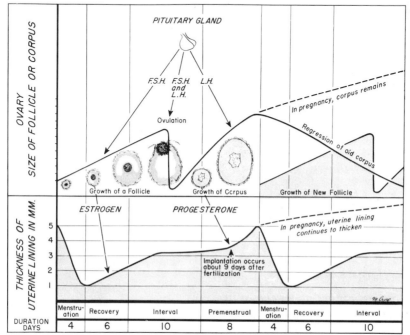

endometrium undergoes changes in structure approximately every 28 days in preparation for the reception of a fertilized ovum. The periodic changes are described as the *menstrual cycle,* for if a fertilized ovum is not implanted within it, the endometrium is sloughed off in a hemorrhagic process termed *menstruation.* The menstrual cycle is a response to, and is regulated by, the amounts of ovarian hormones secreted into the circulating blood by the follicle and the corpus luteum. See illustration at bottom of page 463.

The endometrium passes through four phases related to events occurring in the ovary which lead to ovulation. The menstrual cycle begins with the cessation of the previous menstrual period. During the *follicular phase* the endometrium regenerates. The epithelium is repaired by cellular growth, and the tubular glands whose superficial parts had been lost re-form into straight tubules. Under the influence of estrogenic hormone of the ovarian follicles, the glands begin to widen and assume a wavy appearance as secretions form. Ovulation occurs between the tenth and sixteenth days of the cycle after which the endometrium enters the *progravid phase.* During this period the arteries of the endometrium become coiled, the glands engorged with secretions, and the endometrial tissues boggy with interstitial fluid. Both estrogen and progesterone secreted by the corpus luteum are preparing the endometrium to receive and nourish a fertilized ovum. If fertilization does not occur, the endometrium enters an *ischemic phase* in which the coiled vessels constrict and the endometrium shrinks. The *menstrual phase* occurs as hemorrhage develops from the blood vessels and the superficial epithelium, with parts of the glands, are shed. The *menstrual flow* composed of these elements pours from the uterus to form a bloody vaginal discharge which continues for several days after which a new cycle begins.

If fertilization ensues the endometrium remains in its progravid phase and no menstrual flow occurs. The developing embryo is implanted in the endometrium which becomes involved in the formation of the *placenta.*

External Genital Organs and Vagina

The external genital organs are located in the urogenital triangle of the perineum. These are principally modifications of the skin and sub-cutaneous tissues which surround the openings of the female urethra and the vagina which pass through the perineum from the pelvis.

External Structures. The *mons pubis* is a rounded elevation of skin and fatty tela subcutanea which overlies the front of the symphysis pubis. This pad is continued downward and backward on each side of the midline by the *labia majora* which are similar in composition to the mons pubis. The rounded longitudinal folds of the labia majora merge into the skin of the medial aspect of the thighs laterally. In the anatomical position the medial surfaces of the labia majora touch and enclose between them a crevice, the *pudendal cleft.* The labia majora are homologues of the male scrotum which in the anatomical position, or whenever the thighs are close together, cover and protect the other external genital structures. Both the mons pubis and labia majora become covered by short coarse *pubic hairs* after puberty.

When the thighs are separated the labia majora move apart to open the pudendal cleft. Two folds of skin, the *labia minora,* occupy the pudendal cleft adjacent to the medial surfaces of the labia majora and enclose the *vestibule of the vagina* which presents the openings of the vagina below and the urethra above. The labia minora join superiorly by forming hooded folds about the *clitoris* which forms a small projection above the urethral opening. The clitoris is the female homologue of the penis but differs in that it does not transmit the urethra. The clitoris is composed of two small bodies of erectile tissue, the *corpora cavernosa,* which are attached by crura (as in the penis) to the corresponding ramus of the ischium. The *body* of the clitoris terminates in a small knobby projection, the *glans clitoridis,* which is composed of erectile tissue covered by skin. Like the penis the glans of the clitoris is extremely sensitive to touch and in conditions of sexual excitement becomes filled with blood.

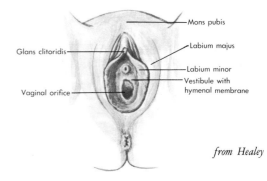

Mons pubis

Glans clitoridis

Labium majus

Labium minor

Vestibule with hymenal membrane

Vaginal orifice

from Healey

Vagina. The *vaginal orifice* is a vertical slit which becomes wider and more rounded with the birth of children. The orifice is partly closed by a thin membrane, the *hymen,* which has a central opening for the flow of the products of menstruation. As sexual relations are established and children are born the hymen becomes reduced to tags of tissue around the margins of the orifice.

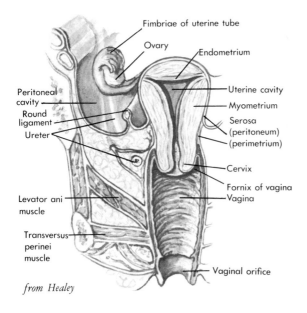

from Healey

At each side of the vaginal opening are the *bulbs of the vestibule* which contain masses of erectile tissue covered by the bulbospongiosus muscle. These structures cover the *greater vestibular glands* whose ducts open into the angle between the margin of the hymen and the labium minor. A mucus secretion forms during sexual activity to lubricate the lower end of the vagina.

The *vagina* is a fibromuscular tube which extends upward and backward through the perineum and into the pelvis from the vestibule. This passageway receives the penis during coitus, acts as the outlet for the menstrual flow, and forms, with the cavity of the uterus, the birth canal. The walls of the vagina are ordinarily collapsed but are capable of tremendous dilatation especially during childbirth. The lumen is lined by a mucosa of the mucous membrane type of stratified squamous epithelium which is thrown up into corrugated ridges. The vagina passes behind the urethra and bladder and in front of the rectum. The upper part of the vagina surrounds the cervix of the uterus. The recesses between the cervix and the upper end of the vagina are called the anterior, lateral, and posterior *fornices of the vagina.* The posterior fornix is deeper than the others and comes into relationship with the recto-uterine pouch.

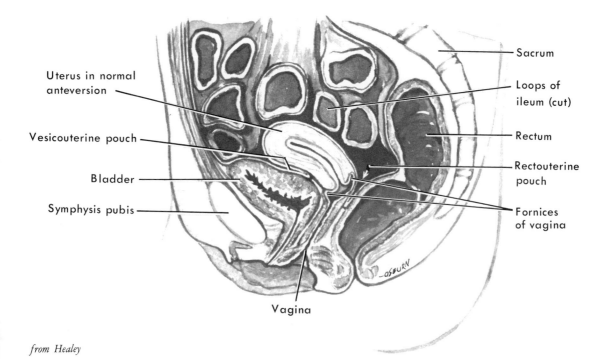

from Healey

Female Urethra

The female urethra has a much less complicated course than in the male. The urethra descends on a straight course downward and forward from the internal urethral orifice of the
bladder through both the pelvic and urogenital
diaphragms. It becomes fused to the anterior wall
of the vagina prior to its termination in the
external urethral orifice at the vestibule.

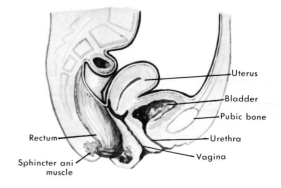

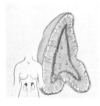

10
The
Endocrine
System

The endocrine system is composed of a group of widely scattered masses of glandular tissue, some of which form separate organs whereas others are part of organs which also have non-endocrine functions. Several factors bring endocrine organs together in a functional system.

1. These glands produce secretions called *hormones* which pass into the blood rather than into excretory ducts.

2. The products of endocrine glands are substances which are needed by many tissues or by certain cells in locations remote from the secretory source. The circulatory system is the most efficient means of distribution for these hormones.

3. Many of the endocrine organs influence the function of other endocrine glands with resultant balance and coördination in the system as a whole.

TYPES OF ENDOCRINE GLANDS

A number of endocrine glands form organs by themselves or become morphologically discrete portions of other organs. These, the *circumscribed endocrine glands,* consist of the hypophysis (pitui-

tary), adrenal, thyroid, and parathyroid glands, which are separate organs. Also included are the islets of Langerhans of the pancreas, the ovarian follicles and corpora lutea of the ovary, and the interstitial tissue of the testicle. Organs of *debated endocrine function* are the pineal gland, thymus gland, and structures called the paraganglia. *Endocrine-like tissues* have a controlling or integrative effect upon other tissues or produce substances required by many organs. Examples of those with a controlling effect are mucous membranes or glands which produce *secretagogues*. The production of secretin and cholecystokinin by the alimentary tract has been mentioned in this respect. The cyclic storage of glucose in the form of glycogen and its release as blood sugar to serve the demand of tissues are considered to be endocrine-like functions of the liver.

HYPOPHYSIS

The hypophysis or *pituitary gland,* barely larger than 1 cm. in all its dimensions, occupies the hypophyseal fossa of the sella turcica portion of the sphenoid bone. This gland is connected to the hypothalamus by the hypophyseal stalk.

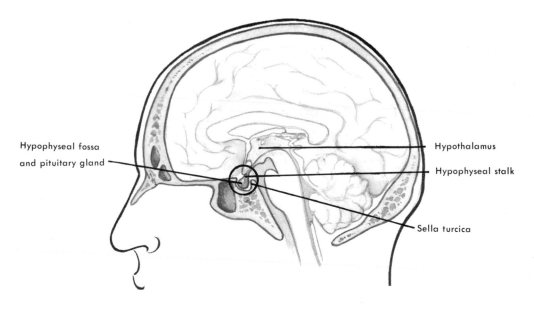

Hypophyseal fossa
and pituitary gland

Hypothalamus

Hypophyseal stalk

Sella turcica

The pituitary gland consists of two portions which differ markedly in microscopic appearance. The *adenohypophysis,* consisting of the anterior part, the narrow intermediate part, and the tuberal (stalk) part, has a distinctly glandular appearance. (See illustration below.) These parts developed as an outgrowth from the roof of the mouth. The *neurohypophysis* is the posterior part of the gland. It grew down from the diencephalon in development and has the appearance of nervous tissue. These two parts of the pituitary gland, more familiarly called the anterior and posterior lobes, are structurally and functionally separate organs.

The functions of both lobes are controlled by the hypothalamus.

The pituitary gland is called the master endocrine gland of the body because its secretions are basic to many body processes and because pituitary hormones stimulate the action of other endocrine tissues.

ANTERIOR LOBE (PARS DISTALIS)

The *anterior lobe* is the largest portion of the pituitary gland and has the greatest number of hormone products and functions. In fact, more

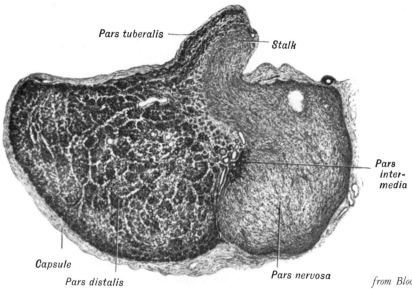

Pars tuberalis

Stalk

Pars inter-media

Capsule

Pars distalis

Pars nervosa

from Bloom and Fawcett

SECTION THROUGH PITUITARY GLAND

hormones are credited to this part, also known as the *adenohypophysis,* than there are clearly demarcated separate cells to produce them.

The anterior lobe is made up of clumps and cords of glandular-appearing cells. These cells lie amidst a connective tissue framework which is penetrated by vascularlike sinusoids filled with blood.

The names given to the cell types of the anterior lobe derive from both the color differentiations in light microscopy and the Greek letter terminology which was devised later as an alternative. Electron microscopy is just beginning to resolve some of the confusion. Histochemistry, through identification of characteristic chemical constituents such as glycoprotein, is beginning to be helpful also. However in this edition, unfortunately, the nomenclature of the pituitary gland still involves a mixture of terms.

Cell Types

Acidophils, also known as *alpha* (α) *cells,* are spherical cells with a prominent Golgi apparatus, mitochondria, and granules. Two subtypes of acidophils are recognized, on the basis of granular size, which are believed to be hormones or their precursors. The cell type with larger granules is believed to form the *growth hormone (somatotropin),* whereas the cell type containing smaller granules is credited with the secretion of *prolactin (lactogenic hormone).* The hormones of the pituitary gland and their function are summarized at the end of this section in the table on page 470.

Basophils (also known as *beta cells*) are similar to acidophils in size and constituents, but the granules are considerably smaller. The names of the cell type and of the granules derive from the process of staining with basic dyes when using light microscopy. Again, two subtypes are recognized.

Beta (β) *basophils* are believed to form the hormone, *thyrotropin* (thyrotropic hormone). The *delta* (Δ) *basophils* secrete *gonadotropins* (gonadotrophic hormones) which are the follicle-stimulating hormone (FSH) and the luteinizing hormone (LH).

Chromophobe cells make up a third group of small cells which have neither granules nor specific staining properties in light microscopy. Recent studies make it apparent that these cells are undifferentiated and may specialize into either

acidophilic or basophilic cells as needed. No specific hormone, therefore, is ascribed to them in their undifferentiated form.

A most important endocrine product of the anterior lobe is the *adrenocorticotropic hormone* (ACTH), which has become very important clinically. As yet, however, none of the cell types of the anterior lobe (pars distalis) have been identified as the cell of origin of this hormone.

INTERMEDIATE PART (*PARS INTERMEDIA*)

There is an intermediate zone in the anterior lobe, just next to the posterior lobe, where a cleft is found. Lining this cleft are cells which are formed into a series of cystic sacs. These cells, similar to the basophils of the anterior lobe in appearance, have been found to secrete the *melanin-stimulating hormone* (MSH) which regulates the production of pigment in the skin of the body.

TUBERAL PART (*PARS TUBERALIS*)

The portion of the anterior lobe which extends upward along the pituitary stalk is the *tuberal part.* Cells here are arranged in long cords with blood vessels coursing between them. Although the tall cuboidal cells of this area have mitochondria, granules, and droplets characteristic of secretory activity, their function is not known at the present time.

POSTERIOR LOBE (*NEUROHYPOPHYSIS*)

The true stalk of the pituitary gland connects the hypothalamus of the brain with the *posterior lobe.* This lobe is sometimes called the *pars nervosa* because it looks like nervous tissue. The bulk of the posterior lobe is made up of nerve fibers whose cell bodies are in the hypothalamus. The fibers course through the stalk to end blindly among the posterior lobe cells, the *pituicytes.* The pituicyte cells resemble the glia cells of the nervous system in forming a connective-tissue-like framework for other components of the lobe. The actual secreting part of the posterior lobe is found, by electron microscopic studies, to be in the nerve fibers which, as axons of the cells of the hypothalamus, traverse the stalk to end amidst vascular sinusoids of the posterior lobe. Definite swollen areas along these fibers have been proved to contain hormone products which are released, presumably, into the blood. The endocrine products of the posterior lobe are the hormones *oxytocin* and *vasopressin* whose properties are indicated in the following table.

PITUITARY HORMONES

Hormone	Cell Location	Cell Type	Function
Growth hormone (somatotropin)	Anterior lobe	Acidophils (**α cells**) a. **large granules**	Controls general body and bone growth
Prolactin (lactogenic h.)	Anterior lobe	b. **small granules**	Stimulates milk secretion Maintains estrogen and progesterone secretion by ovary
Thyrotropin (thyrotropic h.)	Anterior lobe	Basophils (**β cells**) a. **β basophils**	Stimulates thyroid gland and thyroxine production
Gonadotropins (gonadotropic h.) a. FSH (follicle-stimulating h.)	Anterior lobe	b. **Δ basophils**	Stimulates ovarian follicle growth in female Stimulates seminiferous tubule growth in male
b. LSH (luteinizing h.)			Causes follicle maturation and hormone production by ovary in female Causes hormone production by testicle in male
No hormone	Anterior lobe	Chromophobe cells	Undifferentiated: may produce either acidophils or basophils by specialization
MSH (melanin-stimulating h.)	Anterior lobe (Pars intermedia)	Basophil-like epithelial cells	Regulates pigment production
No known hormone	Anterior lobe (Pars tuberalis)	Cords with blood vessels between	Unknown
No hormone	Posterior lobe	Pituicytes	Provides support
Oxytocin	Posterior lobe	Hypothalamic cells via axons to posterior lobe	Stimulates contraction of uterine smooth muscle Stimulates secretion of milk
Vasopressin	Posterior lobe	Hypothalamic cells via axons to posterior lobe	Stimulates smooth muscle contraction, especially in arterial system

CLINICAL ASPECTS OF PITUITARY HORMONES

1. Since the *growth hormone* controls general body growth, including growth of the skeletal system, it may be expected that either underactivity or overactivity of the anterior lobe of the pituitary gland may produce alterations in body form and stature. *Underactivity* during childhood results in *dwarfism. Overactivity,* usually due to diseases such as pituitary tumors, results in excessive growth. If the epiphyseal plates of the skeletal system have not closed, *gigantism* occurs because of excessive growth of the long bones. After epiphyseal closure, the bones thicken and coarsen, producing *acromegaly,* which is characterized by enlargement of the mandible, broadening of facial contours, and spadelike thickening of the hands and feet.

2. Since the thyrotropic hormone stimulates secretion of the thyroid gland, problems seemingly related to overactivity or underactivity of the thyroid gland may not necessarily be caused by disease of the thyroid gland itself. The problem may lie instead in disease of the master controller, the anterior pituitary.

3. In a similar manner, symptoms believed to be related primarily to adrenal gland disease perhaps may be due to a disorder in the production of *adrenocorticotropic hormone* by the anterior pituitary.

4. The development of function leading to ovulation in the follicles of the ovary in the female and the process of spermatogenesis in the male are caused by the *follicle-stimulating hormone* of the anterior pituitary. In the absence of such stimulation from the pituitary gland, the institution of these processes may be impeded or delayed. Such deficiencies may be compounded by the lack of stimulation from the *luteinizing hormone.* This product of the anterior lobe should stimulate the maturation of the ovarian follicles in the female and the development of the corpus luteum after ovulation. The stimulation provided by the luteinizing hormone to the interstitial cells of the testicle results in the secretion of the male androgenic hormone. As may be expected, the subsequent development of secondary sexual characteristics in both sexes is greatly dependent upon the effect of the anterior pituitary hormones on the reproductive organs.

5. If ovarian hormones have affected the growth of the ducts and alveoli of the mammary glands during pregnancy, the anterior pituitary hormone, *prolactin,* will stimulate the glands to secrete milk after a child is born.

6. The posterior lobe hormone, *oxytocin,* is also important at the end of pregnancy, for it stimulates contraction of the uterine muscles and is also related to milk secretion.

7. The other posterior lobe hormone, *vasopressin,* is normally active in causing the contraction of smooth muscle. Sustained contraction of smooth muscle in the walls of the vessels of the arterial system narrows their lumina. The result of the arterial constriction is an increase in blood pressure, a necessary process in the maintainance of a satisfactory blood flow.

THYROID GLAND

The *thyroid gland* is located in the anterior aspect of the neck just below the larynx. It consists of right and left *lateral lobes* connected by the *isthmus* which lies upon the upper part of the trachea. Each lateral lobe is related medially to the larynx, pharynx, and trachea. The lobe presses posterolaterally toward the carotid sheath. The thyroid gland has a lobular structure and is covered by a connective tissue capsule. The gland is intensely vascular and is richly supplied with blood by the prominent superior, middle, and inferior thyroid arteries. (See illustration at top of following page.)

The basic unit of structure is the *thyroid follicle.* (See illustration at bottom of following page.) The follicle is a spherical sac formed by a single layer of cuboidal cells which are intimately associated with a capillary network. The cavity of the follicle is filled with a stiff viscous substance termed *colloid.* The follicular epithelial cells secrete organic forms of iodine which are bound to protein molecules. The main thyroid hormone is termed *thyroxine* although other complex iodine-protein molecules are present. The thyroid gland stores its hormone in the colloid in the form of *thyroglobulin* in addition to secreting it into the blood. When the level of thyroid hormone in the circulating blood falls below normal levels, secretion of the pituitary thyrotropic hormone occurs. The thyroid gland responds by secreting thyroxine or by the converting of stored thyroglobulin to the active hormone form. High

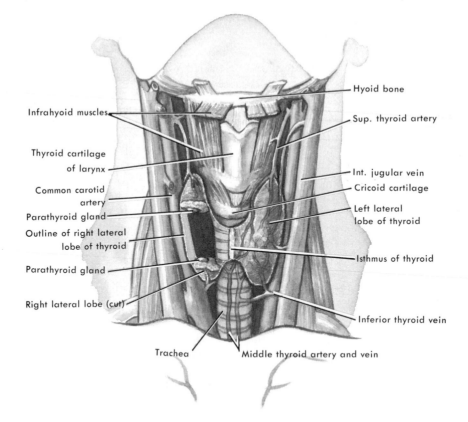

Infrahyoid muscles

Thyroid cartilage
of larynx

Common carotid
artery

Parathyroid gland

Outline of right lateral
lobe of thyroid

Parathyroid gland

Right lateral lobe (cut)

Hyoid bone

Sup. thyroid artery

Int. jugular vein

Cricoid cartilage

Left lateral
lobe of thyroid

Isthmus of thyroid

Inferior thyroid vein

Trachea

Middle thyroid artery and vein

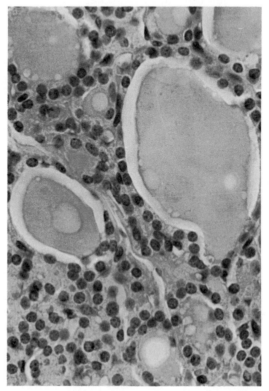

from Leeson and Leeson

MICROSCOPIC APPEARANCE OF THYROID

blood levels of thyroid hormone inhibit the secretion of the pituitary thyrotropic hormone.

Electron microscopic studies show that the structure of the epithelium of the thyroid follicle is composed of the characteristics previously described for glandular epithelia of secretory nature. Many small microvilli project from the cell surface into the lumen of the follicle.

Recent research has demonstrated that there are *parafollicular cells* which lie adjacent to thyroid follicles. These cells have been found to secrete another hormone, *thyrocalcitonin,* which has the effect of lowering the level of calcium in the blood when the process of bone resorption is flooding the vascular system with an excess of calcium ions.

The major function of the thyroid gland is to control the metabolic rate of all the cells of the body. It seems particularly to be an accelerator of cellular activities, for if thyroid function is diminished the metabolism of the body falls far below normal. *Underactivity* of the thyroid gland is called *hypothyroidism.* In this state body functions are slowed, weight accumulates, and the skin is cold and dry. Extreme hypothyroidism in children leads to stunting of growth and retardation (*cretinism*). In adults, untreated hypothyroidism (*myxedema*) is characterized by an exaggera-

tion of the symptoms associated with reduced mental activities, even stupor. *Overactivity* of the thyroid gland (*hyperthyroidism*) is associated with increased cardiac and respiratory rates, perspiration, a hot skin, tremors, and weight loss. Staring bulging eyes (exophthalmos) accompany a severe hyperthyroid state.

PARATHYROID GLANDS

Two small, brownish yellow *parathyroid glands* lie on the capsule of the posterior aspect of each thyroid lobe. The four minute glands consist of densely packed secretory cells without a lobular arrangement. The cells are held in an intimate association with capillaries by a reticular meshwork.

Two cell types, supported by fine, reticular, connective tissue networks, are formed into cords and irregular masses to make up the parathyroid gland. (See illustration on next page.) Larger *oxyphil cells* are grouped together amidst the smaller, more numerous *chief cells* (principal cells). No functional difference can be discerned between the two cell types. Both are believed to be responsible for the production of a hormone of polypeptide nature which is named simply *parathyroid hormone* (parathormone). These glands

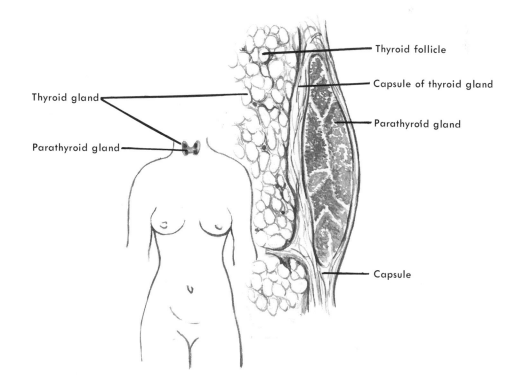

Thyroid follicle

Capsule of thyroid gland

Thyroid gland

Parathyroid gland

Parathyroid gland

Capsule

are essential to life, for they regulate the supply, distribution, and metabolism of calcium and phosphate which are needed by all cells. Without an adequate supply of the parathyroid hormone bone formation and metabolism suffer. In extreme parathyroid deficiency the nervous system becomes hypersensitive, with muscular spasms (tetany) and convulsions resulting.

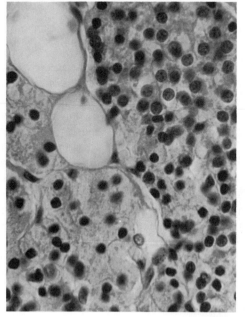

from Leeson and Leeson

MICROSCOPIC APPEARANCE OF PARATHYROID

ADRENAL GLANDS

The *adrenal* (suprarenal) *glands* have been previously noted to be retroperitoneal abdominal organs capping the upper pole of each kidney. They are imbedded in fat through which pass a rich supply of arteries springing directly from the aorta or from many of its branches, such as the inferior phrenic or renal artery or even those within the kidney itself or the renal fascia. The gland is divided into an outer *cortex* and a central *medulla;* these parts are separate endocrine glands.

The *adrenal cortex* is made up of closely packed cords of epithelial cells between which run sinusoidal blood vessels. (See illustration at top of facing page.) The function of the cortex is related to the steroid substances it produces upon stimulation of the pituitary adrenocorticotropic hormone. These substances are grouped together as the *adrenocortical hormones.* Many of them, such as cortisone, desoxycorticosterone, and hydrocortisone, have been identified and produced in pure form. The cortical hormones are essential to life. Atrophy of the cortex or removal of the glands results in a deficiency state called Addison's disease which is fatal unless the hormonal substances are given therapeutically. The major effects of the adrenocortical hormones are in the maintenance of electrolyte and water balance, carbohydrate balance, and connective tissue metabolism. The cortical hormones affect the ability of the body

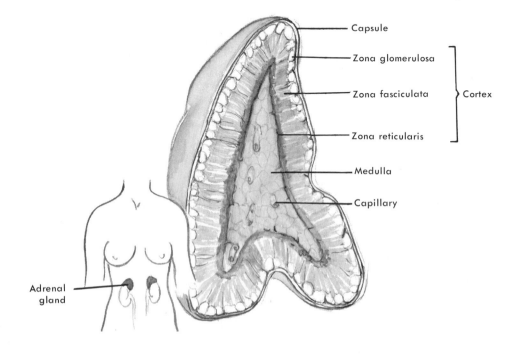

Capsule

Zona glomerulosa

Zona fasciculata — Cortex

Zona reticularis

Medulla

Capillary

Adrenal gland

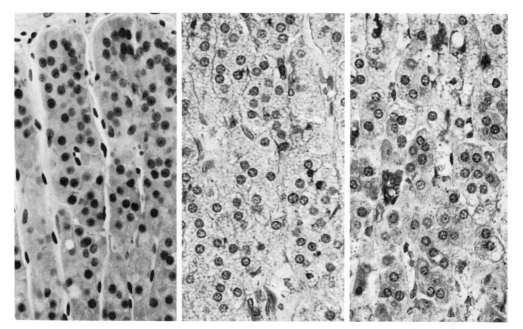

MICROSCOPIC APPEARANCE OF ADRENAL CORTEX   *from Leeson and Leeson*

to react to stress and infections. It has been found that these adrenal steroids provide a means of control and treatment of allergies, diseases of collagen, and some joint diseases.

The *adrenal medulla* makes up the pulpy center of the gland in which groups of irregular cells are located amidst veins which collect the blood

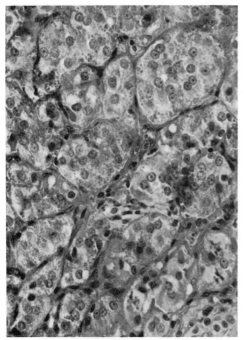

from Leeson and Leeson

MICROSCOPIC APPEARANCE
OF ADRENAL MEDULLA

from the sinusoids. The medulla does not respond to the stimulation of the pituitary hormones as does the cortex, but is stimulated by the thoracolumbar portion of the autonomic nervous system. The medulla secretes two hormones, epinephrine (Adrenalin) and norepinephrine, whose effects upon the body resemble those of thoracolumbar autonomic stimulation. Both hormones increase the blood pressure, epinephrine by an acceleration of the heart rate and an increase in cardiac output and norepinephrine by producing constriction of arterioles. Epinephrine is effective in causing the conversion of glycogen into glucose when there is a demand for carbohydrate. Coupling of this effect with stimulation of the cardiovascular system makes adrenal medullary function important in meeting conditions of stress.

THE PANCREAS

The endocrine activity of the pancreas takes place in the *islets of Langerhans* which are discrete bodies of cells of entirely different nature from the exocrine portion of the gland. (See illustration at top of following page.) The islets, varying in number between several hundred thousand and 1.5 million, are scattered throughout the pancreatic tissue. Each islet consists of several types of cells in which the *alpha* and *beta cells* predominate. A delicate connective tissue

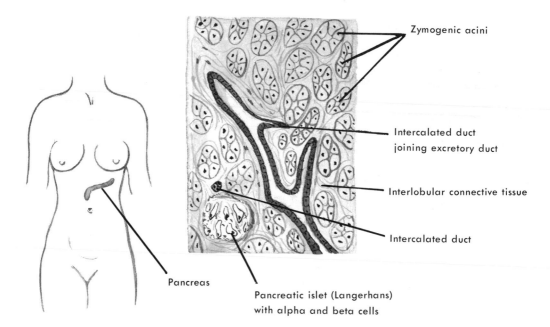

Zymogenic acini

Intercalated duct
joining excretory duct

Interlobular connective tissue

Intercalated duct

Pancreas

Pancreatic islet (Langerhans)
with alpha and beta cells

separates the islets from the surrounding zymo-genic pancreatic acini. The blood supply to each islet is extensive. (See illustration below.)

As may be seen in the diagram, four cell types have been identified in the islet tissue by the use of special stains, by the solubility charac-teristics of granules, and by employment of the electron microscope. The pancreatic hormone,

insulin, is believed to be the secretory product of beta cells.

The alpha cells produce another chemical substance, *glucagon.* The functions of the other two cell types, C and D, is not clear at the present time.

From the endocrine standpoint, the islets of Langerhans, imbedded as thousands of discrete

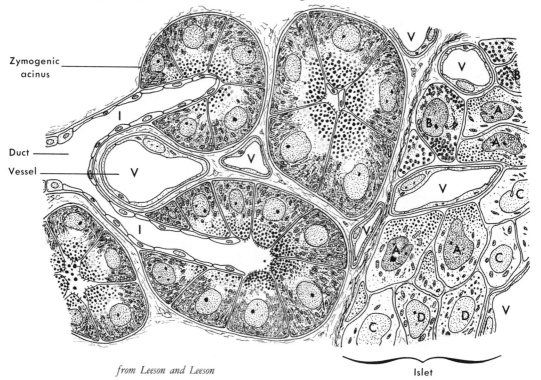

Zymogenic acinus

Duct

Vessel

from Leeson and Leeson

Islet

specks amidst the zymogenic acini of the pancreas, produce the pancreatic hormone, *insulin.* The hormone moves from the beta cells into the rich capillary network, a portion of which is adjacent to the islet cells. Insulin is necessary for the utilization and metabolism of blood glucose by the cells of the body. In health, insulin controls the usage of glucose by the cells and regulates the level of blood sugar. When insufficient insulin is produced the glucose cannot be properly utilized and the level of blood sugar rises. This disease, known as *diabetes mellitus,* is accompanied by the excretion of large amounts of sugar by the kidney into the urine. The energy derived from the utilization of sugar is necessary for the metabolism of fats and proteins. Therefore, in severe diabetes, products of the incomplete metabolism of these substances accumulate to produce the complicating toxic condition of *acidosis.* One of the modern therapeutic triumphs has been the isolation and preparation of insulin and the synthesis of substances which have an insulin effect within the body. These developments have made possible the control of diabetes. Overproduction of insulin occurs when islet tumors, particularly of the beta cells, develop. *Hypo-*

glycemia, a condition in which the level of blood sugar is severely lowered, results when too much insulin is taken by a person with diabetes.

Glucagon enters the circulating blood from the alpha cells of the islets. Its presence in the blood stimulates liver cells to undertake the conversion of stored glycogen into glucose when the blood sugar level is low.

OVARY

The ovarian follicle and the corpus luteum are the endocrine tissues of the ovary. These have been described with the organ itself. The endocrine relationships of the ovary include the following:

1. Pituitary gonadotropic hormones stimulate the growth of ovarian follicles and physiological events related to ovulation.

2. Estrogen and progesterone produced by the follicle and the corpus luteum regulate the cyclic preparation of the uterine endometrium for implantation of a fertilized ovum.

3. Estrogen and progesterone, as the female sex hormones, influence the development of the

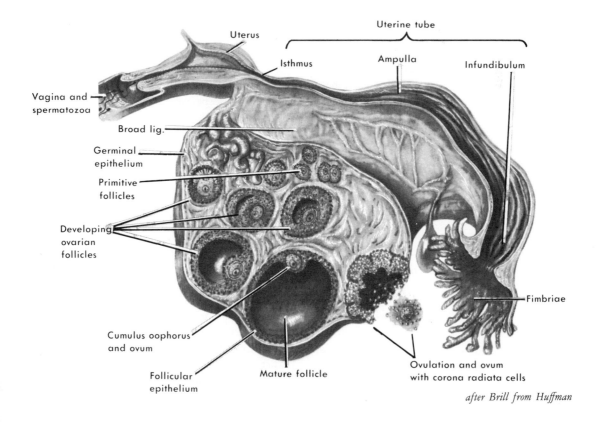

after Brill from Huffman

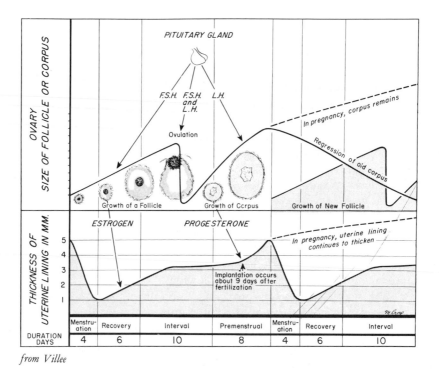

from Villee

female secondary sexual characteristics at puberty and maintain the feminine bodily features thereafter. These characteristics include the rounding of body contours, widening of hips and pelvis, development of fat in the tela subcutanea, hair distribution, and mammary gland development.

4. The ovarian hormones are important in maintaining pregnancy. Increased amounts of estrogen and progesterone are formed by the placenta. Progesterone inhibits ovulation and menstruation during pregnancy. Estrogen stimulates the proliferation of the ducts within the mammary gland and progesterone is responsible for the development of the glandular acini. (See above illustration.)

TESTICLE

The endocrine functions of the testicle are attributed to the *interstitial tissue* which appears as triangular islands between the seminiferous tubules. (See illustration on opposite page.) Large polyhedral cells in these locations produce the male sex hormone, *testosterone.* The production of this hormone is stimulated by the gonadotropic

hormones of the pituitary gland and at puberty is responsible for the development of the male secondary sexual characteristics. These include a heavier, stronger, and squared body form; the development of a male type of hair distribution including the beard; and laryngeal changes which result in a deeper voice.

DEBATED ENDOCRINE ORGANS

The *paraganglia* or chromaffin system consists of scattered bodies of cells, appearing much like those of the adrenal medulla, which are found in the retroperitoneal connective tissues. It is not definitely known whether these secrete epinephrine. The *pineal body* lies at the posterior end of the roof of the third ventricle in the brain. This small body, frequently calcified, consists of cords of epithelial-like cells. It has been claimed to have an effect upon the gonads in early life before involuting at about the age of 14 but the actual function is disputed. The *thymus gland* is an organ of the lymphatic type located in the anterior mediastinum. The thymus is large during infancy

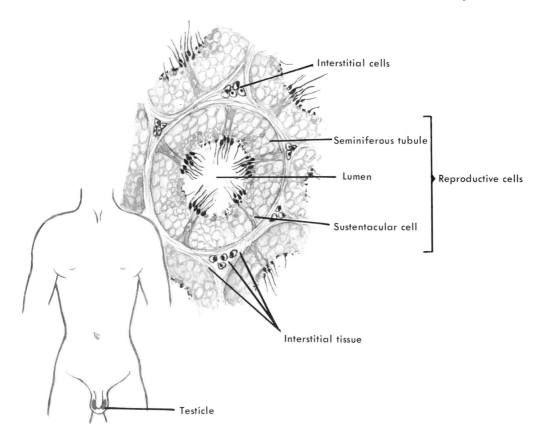

Interstitial cells

Seminiferous tubule

Lumen

Reproductive cells

Sustentacular cell

Interstitial tissue

Testicle

and early childhood after which it undergoes involution. It is not clear whether the thymus gland has an endocrine function. Currently the thymus gland is considered to have a role in the production of lymphocytes in prenatal and early postnatal life. It is believed to be related to the regulation of immune mechanisms of the body prior to its involution.

11

The

Integumentary

System

The skin is the organ which envelops and covers the body. The integumentary system includes this covering membrane, its associated glands and specialized derivatives, and the mammary gland. (See illustration at top of following page.)

THE SKIN

Ectodermal and mesodermal elements are joined to form a surface membrane for the body during embryonic development. An outer cellular layer, the *epidermis,* is of ectodermal origin. The inner fibrous layer, the *dermis,* develops from the mesodermal component.

EPIDERMIS

The *epidermis* is a thin layer composed of squamous epithelial cells which are moved upward as they form. The epidermis is divided into characteristic zones. (See illustration below.) The basal or *germinative zone* consists of cylindrical cells which undergo constant division to form new cells. The new cells are polygonal. As the cells move upward in the epidermis the cells

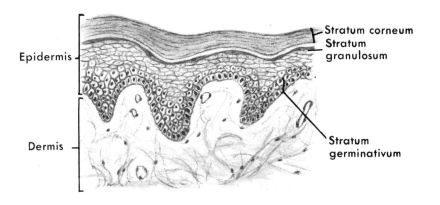

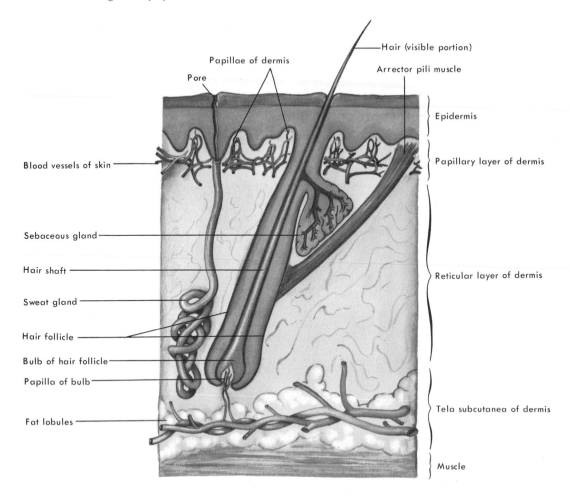

Papillae of dermis

Pore

Hair (visible portion)

Arrector pili muscle

Epidermis

Papillary layer of dermis

Blood vessels of skin

Sebaceous gland

Hair shaft

Sweat gland

Hair follicle

Bulb of hair follicle

Papilla of bulb

Fat lobules

Reticular layer of dermis

Tela subcutanea of dermis

Muscle

gradually flatten and their nuclei degenerate. The superficial layers of the epidermis comprise the *horny zone* or stratum corneum. The cells die and are converted into a protein, keratin. The deeper part of the horny zone forms a barrier to the entrance of water. The outer part of the zone is resistant to abrasion and forms a thick protective layer on surfaces which are subject to pressure or friction. The keratinized cells are worn away or flake off continuously but are replaced by the keratinization of cells which move up from deeper surfaces.

Pigmentation of the skin is a result of the extent of activity of the *melanocyte* cells of the skin.

These cells are displaced from the neural crest in development and settle in the dermis just below its junction with the epidermis. Under the regulation of the melanin-stimulating hormone of the anterior lobe of the pituitary gland, these cells produce *melanin* pigment. The melanin is distributed to some, but not all, of the epidermal cells of the stratum germinativum, as well as to connective cells lying adjacent to the melanocytes in the dermis. The pigment is still visible as cells move up into the stratum spinosum, but it disappears as the process of keratinization proceeds.

Skin coloration is the genetic result of the blending of hues found in naturally occurring skin pigments. True white skin is found only in the *albino,* a person with a congenital absence of skin pigments. Carotene, a yellow pigment, gives the skin a basic yellow appearance, which is always modified by the reddish hue of capillary blood in the dermis to form the basic *flesh pink* color. The shades of yellow-brown, brown, brown-black, blue-black, and black which are found in most of the world's population reflect the degree to which varying amounts of melanin change the basic skin coloration.

DERMIS

The *dermis* or corium is the thicker fibrous portion of the skin which provides strength and elasticity. It is the location of the glands of the skin and the hair follicles. Blood vessels, lymphatic vessels, and nerves ramify within this layer to nourish and innervate the skin. The dermis

is divided into a superficial papillary layer and a deeper reticular layer. The *papillary layer* presses against the epidermis and indents it at intervals with finger-like projections, the *papillae.* (See illustration below.) Papillae consist of an areolar connective tissue core which carries capillaries, nerve fibers, and sensory end organs into more intimate contact with the epidermis. Since the epidermis has no blood vessels, nutriments and oxygen reach it by diffusion through the tissue fluid of the papillae. The papillary layer is also thrown up into papillary ridges which are especially notable on the palmar surfaces of the hands and fingers and the plantar surfaces of the feet. The papillary ridges, covered with epidermis, form raised loops and whorls on the skin surface which aid in touching and grasping objects. The patterns of the ridges are characteristic in each individual and are used in identification as *fingerprints,* palm prints, or sole prints.

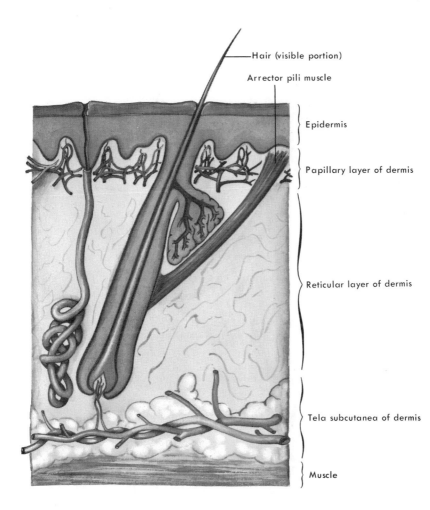

Hair (visible portion)

Arrector pili muscle

Epidermis

Papillary layer of dermis

Reticular layer of dermis

Tela subcutanea of dermis

Muscle

In the *reticular layer* the connective tissue is formed into a dense interlacing network of elastic fibers and bundles of collagenous fibers. The collagenous bundles are very coarse in the depths of the dermis. The boundary of the dermis with the tela subcutanea is marked by the presence of fat and the rearrangement of collagenous bundles into vertical strands. The strands unite the skin to deeper fasciae but yet in most regions of the body permit the skin to move upon underlying structures.

HAIRS

Hairs are keratin shafts formed at the depths of tubular ingrowths of the epidermis called the *hair follicles.* The expanded end of the follicle (bulb), located in the dermis, is indented by a connective tissue papilla which bears blood vessels and sensory receptors. The epithelial cells over the papilla divide and undergo keratinization to form a hair which grows through the slanting tube of the follicle to reach the surface of the skin. Each hair is a solid rod with an expanded *root* at its base where keratin is added to it in the bulb of the follicle. The *shaft* of the hair is formed of an outer *cortex* of hard keratin and an inner *medulla* of soft keratin. Pigment in the cortex gives the hair its color. A small band of smooth muscle fibers, the *arrector pili,* courses between the hair follicle and the papillary layer of the dermis. Sympathetic nerve impulses produce contraction of the muscle fibers which pull the hair erect. "Goose pimples" occur in response to cold or fright when the arrectores pilorum contract.

GLANDS OF THE SKIN

Two types of glands are found in the skin. *Sebaceous glands* cluster in lobules adjacent to hair follicles in the dermis and open into the follicles which are used as excretory ducts. An oily product, *sebum,* is produced by holocrine secretion in which the cells become filled with fat droplets

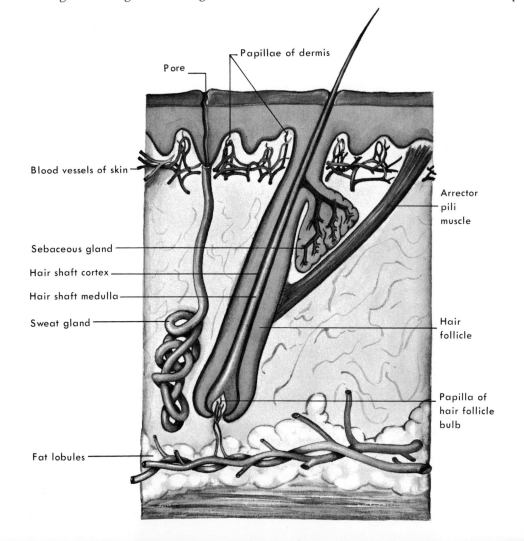

Papillae of dermis

Pore

Blood vessels of skin

Arrector pili muscle

Sebaceous gland

Hair shaft cortex

Hair shaft medulla

Sweat gland

Hair follicle

Papilla of hair follicle bulb

Fat lobules

and die. The oil-filled cells are pushed out into the hair follicles and then to the surface of the skin. The sebum contributes to the pliability of the skin and helps make it impervious to water.

Sweat glands are long tubular glands whose coiled secretory portions are located in the dermis or even in the tela subcutanea. A long straight duct connects the secretory body of the gland to the surface of the skin where an opening forms one of the many *pores* scattered atop the papillary ridges. (See illustration on opposite page.) Perspiration, the secretion of sweat, is a form of elimination of water from the body. It is more significant as a means of reducing body temperature as the sweat is evaporated from the surface of the skin.

NAILS

The *nails* are flat plates of keratin which cover and protect the dorsal surfaces of the tips of the fingers and toes. With their sensitive tips protected the fingers become more useful in grasping, prodding, or scratching. The toes are thrust forward more confidently in walking.

The nails grow forward over the surface of the digit from a narrow fold of skin over the distal phalanx. This fold, the *eponychium,* overlies the *nail bed* which is the germinative zone of the nail. The new part of the nail, which is attached to the formative zone, is the *nail root.* The older keratin forms the *body* of the nail which is pushed forward over the surface of the digit. The body of the nail curves transversely to follow the shape of the digit. The free edge of the nail overhangs a zone of thickened epidermis at the tip of the nail bed called the *hyponychium.*

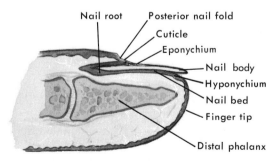

REGIONAL SKIN CHARACTERISTICS

The skin over most of the body surface is loose and movable and closely conforms to the under-lying structures. Keratinization is moderate and the presence of sebum creates a smooth pliable surface. On the palms and soles the skin is adherent to the fascia beneath. The loss of skin movement in these areas increases the efficiency of their contact with environmental structures in work or locomotion. Keratinization is heavy in the skin of the soles where pressure and friction are at a maximum.

The skin is more firmly anchored by dermal fibrous strands where it passes over bony prominences. The fixation of the skin produces a skin indentation commonly called a *dimple.* Tiny lines, marking off geometric areas of the skin, are formed on most surfaces by the elasticity of the dermis. These flatten as the skin moves or stretches. Deeper lines, the *flexure lines,* are creases formed in the skin where one part of the body moves upon another as on the fingers, palm, and wrist. Wrinkles are lines formed by the repeated action of muscles which attach to the skin and, therefore, are common on the face and neck. These become more prominent and permanent as the skin becomes less elastic with age.

FUNCTIONS OF THE SKIN

The many functions of the skin can be summarized as follows:

Protection. The pliable surface intervenes between the tissues of the body and the environment. Its renewable structure is resistant to usual contacts with foreign objects. The mobility of the skin frequently permits the absorption or warding off of externally applied force, but it may be ruptured or lacerated by sharp objects or severe blunt force. An increase in the amount of skin pigments (tanning) protects against exposure to ultraviolet radiation in sunlight.

Environmental Barrier. The skin forms an impermeable barrier to the entrance of water into the body. It also resists the entrance of foreign particulate matter and many but not all microorganisms.

Fluid Retention. The internal fluid environment of body cells and tissues is preserved by the skin which prevents the dissipation of body fluids as long as the surface membrane is intact. When the surface is injured or lost, as in extensive burns or severe skin diseases, enormous amounts of fluids, electrolytes, and proteins are lost from the body.

Sensation. The role of the skin in providing sensory awareness of environmental changes has been previously established.

Regulation of Body Temperature. A constant body temperature is essential to function and health. The body must adjust to fluctuations in the temperature of the environment as well as to heat produced by its own metabolism. Adjustments are made by the regulation of heat loss from the body surface. The skin surface acts to radiate heat from the blood to the environment. Dilation of the arterioles to the skin increases the blood flow to the extensive capillary plexuses of the dermis when heat loss is needed. When heat must be conserved, large units of the capillary bed can be shut off by direct arteriovenous anastomoses which shunt blood away from the surface. The adipose tissue of the tela subcutanea functions as an insulating sheet against external temperature changes. Further cooling of the body can be provided when needed by an increase in sweat production. The pouring out of watery fluid upon the skin surface in perspiration provides cooling by evaporation.

THE MAMMARY GLAND

The paired mammary glands are specialized skin derivatives which produce milk during the lactation period following childbirth.

EXTERNAL APPEARANCE

The female breasts are composite structures in which a radiating compound alveolar gland structure is imbedded in a mound of fat and connective tissue fibers specialized from the tela subcutanea of the pectoral region. The breast, therefore, becomes synonymous with mammary gland. Each mammary gland is suspended by connective tissue strands upon the fascia of the pectoralis major muscle. The skin of the medial aspect of each breast slopes toward the surface of the thorax over the sternum so that the two breasts are separated by the midline *pectoral sinus.* The mammary gland extends vertically between the second and sixth ribs and is rounded laterally over the anterior axillary fold. A modification of the skin, the *nipple,* projects anterolaterally from the surface. The nipple is surrounded by an area of pigmented skin, the *areola,* which usually changes from pink to brown following pregnancy.

STRUCTURE

The skin of the nipple contains circular fibers of smooth muscle which surround a group of 15 to 20 *lactiferous ducts* which open onto the surface. Each lactiferous duct, dilated at its termination to form a *lactiferous sinus,* is the excretory duct for one of the lobes of the mammary gland.

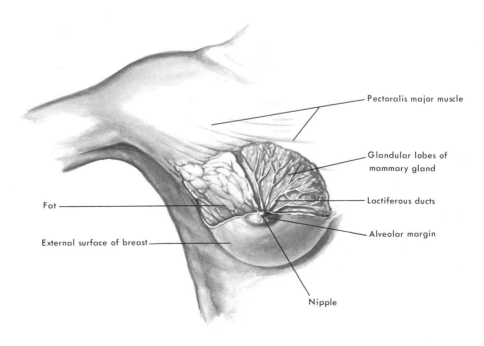

Pectoralis major muscle

Glandular lobes of mammary gland

Lactiferous ducts

Alveolar margin

Fat

External surface of breast

Nipple

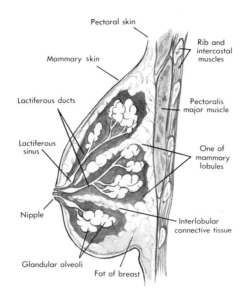

Pectoral skin

Rib and intercostal muscles

Mammary skin

Pectoralis major muscle

Lactiferous ducts

Lactiferous sinus

One of mammary lobules

Nipple

Interlobular connective tissue

Glandular alveoli

Fat of breast

The glandular masses of each lobe radiate from the nipple through the fat of the breast. Smaller lobular subdivisions of each mammary lobe are supported by a framework of fibrous strands. The actual terminal secreting alveoli of the lobule vary in size and development under endocrine control which is related to the menstrual cycle. The alveoli are brought to full functioning state only in the late stages of pregnancy.

The male mammary gland is a rudimentary organ which consists of a nipple with a small areola without glandular development.

Index